Photochemistry of Planetary Atmospheres

Photochemistry of Planetary Atmospheres

Yuk L. Yung

William B. DeMore

New York   Oxford
Oxford University Press
1999

Oxford University Press

Oxford   New York
Athens   Auckland   Bangkok   Bogotá   Buenos Aires   Calcutta
Cape Town   Chennai   Dar es Salaam   Delhi   Florence   Hong Kong   Istanbul
Karachi   Kuala Lumpur   Madrid   Melbourne   Mexico City   Mumbai
Nairobi   Paris   São Paulo   Singapore   Taipei   Tokyo   Toronto   Warsaw

and associated companies in
Berlin   Ibadan

Copyright © 1999 by Oxford University Press, Inc.

Published by Oxford University Press, Inc.
198 Madison Avenue, New York, New York 10016

Oxford is a registered trademark of Oxford University Press

All rights reserved. No part of this publication may be reproduced,
stored in a retrieval system, or transmitted, in any form or by any means,
electronic, mechanical, photocopying, recording, or otherwise,
without the prior permission of Oxford University Press.

Library of Congress Cataloging-in-Publication Data
Yung, Y. L. (Yuk Ling)
   Photochemistry of planetary atmospheres /
  Yuk L. Yung, William B. DeMore.
     p.   cm.
  Includes bibliographical references and index.
  ISBN 0-19-510501-X
   1. Planets—Atmospheres.   2. Earth—Environmental aspects.
3. Atmosphere.   4. Photochemistry.   5. Atmospheric chemistry.
I. DeMore, William B.   II. Title.
QB603.A85Y86   1998
551.51′1′09992—dc21   97-23575

9 8 7 6 5 4 3 2 1

Printed in the United States of America
on acid-free paper

Foreword

One of the greatest triumphs of scientific research during the past several decades has been our achievement of understanding complex macroscopic phenomena, of which combustion and atmospheric chemistry are notable examples. In the field of atmospheric chemistry, our detailed knowledge of elementary chemical reactions and photochemical processes has given us an understanding of phenomena such as the depletion of the ozone layer and creation of the ozone hole. While there are still areas— for example, polyatomic radicals—where our knowledge is incomplete, the use of powerful computers for model calculations allows us to make intelligent predictions. Data from solar system probes, meanwhile, provide a context in which Earth-based observations can be interpreted. The breadth of material relevant to the field means, however, that there is a dire need for a monograph in which this information is sorted and woven into a comprehensive volume. To fill this void, two of the leading experts in the photochemistry of planetary atmospheres, Drs. Yuk Ling Yung and William B. DeMore, have written this well-researched treatise. The book consists of a comprehensive survey of the principal chemical cycles that control the compositions of planetary atmospheres, emphasizing the "big picture" at planetary scales. It gives a detailed account of the history of the various planetary atmospheres, ranging from the comparatively primitive atmospheres of the giant planets to the highly evolved atmospheres of small bodies. It also contains an extensive tabulation of physical constants, astronomical and atmospheric data, and various parameters associated with the eleven planetary atmospheres covered in the book, including material with much significance for evaluating the ecological impact of life on Earth. With its self-contained and lucid writing, this book is useful as a textbook; it also serves well as a reference of source material for experts, containing the basic information needed to understand the complex chemistry of planetary atmospheres. The triumph of understanding that has allowed us to unravel the evolution of planetary atmospheres does ultimately beg the question the authors take pains to avoid: whether Lovelock's controversial Gaia hypothesis, that the Earth is a living organism playing an active role in maintaining its own equilibrium, can be proven or refuted. In spite of the progress made by this book, this will likely remain a mystery. Fortunately, in Einstein's words, mystery, complemented by beauty, is a quintessential element of science.

Dr. Yuan T. Lee
Academia Sinica
Taipei, Taiwan
August, 1997

Preface

This book is based on a course that one of the authors (Y.L.Y.) has taught at the California Institute of Technology since 1977. It is intended as an introduction to photochemistry in the atmospheres of the planets (including Earth) for seniors and first year graduate students in atmospheric science, chemistry, physics, astronomy, environmental science, or engineering.

Most of the material may be traced back to the inspiring course taught by Prof. M. B. McElroy at Harvard and taken by Y.L.Y. in 1972. At that time the scientific community was first arriving at a fundamental understanding of the chemical processes in the atmospheres of the planets (including Earth) and their implications for secular change and atmospheric evolution. It was a time of unparalleled excitement and discovery.

The theoretical foundations of atmospheric chemistry were laid earlier by the pioneers James Jeans, Sidney Chapman, Harold Urey, David Bates, and Marcel Nicolet. The dawn of the space age, heralded by the launching of Sputnik, provided the much-needed data to test these ideas, and to refine and modify some of them. For the first time, we achieved a first-order understanding of Earth's upper atmosphere, space environment, and interactions with solar radiation and solar wind, and how these processes operate on the other planets in the solar system. In retrospect, it is hard to recapture the vibrant "atmosphere" of the period, so full of new possibilities, grand syntheses, previously unsuspected connections, and sheer speculations. The stability and evolution of the thin CO_2 atmosphere of Mars were explained by McElroy, Hunten, and Donahue. Great progress in understanding the chemistry of the atmosphere of Venus was made by Lewis, Prinn, and McElroy and his students, while Strobel did pioneering work on the chemistry of hydrocarbons in the giant planets. Meanwhile, fundamental discoveries were made in Earth's atmosphere. Oxides of nitrogen (a potential pollutant from proposed supersonic aircraft flying in the stratosphere) were shown by Crutzen and Johnston to be capable of destroying the ozone layer. Molina and Rowland demonstrated a similar adverse environmental impact due to the release of chlorofluorocarbons. McElroy proposed that human-induced acceleration of the nitrogen cycle may increase the production of N_2O, with negative impact on stratospheric ozone. Levy first showed that photochemistry in the terrestrial troposphere could be initiated by photolysis of O_3 at long wavelength in the Huggins bands, providing an explanation for the oxidizing power of the troposphere.

It is hardly possible that all these exciting discoveries occurred in isolation. Indeed, the chlorine chemistry on Earth was anticipated by that on Venus, and the sulfur work on Venus by Prinn led Crutzen to propose a similar theory for the origin of the Junge layer of sulfate aerosols in the terrestrial stratosphere. The sudden wealth of data on the atmospheres of Earth and other planets provided a synergy that allowed a more fundamental understanding of all atmospheres in our solar system. For example, this kind of "cosmic vision" created by viewing the Earth in the reference frame of the solar system inspired at least one grand synthesis, Lovelock's Gaia hypothesis.

Noteworthy planetary missions to the inner planets in the 1970s included the Mariners and Vikings to Mars, and the Mariners, Pioneer Venus, and Soviet Venera series to Venus. In the outer solar system, the early Pioneers 10 and 11 to Jupiter and Saturn were followed by the spectacular Voyagers 1 and 2 to all the giant planets and most of their moons. In the 1980s, comet Halley was studied by spacecraft launched by the European Space Agency (Giotto), the Soviet Union (VEGA 2), and Japan (Sakigate and Suisei). In addition, great contributions were made throughout the last few decades by ground-based telescopes, balloon- and aircraft-borne instruments, and Earth-orbiting satellites. Together, these activities resulted in an explosion of knowledge about Earth and its neighbors, and provided a proper perspective for viewing Earth as part of the solar system and the cosmos at large.

Three important ideas guided the development of the field of atmospheric chemistry. The first is the idea of the stability of a planetary atmosphere against thermal escape of its constituents to space, as formulated at the beginning of the century by Jeans. The second is the role of equilibrium chemistry in determining the partitioning of chemical species, as developed by Urey. Third is the role of photochemistry, which produces drastic departures from equilibrium chemistry; a classic example of this is Chapman's theory of the UV-driven formation of the Earth's ozone layer. It is now understood that escape of lighter molecules is important for the terrestrial planets and for satellites, but it is unimportant for the helium- and hydrogen-rich giant planets. Equilibrium chemistry dominates in the deep atmospheres of the giant planets, near the surface of Venus, and in the interior of terrestrial planets, and may also have been important in the solar nebula. Photochemical processes, by contrast, govern atmospheric composition and evolution in the middle atmosphere of all planets, producing phenomena such as the Earth's ozone layer or the hydrocarbon haze layers in the atmospheres of the outer solar system.

To these three ideas we must add a fourth, the role of biochemistry at Earth's surface. While Earth is a part of the solar system and the cosmochemical environment, its atmospheric chemistry appears radically different. Only in Earth's atmosphere do strong reducing and oxidizing species coexist to such a degree: nitrogen species in Earth's atmosphere, for example, span eight oxidation states from ammonia to nitric acid. Consequently, much of the Earth's atmospheric chemistry consists of reactions to relax the disequilibrium caused by such biologically produced molecules. Consideration of the effect of life on the atmosphere brings us inevitably to the question posed by Lovelock in his Gaia hypothesis, of whether life also regulates atmospheric composition in a more proactive biologically beneficial way. The idea is controversial, and it is not the purpose of this book to argue for or against it. However, there is no doubt that the biosphere plays a key role in determining Earth's atmospheric composition. Life provides a means for using solar energy to drive chemical reactions that would otherwise not occur; it represents a kind of photochemistry that is, at least within our solar system, unique to Earth.

We briefly explain the unusual structure of the book. It is common in the study of planets to study Earth first, and then the other planets. In this way we may better understand why the rest of the solar system is different from us. In this book the order of study is reversed. We first try to understand the solar system, and then ask why Earth is unique. With this viewpoint, we can begin to address the issue of the place of life in our (and other) solar system(s). Are there planets around other stars with atmospheres similar to those of our solar system? What are the conditions that can give rise to life and sustain its development? What human perturbations might change our own planet, making it less habitable, or even unhabitable? The diversity of planetary atmospheres in our solar system provides a natural testing ground for our ideas, and the evolutionary history of Earth provides clues about our future. It is clear that our planet and its environment represent a dynamic balance of physical and chemical driving forces. Earth has evolved from initial conditions that are profoundly different from those today, and it will undoubtedly continue to evolve (with or without human intervention) into physical and chemical conditions that may be drastically different from those at present. Yet life has been sustained and has continued to thrive for at least 3.85 billion years. Is there another place in the universe where these conditions are duplicated, even partially? It is a measure of the vitality of a civilization to ask such profound questions, so that posterity may be inspired to search for the answers. Is there a more beautiful vision or a greater technological challenge than this to pass on to the next millennium?

Y.L.Y. is extremely grateful to Academia Sinica in Taiwan, Republic of China, where much of the writing was done. He thanks former President Ta-You Wu of the Academy for his kindness and hospitality, and President Yuan T. Lee of the Academy for his interest in this book and for writing the Foreword. He is especially grateful to his colleagues, Typhoon Lee, Francis Wu, and Leon Teng, for arranging the sabbatical visit and to former Director Yeong Tein Yeh of the Institute of Earth Sciences for hosting the visit.

We thank the following colleagues for valuable help, suggestions, and comments: Mark Allen, Ariel Anbar, Geoffrey Blake, Josh Cheng, James Cho, Lucien Froidevaux, Mimi Gerstell, Randy Gladstone, Andrew Ingersoll, Robert Herman, Heinrich Holland, Kenneth Hsu, Yibo Jiang, David Kass, Zhiming Kuang, Anthony Lee, Ming-Taun Leu,

Lingun Liu, Fred Lo, Franklin Mills, Julie Moses, Elisabeth Moyer, Duane Muhleman, Bruce Murray, Hari Nair, Mitchio Okumura, Joseph Pinto, Charlie Qi, Kelly Redeker-Dewulf, Mark Roulston, Ross Salawitch, Stanley Sander, Charles Shia, Darrell Strobel, Michael Summers, Wing Chi Yung and Hui Zhang. We would like particularly to thank the staff of Oxford University Press for their excellent editing.

Pasadena, California   Y.L.Y.
June 1997   W.B.D.

Contents

Foreword v

1 Introduction 3
   1.1 The Nature of the Problem 3
   1.2 Physical State of Planetary Atmospheres 5
   1.3 Chemical Composition of Planetary Atmospheres 12
   1.4 Stability of Planetary Atmospheres against Escape 13

2 Solar Flux and Molecular Absorption 17
   2.1 Introduction 17
   2.2 Solar Radiation and Other Energy Sources 18
   2.3 Absorption Cross Sections of Atmospheric Gases 28
   2.4 Interaction Between Solar Radiation and the Atmosphere 44
   2.5 Sunlight and Life 53

3 Chemical Kinetics 55
   3.1 Introduction 55
   3.2 Thermochemical Reactions 56
   3.3 Unimolecular Reactions 61
   3.4 Bimolecular Reactions 61
   3.5 Termolecular Reactions 70

xii Photochemistry of Planetary Atmospheres

    3.6   Heterogeneous Reactions   72
    3.7   Miscellaneous Reactions   73

4 Origins 77
    4.1   Introduction   77
    4.2   Cosmic Organization   79
    4.3   The Elements of the Periodic Table   81
    4.4   Molecular Clouds   84
    4.5   The Solar Nebula   91
    4.6   Meteorites   94
    4.7   Comets   99
    4.8   Formation of Planets   102
    4.9   Atmospheres of Terrestrial Planets and Satellites   103
    4.10 Miscellaneous Topics   107

5 Jovian Planets 120
    5.1   Introduction   120
    5.2   Thermosphere   139
    5.3   Hydrocarbon Chemistry   144
    5.4   Nitrogen Chemistry   178
    5.5   Phosphorus Chemistry   182
    5.6   Oxygen Chemistry   183
    5.7   Miscellaneous Topics   186

6 Satellites and Pluto 189
    6.1   Introduction   189
    6.2   Io   191
    6.3   Titan   201
    6.4   Triton   234
    6.5   Pluto   239
    6.6   Unsolved Problems   242

7 Mars 244
    7.1   Introduction   244
    7.2   Photochemistry   247
    7.3   Model Results   256
    7.4   Evolution   273
    7.5   Terraforming Mars   279
    7.6   Unsolved Problems   280

8 Venus 282
    8.1   Introduction   282
    8.2   Photochemistry   287
    8.3   Model Results   294
    8.4   Evolution   311

8.5 Mystery of the Missing Water   313
   8.6 Unsolved Problems   315

9 Earth: Imprint of Life   316
   9.1 Introduction   316
   9.2 Gaia Hypothesis   319
   9.3 Hydrological Cycle   327
   9.4 Carbon Cycle   330
   9.5 Rise of Oxygen   344
   9.6 Nitrogen Cycle   349
   9.7 Minor Elements   351
   9.8 Unsolved Problems   356

10 Earth: Human Impact   357
   10.1 Introduction   357
   10.2 Global Warming   363
   10.3 Tropospheric Chemistry   381
   10.4 Stratospheric Chemistry   389
   10.5 Unsolved Problems   411

Notes and Bibliography   425

Author and Proper Name Index   447

Subject Index   450

Photochemistry of
Planetary Atmospheres

1
Introduction

1.1 The Nature of the Problem

It is usual in the study of planets to consider the Earth first, and then the other planets, so that we can better understand how and why the rest of the solar system is different from us. In this book the order of study will be reversed: we shall first try to understand the solar system, and then we will ask why Earth is unique. We adopt this unconventional approach for two reasons. First, Earth's atmosphere today is the end-point of an evolution that started about 4.6 billion years ago. The pristine materials have all been drastically altered. However, by examining other parts of the solar system that have evolved to a lesser degree, we may deduce what the early Earth might have been like. Second, Earth's atmosphere today is largely determined by the complex biosphere, whose evolution has been intimately coupled to that of the atmosphere. In other words, ours is the only atmosphere in the solar system that supports life, and it is in turn modified by life. Therefore, to appreciate the beauty and the intricacy of our planet, we must start with simpler objects without life.

Chemical composition is intimately connected to evolution, which in turn is driven by chemical change. In this book we attempt to provide a coherent basis for understanding the planetary atmospheres, to identify the principal chemical cycles that control their present composition and past history. Figure 1.1 gives an illustration of the intellectual framework in which our field of study is embedded. The unifying theme that connects the planets in the solar system is "origin"; that is, all planets share a common origin about 4.6 billion years ago. The subsequent divergence in the solar system may be partly attributed to evolution, driven primarily by solar radiation.

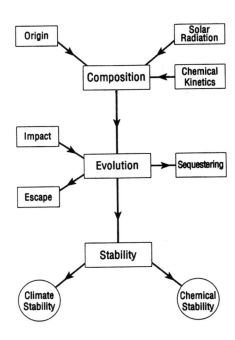

Figure 1.1 Outline of the important concepts and processes addressed in this book.

The bulk of solar radiation consists of photons in the visible spectrum with a mean blackbody radiation temperature of 5800 K. The part that is responsible for direct atmospheric chemistry is a tiny portion (less than 1% of the total flux) in the ultraviolet. In addition, the sun emits a steady stream of corpuscular particles, known as the solar wind. While the sun provides the principal source of energy for change, the time rate of change is crucial, and that is where chemical kinetics and chemical cycles play pivotal roles. The fact that a certain reaction proceeds in a microsecond rather than in a billion years may have the most profound consequences for the observed composition of planetary atmospheres and their evolution. The box in figure 1.1 labeled "origin" represents the source of material to the atmospheres. Solar radiation, both electromagnetic and corpuscular, and chemical kinetics determine the rates of change of the material in the atmospheres. Most chemical reactions in the atmosphere do not result in a net conversion of one species into another, but rather a recycling among different compounds, in specific "chemical cycles." Some processes do result in a net change of the composition of the atmosphere, labeled "Evolution" in figure 1.1. This can happen in two ways: loss of atmospheric species by escape into space, or sequestering in the surface reservoir. In the latter case, the material is not permanently lost from the planet and eventually may be recycled via tectonic or volcanic processes. Material can also be gained or lost from space by impact of comets or meteorites.

The purpose of science is not just to explain the present or reconstruct the past. There is a more utilitarian motive, to contemplate and assess the future. Therefore, we are deeply interested in studying the response of the terrestrial atmosphere to perturbations, that is, the stability (both chemical and climatic) of the atmosphere. The exercise is more than academic, because we now live on a planet whose composition we are seriously perturbing in a potentially irreversible manner. How robust is our planetary environment? What are the consequences of overriding one or more of nature's chemical cycles? These questions are indicated by the two circles in figure 1.1. The answer to our questions will have immediate impact on our welfare and future lifestyle. On a more philosophical level we ask what is the proper place for humans (or whether there is a place) so that we are in harmony with the chemical environment of our solar system and the cosmos. It would be unrealistic to expect complete answers to these questions in the scope of this book, but we state these questions as our motivating forces.

A large number of special topics is included. Many concepts, some of which appear unrelated, are synthesized. To help the reader get an overview of what this book is about, we briefly describe the content of the various chapters.

Chapter 1 gives a preliminary survey of the most fundamental physical and chemical data of planetary atmospheres, the general patterns of speciation, and reasons for stability and instability. Chapter 2 discusses the primary driving force for atmospheric chemistry and change—photochemistry that is driven by solar ultraviolet radiation. Other important drivers of disequilibrium chemistry include solar wind, lightning, and shock waves during impacts. Chapter 3 describes chemical kinetics pertinent to atmospheric modeling. Chapter 4 addresses the question of origins for the solar system; we discuss primitive bodies such as meteorites, asteroids and comets. Having prepared the necessary background, chapter 5 starts with a study of the H_2-dominated atmospheres of the giant planets, Jupiter, Saturn, Uranus, and Neptune. The giant planets have satellites that resemble terrestrial planets in the inner solar system. Chapter 6 is devoted to three such satellites, Titan, Triton, and Io. (Pluto is also included in this chapter.) Chapter 7 deals with the simplest atmosphere in the inner solar system, the Martian atmosphere. Chapter 8 discusses the atmosphere of Venus. The last two chapters are devoted to the jewel of the solar system, the terrestrial atmosphere. Chapter 9 discusses the role of the biosphere (excluding humans), and the human impact on global chemistry is discussed in chapter 10.

1.2 Physical State of Planetary Atmospheres

All the planets except Mercury have substantial atmospheres. The giant planets have very extensive atmospheres that are major portions of the planets themselves. The giant planets also all have a large number of satellites, resembling miniature solar systems. Three such satellites, Titan, Triton, and Io, have atmospheres. For reasons that will become clear later, we group Pluto with the satellites. We are less interested in the tenuous atmospheres of Mercury and the Moon. [The treatment here follows that of Chamberlain and Hunten (1987), cited in section 1.1.]

Tables 1.1 and 1.2 summarize the physical properties of the planetary bodies in the solar system that are of interest to atmospheric studies. Figure 1.2 shows the pressure (P) versus temperature (T) plots for the atmospheres of the giant planets. In this book, we restrict our interest to pressures less than 1 kbar. At this and higher pressures, the atmospheric composition is completely controlled by equilibrium chemistry, which will not be a primary subject of this book. It is convenient to use pressure rather than

Table 1.1 Physical properties of planets

Planet	Mass[a] (Earth = 1)	Equatorial radius (km)	Mean[b] distance from sun (AU)	Length[c] of year	Length[c] of day	Obliquity	Gravity (m s^{-2})	Escape velocity (km s^{-1})
Mercury	0.0554	2439	0.387	88.0^d	176^d	0	3.70	4.2
Venus	0.815	6052	0.723	224.7^d	116.7^d	−177.3	8.87	10.3
Earth	1	6378	1	365.26^d	24^h	23°27′	9.78	11.2
Mars	0.1075	3393	1.524	1.881^{yr}	24^{h}39^{m}35^s	23°59′	3.69	5.0
Jupiter	317.8	71492	5.203	11.86^{yr}	9^{h}55^{m}30^s	3°07′	23.1	59.5
Saturn	95.15	60268	9.539	29.42^{yr}	10^{h}30^m	26°44′	8.96	35.5
Uranus	14.54	25559	19.18	83.75^{yr}	17^{h}14^m	97°55′	8.69	21.3
Neptune	17.2	24766	30.06	163.72^{yr}	16^{h}6^m	28°48′	11.0	23.5
Pluto	0.0021	1137	39.53	248.02^{yr}	6^{d}9^{h}17^m	122°27′	0.66	1.1

[a] 1 Earth mass = 5.98 × 10^{24} kg
[b] 1 astronomical unit (AU) = 1.496 × 10^{11} m
[c] Sidereal

altitude (z) as the vertical coordinate because of the obvious lack of a "surface" for these planets. Pressure and altitude are simply related by the hydrostatic equation

$$\frac{dP}{dz} = -\rho g \tag{1.1}$$

where ρ = mass density and g = acceleration due to gravity (see tables 1.1–1.3 for numerical values). The perfect gas law is

$$P = nkT \tag{1.2}$$

where n = number density and k = Boltzmann constant. The mass density and the number density are related through the definition

$$\rho = mn \tag{1.3}$$

Table 1.2 Physical properties of selected satellites

Satellite	Parent planet	Mass[a] (M_m)	Equatorial radius (km)	distance from planet[b] (R_p)	Sidereal period	Gravity (m s^{-2})	Escape velocity (km s^{-1})
Io	Jupiter	1.21	1821	5.9	42^{h}27^m	1.81	2.56
Titan	Saturn	1.83	2575	20.3	15^{d}22^{h}41^m	1.35	2.64
Triton	Neptune	0.29	1353	14.3	5^{d}21^{h}03^m (retrograde)	0.78	1.46
Moon	Earth	1.00	1738	60.3	27^{d}7^{h}43^m	1.62	2.38

[a] M_m = mass of the Moon = 7.35 × 10^{22} kg
[b] R_p = radius of planet.

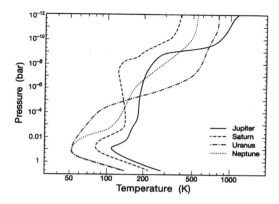

Figure 1.2 Pressure versus temperature for the atmospheres of the giant planets. The lower boundary is arbitrarily set at 6 bar.

where m is the mean molecular mass. Combining these equations, we have

$$\frac{dP}{P} = -\frac{dz}{H} \tag{1.4}$$

where $H = kT/mg$ is the atmospheric scale height. Equation (1.4) states that the pressure of the atmosphere drops off vertically like an exponential function. This can be most easily seen by integrating both sides of (1.4) from z_0 to z (which is above z_0) to yield

$$P(z) = P(z_0) \exp\left(-\int_{z_0}^{z} \frac{dz}{H}\right) \tag{1.5}$$

The result further simplifies to

$$P(z) = P(z_0) \exp\left(-\frac{z - z_0}{H}\right) \tag{1.6}$$

Table 1.3 Physical properties of planetary atmospheres at the surface or 1 bar

Planetary body	Pressure[a] (bar)	Mean temperature (K)	Scale height (km)	Lapse rate[b] (K/km)
Jupiter	1	165	27.0	1.8
Saturn	1	134	59.5	0.7
Uranus	1	76	19.1	0.8
Neptune	1	72	27.7	1.0
Titan	1.5	94	21.5	1.3
Triton	1.6×10^{-5}	38	13.4	0.74
Pluto	$\sim 10^{-5}$	~ 40	~ 22	~ 0.6
Io	$\sim 10^{-8}$	110	7.9	3.1
Mars	$(6.9 - 9) \times 10^{-3}$	210	11.1	4.5
Venus	93	737	15.9	10.5
Earth	1	288	8.4	9.8

[a] Data for the giant planets, which have no surface, are given at a pressure of 1 bar.
[b] adiabatic lapse rate for dry air

for an isothermal atmosphere in which H is approximately constant (the small variation is due to g or m). The physical meaning of the scale height becomes obvious in (1.5) and (1.6). Knowing the scale height allows us to interconvert between pressure and altitude coordinates.

Equations (1.1) and (1.2) will break down when the pressure becomes so small that there are not enough collisions to maintain the hydrostatic equation and the perfect gas law, because their validity depends on the kinetic theory of gases. It is usual to define this level as the critical level or the natural upper boundary of the atmosphere:

$$l = H \tag{1.7}$$

where l is the mean free path for molecular collisions,

$$nlQ = 1 \tag{1.8}$$

with Q = cross section for molecular collisions. The physical meaning of the critical level is that a molecule with sufficient velocity pointing away from the planet has a probability e^{-1} of escaping from the planet. Above the critical level, the molecules can follow individual ballistic trajectories of motion, and the kinetic definitions of static pressure and temperature become ill-defined. Hence, the bound atmosphere generally refers to the region below the critical level. Above this the atmosphere merges with outer space. Combining (1.7) and (1.8), we have the pressure for the critical level:

$$P = \frac{mg}{Q} \tag{1.9}$$

Since Q is of the order of magnitude 10^{-15}cm^2, (1.9) suggests that the critical pressure of the giant planets is of the order of 10^{-12} bar, and is, to first order, independent of temperature. In fact, all the planetary atmospheres in the solar system have critical pressures of this order of magnitude. Unless otherwise stated, the word "atmosphere" will be restricted to a pressure range from about 10^{-12} to 10^3 bar, the lower end being the collision limit and the upper end being the thermodynamic equilibrium chemistry limit.

The thermal structure of the giant planets (figure 1.2) may be divided into three distinct regimes: the troposphere, the middle atmosphere, and the thermosphere, by analogy with what is well known for the terrestrial atmosphere. At pressures higher than about 100 mbar, in the troposphere, convection dominates and the thermal gradient is close to the adiabatic lapse rate, Γ, given in table 1.3. In this region, temperature decreases with height because as an air parcel rises, it has to do work against the environment and it cools due to adiabatic expansion. The adiabatic lapse rate provides a quantitative measure of this cooling effect. At lower pressures, in the middle atmosphere (from about 0.1 to about 10^{-7} bar), the temperature is determined by radiative balance between the heating due to absorbed solar energy and cooling by infrared radiation. The thermal inversion just above the tropopause is caused by the absorption of sunlight by the near infrared bands of CH_4 and aerosols derived from CH_4 photochemistry. The existence of a temperature minimum near what on Earth would be called the mesopause has not been established due to the absence of relevant data. All giant planets are characterized by very hot thermospheres. These high temperatures cannot be explained by conventional theories of solar heating balanced by cooling due to conduction. A hitherto unidentified heating mechanism, most probably related to the planetary magnetospheres, is needed to account for the thermospheric temperature of the giant planets.

The thermal structure of an atmosphere is the result of an intricate interaction between radiation, composition, and dynamics. We state, but do not solve in this book, the one-dimensional equation governing the thermal structure in planetary atmospheres:

$$\rho C_p \frac{dT}{dt} + \frac{d\phi_c}{dz} + \frac{d\phi_k}{dz} = q \qquad (1.10)$$

where C_p = heat capacity (per unit mass) at constant pressure, q = net heating rate = rate of heating − rate of cooling, and the conduction heat flux, ϕ_c, and the convection heat flux, ϕ_k, are given, respectively, by

$$\phi_c = -K \frac{dT}{dz} \qquad (1.11)$$

$$\phi_k = -K_H \rho C_p \left(\frac{dT}{dz} + \frac{g}{C_p} \right) \qquad (1.12)$$

with K = thermal conductivity and K_H = eddy diffusivity. According to the theory of turbulence, ϕ_k as defined by (1.12) vanishes if the quantity in parentheses is positive.

Note that (1.10) is an approximate equation for energy balance in one dimension. The conductive flux, based on the rigorously correct kinetic theory of gases, is important for the thermosphere. The convective flux is, however, based on an empirical theory of turbulence and is not expected to be correct outside the troposphere. Transport of energy in planetary atmospheres is, in general, a three-dimensional process. Equation (1.10) is a globally averaged approximation that provides a first-order picture of the energy budget in the atmosphere.

We give a brief discussion of the significance of the various terms of (1.10). In general, the first term in (1.10) is not important, except for modeling diurnal variations. The third term (convection) dominates in the troposphere; the fourth term (radiation) dominates in the middle atmosphere. The second term (conduction) balances the fourth term in the thermosphere. The thermal structure of a planetary atmosphere depends on the chemical composition of the atmosphere, which in turn may be affected by the temperature through temperature-dependent reactions and condensation of chemical species. However, this coupled interaction, interesting as it seems, does not constitute a significant topic of this book.

The temperatures of the atmospheres of the satellites are presented in figure 1.3. We caution the reader that our understanding of the atmosphere of Io is very poor, and much of our current knowledge is based on indirect evidence and hence is insecure. Little detailed information is available about the atmosphere of Pluto. Conversely, our knowledge of Titan and Triton is relatively secure. The great difference between the atmospheres of the satellites and Pluto and those of the giant planets is the presence of planetary surfaces, which play fundamental roles in regulating these atmospheres. Io's atmosphere is generated by volcanic activity on the surface. Methane in the atmo-

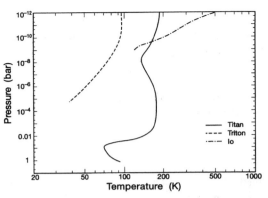

Figure 1.3 Pressure versus temperature for the atmospheres of the satellites (Triton, Titan, Io). No detailed information is available for the atmosphere of Pluto, but its atmosphere is expected to resemble that of Triton.

spheres of Titan, Triton, and Pluto is supplied by a CH_4 reservoir (liquid or ice) on the surface. The surface of Triton has the coldest temperature of all the planetary bodies listed in table 1.3. The reason for the unusual cold is the high albedo of nitrogen ice on the surface that reflects most sunlight to space. The thermospheric temperatures of the satellites are much lower than those of their respective parent bodies and can be adequately accounted for by solar heating alone. This provides further circumstantial evidence that the unusually hot thermospheres of the giant planets are related to their strong magnetic fields or winds. Based on the limited amount of information available, the thermosphere of Pluto is "normal." Among the four small planetary bodies, only Titan has an extensive atmosphere with surface pressure equal to 1.5 bar. The actual column-integrated material per square centimeter is much larger than that in the terrestrial atmosphere because Titan has lower gravity. Titan has well-developed thermal structures that resemble the troposphere, the stratosphere, and the thermosphere of the terrestrial atmosphere. The bulk of stratospheric heating is due to methane and hydrocarbon aerosols, as for the giant planets. Whether Titan has a temperature minimum at the mesopause cannot be established from existing observations.

We briefly comment on the low surface pressures of the other small bodies. Io barely has a collisional atmosphere, despite volcanic sources that supply gases to the atmosphere. Io is immersed in the magnetosphere of Jupiter and bombarded by an intense "Jovian wind" that sputters material from Io. The ejecta from Io remain in the Jovian magnetosphere and ultimately become a source of energetic ions—part of the Jovian wind. This coupled interaction erodes the atmosphere of Io. The low atmospheric pressures of Triton and Pluto are due to the low temperatures at the surface that facilitate condensation of most common atmospheric species.

Figure 1.4 presents the temperature profiles for the atmospheres of the three sister planets in the inner solar system: Mars, Venus and Earth. The thermal structure of the terrestrial atmosphere is the most familiar of all and, indeed, is the basis of the terminology used for describing other atmospheres. In the troposphere, convection dominates and the temperature decreases according to the wet lapse rate, 6.5 K/km. (The wet lapse rate includes the effect of the latent heat released during condensation of water and is less than the dry adiabatic lapse rate of 9.8 K/km.) The thermal inversion in the stratosphere is due to absorption of solar ultraviolet radiation by O_3. Above the

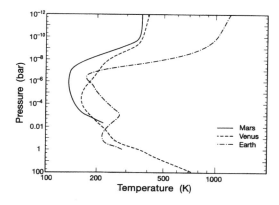

Figure 1.4 Pressure versus temperature for the atmospheres of three planets in the inner solar system: Mars, Venus, and Earth.

stratopause the temperature decreases with height due to cooling in the infrared, mainly by CO_2. The rather hot thermosphere is caused by absorption of extreme ultraviolet radiation by O_2, N_2 and O. In contrast to the Earth, Mars has a thin atmosphere and Venus has a thick atmosphere; this is related to the surface temperatures of these planets. The surface of Mars is sufficiently cold that the major volatiles such as CO_2 and H_2O are frozen out at the poles. The high surface temperature of Venus drives most CO_2 into the atmosphere, which in turn provides a greenhouse effect to keep the surface warm. Earth strikes a happy balance between the frozen Mars and the scalding Venus. That our planet has been able to maintain this condition over the age of the solar system is a most amazing feat and has prompted Lovelock to propose that Gaia (life) has been the planetary homeostat. The tropospheric temperatures of Mars and Venus follow their respective adiabatic lapse rates. There are no well-developed stratospheric inversions, except during dust storms on Mars. The thermospheres of both planets are relatively cold compared with that of Earth, because the major atmospheric constituent (CO_2) is a very efficient radiator of thermal energy. In contrast, the primary homopolar constituents of Earth's atmosphere (N_2, O_2) have no bands in the infrared and are poor radiators.

We briefly comment on the effect of the temperature on atmospheric mixing. Mixing is generally rapid in the troposphere (the Greek word "tropo" means "turning"), as the warmer air (being lighter) at the bottom tends to rise and the colder air at the top tends to sink. In the stratosphere (the Greek word "strato" means "layered") vertical motion is inhibited by the inversion because the warmer air now resides at a higher altitude and tends to remain there. The net result is that a long time constant for exchange between the stratosphere and the troposphere. The very existence of the stratosphere is due to the absorption of sunlight by molecules or aerosols that are chemically produced in the upper atmosphere. Hence, there is a highly nonlinear feedback between chemistry, radiation, and dynamics, such that the absorbers will remain longer and cause a larger thermal inversion, resulting in trapping even more absorbers in the stratosphere. Climatology has an analogy of this kind of positive feedback: when ice melts, the surface albedo decreases, which results in greater amounts of sunlight being absorbed, leading to further melting of ice. The instability of ice is an intricate subject in climatology. Little is known about the stability of the stratosphere.

Table 1.4 Three most abundant gases in each planetary atmosphere

Jupiter	H_2 (0.93)	He (0.07)	CH_4 (3×10^{-3})
Saturn	H_2 (0.96)	He (0.03)	CH_4 (4.5×10^{-3})
Uranus	H_2 (0.82)	He (0.15)	CH_4 (2.3×10^{-2})
Neptune	H_2 (0.80)	He (0.19)	CH_4 ($1 - 2 \times 10^{-2}$)
Titan	N_2 (0.95–0.97)	CH_4 (3.0×10^{-2})	H_2 (2×10^{-3})
Triton	N_2 (0.99)	CH_4 (2.0×10^{-4})	CO (< 0.01)
Pluto	N_2 (?)	CH_4 (?)	CO (?)
Io	SO_2 (0.98)	SO (0.05)	O (0.01)
Mars	CO_2 (0.95)	N_2 (2.7×10^{-2})	Ar (1.6×10^{-2})
Venus	CO_2 (0.96)	N_2 (3.5×10^{-2})	SO_2 (1.5×10^{-4})
Earth	N_2 (0.78)	O_2 (0.21)	Ar (9.3×10^{-3})

Note: Mixing ratios are given in parentheses. All compositions refer to the surface (or 1 bar for the giant planets).

1.3 Chemical Composition of Planetary Atmospheres

A preliminary survey of the composition of planetary atmospheres is given in table 1.4. To limit the scope of discussion, we list only the three most abundant gases, along with concentrations expressed in mixing ratios or mole fractions. It is clear that the composition of the giant planets is dominated by H_2 and He. The dominance of hydrogen is natural given that hydrogen is the most abundant element in the universe, followed by He. The next most abundant element is C, present in the most reducing form as CH_4. The giant planets are so massive that they have preserved the bulk of their atmospheres since the origin of the planets.

The atmospheric compositions of the satellites and Pluto are quite different from those of the giant planets, most importantly in the scarcity of H_2. The major gases are heavy molecules such as N_2, CH_4, and SO_2. The small amount of H_2 in the atmosphere of Titan is derived from the photolysis of CH_4 and is not primordial. As we shall discuss in section 1.4, the gravity of these small bodies is so weak that light species such as hydrogen cannot be retained over geologic time. The overall atmospheric composition of Titan, Triton, and Pluto is mildly reducing. Io's atmosphere is strongly oxidizing, dominated by SO_2 of volcanic origin. Due to its unusual tidal heating by Jupiter and the other Galilean satellites, the interior of Io is probably highly evolved, having lost all of its hydrogen-bearing species.

Mars, Venus, and Earth form a trio of "terrestrial planets" in the inner solar system. The major gases are of neutral oxidation state (N_2) or highly oxidized (CO_2, H_2O). Hydrogen and other reducing compounds are rare. The greatest surprise in the composition of the atmospheres of the inner planets is the large amount of O_2 in the terrestrial atmosphere. Indeed, no other planet in the solar system has this much O_2. This molecule is so reactive chemically that it must be continuously produced at enormous rates to maintain its existence in the atmosphere. In addition, it was shown by Urey (1959) that even N_2 would be unstable in the presence of O_2 and H_2O, and the stable product is HNO_3:

$$2N_2 + 5O_2 + 2H_2O \rightarrow 4HNO_3 \qquad (1.13)$$

Thus, Earth's atmosphere can only be the result of a large input from the biosphere, a view that is championed by Lovelock and expounded in chapter 9.

There is another preliminary survey we can make that gives us further insight into the state of oxidation in the planetary atmospheres. Consider the abundances of the major observed carbon species, CH_4, CO, and CO_2. In the giant planets, the most reduced form, CH_4, dominates, but the more oxidized form, CO, is also present at a concentration that is about 10^5 to 10^6 times less than that of CH_4. In the atmospheres of the satellites, CH_4 is still the dominant carbon species, but the abundance of CO (relative to CH_4) is much higher than in the giant planets. Even CO_2 is present in the atmosphere of Titan and on the surface of Triton. The carbon speciation is fundamentally different in the inner solar system. CO_2 is the dominant carbon species in the atmospheres of Mars and Venus, followed by smaller amounts of CO. In Earth's atmosphere, the bulk of carbon is in the form of CO_2, followed by CH_4 and CO, and this partitioning is the consequence of biospheric input. It is a triumph of atmospheric modeling that the chemical pathways for the interconversion of CH_4, CO, and CO_2 in the planetary atmospheres are now understood. For comparison, we note that CO can dominate the composition of giant molecular clouds and the atmospheres of comets. As we show in chapter 4, a molecular cloud is the precursor of the solar nebula, and comets are primitive bodies left over from the time of formation of the solar system. There are no planetary atmospheres that resemble the composition of molecular clouds or comets, the primitive material for the solar system. Therefore, we conclude that the atmospheres today have undergone extensive evolution since the "beginning."

What was the state of oxidation of Earth's atmosphere just after formation? The answer to this question has profound implications for the origin of life on this planet. Urey (1959) argued that the fact that Titan's atmosphere contains CH_4 proves that the early atmosphere must be massively reducing, dominated by CH_4, NH_3, and H_2S. We show in chapter 9 that this argument is no longer compelling. Based on our current understanding, the primordial terrestrial atmosphere was closer to the present atmospheres of Mars and Venus than that of Titan. This discussion serves to illustrate why the state of oxidation of planetary atmospheres is such a fundamental question, bearing on one of the most profound issues.

1.4 Stability of Planetary Atmospheres against Escape

Why do atmospheres exist? Why do they continue to exist 4.6 billion years after formation? There is no atmosphere (greater than the critical pressure) on the smaller bodies in the inner solar system: Mercury, the Moon, and the asteroids. In contrast, the giant planets possess massive atmospheres. Mars, Venus and the Earth have smaller atmospheres. If we assume that planetary bodies were endowed with atmospheres initially (from either outgassing or subsequent impacts), how stable are the atmospheres against escape to space?

The gas molecules we think of as bound to an atmosphere are in reality particles trapped in the planetary gravitational potential well, as shown in figure 1.5. It is probably not a cosmic coincidence that the order of magnitude of gravitational potential energies for the light gases in the atmospheres of the terrestrial planets is the same

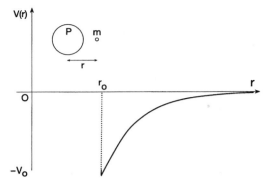

Figure 1.5 Schematic illustration of the planetary gravitational potential well, $V(r) = -GMm/r = -V_0 r_0 / r$, where M and m are, respectively, the mass of the planet (P) and that of the molecule.

as that for the chemical bonds for simple molecules. If this were not so, the atmospheres would be either too unstable or too stable. In the first case there would be no atmospheres (as on the asteroids); in the latter case there would be little planetary evolution (as on the giant planets). We speculate that the development of higher forms of life on Earth is related to planetary evolution. Our planet apparently started in a more reducing state, followed by an inexorable evolution toward a more oxidizing state due to the loss of hydrogen by escape to space. Thus, a chemical arrow of time permeates every chemical reaction on this planet. Is this the ultimate driving force of biological evolution? This subject is further discussed in chapter 9.

There is an analogy to the above speculation. According to John A. Wheeler our entire universe may just be a cosmic accident. If the total mass of the universe were several orders of magnitude higher, the universe would have collapsed long ago, in a "Big Crunch." However, if the total mass were several orders of magnitude smaller, the expansion would have been too fast, and there would be no galaxies or planets. The fact that the total mass of the universe is within an order of magnitude of being either an "open universe" or a "closed universe" is a most profound cosmic puzzle. We can speculate that universes might have existed that were too dense or too tenuous for the formation of planets and the origin of life, just as in our solar system some planetary bodies are too massive or too small to develop and sustain life. The pursuit of this topic, interesting as it is, is beyond the scope of this book.

Table 1.5 gives the binding energies for hydrogen and oxygen atoms in planetary atmospheres. We note that for the giant planets, the binding energies are higher than thermal energies (of the order of 0.1 eV) and chemical bond energies (of the order of 10 eV). Hence, the loss of their atmospheres by massive escape is highly unlikely. Loss of atmosphere from the smaller planetary bodies is possible, even thermally. The theory was developed in its definitive form by James Jeans (1916) to explain why the Moon has no atmosphere (based on the observation of the sharpness of shadows of lunar mountains on the lunar surface), and is based on a straightforward application of the kinetic theory of gases. At temperature T the number density of molecules with velocity between v and $v + dv$ is given by the Maxwell-Boltzmann distribution:

$$n(v) = n_0 \left(\frac{m}{2\pi kT}\right)^{3/2} \exp(-mv^2/2kT) \qquad (1.14)$$

Table 1.5 Gravitational binding energies for hydrogen and oxygen atoms in planetary atmospheres at the critical level (eV)

Planetary body	H	O
Jupiter	17.2	276
Saturn	5.64	90
Uranus	2.32	37
Neptune	2.84	45.0
Titan	0.036	0.58
Triton	0.011	0.18
Pluto	0.008	0.13
Io	0.034	0.55
Mars	0.13	2.1
Venus	0.55	9.0
Earth	0.66	10.6

where n_0 is the total number density of molecules and k and m refer to the Boltzmann constant and molecular mass, respectively. Although the mean kinetic energy of each molecule is $\frac{3}{2}kT$, there is a "Maxwellian tail" consisting of a small number of highly energetic particles, some of which can escape if they reach the critical level. The crucial physical quantity in the problem is the dimensionless parameter

$$\lambda = \frac{GMm}{rkT} \tag{1.15}$$

where GMm/r is the gravitational potential energy and kT is a measure of the thermal kinetic energy of the molecule. λ is thus the the gravitational energy measured in units of kT. A formula that quantitatively gives the rate of escape of gas from a planet can be derived from the kinetic theory of gases:

$$\phi = \frac{n_c v}{2\sqrt{\pi}}(1+\lambda)\exp(-\lambda) \tag{1.16}$$

where n_c is the number density of the escaping gas at the critical level, $v = \sqrt{2kT/m}$, and ϕ is in units of cm^{-2} s^{-1}. This formula is known as the Jeans escape formula. Note that when the gravitational binding energy is much larger than kT, λ is large and the exponential factor in (1.16) ensures a small escape rate; thus, the atmosphere is stable. When the gravitational binding energy is smaller than kT, we can ignore the exponential factor in (1.15). The escape rate is then roughly the rate of thermal expansion of the gas into a vacuum; thus, the atmosphere would be unstable. Table 1.6 gives the escape parameter and escape velocities for hydrogen and oxygen in planetary atmospheres. From the escape rates, we can estimate a lifetime for an atmosphere. This provides a simple way to classify all planetary atmospheres according to their rates of evolution:

1. no evolution: giant planets
2. moderate evolution: Mars, Venus, Earth
3. extensive evolution: Titan, Triton, Pluto, Io
4. catastrophic evolution: Mercury, Moon, comets

Table 1.6 Jeans escape parameters for hydrogen and oxygen atoms

Planetary body	Temperature[a] (K)	λ_H[b]	λ_O[b]	V_H[c] (cm s^{-1})	V_O[c] (cm s^{-1})
Jupiter	1200	165	2665	0	0
Saturn	400	162	2614	0	0
Uranus	810	33	531	1.6×10^{-8}	0
Neptune	540	61	977	1.7×10^{-20}	0
Titan	185	2.3	36	1.6×10^4	1.1×10^{-10}
Triton	95	1.3	22	2.2×10^4	5.7×10^{-5}
Pluto	100	0.9	14	2.8×10^4	0.11
Io	500	0.8	13	6.6×10^4	0.64
Mars	365	4.1	67	5.9×10^3	0
Venus	400	16.2	260	0.11	0
Earth	1200	6.3	100	1.7×10^3	0
Moon	390	0.9	14	5.5×10^4	0.22
Mercury	700	1.6	25	5.0×10^4	2.4×10^{-5}

[a] Representative exospheric temperature.
[b] Jeans escape parameters for hydrogen and oxygen atoms, computed using (1.15).
[c] Jeans escape velocity for hydrogen and oxygen atoms. The product of this velocity and the density of the escaping species equals the escape flux (1.16).

We should point out that the Jeans formula is not rigorously correct for large escape rates, because it is derived for an equilibrium distribution of gas velocities. Nor is it the only process for escape from an atmosphere; nonthermal escape can be important in many atmospheres. Limiting factors such as diffusion also play a fundamental role in regulating the rate of escape. But for the purpose of this chapter, Jeans escape gives us a preliminary view of the stability of planetary atmospheres.

2

Solar Flux and Molecular Absorption

2.1 Introduction

In this book we are concerned primarily with disequilibrium chemistry, of which the sun is the principal driving force. The sun is not, however, the only source of disequilibrium chemistry in the solar system. We briefly discuss other minor energy sources such as the solar wind, starlight, precipitation of energetic particles, and lightning. Note that these sources are not independent. For example, the ultimate energy source of the magnetospheric particles is the solar wind and planetary rotation; the energy source for lightning is atmospheric winds powered by solar irradiance. Only starlight and galactic cosmic rays are completely independent of the sun.

While the sun is the energy source, the atoms and molecules in the planetary atmospheres are the receivers of this energy. For atoms the interaction with radiation results in three possibilities: (a) resonance scattering, (b) absorption followed by fluorescence, and (c) ionization. Ionization usually requires photons in the extreme ultraviolet. The interaction between molecules and the radiation field is more complicated. In addition to the above (including Rayleigh and Raman scattering) we can have (d) dissociation, (e) intramolecular conversion, and (f) vibrational and rotational excitation. Note that processes (a)–(e) involve electronic excitation; process (f) usually involves infrared radiation that is not energetic enough to cause electronic excitation. The last process is important for the thermal budget of the atmosphere, a subject that is not pursued in this book. Scattering and fluorescence are a source of airglow and aurorae and provide valuable tools for monitoring detailed atomic and molecular processes in the atmosphere. Processes (c) and (d) are most important for determining the chemical composition of planetary atmospheres. Interesting chemical reactions are initiated when the absorption of solar energy leads to ionization or the breaking of chemical bonds. In this chapter we provide a survey of the absorption cross sections of selected atoms and molecules. The selection is based on the likely importance of these species in planetary atmospheres.

2.2 Solar Radiation and Other Energy Sources

2.2.1 Solar Energy Flux

The sun radiates with a total luminosity of 3.8×10^{26} W, with its energy source ultimately derived from nuclear fusion in the core. The nuclear fuel consists of the light elements hydrogen and helium, material that was created in the Big Bang, the origin of the universe, about 15 billion years ago. The sun is a typical main sequence star of type G2V, formed about 5 billion years ago, and has enough fuel for another 5 billion years. This is the largest source of energy in the solar system, since the next nearest star is 270,000 AU away. The contribution of stellar radiation is generally small, except perhaps for brief periods following a supernova explosion. The sun's average energy flux per unit area incident on Earth is 1373 W m^{-2}, and this is known as the solar constant, F_0, which is actually not constant with time (see section 2.2.2).

The sun's energy has a characteristic spectral distribution, which to first order is blackbody radiation with emission temperature $T = 5785$ K. The expression for this distribution is the Planck function,

$$B_\nu(T) = \frac{2h\nu^3}{c^2(e^{h\nu/kT} - 1)} \tag{2.1}$$

where ν = frequency (Hz), h = Planck's constant, k = Boltzmann's constant, and c = velocity of light. B_ν is known as the specific intensity of radiation, with units of energy per unit area per second and per hertz per steradian. There is a straightforward integral over ν,

$$\int_0^\infty B_\nu d\nu = \frac{\sigma T^4}{\pi} \tag{2.2}$$

where σ is the Stefan-Boltzmann constant. It can be shown that the energy flux from the surface of the sun is just π times integral (2.2):

$$F_{\text{sun}} = \sigma T^4 \tag{2.3}$$

The solar constant F_0 is related to F_{sun} by

$$F_0 = F_{\text{sun}} \left(\frac{R_{\text{sun}}}{d}\right)^2 \tag{2.4}$$

where R_{sun} is the radius of the sun and d is the distance from the sun to Earth (1 AU). We sometimes prefer to express the blackbody radiation as a function of wavelength $\lambda = c/\nu$. In this case we have

Solar Flux and Molecular Absorption   19

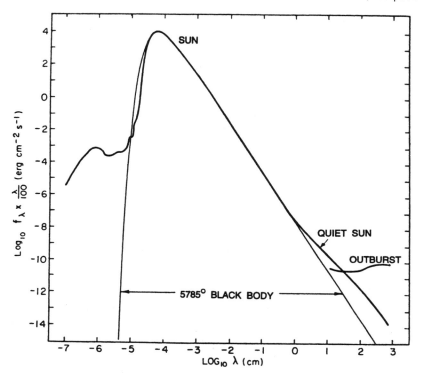

Figure 2.1 The solar irradiance from 10^{-7} to 10^3 cm. After Allen, C. W., 1958, "Solar Radiation." *Quart. J. Roy. Meteorol. Soc.* **84**, 307.

$$B_\lambda = \frac{C_1 \lambda^{-5}}{\pi(\exp(C_2/\lambda T) - 1)} \qquad (2.5)$$

where $C_1 = 2\pi hc^2$ and $C_2 = hc/k$.

The observed spectral distribution of solar flux is shown in figure 2.1, along with a theoretical Planck curve for comparison. A broad-band smoothing has been applied to the observed spectrum to remove the sun's Fraunhofer absorption features. Note that the bulk of radiation under the main peak from 3000 Å (soft ultraviolet) to 1 cm (microwave) is very well approximated by a blackbody radiation at 5785 K. The agreement is good over an impressive five orders of magnitude in wavelength. However, under 3000 Å, the agreement with the Planck curve breaks down. The sun radiates far more energy in the ultraviolet than that predicted by the Planck distribution, because the bulk of the solar energy originates from the visible disk of the sun, the photosphere, where the solar atmosphere is in thermodynamic equilibrium and the assumptions of the blackbody are valid. However, the ultraviolet radiation is derived from the upper atmosphere of the sun, the corona, which radiates at 10^6 K and is out of equilibrium with the photosphere. We should point out that the mechanism that sustains the extremely hot corona of the sun is not well understood. This problem is

Table 2.1 Solar spectral irradiance at 1 AU

λ (Å)	Energy flux (fraction of solar constant $< \lambda$)	Photon flux (photons cm^{-2}s^{-1})
1200	4.4×10^{-6}	1.97×10^{11}
2000	8.1×10^{-5}	1.14×10^{13}
2500	1.9×10^{-3}	2.49×10^{14}
3000	1.2×10^{-2}	2.26×10^{15}
3500	4.5×10^{-2}	9.14×10^{15}
4000	8.7×10^{-2}	1.98×10^{16}
5000	2.3×10^{-1}	6.44×10^{16}
8000	5.6×10^{-1}	2.15×10^{17}

reminiscent of that of the hot thermospheres of the giant planets discussed in chapter 1. The variation in the corona is largely responsible for variations of the solar ultraviolet flux over a solar cycle.

We also note the large excess emission of the sun in the radio region compared with blackbody emission (see figure 2.1). The reason for the excess is the same as that of the ultraviolet excess: Both emissions are from the corona. Hence, even though the radio emissions play no role in the photochemistry of planetary atmospheres, they are easily observed from the ground and the data can serve as an indicator of coronal activity. Since there is no simple way to monitor the variability of solar ultraviolet radiation from the ground, the 10.7 cm emission from the sun serves as a useful proxy.

A coarse resolution tabulation of the solar flux is given in table 2.1. For convenience we also list the photon flux, which is a particularly useful quantity when we study photochemistry. The total number of chemical bonds broken is equal to the total number of photons absorbed times the probability of dissociation. Table 2.1 shows that the bulk of solar energy is in the visible and near infrared region. Ultraviolet radiation, which is crucial for photochemistry, amounts to a tiny fraction of the solar constant. Thus, all the solar flux below 3000 Å contributes about 1% to the integrated flux, or 10^{15} photons cm^{-2} s^{-1}. The portion of solar energy that is capable of ionizing simple molecules, at Lyman α and shorter wavelengths, is only 10^{-5} of the solar constant; the total photon flux is 10^{12} cm^{-2} s^{-1}. Hence, the total column-integrated rate of ionization in the terrestrial atmosphere cannot exceed this value (10^{12} cm^{-2} s^{-1}), which determines the maximum state of ionization in Earth's ionosphere.

2.2.2 Solar Variability

The bulk of solar energy originates from the photosphere, where the ambient pressure of the solar atmosphere is high and variability is low. It is difficult to detect from the ground such small variations in the solar constant, despite a century of such effort. The only reliable data were taken from space-borne platforms in the last decade. Figure 2.2 shows the variability of solar irradiance from 1980 to 1989. The short-term variability could be as high as 0.3%, but the long-term variability is less than 0.1% over this period. Variations of this magnitude cannot account for substantial climatic changes based on known mechanisms.

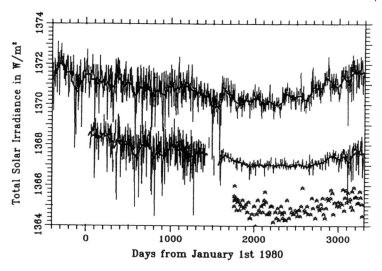

Figure 2.2 Time series of total solar irradiance taken by satellites studies: ERB (top), ACRIM (middle), and ERBE (bottom). The lines are 81 day running means. After Frohlich, C. et al., 1991, "Solar Irradiance Variability from Modern Measurements," in Sonett et al. (1991; cited in section 2.2), p. 11.

The most striking evidence of the sun's variability is the visual disturbance of the photosphere caused by sunspots. There is a continuous record of the observed number of sunspots since their discovery by Galileo. The record from 1749 to 1991, as shown in figure 2.3, reveals a sunspot cycle with an average period of 11 yr. The amplitude of each cycle is not constant and can vary between zero and 200 sunspots. There appears to be another modulation of the amplitude with period of about 88 yr. In the seventeenth century (not shown), there was a period of over 50 yr during which the sunspots disappeared. This interval, known as the Maunder Minimum, corresponds to an unusually cold period in the northern hemisphere. There are speculations that the solar irradiance was reduced at that time, but this has not been proved.

A number of other phenomena are related to or modulated by the sunspot cycle. The solar magnetic field reverses polarity every 11 yr, completing a full cycle in 22 yr. Since galactic cosmic rays are deflected by the solar magnetic field, this results in a modulation of the galactic cosmic ray flux incident on the planets. Galactic cosmic rays are the primary source of ^{14}C in the atmosphere, so the ancient record of ^{14}C (e.g., in tree rings) provides a useful proxy for the sunspot cycle in prehistoric times. This record shows that in addition to the 11 and 88 yr periods evident in figure 2.3, there are other prominent periods at 208 and 2300 yr. Another solar output that varies strongly with the solar cycle is the solar wind (see section 2.2.4). By far the most important for the atmospheres of the planets, the ultraviolet radiation varies with the sunspot cycle.

The ultraviolet portion of the solar radiation originates from the corona of the sun, where the pressure is low and the temperature is high. This region is strongly disturbed

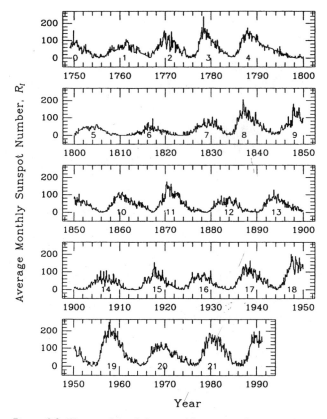

Figure 2.3 Time series of the monthly averaged sunspot numbers from 1749 to 1991. The solar cycle number is given under each peak. After Donahue, R. A., and Baliunas, S. L., 1992, "Periodogram Analysis of 240 Years of Sunspot Records." *Solar Physics* **141**, 181.

by solar magnetic activity, of which the visible sunspots are only one manifestation. Variability of ultraviolet radiation over a solar cycle is shown in figure 2.4. Note that there is little variation above 2000 Å. At Lyman α, the variations are about a factor of 2; in the extreme ultraviolet (EUV) region, the variations approach a factor of 10. Most of the solar energy in the EUV is absorbed in the thermospheres of planets. The large variability in the solar EUV results in great variations over a solar cycle in the temperature of planetary thermospheres, the chemistry of the ionosphere, and even the dynamics of the upper atmosphere. Although the influence of the solar cycle on the upper atmosphere is well documented, there is no convincing evidence that the troposphere is affected by the solar cycle, despite occasional claims to the contrary.

The sun's variability over a longer time period is not known from observations but may be inferred from the theory of evolution of stars that are similar to the sun. There seems to be general agreement in the astrophysical community that the sun was

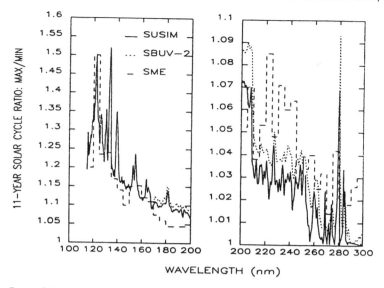

Figure 2.4 Comparison of UV irradiance variability between solar maximum and solar minimum using satellite data from SUSIM (solid line), SBUV-2 (short dashed line), and SME (long dashed line). After Lean, J. et al., 1992, "SUSIM/UARS Observations of the 120 to 300 nm Flux Variations during the Maximum of the Solar Cycle: Inferences for the 11-Year Cycles." *Geophys. Res. Lett.* **19**, 2203.

smaller in radius and cooler in the past. The estimated solar irradiance 5 billion years ago was 40% less than the present value, and it will continue to increase beyond the present value in the next few billion years. This magnitude of change over the age of the solar system must have had a profound impact on the climate of the planets. How the surface of Earth could maintain conditions suitable for the continuation of life in face of such substantial changes in the solar radiation is a puzzle. The composition of the terrestrial atmosphere must have been different. Did Gaia (the biosphere) play a pivotal role in the planetary transformation? We return to these questions in chapter 9.

Recent astronomical observations show that stars like the sun go through an initial phase of enormous activity accompanied by enhanced ultraviolet emissions. The enhancement factor over the current brightness in the EUV could be as high as 10^4 but decreases rapidly with time. This period, the T-Tauri phase, occurred at about the time of the formation of the solar system. The primitive atmospheres of the planets could have been subjected to this intense ultraviolet radiation (see figure 2.5). The resulting photochemistry and rate of evolution would have been different from any subsequent period in the solar system. One interesting possibility is that the huge EUV flux could have induced hydrodynamic escape of gases from planetary atmospheres, a process that differs from the gentler Jeans escape described in chapter 1. Heavy inert gases such as xenon could have escaped from Earth via this mechanism. However, it is not known how transparent the solar system was for the transmission of EUV radiation during this epoch.

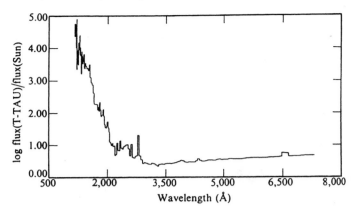

Figure 2.5 Enhancement of UV flux in T-Tauri stars relative to the sun derived from 14 spectra of three T-Tauri stars. After Canuto, V. M. et al., 1983, "The Young Sun and the Atmosphere and Photochemistry of the Early Earth." *Nature* **305**, 281.

2.2.3 Solar Entropy Flux

The total solar power intercepted by Earth is 10^{17} W, of which about 30% is reflected into space. The remaining 70% is absorbed by the planet. But Earth does not store this energy from the sun, and an equal amount of thermal energy is radiated back to space. There is a fundamental difference in the quality of the incident radiation with radiation temperature at 5785 K and that of the emergent radiation from Earth with radiation temperature of about 255 K. There is a quantitative measure of the quality of energy flux, the entropy flux. The mean entropy flux emitted from the top of the atmosphere is about 22 times larger than the mean entropy flux of the incident solar radiation. The sun is thus an enormous source of negative entropy based on its higher radiation temperature alone. This source is required to maintain the chemistry of the atmosphere so far out of equilibrium with thermodynamics. Since entropy and information are different only by a negative constant, we can regard the sun as the source of 10^{37} bits of information per second to Earth. To first order, this is just the total number of photons intercepted by our planet. This may be regarded as the ultimate source of life and intelligence, the ability to create and sustain a high degree of organization.

We should point out that there are, in addition to the radiation entropy, two other "geometric" sources of negative entropy. Due to its large distance from the sun, sunlight reaches Earth in an almost parallel beam. The thermal radiation of the planet is roughly isotropic. Thus, the solar beam can be focused (e.g., using mirrors), whereas the planetary radiation cannot. The former obviously has less entropy. Finally, Earth being a sphere, there is more solar power per unit area incident on the tropics than on the higher latitudes. This unevenness in incident solar flux is a source of negative entropy that maintains highly organized atmospheric motions (e.g., Hadley cells) against dissipation (entropy production).

Table 2.2 Disequilibrium energy fluxes for Earth

Physical process	Energy flux (W/m^2)
Electromagnetic radiation	
Total solar irradiance	1370
Irradiance < 3000 Å	16
Solar wind	3.0×10^{-4}
Visible aurorae (variable)	
International brightness coefficient I	6.0×10^{-4}
International brightness coefficient IV	0.6
Auroral particles (variable)	
Bright auroral arc	1.0
Proton aurora, very disturbed conditions	3.6×10^{-2}
Polar cusp electrons	1.0×10^{-3}
Polar cusp protons	1.0×10^{-4}
Joule heating of the thermosphere	
(E = 100 mV/m)	0.14
Solar protons (polar cap absorption, variable)	2.0×10^{-3}
Galactic cosmic rays	7.0×10^{-6}
Downward conduction of heat from magnetosphere	3.2×10^{-5}
Lightning	3.0×10^{-3}
Starlight	2.0×10^{-5}
Internal heat of earth	8.7×10^{-2}

The above discussion is directed at Earth but can obviously be generalized to all planetary atmospheres with minor modifications.

2.2.4 Other Sources of Energy

Table 2.2 lists the major energy sources for Earth's atmosphere. Note that all other sources are very small compared with solar radiation. However, in the outer solar system, the sun's influence falls off rapidly with distance, but the galactic sources such as starlight or galactic cosmic rays remain constant and become relatively more important. We briefly discuss the characteristics of each of the energy sources.

The solar wind, discovered in the 1950s, is a stream of corpuscular radiation emanating from the sun. It is an ionized plasma and carries with it its own magnetic field. At 1 AU the the speed of the solar wind is approximately 500 km/sec and the mean density is 5 particles/cm^3. The energy flux and particle flux are 3×10^{-4} W m^{-2} and 2.5×10^8 cm^{-2} s^{-1}, respectively. The most abundant elements in the solar wind are H ($\sim$0.96), He ($\sim$0.04), and O ($\sim 5 \times 10^{-4}$), where the numbers refer to mole fraction. Though the energy flux carried by the solar wind is negligible compared with the photon flux, the momentum flux is higher than that due to photons. Thus, in planetary atmospheres that are not shielded by magnetic fields, the solar wind interacts directly with the exosphere, causing sputtering of the atmospheres. Secondary processes involve local acceleration of ions by the solar wind magnetic field and may enhance the direct sputtering process. Direct and induced solar wind sputtering may have played an important role in the evolution of the atmospheres of Mars and Venus and may have contributed also to the loss of primordial atmospheres on Mercury and the Moon.

26   Photochemistry of Planetary Atmospheres

On planets with intrinsic magnetic fields, the solar wind is repelled at distances many planetary radii away from the planet. For example, the solar wind at Earth is stopped ~10 Earth radii away, and at Jupiter this occurs at ~60 Jovian radii. The exact distance varies with solar activity. The solar wind in fact defines the outer limits of planetary magnetospheres. Thus, the total solar wind energy flux that is intercepted by a magnetic planet far exceeds that of a nonmagnetic planet. The solar wind is a major energy source for Earth's magnetosphere. The precipitation of energetic particles from the magnetosphere into the upper atmospheres of planets (usually in the polar regions) gives rise to spectacular displays known as aurorae. The associated energy fluxes are highly variable. Representative values are given in table 2.2.

The solar system is bathed in starlight—we can observe stars at night. In the inner solar system the intensity of this radiation is trivial compared with that of the sun, but in the outer solar system starlight is not negligible. Starlight is essentially isotropic (there is little diurnal or latitude effect). The spectrum is known and can be approximated by a superposition of several blackbody emissions, as shown in

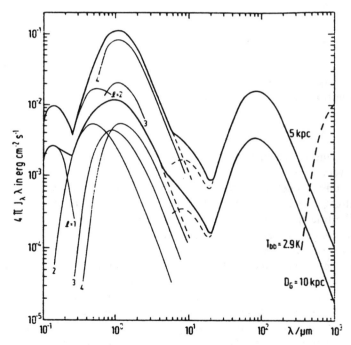

Figure 2.6 The interstellar radiation field at galactocentric distances $D_G = 5$ and 10 kpc (kiloparsecs), respectively. Curves labeled $l = 1, 2, 3, 4$ relate to contributions of four stellar components to the interstellar radiation field. The sun is in the vicinity of $D_G = 10$ kpc. T_{bb} = black body temperature. After Mathis, J. S., Mezger, P. G., and Panagia, N., 1983, "Interstellar Radiation Field and Dust Temperatures in the Diffuse Interstellar Matter and in Giant Molecular Clouds." *Astron. Astrophys.* **128**, 212.

figure 2.6. The hotter blackbody emits more ultraviolet radiation, which could be an important source of ionization for Earth's atmosphere at night or in the polar night. The integrated intensity of radiation from the interstellar radiation field is 2.17×10^{-5} W m^{-2}.

The sun is a potent source of hydrogen Lyman α emission. Part of it reaches the planets directly, but a smaller fraction can reach the planets indirectly after being scattered by hydrogen atoms in the interstellar medium. The backscattered Lyman α is not important for the inner solar system but becomes comparable to the direct solar Lyman α at Neptune.

The solar system is bombarded by galactic cosmic rays, the majority of which are deflected by the solar magnetic field and the solar wind. However, the most energetic particles can penetrate into the solar system. Figure 2.7 shows the energy spectrum of galactic cosmic rays at 1 AU in 1977 (solar minimum) and 1970 (solar maximum). For comparison, the inferred spectrum in the local interstellar medium is given by the dashed curve. Note the cutoff of low-energy particles below 10 GeV by the sun at the minimum and maximum of a solar cycle. Particles with energies above 10 GeV are not modulated by the sun. The galactic cosmic rays have great penetrating power. In Earth's atmosphere, they are stopped in the stratosphere and the upper troposphere. They provide an important source of NO and ^{14}C to the atmosphere. In the outer solar system, ionization induced by the absorption of galactic cosmic rays results in organic synthesis.

Lightning is common in the terrestrial atmosphere and has been detected in the atmosphere of Jupiter via optical observations. Reports of detections of lightning in planetary atmospheres using plasma waves are controversial. In a lightning bolt an

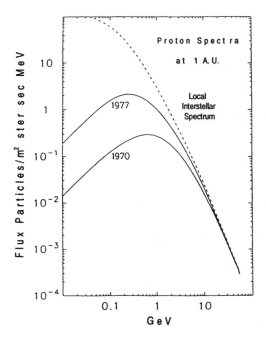

Figure 2.7 Cosmic ray proton spectra in 1970 (solar maximum) and 1977 (solar minimum). The dashed line is an estimate of the unmodulated spectrum. After Beer, J., Raisbeck, G. M., and Yiou, F., 1991, "Time Variations of ^{10}Be and Solar Activity", in Sonett et al. (1991; cited in section 2.2), p. 343.

Table 2.3 Ionization potentials of selected molecules of atmospheric interest

Species	Ionization potential (eV)	Equivalent wavelength (Å)
Na	5.1	2412
Mg	7.7	1621
Si	8.2	1520
NO	9.3	1350
CH_3	9.8	1262
S	10.4	1196
C	11.3	1100
O_2	12.1	1026
H_2O	12.6	987
O_3	12.8	970
CH_4	13.0	954
SO_2	13.1	946
H	13.6	912
O	13.6	911
CO_2	13.8	899
HCN	13.9	892
CO	14.0	885
N	14.5	852
H_2	15.4	804
N_2	15.6	799
Ar	15.8	787
He	24.6	504

Compiled from Franklin, J. L. et al. (1969).

air parcel is initially heated up to 2000 K, followed by rapid cooling due to adiabatic expansion. The composition of the parcel equilibrates with the high temperature until the "freezing temperature" (about 1000 K) is reached. After that, as the temperature gets even lower the composition will not undergo further change. The most important nonequilibrium product of lightning in Earth's atmosphere is NO.

2.3 Absorption Cross Sections of Atmospheric Gases

As pointed out in section 2.1, the two most important processes in planetary atmospheres are ionization and dissociation of atmospheric species following the absorption of a photon. These processes require energies above certain thresholds, known respectively as the ionization potential and the dissociation energy. Tables 2.3 and 2.4 give these threshold energies for atoms and simple molecules that are likely to be important in planetary atmospheres. Note that ionization potentials are generally high, of the order of 10 eV (except for metals), and would require EUV photons. Since the solar output in the EUV is relatively small, this limits the rate of ionization in the planetary thermospheres. Dissociation energies for most common molecules are much lower, and at longer wavelengths more ultraviolet photons are available. The absorption cross sections, σ, are defined by the Beer's law expression,

$$I = I_0 \exp(-\sigma n l)$$

Table 2.4 Dissociation energies of selected molecules of atmospheric interest

Species	Dissociation energies (eV)	Equivalent wavelength (Å)
H_2	4.52	2743
CH_4	4.55	2722
C_2H_6	4.36	2844
C_2H_4	4.80	2583
C_2H_2	5.76	2153
PH_3	3.40	3650
NH_3	4.70	2637
H_2S	3.96	3134
H_2O	5.17	2398
H_2O_2	2.22	5582
O_2	5.17	2400
O_3	1.10	11222
CO	11.14	1113
CO_2	5.52	2247
H_2CO	3.82	3244
N_2	9.80	1265
N_2O	1.73	7152
NO	6.55	1893
NO_2	3.18	3903
NO_3	2.16	5731
N_2O_5	0.99	12536
HNO_3	2.15	5774
HO_2NO_2	1.01	12320
HCl	4.47	2773
$CFCl_3$	3.25	3811
CF_2Cl_2	3.46	3582
$ClONO_2$	1.16	10665
Cl_2O_2	0.77	16057
SO	5.40	2296
SO_2	5.72	2168
COS	3.20	3873
HCN	5.37	2309
HC_3N	6.34	1955
C_2N_2	5.84	2123

where I_0 and I are the incident and transmitted light intensity, respectively, σ is the absorption cross section (cm^2 molecule^{-1}), n is the molecular density (cm^{-3}), and l is the pathlength (cm). The cross sections presented in figures 2.8–2.22 in this section are smoothed over 50 Å (for photolysis calculations), except in the EUV, where solar resonance lines are important.

2.3.1 Hydrogen and Helium

H and He are the simplest atoms. In EUV the most important transitions are the excitation by Lyman α (1216 Å) and Lyman β (1026 Å) for hydrogen, and by the He resonance line at 584 Å. These excitations are followed by emission of photons

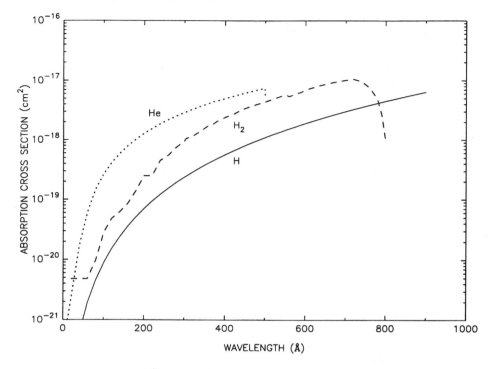

Figure 2.8 Absorption cross section of H, He, and H_2.

of the same wavelengths, a process known as resonance scattering. Since the solar atmosphere contains a large number of H and He atoms, radiation at the resonant wavelengths is greatly enhanced. Scattering of the solar lines by H and He provides valuable information about the physics of planetary thermospheres. The ionization continua for H and He start, respectively, at 912 and 504 Å.

H_2 is the simplest molecule, with ionization potential of 15.42 eV (804 Å). Absorption of radiation below 804 Å leads entirely to ionization with cross sections given by figure 2.8. The dissociation energy of H_2 is 4.52 eV. Therefore, in principle, photons under 2743 Å are capable of dissociating H_2, but there is no absorption until 1108 Å. The reason can be found in the potential diagram for H_2 in figure 2.9. H_2 has no dissociating or predissociating states near the threshold of dissociation. In fact, the lowest excited state is the B state, which lies at 11.4 eV above the ground state. The B state lies close to the C state. Above these are the D, E, and F states, and finally the ionization continuum. The excitation of the H_2 molecule into the upper states is followed by fluorescence known as the Lyman bands (from B state) and the Werner bands (from C state). A small fraction of the Lyman bands (21%) and the Werner Bands (1%) radiates into the dissociating continuum and contributes to the dissociation of the molecule. The absorption cross sections of H, He, and H_2 are shown in figure 2.8.

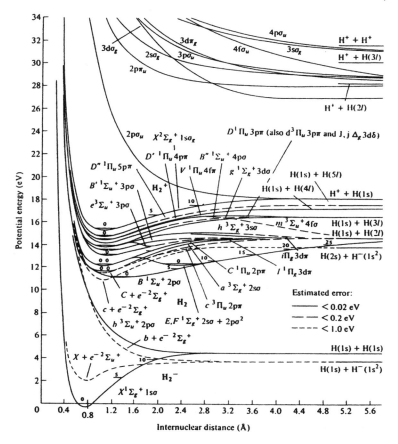

Figure 2.9 Potential energy diagram for H_2^-, H_2, and H_2^+. After Sharp, T. E., 1971, "Potential-Energy Curves for Molecular Hydrogen and Its Ions." *Atomic Data* **2**, 119.

2.3.2 Simple Hydrocarbons

The most abundant hydrocarbon in the outer solar system is methane, followed by its primary photochemical products, C_2H_6, C_2H_2, and C_2H_4. The absorption cross sections are shown in figure 2.10. Ionization at shorter wavelengths is important for planetary ionospheres. The most striking feature of the cross section of CH_4 is that it becomes negligible at wavelengths beyond 1450 Å. This is surprising in view of the relatively weak CH_3-H bond, 4.55 eV (2722 Å). The reason is the same as that given for H_2: Methane does not have low-lying dissociating states. The second simplest alkane is C_2H_6. Its absorption spectrum is similar to CH_4 but shifted slightly to higher wavelengths. C_2H_2 and C_2H_4 have absorptions that are very different from the simple alkanes. The cross sections extend into the region of 2000 Å. Since CH_4 is the most abundant carbon species in the atmosphere, absorption by CH_4 provides shielding for ethane, though the shield is not perfect. Acetylene and ethylene are not

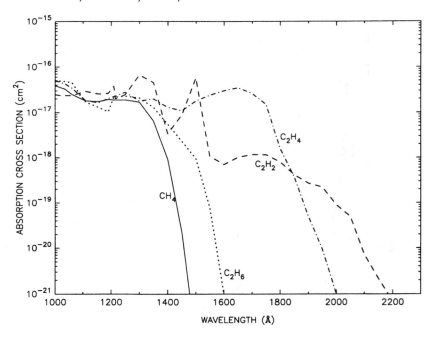

Figure 2.10 Absorption cross section of CH_4, C_2H_6, C_2H_4, and C_2H_2.

shielded at all in the long-wavelength region, where the solar flux is rapidly increasing. Thus, the initial absorption of photons under 1450 Å by CH_4 leads to the production of simple photochemical products, which can then absorb at higher wavelengths. The latter process is particularly interesting for the case of C_2H_2, since the photolysis of C_2H_2 initiates new photochemical reactions that can result in the breaking up of the CH_4 molecule:

$$\begin{aligned} C_2H_2 + h\nu &\to C_2H + H \\ C_2H + CH_4 &\to C_2H_2 + CH_3 \\ \hline \text{net} \quad CH_4 &\to CH_3 + H \end{aligned}$$

This process is known as photosensitized dissociation of CH_4 and could, in principle, be more important than the primary photolytic process for reasons discussed in the previous paragraph. In the laboratory, a classic example is the photosensitized dissociation of CH_4 at 2537 Å, catalyzed by mercury atoms. At this wavelength, CH_4 is transparent but Hg has a resonance line. The excited Hg atom transfers its energy to CH_4, resulting in its dissociation into $CH_3 + H$.

2.3.3 NH_3, PH_3, and H_2S

The bond energy for ammonia (NH_2-H) is 4.70 eV (2637 Å). Absorption becomes important under 2300 Å, reaching peak values near 2000 Å, as shown in figure 2.11. The diffuse bands superposed on the continuum are due to out-of-plane vibration of the excited molecule. Phosphine has a smaller bond energy (PH_2-H) of 3.4 eV

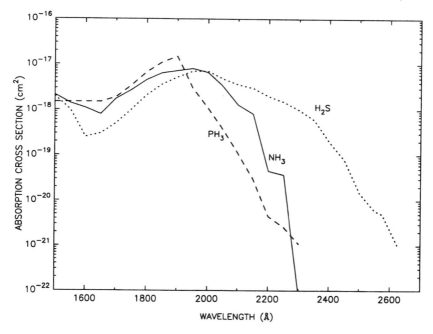

Figure 2.11 Absorption cross section of NH_3, PH_3, and H_2S.

(3650 Å). The absorption cross section, as shown in figure 2.11, closely follows the continuum part of NH_3 absorption. Hydrogen sulfide has a bond energy (HS—H) of 3.96 eV (3134 Å). Absorption starts at about 2500 Å, reaching a maximum at about 2000 Å. The absorption cross section is shown in figure 2.11. Note that NH_3, PH_3, and H_2S all absorb in the ultraviolet at wavelengths above the absorption by the simple hydrocarbons. Hence, there is no shielding by the simple hydrocarbons. Due to the large solar flux at longer wavelengths, these three molecules are less stable than the simple hydrocarbons and cannot survive long enough to be transported to the upper atmosphere of the giant planets.

2.3.4 H_2O and H_2O_2

The water molecule has a bond energy (HO—H) of 5.17 eV (2398 Å). The absorption cross section from about 1900 Å to 1000 Å is shown in figure 2.12. The long-wavelength portion from 1900 to 1400 Å consists of a smooth continuum. Below 1400 Å, diffuse bands become important. Ionization occurs below 1000 Å. The photolysis of H_2O readily leads to the production of H_2O_2, a temporary reservoir of the reactive radical family (H, OH, HO_2). The weaker bond (HO—OH) has energy of 2.22 eV (5582 Å). Absorption starts at 3500 Å (see figure 2.12). Water is shielded by CO_2, O_2, and O_3; hydrogen peroxide is not completely shielded by any of these three molecules.

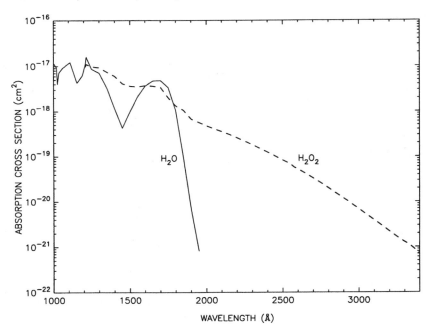

Figure 2.12 Absorption cross section of H_2O and H_2O_2.

There are other important photochemical products that can be generated from the dissociation of H_2O, such as OH and HO_2. However, these radicals are extremely reactive toward other molecules, and interaction with solar radiation is less important.

2.3.5 Oxygen

Oxygen has three important allotropes: O, O_2, and O_3. Oxygen dimer (O_4) is a van der Waals molecule that has been detected in the terrestrial atmosphere but is not known to play a significant role in atmospheric chemistry. Atomic oxygen has two low-lying excited metastable states at 1.97 and 4.19 eV that give rise to the nightglow lines near 6300 and at 5577 Å, respectively. The first resonance transition in the ultraviolet is a triplet at 1302, 1304, and 1306 Å. Ionization occurs below 911 Å. (see figure 2.13).

Figure 2.13 gives the absorption cross sections of molecular oxygen and reveals the complexity and richness of the interaction between O_2 and radiation. The interpretation of the absorption features can be facilitated with reference to the potential diagram for molecular oxygen, as given in figure 2.14. The bond energy of molecular oxygen is 5.17 eV (2400 Å).

The molecule has two low-lying metastable states: the ($a^1\Delta$) state at 0.977 eV, and the ($b^1\Delta$) state at 1.626 eV. These states are important in airglow studies, giving rise to the infrared atmospheric bands and the atmospheric bands, respectively. Perhaps the most interesting feature of O_2 is the excited A state lying just beneath the dissociation threshold. Absorption of ultraviolet radiation to this state results in molecular dissociation. This corresponds to the Herzberg continuum from 1900 to 2424 Å in

Solar Flux and Molecular Absorption   35

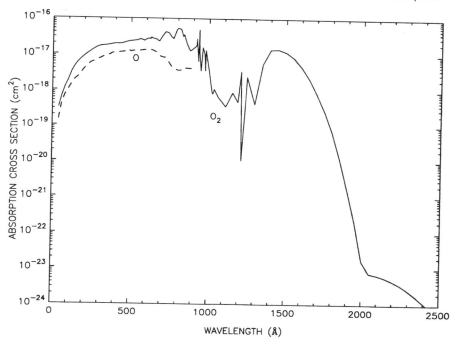

Figure 2.13 Absorption cross section of O_2 and O.

figure 2.13. Between 1900 and 1750 Å the absorption goes through a series of bands known as the Schumann-Runge bands due to the X-B transitions (see figure 2.14). Both of the oxygen atoms are produced in the ground state by predissociation. Between 1750 Å and 1300 Å the absorption goes through another continuum known as the Schumann-Runge continuum. Here one of the oxygen atoms produced is in the excited state, $O(^1D)$. Below 1300 Å there are numerous Rydberg states. Ionization becomes important below 1026 Å.

The bond energy of ozone (O_2-O) is 1.10 eV (11,222 Å). Ozone has three absorption bands, as shown in figure 2.15a. The Chappuis bands start at 9000 Å, reach a peak at 6000 Å, and fall off at 4500 Å. The cross sections are small. The products of dissociation are O and O_2 in the ground states. The Huggins bands range from 3000 to 3600 Å. There is a strong temperature dependence in the cross sections of O_3 in the Huggins bands, as shown in high resolution (1 Å) in figure 2.15b. At lower temperatures, the valleys of the bands become much deeper; the peaks remain roughly constant. The products of dissociation include $O_2(^1\Delta)$ and $O(^1D)$. The $O(^1D)$ yield declines rapidly above 3080 Å. The primary absorption of O_3 is due to the Hartley bands from 2000 Å to 3000 Å, with maximum cross section approaching 10^{-17} cm^2. The products of photolysis are mostly $O_2(^1\Delta)$ and $O(^1D)$, although some fraction (5–10%) of the O is in the ground state. Note that the bulk of the ultraviolet shielding of the terrestrial biosphere is via the Hartley and the Huggins bands of ozone. That the efficiency of these bands for filtering out the actinic rays is high is demonstrated by

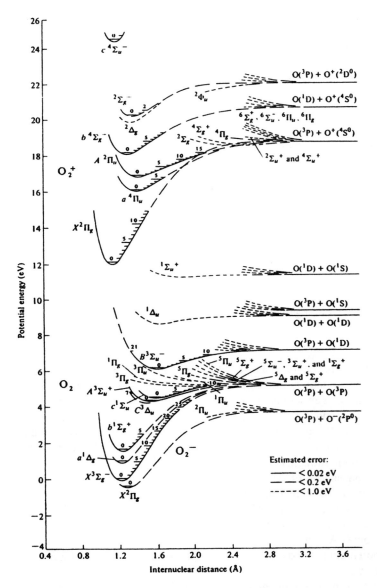

Figure 2.14 Potential energy diagram for O_2. After Gilmore (1971), as quoted in Steinfeld, J. I., 1979, *Molecules and Radiation: An Introduction to Modern Molecular Spectroscopy* (Cambridge, Mass.: MIT Press).

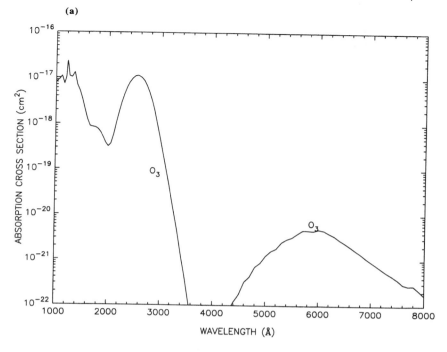

Figure 2.15 (a) Absorption cross section of O_3.

noting that if the ozone layer were condensed into a sheet of solid matter, it amounts to just a few microns in thickness.

2.3.6 CO_2, CO, and H_2CO

The bond energy of carbon dioxide (CO—O) is 5.52 eV (2247 Å). As shown in figure 2.16, absorption begins at about 2300 Å. The absorption near the threshold is weak and goes through a spin-forbidden transition to produce CO and O in the ground states. The absorption spectrum is characterized by three peaks at 1474, 1332, and 1119 Å that correspond to excitation to the various excited states of the molecule. Ionization becomes important below 899 Å. At longer wavelengths the absorption cross section of CO_2 has a strong temperature dependence. This has significant consequences for the photochemistry of the atmospheres of Mars and Venus.

Carbon monoxide has one of the strongest bonds among the simple molecules, with bond energy (C—O) equal to 11.14 eV (1113 Å). A weak discrete absorption between 1765 and 2155 Å is the Cameron system. The absorption bands between 1280 and 1600 Å correspond to the fourth positive system. Excitation into these transitions are followed by fluorescence. Absorption followed by dissociation occurs at much shorter wavelengths (see figure 2.16). Ionization becomes important below 885 Å.

One of the photochemical products of CH_4 oxidation in an oxidizing atmosphere is formaldehyde. The bond energy (HCO—H) is estimated to be 3.82 eV (3244 Å). The primary absorption is in the visible wavelengths (see figure 2.16). The products

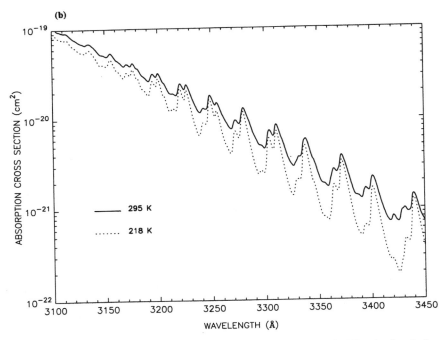

Figure 2.15 (b) High-resolution absorption cross section of O_3 in the Huggins bands for 295 and 218 K, after Malicet, J, et al., 1995, "Ozone UV Spectroscopy. II. Absorption Cross-Section and Temperature Dependence." *J. Atmos. Chem.* **21**, 263.

of dissociation consist of two channels, H + HCO and H_2 + CO. There are absorptions at shorter wavelengths, but these are not important compared with those at long wavelengths for atmospheric applications and are not shown.

2.3.7 The Nitrogen Family

Molecular nitrogen is one of the most stable molecules, and its chemistry is somewhat limited. However, nitrogen has a large number of oxides that play fundamental roles in the photochemistry of planetary atmospheres. This limited survey will include N_2, N_2O, NO, NO_2, NO_3, N_2O_5, HNO_3, and HO_2NO_2.

The bond energy for nitrogen (N–N) is 9.80 eV (1265 Å). The Lyman-Birge-Hopfield bands are the main absorption bands in the 1000–1500 Å region. These absorptions are followed by fluorescence. Dissociation occurs between 600 and 1000 Å. Ionization is possible below 799 Å. The dissociative and ionization cross sections are shown in figure 2.17.

The (N_2–O) bond of nitrous oxide has energy of 1.73 eV (7152 Å). However, the photodissociation threshold is at 2500 Å. The products are N_2 and $O(^1D)$, the yield of N + NO being less than 1%. The absorption peaks at 1809, 1455, and 1291 Å correspond to three distinct excitations of the molecule.

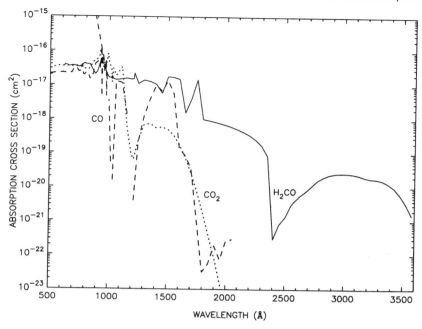

Figure 2.16 Absorption cross section of CO_2, CO, and H_2CO.

The bond energy of nitric oxide (N—O) is 6.55 eV (1893 Å). Absorption begins at 1910 Å via the β, γ, δ, and ϵ bands. Predissociation occurs with the absorption of radiation below 1910 Å. This includes the states $v' > 6$ for β bands, $v' > 3$ for γ bands, $v' > 3$ for ϵ bands, and all of δ bands. Figure 2.18 shows the absorption cross sections in the δ (0-0) bands.

The bond energy of nitrogen dioxide (ON—O) is 3.18 eV (3903 Å). At long wavelengths beyond the dissociation limit, NO_2 absorption is followed by fluorescence. Figure 2.19 shows the absorption cross sections of NO_2 in the ultraviolet. The absorption is known to be slightly dependent on temperature. The quantum yield of dissociation is unity below 3130 Å but falls off rapidly to about 1% at 4358 Å.

Nitrogen trioxide has a bond energy (O—NO_2) equal to 2.16 eV (5731 Å). Figure 2.19 shows that there is strong absorption in the visible spectrum, between 4500 and 7000 Å. The principal products are $NO_2 + O$, with only about 10% contribution of the product channel $NO + O_2$.

Dinitrogen pentoxide has an extremely weak bond (O_2N—NO_3) of 0.99 eV (1.25 μm). The molecule is barely stable against thermal decomposition in the terrestrial atmosphere. The absorption cross sections are shown in figure 2.19. There is a significant temperature dependence.

The weakest bond in nitric acid (HO—NO_2) has energy of 2.15 eV (5774 Å). The primary process occurs from 3100 to 2000 Å (see figure 2.19), with quantum yield of unity for products $OH + NO_2$. Below 1500 Å there is another well-defined region of absorption. The bond (HO_2—NO_2) in pernitric acid has energy 1.0 eV. The

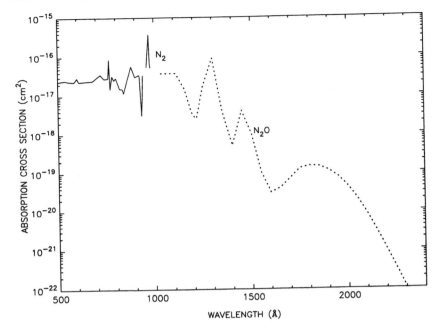

Figure 2.17 Absorption cross section of N_2 and N_2O.

molecule dissociates readily by absorption of solar radiation in the near ultraviolet. The absorption cross sections of HO_2NO_2 are shown in figure 2.19.

2.3.8 HCl, CFC, ClONO$_2$, and Cl$_2$O$_2$

The bond energy of hydrogen chloride (H—Cl) is 4.47 eV (2773 Å). Absorption starts at 2000 Å and reaches a maximum at 1500 Å (see figure 2.20). The primary process produces H + Cl with quantum yield of unity.

There are a large number of chlorofluorocarbons (CFCs). Only $CFCl_3$ (CFC-11) and CF_2Cl_2 (CFC-12) will be discussed here. The weakest bond energy for $CFCl_3$ (Cl—$CFCl_2$) is 3.25 eV (3811 Å). The bond energy for CF_2Cl_2 (Cl—CF_2Cl) is 3.46 eV (3582 Å). The absorption cross sections in the near ultraviolet are shown in figure 2.20. Note that these two CFCs are completely shielded by O_3; there is no destruction until the molecules enter the terrestrial stratosphere.

Chlorine nitrate is a weakly bound molecule. The weaker bond (ClO—NO_2) is 1.16 eV; the stronger bond (Cl—ONO_2) is 1.16 eV. The absorption cross sections in the ultraviolet are given in figure 2.20. The dissociation products are believed to include both channels: ClO + NO_2 and Cl + NO_3. The quantum yield may have an unusual pressure dependence. At high pressure (>50 torr) its value is decreased.

Cl_2O_2 is a dimer of ClO. The weaker bond (ClO − OCl) is 0.77 eV; the stronger bond (Cl—OClO) is 0.92 eV. The absorption cross sections are shown in figure 2.20.

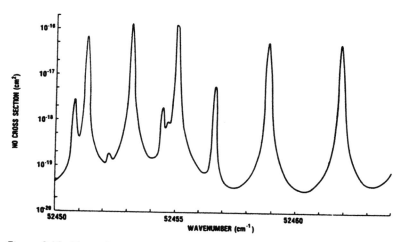

Figure 2.18 Absorption cross section for NO in the delta (0-0) band (52455 cm^{-1} corresponds to 1906.64 Å). After Frederick, J. E., and Hudson, R. D., 1979, "Predissociation of Nitric Oxide in the Mesosphere and Stratosphere." *J. Atmos. Sci.* **36**, 737.

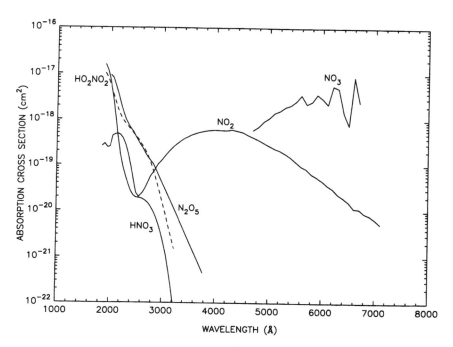

Figure 2.19 Absorption cross section of NO_2, NO_3, N_2O_5, HNO_3, and HO_2NO_2.

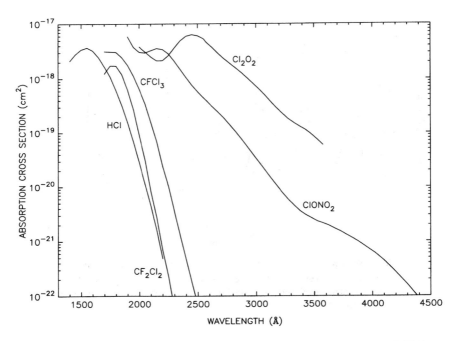

Figure 2.20 Absorption cross section of HCl, CFCl$_3$, CF$_2$Cl$_2$, ClONO$_2$, and Cl$_2$O$_2$.

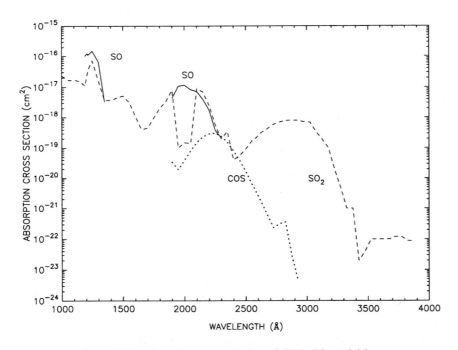

Figure 2.21 Absorption cross section of COS, SO, and SO$_2$.

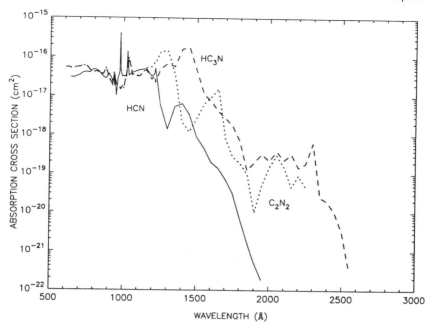

Figure 2.22 Absorption cross section of HCN, HC$_3$N, and C$_2$N$_2$.

2.3.9 COS, SO, and SO$_2$

Carbonyl sulfide has bond energy (OC—S) of 3.2 eV (3873 Å). Absorption starts at about 2500 Å in a continuum to about 2000 Å (see figure 2.21). The primary products are CO and S. At shorter wavelengths, the products include S(^{1}D) and O(^{1}D).

The bond energy for sulfur monoxide (S—O) is 5.40 eV (2296 Å). Figure 2.21 gives the absorption cross sections. The quantum yield for dissociation is not known but may be estimated from the detailed structure of the absorption structure to be close to unity.

The bond energy for sulfur dioxide (O—SO) is 5.72 eV (2168 Å). There is an extremely weak absorption from 3400 to 3900 Å and a weak absorption from 2200 to 3400 Å (see figure 2.21). These absorptions result in fluorescence and phosphorescence. Below 2168 Å absorption is followed by dissociation into products SO + O.

2.3.10 HCN, HC$_3$N and C$_2$N$_2$

The bond energy for hydrogen cyanide (H—CN) is 5.37 eV (2309 Å). Weak absorption starts at about 1900 Å, as shown in figure 2.22. The products of dissociation are H + CN. In general, photolysis of nitrile compounds does not split the strong C—N bond (7.8 eV). The weaker bonds are preferentially broken.

The bond energy for cyanoacetylene (NC—C$_2$H) is 6.34 eV (1955 Å). Two absorption regions are known, from 2300 to 2715 Å and from 2300 to 2100 Å. These are probably transitions to excited states that do not result in dissociation. The absorption cross sections below 2000 Å are shown in figure 2.22. The photolysis products are most likely $C_2H + CN$. The quantum yields are unknown.

The bond energy for cyanogen (NC–CN) is 5.84 eV (2123 Å). Weak absorption bands are from 3020 to 2400 Å and from 2260 to 1820 Å. Strong absorption occurs below 1800 Å (see figure 2.22). The products of dissociation are $CN + CN$.

2.4 Interaction Between Solar Radiation and the Atmosphere

2.4.1 Absorption

For a plane parallel atmosphere containing a gas mixture with number densities $n_i(z)$, we can define an absorption optical depth at wavelength λ by

$$\tau(z, \lambda) = \sum_i \int_z^\infty n_i(z)\sigma_i(\lambda)dz \tag{2.6}$$

where $\sigma_i(\lambda)$ is the absorption cross section at wavelength λ and z is the altitude ($z = 0$ usually refers to the surface). The attenuation of the solar flux that enters the atmosphere at zenith angle θ is given by the expression

$$F(z, \lambda) = F(\infty, \lambda)e^{-\tau(z,\lambda)/\mu} \tag{2.7}$$

where $F(\infty, \lambda)$ is the solar flux at the top of the atmosphere and $\mu = \cos\theta$. The exponential factor in (2.7) suggests that the solar flux is rapidly attenuated as it penetrates into the atmosphere when the absorption is strong and the optical depth is large. The volume absorption rate by species i is given by

$$r_i(z) = J_i(z)n_i(z) \tag{2.8}$$

where

$$J_i(z) = \int \sigma_i(\lambda) F(z, \lambda)d\lambda \tag{2.9}$$

is known as the photodissociation coefficient. The units of $F(z, \lambda)$, $r_i(z)$, and $J_i(z)$ are photons cm^{-2} s^{-1}, photons cm^{-3} s^{-1}, and s^{-1}, respectively. Equation (2.6) ignores the curvature of the planet, an approximation that holds in the terrestrial atmosphere until θ exceeds 87°.

To gain some insight into equations (2.6)–(2.9), we examine the special case of a monochromatic beam of light incident on an isothermal atmosphere containing a single absorbing species whose concentration is given by

$$n(z) = n_0 e^{-z/H} \tag{2.10}$$

where H is the atmospheric scale height. In this case the volume absorption rate is

$$r(z) = n_0 \sigma F(\infty) e^{-z/H - \alpha e^{-z/H}} \tag{2.11}$$

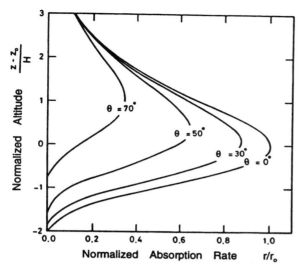

Figure 2.23 Normalized Chapman function as a function of normalized altitude for different values of the solar zenith angle. After Brasseur, G., and Solomon, S., 1984, *Aeronomy of the Middle Atmosphere: Chemistry and Physics of the Stratosphere and the Mesosphere.* (Dordrecht: Reidel).

where $\alpha = \sigma H n_0 / \mu$. If $\alpha > 1$, it can be shown that $r(z)$ has a maximum located at $z_0 = H \log(\alpha)$. At this position the optical path $\tau(z)/\mu$ equals unity. We can rewrite (2.11) as

$$r(z) = r_o \exp\left[1 - \frac{z - z_o}{H} - \exp\left(-\frac{z - z_o}{H}\right)\right] \qquad (2.12)$$

The function $r(z)/r_o$ is known as the Chapman function, where $r_o = \mu F(\infty)/e\sigma H$. Figure 2.23 shows plots of this function for various zenith angles. The function exhibits a well-defined layered structure, known as the Chapman layer. The physical reason for the existence of the Chapman layer is simple. From (2.8) $r(z)$ consists of the product of two terms, $J(z)$ and $n(z)$. At the top of the atmosphere $J(z)$ is a constant but $n(z)$ falls off as a decreasing exponential function, hence, the product is small. Deep down in the atmosphere $n(z)$ increases but $J(z)$ decreases more rapidly; again, the product is small. The maximum occurs at optical path equal to unity. The Chapman function falls off with the characteristic length scale of about H, the atmospheric scale height.

The preceding discussion reveals the special significance of the region of the atmosphere where the optical depth $\tau(z)$ is unity. Figure 2.24 shows the altitude of the optical depth unity in the terrestrial atmosphere. The EUV fluxes are absorbed at high altitudes, resulting in the dissociation and ionization of the major constituents in the thermosphere, leading to the formation of the layers of the ionosphere. At longer wavelengths, the solar flux penetrates deeper into the atmosphere and is absorbed by O_2 in the Schumann-Runge bands and the Herzberg bands, leading to the production of O_3. Beyond 2400 Å the major absorber is O_3 in the stratosphere. Above 3100 Å the atmosphere is transparent except for Rayleigh scattering and aerosol (and cloud) scattering.

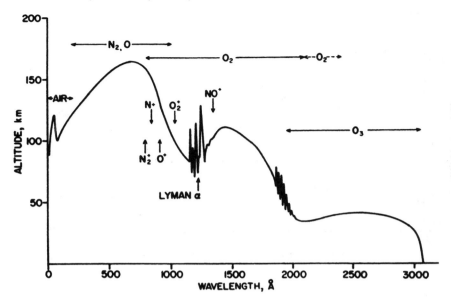

Figure 2.24 Depth of penetration of solar UV radiation in Earth's atmosphere. The line indicates the altitude where optical depth is unity. The vertical arrows show ionization limits. After Herzberg, L., 1965, "Solar Optical Radiation and Its Role in Upper Atmospheric Processes," in *Physics in the Earth's Upper Atmosphere*, C. O. Hines et al., editors (Englewood Cliffs, N.J.: Prentice Hall, 1965), p. 31.

2.4.2 Rayleigh Scattering

Rayleigh scattering is important in planetary atmospheres. Photons are not destroyed in this process but reemitted, with a definite angular distribution known as the phase function. The cross section for Rayleigh scattering is given by

$$\sigma_R = \frac{8\pi^3}{3\lambda^4} \frac{(m^2 - 1)^2}{n^2} D \tag{2.13}$$

where m is the index of refraction of the gas and n is the number density. Note that $m - 1 \ll 1$ and is proportional to n. The depolarization factor D is given by

$$D = \frac{3(2 + \delta)}{(6 - 7\delta)} \tag{2.14}$$

with $\delta = 0.0305$ (N_2), 0.054 (O_2), 0.0805 (CO_2), and 0.0 (Ar). Expression (2.13) contains the famous inverse fourth-power law in wavelength for scattering and is the basis of Rayleigh's explanation of the blue sky.

The attenuation and redistribution of solar radiation in an atmosphere containing both absorbers and scatterers are much more complicated than (2.7). It is necessary to define a few new quantities. The extinction cross section is

$$\sigma_e = \sigma_a + \sigma_s \tag{2.15}$$

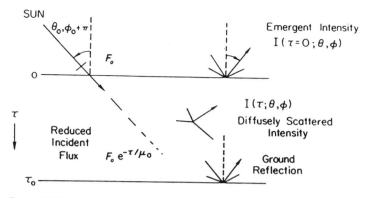

Figure 2.25 Schematic diagram of radiative transfer of solar radiation in an atmosphere, including absorption, scattering, ground reflection, and escape to space. After Chamberlain, J. W., and Hunten, D. M., 1987, *Theory of Planetary Atmospheres* (New York: Academic Press).

where the suffices e, a, and s refer to extinction, absorption, and scattering, respectively. The single scattering albedo is defined as

$$a = \frac{\sigma_s}{\sigma_e} \qquad (2.16)$$

The extinction optical depth is defined as in (2.6) but with σ_e in the place of $\sigma(\lambda)$. The radiation field is completely determined by a fundamental quantity, the specific intensity, $I(\tau, \theta, \phi)$. This function has units of photons cm^{-2} s^{-1} sr^{-1} (sr = steradian) and depends on two angles as well as τ (see figure 2.25). We state, but do not solve, the equation of radiative transfer for the specific intensity for monochromatic unpolarized light:

$$\mu \frac{dI(\tau, \theta, \phi)}{d\tau} = I(\tau, \theta, \phi) - \frac{F_0}{4\pi} e^{-\tau/\mu_0} P(\tau, \theta, \phi; \theta_0 + \pi, \phi_0)$$

$$- \frac{a}{4\pi} \int_0^{2\pi} \int_0^{\pi} I(\tau, \theta', \phi') P(\tau, \theta, \phi; \theta', \phi') \sin\theta' d\theta' d\phi' \qquad (2.17)$$

where $P(\tau, \theta, \phi; \theta', \phi')$ is the phase function for scattering. Figure 2.25 gives a schematic diagram of radiative transfer of solar radiation in an atmosphere, including absorption, scattering, ground reflection, and escape to space, as described equation (2.17) (with appropriate boundary conditions). For Rayleigh scattering, the phase function, ignoring polarization, is

$$P(\cos\Theta) = \frac{3}{4}(1 + \cos^2\Theta) \qquad (2.18)$$

where $\cos(\Theta) = \cos\theta \cos\theta' + \sin\theta \sin\theta' \cos(\phi - \phi')$ and Θ is the angle between the directions of the photon before and after scattering. The phase function is normalized:

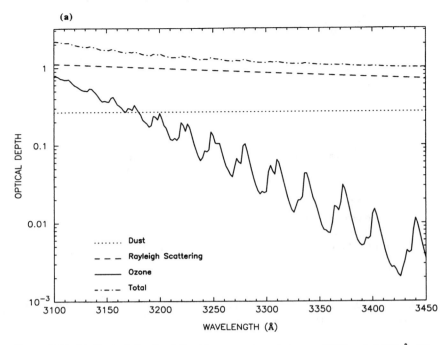

Figure 2.26 (a) Optical depth of Earth's atmosphere between 3100 and 3450 Å. The principal sources of opacity are Rayleigh scattering, ozone absorption, and dust scattering. After Demerjian, K. L., Schere, K. L., and Peterson, J. T., 1980, "Theoretical Estimates of Actinic (Spherically Integrated) Flux and Photolytic Rate Constants of Atmospheric Species in the Lower Troposphere." *Adv. Environ. Sci. Technol.* **10**, 369.

Its integral over 4π equals unity. After $I(\tau, \theta, \phi)$ has been obtained by solving (2.17), we can evaluate the irradiance (energy) flux and the actinic flux by

$$F_{\text{irr}}(\tau) = \int_0^{2\pi} \int_0^{\pi} I(\tau, \theta, \phi) \cos\theta \sin\theta \, d\theta \, d\phi \tag{2.19}$$

and

$$F_{\text{act}}(\tau) = \int_0^{2\pi} \int_0^{\pi} I(\tau, \theta, \phi) \sin\theta \, d\theta \, d\phi \tag{2.20}$$

For computation of the photodissociation coefficient (2.9), we have to use F_{act}. As an illustration, figure 2.26a shows the effect of Rayleigh scattering in the terrestrial atmosphere. At short wavelengths, absorption by O_3 dominates and scattering is not important. At longer wavelengths, Rayleigh scattering becomes important. Aerosols (see section 2.4.3) are less important in the unpolluted atmosphere. Figure 2.26b presents the photodissociation coefficient for O_3 and NO_2 for a "standard" atmosphere in spring. The importance of multiple scattering is clearly illustrated by these results.

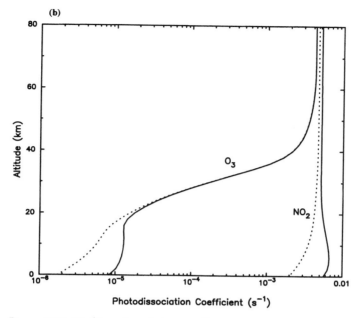

Figure 2.26 (b) Photodissociation coefficient for O_3 [$O_3 + h\nu \to O_2 + O(^1D)$] and NO_2 ($NO_2 + h\nu \to NO + O$) in Earth's atmosphere. The dotted line gives the contribution of the direct attenuated solar flux; the solid line gives the sum of the direct and diffuse fluxes. The calculations are diurnally averaged for a midlatitude atmosphere in spring with surface reflectivity of 0.25 and without aerosols. The total ozone column in the model is 341 DU (Dobson units). From the authors' Caltech/Jet Propulsion Lab model of Earth's atmosphere. See, for example, Michelangeli, D. V., Allen, M., and Yung, Y. L., 1989, "El Chichon Volcanic Aerosols: Impact of Radiative, Thermal and Chemical Perturbations." *J. Geophys. Res.* **94**, 18429.

2.4.3 Aerosols

There are at least four types of particulate matter in the atmosphere. In the lower atmosphere we have sea-salt particles, plus dust blown by wind from the surface and condensate from the freezing of an atmospheric constituent. An example of the former is a dust storm on Mars. A water cloud in the terrestrial atmosphere is an example of the latter. In the upper atmosphere, a source of solid particles is the debris from the ablation of micrometeoroids. Although too small to have an effect on the optics of the atmosphere, these particles may serve as condensation nuclei and thereby exert an indirect effect. Perhaps the most interesting form of particulate matter in the upper atmosphere is aerosols that are photochemically produced in situ.

Photochemical aerosols are ubiquitous in planetary atmospheres. The terrestrial stratosphere has the Junge layer consisting of H_2SO_4 aerosols. The clouds of Venus are also mostly H_2SO_4 particles and a dark ultraviolet absorber that is mostly likely polysulfur. The origin of these aerosols is photochemical, with the "parent molecules" such as COS and SO_2 supplied from the surface. In the outer solar system the presence of a dark hydrocarbon aerosol seems to be well established in the stratospheres of the giant planets and Titan. The ultimate source of this material, known as Axel-Danielson dust, is CH_4.

The presence of aerosols in the upper atmosphere profoundly affects the radiation field. The theory of the interaction between an electromagnetic field and an aerosol has been well developed only for spherical particles, and our discussion is restricted to this theory, known as Mie theory. For a sphere of radius r made of material with index of refraction m (a complex number), the cross sections are given by

$$\sigma_e = \pi r^2 Q_e \quad (2.21)$$

$$\sigma_a = \pi r^2 Q_a \quad (2.22)$$

$$\sigma_s = \pi r^2 Q_s \quad (2.23)$$

where Q_e, Q_a, and Q_s denote efficiency of extinction, absorption, and scattering, respectively. Mie theory provides exact expressions for the efficiency factors. Note that $Q_e = Q_a + Q_s$, and in the case of a nonabsorbing material (m = a real number), $Q_a = 0$. Figure 2.27 shows the values of Q_e for $m = 1.33$ (water) $+ ik$, where k = the imaginary part of the index of refraction and is plotted against the size parameter $x = 2\pi r/\lambda$. It is clear that for small particles ($x \ll 1$), the extinction efficiency is small. In fact it will approach the limit of Rayleigh scattering and we will recover the formula in (2.13). For large particles $x \gg 1$, the extinction efficiency approaches an asymptotic value of 2. This is surprising because we expect the limit to be 1 from geometric optics. The reason is that Mie theory includes the extremely narrow forward diffraction peak, which is not accounted for in geometric optics. The high-frequency oscillations are caused by wave interference in the sphere and will be filtered when we increase the imaginary part of the index of refraction (k). The wave interference features are also suppressed when we carry out an averaging over spheres of various sizes.

The single scattering albedo can be defined using (2.16), with the appropriate cross sections given by (2.21)–(2.23). The phase function can also be obtained from Mie theory in closed form. For small particles, the Mie phase function is the same as the Rayleigh phase function (2.18). For large dielectric particles, the Mie phase function is strongly peaked in the forward direction (the diffraction peak). The asymmetry parameter (g) provides a measure of this behavior:

$$g = \int_0^{2\pi} \int_0^{\pi} P(\theta, \phi) \cos\theta \sin\theta \, d\theta \, d\phi \quad (2.24)$$

For isotropic and Rayleigh scattering, $g = 0$; g is positive or negative as the particle scatters more or less energy into the forward or backward direction. The equation of radiative transfer (2.17) may now be solved, and the relevant quantities, the irradiance and actinic fluxes, may be derived from the solution.

It can be shown that for a given amount of material, the greatest opacity is obtained if the material is used to form aerosols with scattering parameter x equal to about 1. That is, the particle size is roughly the same as the wavelength of the incident light.

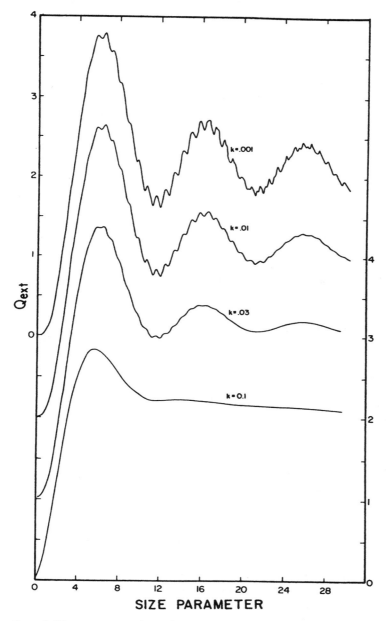

Figure 2.27 Extinction efficiency for a dielectric sphere as a function of the size parameter (x-axis). The real part of the index of refraction is 1.33, and the imaginary part of the index of refraction (k) is given in the figure. Each curve (except the bottom one) has been shifted up by 1 unit relative to the one beneath it. After Bohren, C. F., and Huffman, D. R., 1983, *Absorption and Scattering of Light by Small Particles* (New York: Wiley).

The reason is as follows: When the particle sizes are small, in the Rayleigh limit, the extinction efficiency is too small. For large particles, the extinction cross section varies as r^2, but the volume varies as r^3. Therefore, large aerosols are not efficient in generating atmospheric opacities.

The aerosols in the upper atmospheres of planets have scattering parameters of the order of unity, and hence are extremely effective in using a limited amount of material for altering the radiation field. The total fractional abundance of material is often in the range of 10^{-7} to 10^{-9}, and yet in the form of these aerosols they have a profound optical effect. The H_2SO_4 particles in the atmospheres of Venus and Earth are white (conservative) scatterers, and their first-order effect is to redistribute radiation. The Axel-Danielson dust in the atmospheres of the outer solar system is dark and is a good absorber of solar radiation.

2.4.4 Radiative Transfer in Planetary Atmospheres

The general problem of radiative transfer in planetary atmospheres involves the combined interaction of absorption, Rayleigh scattering, and absorption and scattering by aerosols. The combined problem is a straightforward extension of what has been discussed above. The total optical depth is the sum of the optical depths for each process. The overall albedo and phase function are the weighted average of the individual processes.

The interaction between solar radiation and the atmosphere gives rise to a number of interesting feedbacks, which can profoundly alter the composition and the structure of the atmosphere. An example is the absorption of ultraviolet radiation, $\lambda < 2400$ Å, by O_2. This results in the dissociation of O_2 and production of O_3, which now absorbs all ultraviolet radiation to 3000 Å. Heating of the atmosphere by O_3 maintains the thermal inversion in the stratosphere, which inhibits exchange of air masses between the troposphere and the stratosphere. Thus, the absorption of solar ultraviolet radiation by O_2 initiates a sequence of events that ultimately controls the thermal structure and dynamics of the middle atmosphere. Another example is the absorption of far UV radiation ($\lambda < 1450$ Å) by CH_4 in the atmospheres of the outer solar system, leading to the production of C_2H_2. The latter can absorb solar radiation to $\lambda < 2000$ Å, and its subsequent photochemistry results in the production of the Axel-Danielson dust. These dark aerosols then become the major absorber of solar energy in the upper atmosphere, creating a thermal inversion, a stratosphere. Note that the dynamical stability of the stratosphere (caused by the thermal inversion) is in turn important for trapping the Axel-Danielson dust in the upper atmosphere.

In most modeling studies of planetary atmospheres, the interactions outlined above between radiation, chemistry, and dynamics are conveniently decoupled. Chemical studies are carried out by using an observed temperature profile and empirical transport coefficients. Thermal modeling is based on observed chemical composition. A new kind of modeling that includes the coupled effects of radiation, chemistry, and dynamics is needed. An analogy is the ab initio calculation in quantum chemistry. We may ask why this kind of modeling is interesting at all. After all, the atmospheres have been observed, and even the most sophisticated models have to satisfy the observational constraints. The answer is that the coupled model allows us to study the stability of planetary atmospheres. Are the present states stable? What is the range

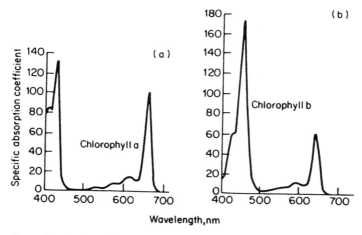

Figure 2.28 Absorption spectrum of chlorophyll a and b in 80-acetone solution. The unit is approximately equal to 1000 liter mole^{-1} cm^{-1}. After Gregory, R. P. F., 1989, *Biochemistry of Photosynthesis* (New York: Wiley).

of stability if we vary the key parameters such as sunlight, speciation, and transport coefficients? The chemical and climatic stability of Earth's atmosphere is an active area of research today. The comparative study of similar problems in planetary atmospheres would add greatly to our understanding of Earth and will undoubtedly be important for our long-term survival and well-being on the planet.

2.5 Sunlight and Life

A unique feature of Earth is the presence of life, which is maintained against decay by solar energy. The primary absorber of solar energy by the biosphere is chlorophyll, whose absorption spectrum is shown in figure 2.28. The spectrum is characterized by three maxima at 2500 (not shown), 4250, and 6750 Å. There is a broad minimum between 4500 and 6500 Å. The preferential absorption of the blue and red light gives vegetation its characteristic green color. The mean cross section of chlorophyll-a in the visible spectrum, equal to about a third of the red peak, is 5×10^{-17} cm^2. No other molecule discussed in this chapter has an absorption cross section of this magnitude in the visible range. The chlorophyll molecule is a marvel of biochemistry, rendering it a most efficient light-gathering antenna for solar energy. The energy from the sun is stored chemically by the synthesis of organic matter, carbohydrate, from CO_2. The overall reaction may be schematically written as

$$CO_2 + H_2O + h\nu \rightarrow CH_2O + O_2 \tag{2.25}$$

where $(CH_2O)_n$ represents carbohydrate. If reaction (2.25) were to be driven by a single photon, the threshold energy is 5.39 eV, or at the wavelength of 2300 Å. The total fraction of the solar irradiance that lies below this wavelength is less than 0.1%.

However, the plant is able to carry out this synthesis using eight photons of much lower energy so as to take advantage of a greater fraction of the sun's energy that is in the visible part of the spectrum. How this is achieved need not concern us here. The key step is the temporary storage of solar energy in complex molecules known as adenosine triphosphate (ATP) and nicotinamide adenine dinucleotide (NADP). These molecules can later deliver the necessary energy to the reaction center where the final synthesis takes place, with the absorption of more photons. The beauty of the scheme is threefold. First, all the energy required to drive the reaction does not need to be absorbed at once at the reaction center. Second, the energy efficiency of the multiquantum process is as high as 33%. Third, since eight photons must be used for the synthesis of carbohydrate, the product is extremely stable against photolytic destruction (a single-photon process) by the same photons. No manmade chemical scheme has been able to duplicate this truly elegant solar energy collector of Nature, the ultimate source of almost all life on this planet. The only exceptions are probably the bacteria that use the Earth's internal heat (see table 2.2) and chemical energy in underwater thermal vents.

In section 2.4.4 we point out the curious consequence of photochemistry in planetary atmospheres—the synthesis of chemicals that absorb at increasingly longer wavelengths (where there is more solar energy). Chlorophyll seems an extreme extension of this concept. It is a biochemical triumph to be able to tap into the visible part of the sun's radiation and turn its enormous energy flux into chemical energy.

What is life but an excited state of matter that can extend itself in space (by growth) and continue itself in time (by reproduction)? Since life is an excited state it is out of equilibrium with its natural surroundings. Coming into an equilibrium would mean decay (the opposite of growth) and death (the opposite of reproduction). How do we measure life quantitatively? Here are some possibilities:

1. number of carbon bonds made
2. amount of solar energy converted into chemical energy
3. amount of negative entropy needed
4. number of cells reproduced (equal to number of deaths in steady state)
5. number of genetic mutations

Sunlight is the ultimate quantitative limit of life. But electromagnetic energy is hard to store, transport, and use. The synthesis of carbohydrates converts an elusive form of energy into a more suitable form for living organisms. We may take this as a good planetary definition of the biosphere. Here is the beauty of it—as an extension of photochemistry.

If we regard the biosphere as a planetary disequilibrium phenomenon driven by the sun, it would be of interest to compare it with other similarly driven planetary phenomena: the ionosphere, the ozone layer, and the Hadley circulation of the troposphere. In a generalized sense, the biosphere may be regarded as a global photochemical extension.

3

Chemical Kinetics

3.1 Introduction

It is convenient to distinguish two types of chemistry in the solar system. The first is thermochemical chemistry driven by the thermal energy of the atmosphere. This type of chemistry is important, for example, in the interior of the giant planets where pressures and temperatures exceed 1000 bar and 1000 K, respectively. The second type is disequilibrium chemistry, driven by an external energy source, of which solar radiation is the most important in the solar system.

Chemical reactions between stable molecules are very slow at pressures less than 1 bar in planetary atmospheres. Sunlight is the ultimate source of greater chemical activity in the middle and upper atmospheres of planets. As shown in chapter 2, the absorption of solar ultraviolet radiation by atmospheric gases leads to the production of radical species (e.g., atoms, ions, excited molecules) that are extremely reactive. The bulk of atmospheric chemistry involves the reaction between the radicals themselves and between the radicals and stable molecules.

In this chapter we briefly survey the chemical kinetics that are important for understanding the chemistry of planetary atmospheres. All atmospheric reactions may be classified into three types:

1. Unimolecular reactions

$$A \to B + C \tag{3.1}$$

2. Bimolecular reactions

$$A + B \to C + D \tag{3.2}$$

3. Termolecular reactions

$$A + B + M \to C + M \tag{3.3}$$

where M is a third body, the concentration of which is that of the ambient atmosphere. The rates of reaction (in units of molecules cm^{-3} s^{-1}) for reactions (3.1)–(3.3) are given, respectively, by

$$r_1 = k_1[A] \tag{3.4}$$

$$r_2 = k_2[A][B] \tag{3.5}$$

$$r_3 = k_3[A][B][M] \tag{3.6}$$

where the quantities in brackets denote concentrations in units of molecules per cm^3, and k_1, k_2, and k_3 are rate coefficients for first-order, second-order, and third-order reactions, respectively, in units of s^{-1}, cm^3 s^{-1}, and cm^6 s^{-1}. The rate coefficients have a simple physical meaning. Consider the rate of loss of species A due to each of the reactions (3.1)–(3.3):

$$\frac{d[A]}{dt} = -\alpha_i[A] \tag{3.7}$$

where $\alpha_1 = k_1$, $\alpha_2 = k_2[B]$, and $\alpha_3 = k_3[B][M]$. Solving (3.7) yields

$$[A](t) = [A](0)e^{-\alpha_i t} \tag{3.8}$$

Therefore, species A will decay to e^{-1} of its initial value in a time interval equal to $1/k_1$, $1/k_2[B]$, and $1/k_3[B][M]$ due to reactions (3.1), (3.2), and (3.3), respectively.

In the following sections we discuss measurements and theories of rate coefficients. In general, rate coefficients are complex functions of pressure and temperature and are measured in the laboratory. Only the simplest reaction coefficients have been computed successfully. Nevertheless, the theoretical ideas are important for understanding the order of magnitude of rate coefficients and certain regularities in their behavior.

3.2 Thermochemical Reactions

Chemical reactions involve exchange of energy as well as exchange of mass. A convenient measure of the energy content of a molecule is the enthalpy of formation, H. The enthalpy of the most stable form of the elements is usually taken as zero. For example, the enthalpy of H_2 is zero; that for H is 52.1 kcal mole^{-1}. The enthalpy change in the reaction

$$H_2 + M \to H + H + M \tag{3.9}$$

is 104.2 kcal mole^{-1}. This is the amount of energy required to break the H—H bond. Since average collisional energies are much smaller in most atmospheres, reaction (3.9) will not occur at a significant rate. The reverse of (3.9),

$$H + H + M \rightarrow H_2 + M \qquad (3.10)$$

releases energy and can proceed spontaneously. Table 3.1 gives enthalpies of formation for molecules of interest in planetary atmospheres. The thermodynamic properties strongly influence the reactivity of molecules. Molecules such as N_2, H_2, and CH_4 have low values of enthalpy and are thermally stable. Radicals such as O, H, and Cl have high values of enthalpy and are extremely reactive.

Reaction (3.10) proceeds spontaneously to the right because of the natural tendency of a system to move toward a state of lower energy. Indeed, in a static state the laws of mechanics require that the physical system be in a state of minimum energy. However, in a dynamical system with many gas molecules, there is another fundamental principle, the principle of maximum entropy, or the tendency for a system to reach maximum disorder. Thus, each molecule is characterized by a thermodynamic quantity known as entropy (S), which is a measure of the number of states available to it. By the laws of probability, the configurations with more states are favored. The entropies (S) of simple molecules are listed in table 3.2. Referring to reaction (3.9), the entropy on the left side is 31.2 cal mole^{-1} K^{-1} and on the right side is $2 \times 27.39 = 54.8$ cal mole^{-1} K^{-1}. Thus, according to the maximum entropy principle, (3.9) will proceed so as to increase the entropy of the system by 23.6 cal mole^{-1} K^{-1}. Physically, 2 moles of H atoms represent more disorder than 1 mole of H_2 and in the absence of the energy principle (3.9) would prevail.

The combination of these two principles is formulated in terms of Gibbs free energy:

$$G = H - TS \qquad (3.11)$$

where T is temperature. For a closed thermodynamic system, the equilibrium state is attained when G is a minimum. This is analogous to a mechanical system, which seeks minimum potential energy. Note that both the minimum energy principle and the maximum entropy principle are incorporated as special cases of the Gibbs principle. As an application of the Gibbs principle, consider the change in G in (3.9). At 300 K (representative of planetary atmospheres) the change in G from the left side to the right side is 97.1 kcal mole^{-1}. Therefore, equilibrium favors the molecular form, H_2. However, at 6000 K (representative of the solar atmosphere), the change in Gibbs free energy is -37.4 kcal mole^{-1} (for simplicity we use the enthalpies and entropies at 300 K). Equilibrium is now shifted to the right. This explains why H_2 is the more stable form of hydrogen in planets, whereas in the solar atmosphere H atoms dominate over H_2.

Referring to the definition of G in (3.11), the energy term H is more important than the entropy term TS at 300 K, but the reverse is true at 6000 K. Note that enthalpy is a measure of the energy involved in making or breaking chemical bonds— an internal energy that depends, to first order, on the electronic configurations of the molecules and is largely independent of external parameters. Conversely, the entropy part of G is proportional to temperature, which is a macroscopic property of the gas. The Gibbs principle provides a beautiful and simple connection between the micro-

Table 3.1 Enthalpies of simple molecules

Species	ΔH_f(298 K) (Kcal mole^{-1})	Species	ΔH_f(298 K) (Kcal mole^{-1})	Species	ΔH_f(298 K) (Kcal mole^{-1})	Species	ΔH_f(298 K) (Kcal mole^{-1})
H	52.1	C_2H_4	12.45	FNO_2	-26 ± 2	$CFCl_3$	-68.1
H_2	0.0	C_2H_5	28.4	$FONO_2$	2.5 ± 7	CF_2Cl	-64 ± 3
O	59.57	C_2H_6	-20.0	CF		CF_2Cl_2	-117.9
O_2	0.0	CH_2CN	57 ± 2	CF_2	-44 ± 2	CF_3Cl	-169.2
O_3	34.1	CH_3CN	15.6	CF_3	-112 ± 1	$CHFCl_2$	-68.1
HO	9.3	CH_2CO	-11 ± 3	CF_4	-223.0	CHF_2Cl	-115.6
HO_2	3 ± 1	CH_3CO	-5.8	CHF_3	-166.8	COFCl	-102 ± 2
H_2O	-57.81	CH_3CHO	-39.7	CHF_2	-58 ± 2	CH_3CH_2F	-63 ± 2
H_2O_2	-32.60	C_2H_5O	-4.1	CH_2F_2	-107.2	CH_3CHF	-17 ± 2
N	113.00	CH_2CH_2OH	10 ± 3	CH_2F	-8 ± 2	CH_2CF_3	-124 ± 2
N_2	0.0	C_2H_5OH	-56.2	CH_3F	-55.9 ± 1	CH_3CHF_2	-120 ± 1
NH	85.3	CH_3CO_2	-49.6	FCO	-41 ± 14	CH_3CF_2	-71 ± 2
NH_2	45.3	$C_2H_5O_2$	-6 ± 2	COF_2	-153 ± 2	CH_3CF_3	-179 ± 2
NH_3	-10.98	CH_3COO_2	-41 ± 5	CF_3O		CF_2CF_3	-213 ± 2
NO	21.57	CH_3OOCH_3	-30.0	CF_3O_2		CHF_2CF_3	-264 ± 2
NO_2	7.9	C_3H_5	39.4	CF_3OH		CH_3CF_2Cl	-127 ± 2
NO_3	17 ± 2	C_3H_6	4.8	CF_3OOCF_3		CH_2CF_2Cl	-75 ± 2
N_2O	19.61	$n-C_3H_7$	22.6 ± 2	CF_3OOH		C_2Cl_4	-3.0
N_2O_3	19.8	$i-C_3H_7$	19 ± 2	CF_3OF		C_2HCl_3	-1.9
N_2O_4	2.2	C_3H_8	-24.8	Cl	28.9	CH_2CCl_3	17 ± 2
N_2O_5	2.7 ± 2	C_2H_5CHO	-44.8	Cl_2	0.0	CH_3CCl_3	-34.0
HNO	23.8	CH_3COCH_3	-51.9	HCl	-22.06	CH_3CH_2Cl	-26.8
HONO	-19.0	$CH_3COO_2NO_2$	-62 ± 5	ClO	24.4	CH_2CH_2Cl	22 ± 2
HNO_3	-32.3	S	66.22	ClOO	23 ± 1	CH_3CHCl	17.6 ± 1
HO_2NO_2	-11 ± 2	S_2	30.72	OClO	23 ± 2	Br	26.7
C	170.9	HS	34 ± 1	ClO_2	21.3	Br_2	7.39
CH	142.0	H_2S	-4.9	$ClOO_2$	16.7	HBr	-8.67
CH_2	93 ± 1	SO	1.3	ClO_3	52 ± 4	HOBr	-19 ± 2
CH_3	35 ± 0.2	SO_2	-70.96	Cl_2O	19.5	BrO	30
CH_4	-17.88	SO_3	-94.6	Cl_2O_2	31 ± 3	BrNO	19.7
CN	104 ± 3	HSO	-1 ± 3	Cl_2O_3	34 ± 3	BrONO	25 ± 7
HCN	32.3	HSO_3	-92 ± 2	HOCl	18 ± 3	$BrNO_2$	17 ± 2
CH_3NH_2	-5.5	H_2SO_4	-176	NOCl	12.36	$BrONO_2$	12 ± 5
NCO	38.0	CS	67 ± 2	ClNO	12.4	BrCl	3.5
CO	-26.42	CS_2	28.0	$ClNO_2$	3.0	CH_2Br	40 ± 2
CO_2	-94.07	CS_2OH	26.4	ClONO	13	$CHBr_3$	6 ± 2
HCO	10 ± 1	CH_3S	33 ± 2	$ClONO_2$	5.5	$CHBr_2$	45 ± 2
CH_2O	-26.0	CH_3SOO	0.0	$ClNO_3$	6.28	CBr_3	48 ± 2
COOH	-53 ± 2	CH_3SO_2	-57	FCl	-12.1	CH_2Br_2	-2.6 ± 2
HCOOH	-90.5	CH_3SH	-5.5	CCl_2	57 ± 5	CH_3Br	-8.5
CH_3O	4 ± 1	CH_2SCH_3	36 ± 3	CCl_3	18 ± 1	CH_3CH_2Br	-14.8
CH_3O_2	4 ± 2	CH_3SCH_3	-8.9	CCl_4	-22.9	CH_2CH_2Br	32 ± 2
CH_2OH	-6.2	CH_3SSCH_3	-5.8	C_2Cl_4	-2.97	CH_3CHBr	30 ± 2
CH_3OH	-48.2	OCS	-34	$CHCl_3$	-24.6	I	25.52
CH_3OOH	-31.3	F	18.98	$CHCl_2$	23 ± 2	I_2	14.92
CH_3ONO	-15.6	F_2	0.0	CH_2Cl	29 ± 2	HI	6.3
CH_3NO_2	-17.86	HF	-65.34	CH_2Cl_2	-22.8	CH_3I	3.5
CH_3ONO_2	-28.6	HOF	-23.4 ± 1	CH_3Cl	-19.6	CH_2I	52 ± 2
CH_3NO_3	-29.8	FO	26 ± 5	ClCO	-5 ± 1	IO	41.1
$CH_3O_2NO_2$	-10.6 ± 2	F_2O	5.9 ± 0.4	$COCl_2$	-52.6	INO	29.0
C_2H	-133 ± 2	FO_2	6 ± 1	CHFCl	-15 ± 2	INO_2	14.4
C_2H_2	54.35	F_2O_2	5 ± 2	CH_2FCl	-63 ± 2		
C_2H_2OH	30 ± 3	FONO	-15 ± 7	CFCl	7 ± 6		
C_2H_3	72 ± 3	FNO	-16 ± 2	$CFCl_2$	-22 ± 2		

Table 3.2 Entropies of simple molecules

Species	$S°(298\ K)$ (cal^{-1} mole^{-1} deg^{-1})	Species	$S°(298\ K)$ (cal^{-1} mole^{-1} deg^{-1})	Species	$S°(298\ K)$ (cal^{-1} mole^{-1} deg^{-1})	Species	$S°(298\ K)$ (cal^{-1} mole^{-1} deg^{-1})
H	27.4	C_2H_4	52.5	FNO_2	62.3	$CFCl_3$	74.0
H_2	31.2	C_2H_5	58.0	$FONO_2$	70.0	CF_2Cl	68.7
O	38.5	C_2H_6	54.9	CF	50.9	CF_2Cl_2	71.8
O_2	49.0	CH_2CN	58.0	CF_2	57.5	CF_3Cl	68.3
O_3	57.0	CH_3CN	58.2	CF_3	63.3	$CHFCl_2$	70.1
HO	43.9	CH_2CO	57.8	CF_4	62.4	CHF_2Cl	67.2
HO_2	54.4	CH_3CO	64.5	CHF_3	62.0	COFCl	66.2
H_2O	45.1	CH_3CHO	63.2	CHF_2	61.7	CH_3CH_2F	63.3
H_2O_2	55.6	C_2H_5O	65.3	CH_2F_2	58.9	CH_3CHF	
N	36.6	CH_2CH_2OH		CH_2F	55.9	CF_2CF_3	71.8
N_2	45.8	C_2H_5OH	67.5	CH_3F	53.3	CH_3CHF_2	67.6
NH	43.3	CH_3CO_2		FCO	59.4	CH_3CF_2	69.9
NH_2	46.5	$C_2H_5O_2$	75.0	COF_2	61.9	CH_3CF_3	68.6
NH_3	46.0	CH_3COO_2		CF_3O		CF_2CF_3	81.6
NO	50.3	CH_3OOCH_3	74.1	CF_3O_2		CHF_2CF_3	
NO_2	57.3	C_3H_5	62.1	CF_3OH		CH_3CF_2Cl	68.7
NO_3	60.3	C_3H_6	63.8	CF_3OOCF_3		CH_2CF_2Cl	
N_2O	52.6	n-C_3H_7	68.5	CF_3OOH		C_2Cl_4	81.4
N_2O_3	73.9	i-C_3H_7	66.7	CF_3OF	77.1	C_2HCl_3	77.5
N_2O_4	72.7	C_3H_8	64.5	Cl	39.5	CH_2CCl_3	80.6
N_2O_5	82.8	C_2H_5CHO	72.8	Cl_2	53.3	CH_3CCl_3	76.4
HNO	52.7	CH_3COCH_3	70.5	HCl	44.6	CH_3CH_2Cl	65.9
HONO	59.6	$CH_3COO_2NO_2$		ClO	54.1	CH_2CH_2Cl	
HNO_3	63.7	S	40.1	ClOO	64.0	CH_3CHCl	
HO_2NO_2		S_2	54.5	OClO	61.5	Br	41.8
C	37.8	HS	46.7	ClO_2	63.0	Br_2	58.6
CH	43.7	H_2S	49.2	$ClOO_2$	73.0	HBr	47.4
CH_2	46.3	SO	53.0	ClO_3	67.0	HOBr	59.2
CH_3	46.4	SO_2	59.3	Cl_2O	64.0	BrO	56.8
CH_4	44.5	SO_3	61.3	Cl_2O_2	72.2	BrNO	65.3
CN	48.4	HSO		Cl_2O_3		BrONO	
HCN	48.2	HSO_3		HOCl	56.5	$BrNO_2$	
CH_3NH_2	58.0	H_2SO_4	69.1	NOCl	62.5	$BrONO_2$	
NCO	55.5	CS	50.3	ClNO	62.5	BrCl	57.3
CO	47.3	CS_2	56.9	$ClNO_2$	65.1	CH_2Br	
CO_2	51.1	CS_2OH		ClONO		$CHBr_3$	79.1
HCO	53.7	CH_3S	57.6	$ClONO_2$		$CHBr_2$	
CH_2O	52.3	CH_3SOO		$ClNO_3$		CBr_3	80.0
COOH	61.0	CH_3SO_2		FCl	52.1	CH_2Br_2	70.1
HCOOH	59.4	CH_3SH	61.0	CCl_2	63.4	CH_3Br	58.7
CH_3O	55.0	CH_2SCH_3		CCl_3	71.0	CH_3CH_2Br	68.6
CH_3O_2	65.3	CH_3SCH_3	68.4	CCl_4		CH_2CH_2Br	
CH_2OH	58.8	CH_3SSCH_3	80.5	C_2Cl_4	81.4	CH_3CHBr	
CH_3OH	57.3	OCS	55.3	$CHCl_3$	70.7	I	43.2
CH_3OOH	67.5	F	37.9	$CHCl_2$	66.5	I_2	62.3
CH_3ONO	68.0	F_2	48.5	CH_2Cl	58.2	HI	49.3
CH_3NO_2	65.7	HF	41.5	CH_2Cl_2	64.6	CH_3I	60.6
CH_3ONO_2	72.1	HOF	54.2	CH_3Cl	56.1	CH_2I	
CH_3NO_3	76.1	FO	51.8	ClCO	63.3	IO	58.8
$CH_3O_2NO_2$		F_2O	59.1	$COCl_2$	67.8	INO	67.6
C_2H	49.6	FO_2	61.9	CHFCl		INO_2	70.3
C_2H_2	48.0	F_2O_2	66.3	CH_2FCl	63.3		
C_2H_2OH		FONO	62.2	CFCl	62.0		
C_2H_3	56.3	FNO	59.3	$CFCl_2$	71.5		

scopic properties of the individual molecules and the macroscopic properties of the gas as a whole.

Consider a general chemical reaction,

$$m_1X_1 + m_2X_2 + m_3X_3 \rightarrow n_1Y_1 + n_2Y_2 + n_2Y_3 \qquad (3.12)$$

with rate coefficient k_f. Note that by appropriate choice of the values m_i and n_i (including zero) we have all reactions of types (3.1)–(3.3). All our arguments can obviously be extended to reactions involving more than three species. The reverse of reaction (3.12) is

$$n_1Y_1 + n_2Y_2 + n_3Y_3 \rightarrow m_1X_1 + m_2X_2 + m_2X_3 \qquad (3.13)$$

with rate coefficient k_r. For a closed chemical system that has reached equilibrium, the forward and the backward reaction rates must be equal:

$$k_f[X_1]^{m_1}[X_2]^{m_2}[X_3]^{m_3} = k_r[Y_1]^{n_1}[Y_2]^{n_2}[Y_3]^{n_3} \qquad (3.14)$$

Rearranging (3.14) we have

$$\frac{[Y_1]^{n_1}[Y_2]^{n_2}[Y_3]^{n_3}}{[X_1]^{m_1}[X_2]^{m_2}[X_3]^{m_3}} = \frac{k_f}{k_r} \qquad (3.15)$$

It can be shown that ratio (3.15) can be expressed as a function of the Gibbs free energies:

$$\frac{k_f}{k_r} = e^{(G_f - G_r)/RT} \qquad (3.16)$$

where R is the molar gas constant and

$$G_f = m_1 G(X_1) + m_2 G(X_2) + m_3 G(X_3) \qquad (3.17)$$

$$G_r = n_1 G(Y_1) + n_2 G(Y_2) + n_3 G(Y_3) \qquad (3.18)$$

Although the Gibbs free energy gives no direct prediction of the values of individual rate coefficients, the ratio of the forward and the reverse rate coefficients can always be computed using Gibbs free energies. This is an important result because it is often hard to measure both k_f and k_r. By (3.16) we only have to measure one and can obtain the other using thermodynamic considerations.

It can be shown that for a system that has reached equilibrium, the partitioning of the species is such that the total Gibbs free energy of the system is a minimum. Note that the thermodynamic properties of the molecules now completely determine the chemical composition of the system without any reference to the rate coefficients.

As an example of thermoequilibrium chemistry, consider a parcel of air containing N_2 and O_2, heated up to thousands of degrees, as in a lightning bolt or the combustion chamber of an automobile. The reaction

$$N_2 + O_2 \leftrightarrow 2NO \qquad (3.19)$$

repartitions the nitrogen and oxygen atoms into the molecules N_2, O_2, and NO. The composition of the air parcel is uniquely determined by specifying the initial partial pressures of N_2 and O_2 and the final temperature of the parcel.

There is one great deficiency in this thermodynamic scheme. Thermodynamics offers no clue to the time constant that is needed for the equilibrium to be reached. For instance, in reaction (3.19) we do not know whether the equilibrium concentrations are attained. We do not know, as the air parcel cools down, at what temperature the parcel is "quenched" so that its composition will not undergo further change with lowering temperature. For this we need the kinetic information, discussed in the following sections.

3.3 Unimolecular Reactions

Once a chemical bond is formed, it is generally difficult to break. The rate coefficient for the spontaneous decomposition of A into products B and C by (3.1) is

$$k_1 = \frac{kT}{h} e^{-\Delta G/RT} \tag{3.20}$$

where k is the Boltzmann constant, h is the Planck constant, and $\Delta G = G(B) + G(C) - G(A)$. At room temperature the preexponential factor is of the order 10^{13} s^{-1} (~molecular vibrational frequency). For stable molecules we often have $\Delta G \gg RT$, and the exponential factor tends to make k_1 very small.

By far the most important first-order (unimolecular) processes in the atmosphere are ones that are excited by the absorption of a photon, as discussed in chapter 2:

$$A + h\nu \rightarrow A^\dagger \tag{3.21}$$

followed by

$$A^\dagger \rightarrow B + C \tag{3.22}$$

The second step is now fast because the bond energy barrier has been overcome by the absorption of a photon. The rate-limiting step is (3.21). Hence, for most applications, we have

$$k_1 = J \tag{3.23}$$

where J is the photodissociation coefficient defined by (2.9) for species A.

Process (3.21) usually occurs with the absorption of a single ultraviolet photon with sufficient energy to break the chemical bond of A. There is another process using multiphoton absorption in the infrared. The molecule is pumped to the dissociation continuum via multiphoton excitation of the vibrational levels. However, this can only be done using a laser in the laboratory, and is unimportant in the atmosphere.

3.4 Bimolecular Reactions

3.4.1 Molecular Collisions

Consider the bimolecular reaction (3.2). Molecules A and B have to collide with each other before they can react. For simplicity let us assume that molecules A and B are spheres with masses m_A and m_B, respectively, and have radii r_A and r_B, respectively. The cross section for a binary collision is

$$\sigma = \pi(r_A + r_B)^2 \tag{3.24}$$

The number of binary collisions cm^{-3} s^{-1} is given by (3.5) with

$$k_2 = k_c = \int\int \sigma(w) f(v_A) f(v_B) w d^3 v_A d^3 v_B \tag{3.25}$$

where f represents the normalized distributions for molecular velocities and $w = |\vec{v}_A - \vec{v}_B|$. For a gas in equilibrium at temperature T, the velocity distribution is given by the Maxwell-Boltzmann distribution. In this case the double integral in (3.25) may be evaluated to yield

$$k_c = \sigma v_r \tag{3.26}$$

where the mean speed v_r is given by

$$v_r = \sqrt{\frac{8kT(m_A + m_B)}{\pi m_A m_B}} \tag{3.27}$$

Note that v_r is just the mean speed of a particle in thermal equilibrium at temperature T with reduced mass $\mu = m_A m_B/(m_A + m_B)$. We may evaluate k_c for the major molecules in the terrestrial atmosphere. The cross section for collision is about 5×10^{-15} cm^2; v_r at 300 K is about 4×10^4 cm s^{-1}. Hence, the gas kinetic rate coefficient is approximately

$$k_c = 2 \times 10^{-10} \sqrt{\frac{T}{300}} \text{ cm}^3 \text{s}^{-1} \tag{3.28}$$

Since the rate of chemical reactions normally cannot exceed that of kinetic collisions, the value of k_c given by (3.28) must be the upper limit for the rate coefficient of binary reactions between neutral molecules. Ions can react via the long-ranged Coulomb force and are not subjected to the limitation of (3.28). (See discussion in section 3.4.3.)

There are a number of reactions whose rate coefficients do approach that of (3.28):

$$O + OH \rightarrow O_2 + H \quad k = 2.2 \times 10^{-11} e^{120/T} \tag{3.29}$$

$$O + ClO \rightarrow O_2 + Cl \quad k = 3 \times 10^{-11} e^{70/T} \tag{3.30}$$

$$CH + CH_4 \rightarrow C_2H_4 + H \quad k = 1.0 \times 10^{-10} \tag{3.31}$$

where the units for the rate coefficients are cm^3 s^{-1}. Unfortunately, this collisional theory is too simplistic to account for the detailed molecular interactions involved in a chemical reaction. In sections 3.4.2–3.4.4, we present more realistic theories of chemical kinetics.

3.4.2 Forming Chemical Bonds

In the previous section we obtained the kinetic collision rate, without further regard for the internal structure of each molecule. How effective is each collision at inducing a chemical reaction, that is, forming a new chemical bond? To answer this question we need detailed quantum mechanical knowledge of the reactants and the products. (The discussion here is based on unpublished material by G. Blake.)

Consider two atoms A and B undergoing a collision to form a stable molecule AB. In section 3.3 we studied the unimolecular dissociation of a stable molecule. The breaking of a chemical bond is difficult because a minimum amount of energy (the bond energy) must be supplied to the molecule in appropriate form. The reverse process, the forming of a new bond, is also difficult. The bond energy (D_0) is of the order of 5 eV, which must be removed in the brief interval of time during collision. The size of a molecule (d) is of the order of 5 Å, and at thermal velocity $v = 3 \times 10^4$ cm s^{-1}, the "collision time" is about 2×10^{-12} s. The power density of dissipation is vD_0/d^3, or about 10^8 W cm^{-2}. It is difficult to remove energy at this rate. Hence, in general, the transient molecule AB would not be stabilized but would fly apart as A and B.

The formation of a new chemical bond from atoms is an inefficient process, as shown above. There are three mechanisms for removing the excess energy from the newly formed molecule: (a) termolecular reaction, using collisions with a third body; (b) radiative association, radiating the energy away; and (c) heterogeneous reaction, using a surface to absorb the energy. These processes are discussed in detail in sections 3.5, 3.7.3, and 3.6, respectively.

3.4.3 Transition State Theory

While it is difficult in the atmosphere to create or break a chemical bond, it is relatively easy to exchange chemical bonds. For example, in a metathetical reaction such as

$$A + BC \rightarrow AB + C \quad (3.32)$$

there is a preexisting chemical bond, B—C, and a new bond, A—B, is formed. The excess energy from the bond formation is dissipated in the breaking of the old B—C bond, and the overall reaction exothermicity can be taken up as relative translational energy of the products. Thus, the original B—C bond is exchanged for the new A—B bond. This type of reaction makes up the bulk of chemical reactions in the atmosphere.

We now examine (3.32) in more detail. The first step is the formation of a transient triatomic complex, known as the activated complex or transition state:

$$A + BC \rightarrow ABC^\dagger \quad (3.33)$$

There are two possible fates for the transition state. One choice is going back to the reactants:

$$ABC^\dagger \rightarrow A + BC \quad (3.34)$$

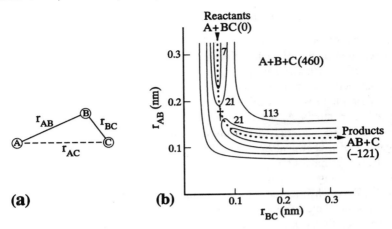

Figure 3.1 (a) Schematic diagram showing the distances between the atoms A, B, and C. (b) Typical potential energy surface for a particular ABC angle. The lowest-energy path for the reaction to proceed from A + BC to AB + C is indicated by the dotted line, which represents the reaction coordinate. On this path, the highest energy is reached at the transition state, marked by the cross. This potential surface is computed for the reaction $F + H_2 \rightarrow H + HF$. The numerical values are energies in kJ mole^{-1}. After Muckerman, J. T., *J. Chem. Phys.* **54**, 115, 1971.

In this case the net result of (3.33) plus (3.34) is a null reaction. Indeed, transition state theory is based on the hypothesis of equilibrium between reactants and the transition state intermediate. Another possibility is decaying into a different product channel.

$$ABC^\dagger \rightarrow AB + C \qquad (3.35)$$

The net result of (3.33) and (3.35) is equivalent to (3.32).

It is illuminating to view the interaction as a trajectory on the potential energy surface of the triatomic system using as coordinates the internuclear distances between A and B, and B and C, r_{AB} and r_{BC} as shown in figure 3.1, a and b. The reaction trajectory is represented by the dotted line in this figure. At the starting point we have reactant A far away from reactant BC. Therefore, the value for r_{AB} is large and the value for r_{BC} is small. As A approaches BC, a transition state $ABC^\dagger$ is formed and the values of both r_{AB} and r_{BC} become small. The transition state then decays into products AB + C. The distance r_{BC} now increases while r_{AB} remains constant.

We may also study this interaction from another (equivalent) point of view using the "interaction coordinate." Consider the potential for atom A interacting with molecule BC, as shown in figure 3.2a. For the reaction to take place, A must be attracted to BC. Hence, the potential is attractive as A approaches BC (as r_{AB} gets small). At very small distances there is a repulsive barrier between A and BC. A similar potential for the interaction between molecule AB and the atom C is shown in figure 3.2b. If we now join the two potential curves smoothly together, we have the transition state in the middle, $ABC^\dagger$. Note that in going from the reactants to the products, the reaction trajectory has to traverse a local potential maximum, an energy barrier. Two types

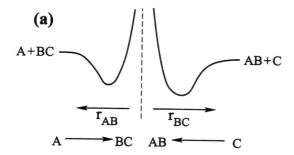

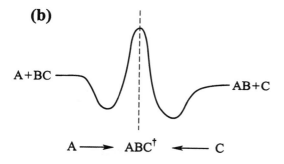

Figure 3.2 (a) Schematic diagram showing the attractive potentials for A + BC and AB + C. (b) Schematic diagram of the potential surface along the reaction coordinate of figure 3.1b (dotted line) for the reaction A + BC → AB + C. Based on unpublished material by G. Blake.

of situations can arise. In the first case, as shown in figure 3.3, the potential energy of reactants A + BC is lower than the transition state ABC†. Even though the net reaction from A + BC to AB + C is exothermic, an extra amount of energy, known as the activation energy (E_a), must be supplied to overcome the barrier. The bimolecular rate coefficient k_2 may now be expressed by the Arrhenius expression:

$$k_2 = Ae^{-E_a/RT} \tag{3.36}$$

where A is the gas kinetic rate coefficient k_c given in (3.28), and the exponential function takes into account of the activation energy. It is usual to measure k_2 empirically at various temperatures and determine E_a empirically from the data. The temperature dependence of A is weak compared with the exponential factor, and A may be regarded as a constant. In the second case, as shown by the reverse reaction in figure 3.3, the energy barrier is $E_a + \Delta H$.

The transition state theory predicts more than the empirical formula (3.36). By assuming quasiequilibrium between the reactants and the transition state, the theory predicts a rate coefficient given by

$$k_2 = \frac{kT}{h} \frac{Q(\text{ABC}^\dagger)}{Q(\text{A})Q(\text{BC})} e^{-E_a/RT} \tag{3.37}$$

where h is Planck's constant, Q is the partition function, and E_a is the difference in the enthalpies of the reaction. Note that the ratio of the partition functions in (3.37) is the equilibrium constant for the formation of the transition state. In applying formula (3.37), we usually know $Q(A)$ and $Q(B)$. Computation of $Q(\text{ABC}^\dagger)$ requires knowledge of the structure of the transition state (e.g., whether it is linear, bent, or cyclic)

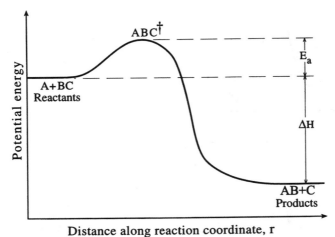

Figure 3.3 Barrier energies for the forward reaction (E_a) and the reverse reaction ($E_a + \Delta H$).

and its modes of vibration and rotation. The theory is not advanced enough to derive rate coefficients from first principles but can be helpful for understanding the mechanism of chemical reactions as well as the order of magnitude of the reaction rates. Using the thermodynamic formulation of the equilibrium constant for the formation of the transition state, the rate coefficient may be expressed in the form

$$k_2 = k_c e^{\Delta S/R} e^{-E_a/RT} \qquad (3.38)$$

where ΔS is the change in entropy, $S(ABC^\dagger) - S(A) - S(B)$. This entropy difference is a measure of the molecular rearrangement involved in the formation of the transition state. It accounts for the magnitude of the preexponential factor and may differ substantially among different reaction types.

Rate constants are most strongly affected by the activation energy, E_a. This quantity is, unfortunately, the most difficult to calculate *a priori*. Two examples of reactions with large activation energy are:

$$H + CH_4 \rightarrow CH_3 + H_2 \quad k = 3.0 \times 10^{-10} \, e^{-6560/T} \qquad (3.39)$$

$$O + H_2 \rightarrow OH + H \quad k = 1.6 \times 10^{-11} \, e^{-4570/T} \qquad (3.40)$$

The rate coefficients are in units of $cm^3 \, s^{-1}$. These reactions are quite slow, except at very high temperatures.

For most reactions of interest in the atmosphere, the theory given by (3.37) is currently not capable of providing rate constants accurate enough for atmospheric modeling. The bulk of rate constants must usually be determined experimentally. However, there exist many correlations between them that are very precise. These correlations can often be used to estimate unmeasured rate constants, based on experimental data for related reactions. This is particularly true in second-order radical-molecule reactions. As experimental data become progressively more accurate, the precision of

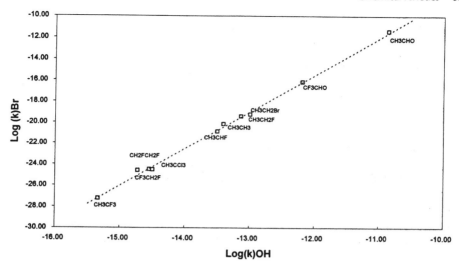

Figure 3.4 Plot of rate constants for reactions between Br and a series of ethane derivatives versus those of OH with the same compounds.

the correlations among them becomes increasingly apparent. The reason for the correlations is that the same bond is attacked by the radicals as, for example, in the abstraction of an H atom by various radical species such as halogen atoms or hydroxyl radicals. An example of a particularly good correlation is shown in figure 3.4, relating to the abstraction reactions of OH and Br with a series of ethane derivatives. Note the wide range of rates, as well as the widely disparate absolute rate constant values, over which the logarithm of the Br rate constants is proportional to the logarithm of the corresponding OH rate constants. The basis for the correlation can be traced simply. Both reactions appear to follow the Evans-Polanyi relationship, E_a/R = a(BDE) + b, where E_a/R is the activation temperature in K, a and b are constants for a given reaction class, and BDE is the bond dissociation energy of the bond that is attacked. A further correlation exists between the preexponential factors, A, in the Arrhenius equation for the rate constant (3.36). For reactions within a given type, such as abstraction of atomic hydrogen by OH radicals, the A-factors per bond attacked are approximately constant. This constancy stems from the fact that reactant approach is closely determined by the requirement to minimize energy, resulting in a well-defined reaction transition state. Using transition state theory, which postulates equilibrium between reactants and transition state, A-factors can be calculated approximately by taking a model to estimate the entropy of the transition state. For example, in the OH reaction with CH_4, the model transition state would be methanol, CH_3OH. A small correction is necessary to account for the fact that the transition state is not quite as rigidly defined as the reactant molecule (i.e., it is "looser" and therefore has a slightly higher entropy). When reactions obey these two relations, it can be shown that the observed linear relation between the logarithms of the rate constants is expected.

Radical-radical reactions, such as the important atmospheric reaction

$$OH + HO_2 \rightarrow H_2O + O_2 \tag{3.41}$$

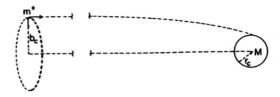

Figure 3.5 Capture cross section for ion-molecule collisions. After Su, T., and Bowers, M. T., 1979, "Classical Ion-Molecule Collision Theory," in Bowers (1979; cited in section 3.1), p. 83.

are not as well understood as radical-molecule reactions. These reactions often have zero or negative activation energies, since the energy input associated with bond breaking is very small. Higher energy in the reactants may actually reduce the reaction probability. Another difference exists is that, though reaction (3.41) is nominally an OH abstraction of an H atom, the limited approach geometry related to energy minimization is not required. As a result, the effective preexponential factor is very high and is difficult to estimate based on any model structure of the transition state.

3.4.4 Ion-Molecule Reactions

For ions colliding with neutrals, the effective cross section is much larger than that for neutral-neutral collisions, because atoms and molecules are polarizable, and the ion with charge q at distance r can induce a dipole. The subsequent interaction between the ion and the induced dipole can be described by an attractive potential:

$$V(r) = -\frac{\alpha q^2}{2r^4} \qquad (3.42)$$

where α is the polarizability (in units of cm^3) of the atom or molecule. This attractive potential has two important effects on the rate coefficient: (a) increase the collision cross section, and (b) increase the energy of the collision such that the activation energy is easily overcome. Note that (a) is due to the long-ranged Coloumb force and (b) is from the conversion of potential energy to kinetic energy.

The mechanics for the interaction between an incident ion (A) and a neutral particle (B) are shown in figure 3.5. The masses are, respectively, m_A and m_B. The impact parameter is b, and the initial velocity is v_0. The effective potential is

$$V_{\text{eff}}(r) = -\frac{\alpha q^2}{2r^4} + \frac{L^2}{2\mu r^2} \qquad (3.43)$$

where we have added the repulsive centrifugal potential due to the angular momentum (L) of the system and μ is the reduced mass $m_A m_B/(m_A + m_B)$. Since angular momentum is conserved, L is $\mu v_0 b$, a constant. V_{eff} has two parts. At large distance the repulsive (centrifugal) part dominates; at small distance the attractive (induced dipole) part becomes more important. An example for $N_2^+ + N_2$ is given in figure 3.6. The distance at which the repulsive and the attractive forces balance is the critical distance r_c, given by solving the equation

$$\frac{dV_{\text{eff}}(r)}{dr} = 0 \qquad (3.44)$$

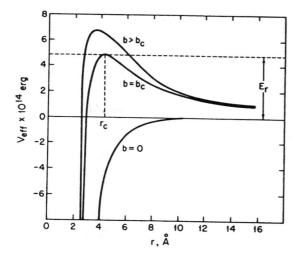

Figure 3.6 Plot of V_{eff} versus r for N_2^+ colliding with N_2. After Su, T., and Bowers, M. T., 1979, "Classical Ion-Molecule Collision Theory," in Bowers (1979; cited in section 3.1), p. 83.

The result is

$$r_c = \frac{q}{L}\sqrt{2\alpha\mu} \tag{3.45}$$

The critical distance has a special physical meaning. All collisions that pass within r_c result in a capture. At $r = r_c$ the energy needed to get over the centrifugal barrier is

$$E_c = V_{\text{eff}}(r_c) = \frac{L^4}{8\alpha\mu^2 q^2} \tag{3.46}$$

However, at r_c all the energy is in the tangential motion. E_c is then the total energy $\mu v_0^2/2$. Hence, we can solve for the critical impact parameter b_c:

$$b_c = \left(\frac{4\alpha q^2}{\mu v_0^2}\right)^{\frac{1}{4}} \tag{3.47}$$

The rate coefficient is given by

$$k_2 = k_L = v_0 \pi b_c^2 = 2\pi q \left(\frac{\alpha}{\mu}\right)^{\frac{1}{2}} \tag{3.48}$$

This is known as the Langevin rate constant for ion-molecule reactions. Ion-molecule reactions are usually faster than neutral-neutral reactions, as shown in the following examples:

$$H_2^+ + H_2 \rightarrow H_3^+ + H \quad k = 2.1 \times 10^{-9} \tag{3.49}$$

$$O^+ + CO_2 \rightarrow O_2^+ + CO \quad k = 1.1 \times 10^{-9} \tag{3.50}$$

$$O_2^+ + Mg \rightarrow Mg^+ + O_2 \quad k = 2 \times 10^{-9} \tag{3.51}$$

where the units of the rate coefficients are $cm^3\ s^{-1}$.

3.5 Termolecular Reactions

In section 3.4.2 we pointed out the difficulty of forming a new chemical bond because of the need to dissipate a large amount of energy (the bond energy) in a short time (the collision time) over a small amount of space (the size of the molecule). One way to stabilize the new bond is to invoke collision with a third body. Consider the association of the reactants A and B forming an association complex $AB^\dagger$:

$$A + B \rightarrow AB^\dagger \qquad k_a \qquad (3.52)$$

The complex can decay via a unimolecular reaction:

$$AB^\dagger \rightarrow A + B \qquad k_b \qquad (3.53)$$

or can be stabilized by collision with the background gas M:

$$AB^\dagger + M \rightarrow AB + M \qquad k_c \qquad (3.54)$$

The net result may be summarized as

$$A + B + M \rightarrow AB + M \qquad k_3 \qquad (3.55)$$

with an effective three-body rate coefficient

$$k_3 = k_{LH} = \frac{k_a k_c}{k_b + k_c[M]} \qquad (3.56)$$

This is known as the Lindemann-Hinshelwood rate coefficient. At low pressure we have

$$\lim_{M \to 0} k_{LH} \equiv k_0 = \frac{k_a k_c}{k_b} \qquad (3.57)$$

In this case the most critical parameter that determines the value of k_{LH} is the lifetime $1/k_b$ of the association complex $AB^\dagger$. The lifetime of the association complex may be estimated using the Rice-Ramsperger-Kassel-Marcus theory. In general, a simple association complex like $H_2^\dagger$ or $O_2^\dagger$ has only one vibrational mode, and all the excess energy in the excited molecule is available for breaking the bond. The lifetime is short, resulting in small values of rate coefficients:

$$H + H + M \rightarrow H_2 + M \qquad k = 8.1 \times 10^{-33} \qquad (3.58)$$

$$O + O + M \rightarrow O_2 + M \qquad k = 4.8 \times 10^{-33} \qquad (3.59)$$

where the k values refer to room temperature in units of $cm^6\ s^{-1}$.

Compare the above rate coefficients with those for more complex reactions such as

$$H + CH_3 + M \rightarrow CH_4 + M \qquad k = 5.6 \times 10^{-29} \qquad (3.60)$$

$$CH_3 + CH_3 + M \rightarrow C_2H_6 + M \qquad k = 1.8 \times 10^{-25} \qquad (3.61)$$

These rate coefficients are many orders of magnitude greater than those for (3.58) and (3.59). The reason is as follows: The activated complexes $CH_4^\dagger$ and $C_2H_6^\dagger$ have

8 and 18 vibrational modes, respectively. The excess bond energy is distributed over many degrees of freedom. It would take thousands of vibrations before the excess energy could be concentrated into one particular bond that breaks. The lifetime of the association complex is long, and by (3.57), this implies a fast rate of formation. The high rate coefficients for the synthesis of complex organic molecules explain the abundance of these compounds in the atmosphere of Titan and have implications for the origin of life.

The high-pressure limit of (3.56) is

$$\lim_{M \to \infty} k_{LH} \equiv \frac{k_\infty}{[M]} = \frac{k_a}{[M]} \qquad (3.62)$$

In this case the rate of reaction (3.55) is

$$r_3 = k_{LH}[A][B][M] = k_a[A][B] = k_\infty[A][B] \qquad (3.63)$$

In this limit the three-body reaction reduces to a two-body reaction, with rate given by the rate-limiting reaction (3.52). Physically, at high pressure the quenching is so efficient that every association complex formed by (3.52) is eventually stabilized by collisions.

The Lindemann-Hinshelwood mechanism does not give a good approximation to the experimental data in the "falloff" region, that is, in the transition region between two-body and three-body limits. An improvement to the Lindemann-Hinshelwood theory is given by Troe; we state without derivation his formula for three-body rate coefficient:

$$k_{Troe} = k_{LH} 0.6^{(1+X^2)^{-1}} \qquad (3.64)$$

where k_{LH} is $k_0/(1 + (k_0[M]/k_\infty)$ from (3.56) and $X = \log (k_0[M]/k_\infty)$. Note that the correction factor from Troe's equals unity when $M \to 0$ or ∞. Figure 3.7 shows a

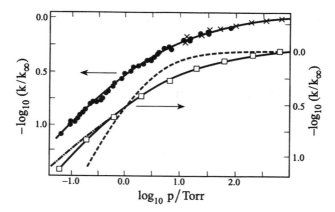

Figure 3.7 Experimental and calculated falloff curves for cyclopropane isomerization at 490°C. Upper curve: experimental data. Lower curves: dashed, based on Lindemann-Hinshelwood theory; solid and dot-dashed, based on modified Lindemann-Hinshelwood theory. After Robinson, P. J., and Holbrook, K. A., 1972, *Unimolecular Reactions* (New York: Wiley).

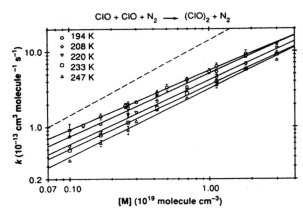

Figure 3.8 Rate coefficient for $ClO + ClO + N_2 \rightarrow (ClO)_2 + N_2$ as a function of N_2 concentrations over the temperature range from 194 to 247 K. The dashed line is extrapolated from earlier work. After Sander, S. P., Friedl, R. R., and Yung, Y. L., 1989, "Rate of Formation of ClO Dimer in Polar Stratospheric Chemistry: Implications for Ozone Loss." *Science* **245**, 1095.

comparison of the Lindemann-Hinshelwood theory and its modified form for providing an empirical fit to experimental data.

Figure 3.8 shows a laboratory determination of the rate coefficient for the reaction $ClO + ClO + N_2 \rightarrow (ClO)_2 + N_2$ as a function of N_2 concentrations. The results suggest that it is difficult to extrapolate from a limited database for atmospheric applications. Accurate measurements must be carried out to determine the rate constants as a function of temperature, pressure, and the third body (in the case of termolecular reactions) over the range of conditions expected in the atmosphere under study.

3.6 Heterogeneous Reactions

Consider an atmosphere containing species X and aerosol particles. X can collide with and stick to the aerosols:

$$X + \text{aerosol} \rightarrow X(\text{ads}) \tag{3.65}$$

X(ads) denotes the adsorbed species. Let n_a and r_a be the number density and radius of the aerosols, respectively. The rate at which X is adsorbed is

$$r_h = k_h[X] \tag{3.66}$$

with the unimolecular rate coefficient given by

$$k_h = \frac{1}{4}\gamma v \pi r_a^2 n_a \tag{3.67}$$

where $v = (8kT/\pi m)^{\frac{1}{2}}$ is the mean thermal speed of X and γ is the accommodation (or sticking) coefficient. This formula can obviously be generalized to a distribution of aerosol particles with different radii. An alternative way to write (3.67) is

$$k_h = \frac{1}{4}\gamma v A \tag{3.68}$$

where A is the area of aerosol per unit volume. The most difficult quantity to determine in the formulas (3.67) and (3.68) is the dimensionless parameter γ. We give an example of an important heterogeneous reaction in the terrestrial stratosphere:

$$HCl + \text{aerosol} \rightarrow HCl(ads) \qquad (3.69)$$

$$ClONO_2 + HCl(ads) \rightarrow Cl_2 + HNO_3 \qquad (3.70)$$

where the "aerosol" may be a H_2SO_4 particle or an ice particle. The values of γ are of the order 10^{-2} and 10^{-1}, respectively, for the above two types of particles. Note that heterogeneous reactions (3.69) and (3.70) turn two rather stable reservoir species HCl and $ClONO_2$ into Cl_2, from which reactive chlorine is readily released.

A famous heterogeneous reaction in astrophysics is the formation of H_2 from H atoms on the surface of grains in the interstellar medium:

$$H + \text{grain} \rightarrow H(ads) \qquad (3.71)$$

$$H(ads) + H \rightarrow H_2 \qquad (3.72)$$

This is believed to be the principal pathway for the recombination of H atoms.

3.7 Miscellaneous Reactions

3.7.1 Reaction with Excited Species

Photolysis and chemical reactions often produce species that are vibrationally or electronically excited. These species have more energy than their counterparts in the ground state. The electronically excited species may have different spin states that facilitate chemical reactions. The excited states of common molecules are listed in table 3.3. The theory of excited state reactions is the same as that for the ground state. We give examples in the following paragraphs.

The reaction

$$H^+ + H_2 \rightarrow H + H_2^+ \qquad (3.73)$$

is endothermic by 41 kcal mole^{-1}, and hence is not important in the ionospheres of the giant planets. However, if the H_2 molecules are in vibrationally excited states with $v' > 4$, then this reaction is exothermic and is expected to be fast.

Photolysis of CH_4 produces methylene radicals in the ground state ($^3CH_2$) and in an excited state ($^1CH_2$), the two states being isoelectronic to $O(^3P)$ and $O(^1D)$, respectively. The reactivities of the two states are very different. For example, the reaction

$$^1CH_2 + H_2 \rightarrow CH_3 + H \qquad (3.74)$$

has rate coefficient $k = 7 \times 10^{-12}$ cm^3 s^{-1}, but the corresponding reaction between $^3CH_2$ and H_2 is negligible.

The reaction

$$O + H_2O \rightarrow OH + OH \qquad (3.75)$$

Table 3.3 Some metastable excited states of atoms and molecules[a]

Ground state	Excited state	Energy (eV)	Radiative lifetime (s)[b]
$H(^2S)$	$H(^2S)$	10.2	0.12
$C(^3P)$	$C(^1D)$	1.26	3.2(3)
	$C(^1S)$	2.68	2
$CH_2(\tilde{X}^3B_1)$	$CH_2(\tilde{a}^1A_1)$	0.1–1	—
$O(^3P)$	$O(^1D)$	1.97	110
	$O(^1S)$	4.19	0.74
$N(^4S)$	$N(^2D)$	2.38	9.4
	$N(^2P)$	3.58	12
$O_2(X^3\Sigma_g^-)$	$O_2(a^1\Delta_g)$	0.98	2.7(3)
	$O_2(b^1\Sigma_g^+)$	1.63	12
$SO_2(\tilde{X}^1A_1)$	$SO_2(\tilde{a}^3B_1)$	3.19	2.7(−3)

[a] Data are from Okabe (1978).
[b] The notation for radiative lifetime a(b) reads a × 10^b.

is endothermic by 17 kcal mole^{-1} when the oxygen atom is in the ground (^3P) state. However, if the O atom is in the first excited state (^1D), the reaction is exothermic by 28.7 Kcal mole^{-1} and is exceedingly rapid:

$$O(^1D) + H_2O \rightarrow OH + OH \qquad k = 1 \times 10^{-11} \qquad (3.76)$$

The primary source of $O(^1D)$ in the terrestrial atmosphere is photolysis of O_3:

$$O_3 + h\nu \rightarrow O(^1D) + O_2(^1\Delta) \qquad (3.77)$$

The bulk of $O(^1D)$ produced in (3.77) is quenched to the ground state by

$$O(^1D) + M \rightarrow O + M \qquad (3.78)$$

but enough survives to drive the reaction (3.76) and make it the principal source of the hydroxyl radical in the terrestrial troposphere and stratosphere.

3.7.2 Dissociative Recombination Reactions

We pointed out in section 3.4.4 that ion-molecule reactions are generally faster than neutral-neutral reactions. Reactions between electrons and molecular ions such as

$$AB^+ + e \rightarrow A + B \qquad (3.79)$$

are extremely fast, with rate coefficients of the order of 10^{-6} to 10^{-8} cm^3 s^{-1}. The reasons are the strong Coulomb attraction between an ion and an electron, and the greater mobility of the electron. We give a few examples of this type of reaction that are important in planetary atmospheres:

$$H_3^+ + e \rightarrow H_2 + H \text{ (or 3H)} \quad k = 2.8 \times 10^{-7}(200/T_e)^{0.7} \quad (3.80)$$

$$O_2^+ + e \rightarrow O + O \quad k = 3 \times 10^{-7} \quad (3.81)$$

$$CH_5^+ + e \rightarrow CH_4 + H \quad k = 3.9 \times 10^{-6} \quad (3.82)$$

where the units for the rate coefficients are cm^3 s^{-1} and the rate coefficient for (3.80) refers to electron temperature T_e, which can be higher than the neutral temperature T in the ionosphere. The fast rates of dissociative recombination ultimately limit the concentrations of molecular ions in planetary ionospheres.

3.7.3 Radiative Association Reactions

The reaction between an electron and an atomic ion of the type

$$A^+ + e \rightarrow A + h\nu \quad (3.83)$$

is much slower than (3.79), because the excess energy produced in (3.79) can all be carried away in the breaking of the A—B bond and the kinetic energy of the fragments. However, in reaction (3.83) all the excess energy associated with the capturing the electron into a bound state must be radiated away, a relatively inefficient process for energy loss. Consequently, the rate coefficients for radiative association are small, of the order of 10^{-12} cm^3 s^{-1}, as shown in the following examples:

$$H^+ + e \rightarrow H + h\nu \quad k = 4.0 \times 10^{-12}(200/T_e)^{0.7} \quad (3.84)$$

$$O^+ + e \rightarrow O + h\nu \quad k = 3.0 \times 10^{-12} \quad (3.85)$$

The rate coefficients are in units of cm^3 s^{-1}. Note that the difference between the rates for (3.79) and (3.83) is five orders of magnitude. The slow rate of radiative association is the main reason for the long lifetime of atomic ions in the upper ionospheres of Earth and the giant planets.

Neutral atoms can also recombine by radiative association:

$$A + B \rightarrow AB + h\nu \quad (3.86)$$

This is also an inefficient process and requires the existence of a low-lying electronic state that can radiate into the ground state. We may make a rough estimate as follows. The radiative association rate coefficient may be computed from

$$k_{ra} = k_c f \quad (3.87)$$

where k_c is the gas kinetic rate coefficient (3.28) and f is an efficiency factor that measures the probability that the emission takes place during the encounter period. The time for a potential curve crossing is of the order of 10^{-12} s. The Einstein coefficient for a fast radiative rate is 10^8 s^{-1}. Therefore, the probability of emission is 10^{-4}. This would yield an upper limit of 10^{-14} cm^3 s^{-1} for k_{ra}. The actual rate coefficients are much smaller, as shown in the following examples

$$C + H \rightarrow CH + h\nu \quad k = 10^{-17} \text{ cm}^3 \text{ s}^{-1} \quad (3.88)$$

$$N + O \rightarrow NO + h\nu \quad k = 6.4 \times 10^{-17} \text{ cm}^3 \text{ s}^{-1} \quad (3.89)$$

In general, radiative association of neutral atoms is too slow to compete with termolecular reaction and recombination on the surface of aerosols and grains. However, radiative association between an ion and a neutral is expected to be faster and the reaction

$$C^+ + H_2 \rightarrow CH_2 + h\nu \quad (3.90)$$

is expected to have a rate coefficient of 10^{-15} to 10^{-16} cm^3 s^{-1}. At this rate (3.90) may be important in the interstellar medium.

4
Origins

4.1 Introduction

Cosmology is a subject that borders on and sometimes merges with philosophy and religion. Since antiquity, the deep mysteries of the universe have intrigued mankind. Who are we? Where do we come from? What are we made of? Is the development of advanced intelligence capable of comprehending the grand design of the cosmos, the ultimate purpose of the universe? Is there life elsewhere in the universe? Is ours the only advanced intelligence or the most advanced intelligence in the universe? These questions have motivated great thinkers to pursue what Einstein called "the highest wisdom and the most radiant beauty." In the fourth century B.C., the essence of the cosmological question was formulated by the philosopher Chuang Tzu:

> If there was a beginning, then there was a time before that beginning. And a time before the time which was before the time of that beginning. If there is existence, there must have been non-existence. And if there was a time when nothing existed, then there must be a time before that—when even nothing did not exist. Suddenly, when nothing came into existence, could one really say whether it belonged to the category of existence or of nonexistence? Even the very words I have just uttered, I cannot say whether they have really been uttered or not. There is nothing under the canopy of heaven greater than the tip of an autumn hair. A vast mountain is a small thing. Neither is there any age greater than that of a child cut off in infancy. P'eng Tsu [a Chinese Methuselah] himself died young. The universe and I came into being together; and I, and everything therein, are one.

78 Photochemistry of Planetary Atmospheres

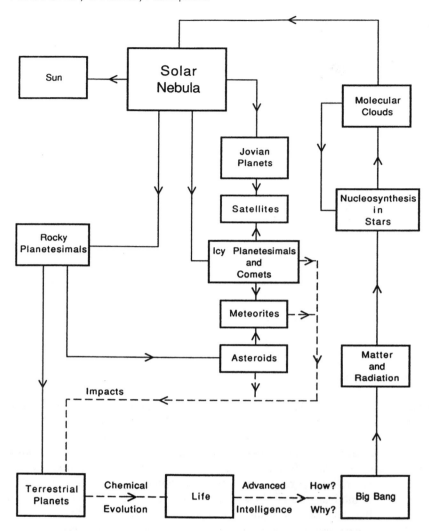

Figure 4.1 Schematic diagram showing the principal pathways of the origin and evolution of the solar system. After Yung, Y. L., and Dissly, R. W., 1992, "Deuterium in the Solar System," in *Isotope Effects in Gas-Phase Chemistry*, J. A. Kaye, editor (Washington, D.C.: American Chemical Society), pp. 369–389.

Fortunately, our subject matter, solar system chemistry, is less esoteric than the questions asked by Chuang Tzu. A schematic diagram showing the principal pathways by which our solar system is formed is given in figure 4.1. The great triumphs of modern science have been summarized in this figure as fundamental contributions to the five "origins": (a) origin of the universe, (b) origin of the elements, (c) origin of the solar system, (d) origin of life, and (e) origin of advanced intelligence. Note that almost all the concrete advances have been made during the last hundred years. We now stand at the epochal point in the history of the solar system, when the development of the human consciousness arrives at the level of a fundamental understanding of the planetary and cosmic environments that give rise to it. It is an exhilarating and intoxicating experience to recognize that the causal sequence from creation (Big Bang) to planets is a logical and inevitable consequence of the known laws of physics. It is not necessary to invoke supernatural or unnatural assumptions.

In this chapter we briefly outline the part of cosmochemistry that is important for determining the chemical environment of the origin of the solar system. What was the source of solar system material? What were the principal events leading to the formation of the solar system? How did planets form? How did the terrestrial planets acquire their atmospheres? Most of the evidence from the time of the formation of the solar system is lost. We have to reconstruct the events from the few fragments that survive to the present.

The topics of this chapter have been chosen not on the basis of what we would like to know, but on the basis of what limited information is available. Observations of the sun and the local interstellar medium provide us with fundamental data on the "cosmic abundances" of the elements. Molecular clouds are precursors of solar system material. There is a considerable body of data on the chemical speciation, processing, and evolution of molecular clouds. Within the solar system, it is convenient to divide the materials into three types: (a) gas, (b) rock, and (c) ice. The giant planets contain gases that were remnants of the solar nebula. In the inner solar system we have more "rocky" material. Meteorites are the best samples of this type of material. The materials that are in the terrestrial planets are not "primitive." The planets are large enough that extensive differentiation has occurred, and it is difficult to infer the original composition by studying the present surface composition. In the outer solar system we have more "icy" material, of which comets are examples. Thus, the giant planets are representative of the bulk of the material in the solar system. The meteorites and comets are probes to the inner and outer parts, respectively, of the solar nebula that has since dissipated. In addition, the study of radioactive nuclei found in meteorites yields important time constraints for the duration of the solar nebula. Noble gases and isotopic fractionations provide clues to the origin of atmospheres.

4.2 Cosmic Organization

When one contemplates the wonders of the universe from the large astronomical scales to the small atomic scales, nothing is more wonderful than the hierarchy of scales of organization. In other words, the mass of the universe is neither uniformly nor randomly distributed over all sizes, but seems to be concentrated in certain preferred ranges. For example, in the large scale we have well-known highly organized units such as planets, stars, molecular clouds, and galaxies. In the small scale we have atoms, molecules, proteins, aerosols, and cells. Our wonder at this intricate pattern of organization is not diminished by the fact that the fundamental laws of physics that govern the distribution of matter are known. On the large scale, the agent of organization is gravity; on the small scale, the corresponding agent is the Coulomb force. Both are long-range forces whose intensity falls off as the inverse square of distance. Of the four fundamental forces of nature, only these two are mediated by

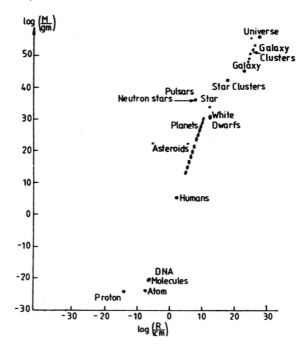

Figure 4.2 Observed objects in the universe plotted in a size-mass diagram. After Barrow, J. D., and Tipler, F. J., 1986, *The Anthropic Cosmological Principle* (New York: Oxford University Press).

massless particles (photon and graviton), which account for their infinite range. Even though the electromagnetic force is many times stronger than that of gravity, it can be both attractive and repulsive and can be canceled. Gravity is always attractive and will dominate at large distances. The human being occupies an intermediate position in this hierarchy of scales. Structurally we are built by electromagnetic forces, but our mobility is severely limited by gravity. The terrestrial planets are unusual in that the gravitational binding energy of a molecule to the planet is of the same order of magnitude as the chemical bond energy (see section 1.4). This determines a rate of chemical evolution that is neither too fast (with catastrophic consequences) nor too slow (with no evolution).

The masses and sizes (linear dimension) of the important organized units in the universe are shown in figure 4.2. The mass of the known universe is very large, estimated to be of the order of 10^{56} g. The bulk of the universe, stretched over 10^{28} cm, is empty, and in this vast monotony of space are scattered concentrated populations of stars, known as galaxies. The mass of a galaxy is of the order 10^{45} g. The stars have masses in the range of 0.1–100 solar masses (1 solar mass = 2×10^{33} g). In the intermediate mass range, between a galaxy and a star, we have giant molecular clouds, of the order of 10^5 to 10^6 solar masses. As we shall show, the cosmic chemical environment is largely determined by processes at this scale.

In our solar system, the bulk of the mass is concentrated in the sun. The mass of Earth is 6×10^{27} g. The mass of Jupiter, the largest planet, is 318 Earth masses.

The giant planets consist mostly of gases, with small cores made of rocky material. The terrestrial planets are composed primarily of rocky material, with thin envelopes of gases. The mass of Earth's atmosphere is 5×10^{21} g, and that of the ocean is 1.5×10^{24} g. All life on Earth is in 10^{18} g of carbon (living biomass). This hierarchy of masses suggests that the various "origins" discussed above are closely connected to each other. The origin of the planets is part of the origin of the solar system. The origin of planetary atmospheres is part of the origin of the planets.

The ultimate source of all energy and matter in our universe is the Big Bang, which occurred about 15 billion years ago. Why should the initially hot and nearly homogeneous gaseous universe evolve into highly organized structures such as galaxies, molecular clouds (with organic molecules), stars, solar systems, planets, and eventually (on at least one planet) life and advanced intelligence? This is all the more surprising because according to the second law of thermodynamics, matter in a closed box tends to evolve toward equilibrium, a state of maximum entropy known to the nineteenth-century cosmologists as the "heat death of the universe." That the universe has evolved along a more interesting alternative path, as shown in figure 4.1, rather than toward heat death is now understood. There are at least three reasons: (a) the expansion of the universe, (b) gravity, and (c) relaxation time. We briefly comment on these reasons. First, the universe is expanding. It is a dynamic, not a static (equilibrium), system. Second, gravity is a long-ranged force that is capable of creating and maintaining complex structures on all macroscopic scales from galaxies to planetary systems. An extreme example of the effect of gravity is the formation of a black hole, the ultimate compact and superdense form of matter. Third, as pointed out in chapter 3 on chemical kinetics, the time needed for a chemical system to reach thermodynamic equilibrium may be too long for the process to be relevant in an evolving system. For example, in the first few minutes of the expansion of the universe, the most thermodynamically stable form of the nucleons was Fe. However, the neutron flux during this period was too low for the equilibrium composition to be attained. As discussed in section 4.3, this is the ultimate source of the nuclear fuel that was left over from the origin of the universe. It would be a major source of energy for stars later.

4.3 The Elements of the Periodic Table

A major triumph of astrophysics is its ability to account for the origin of the elements as part of the origin of the universe and stellar evolution. The cosmic abundances of the elements determine the "initial conditions" for the chemistry of the solar system. The distribution of the elements in the solar system and the planets depends on their "geochemical tendencies," and on this basis they are classified. A most important property, for example, is volatility. The noble gases are very volatile and always remain in the gas phase. Iron, on the other hand, is nonvolatile and prefers to be in the solid state. The classical scheme of Goldschmidt divides all elements into four groups: lithophile, siderophile, chalcophile, and atmophile (meaning, respectively, rock loving, iron loving, sulfide loving, and atmosphere loving). We adopt the more recent classification scheme of Larimer (1987).

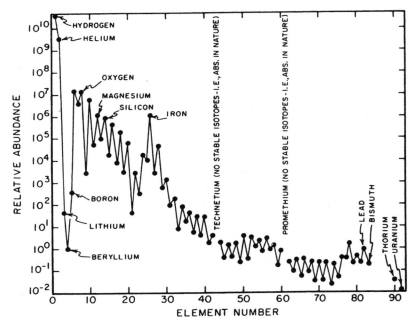

Figure 4.3 Relative abundance of the elements in our sun. ABS. = absent. The abundance of Si is taken to be 10^6. After Broecker, W. S., 1985, *How to Build a Habitable Planet* (New York: Eldigio Press).

4.3.1 Cosmic Abundances of the Elements

In the first few minutes of the origin of the universe, the temperature of the fireball exceeded 10^9 K and the simplest nuclei H, D, He, and Li were synthesized. The nascent universe cooled rapidly due to adiabatic expansion, and there was not enough time for the synthesis of the heavier elements beyond Li. Matter consisted of 90% H and 10% He by mass, with smaller amounts of D (10^{-4}) and Li (10^{-9}). The conversion of the simplest elements into the approximately 100 heavier elements of the periodic table took place in the interior of stars, where the heating (due to gravitational compression) is intense enough to ignite the nuclear fuel. The details of how the originally diffuse matter of the universe organized into dense self-gravitating balls of gas is not pursued here. In the interior of stars, hydrogen and helium are "burned," resulting in the formation of heavier elements in a process known as nucleosynthesis, which is in fact the principal source of energy for stars. Stars have finite lifetimes, and on death, eject the synthesized heavy elements into the interstellar medium.

The material of which the bulk of the universe is made is thus a mixture of primordial material and material of secondary origin. This distribution of matter is known as the cosmic abundances of elements presented in figure 4.3. Appendix 4.1 gives the same information in a table and provides data on the relative abundances of important isotopes of the elements. This is the ultimate chemical yardstick on which all partitioning and speciation is based. The sun and the interstellar medium in the solar neighborhood have roughly this composition.

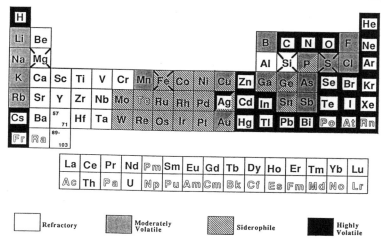

Figure 4.4 Cosmochemical classification of the elements. Elements with crosses are lithophiles. After Larimer, J. W., 1987, "The Cosmochemical Classification of the Elements," in Kerridge and Matthews (1988; cited in section 4.1), pp. 375–389.

4.3.2 Classification of Elements

It is convenient in cosmochemistry to divide the elements into four groups: refractory, siderophile, moderately volatile, and highly volatile, as shown in figure 4.4. This scheme is proposed by Larimer (1987) on the basis of Urey's (1959) recognition that the most important fractionation process in the formation of the solar system is the separation of dust and gas. Thus, the primary motivation of this classification is the equilibrium condensation sequence in the solar nebula as it cools. The refractory group consists of most of the elements to the left of the periodic table, plus lanthanides and actinides. The refractory elements are the first to condense in the solar nebula at temperatures above 1000 K. The siderophile elements as a group all occur as metals usually alloyed with Fe and Ni. As a result of this affinity, they are concentrated in the cores of differentiated planetary bodies. The elements in the moderately volatile group include most of the first column (the alkali metals) and the right side of the periodic table (figure 4.4). These elements condense at significantly lower temperatures. The last group, the highly volatile elements, condense at low temperatures, forming ices in the outer solar system. The noble gases, which belong to this group, will not condense even in the outer solar system. However, they can be trapped in ices and in carbonaceous material.

This classification is not mutually exclusive and depends on ambient pressure, temperature, and oxidation state. A key variable in the solar nebula is the C/O ratio, which is of the order of unity. A slight excess of C over O would imply that the nebula favors the production of "refractories" such as graphite, SiC, AlN, and TiN. The alternative possibility, with an excess of O over C, would imply the dominance of CO, a highly volatile gas, and the instability of organics against oxidation.

To the Larimer scheme we can add another group, the biogeochemical elements C, H, O, N, S, and P. Interest in the latter group is connected to the origin of life and the history of Earth. We note that biology has the ability to mobilize oxygen and sulfur as gaseous compounds. The geochemistry and atmospheric chemistry of Earth are profoundly affected by the biosphere. The synthesis of complex organic compounds is not, however, the monopoly of living organisms. As we show in section 4.4, organic synthesis is prevalent on a cosmic scale.

4.4 Molecular Clouds

From figure 4.2 we note that on a size scale much smaller than a planet, the agent of organization is the Coulomb force, and the most remarkable product is life. The fundamental unit of matter is an atom, with mass of the order of 10^{-24} g and size 10^{-8} cm. A primitive living organism (a cell) consists of 10^{10} atoms. A complex form of life (a human being) is about 10^5 g. The basis of this organizational hierarchy is the organic chemistry based on carbon that allows the construction of molecules of arbitrary complexity, given a suitable source of energy. Today, the rate of organic synthesis on Earth is greater than that on any other planet in the solar system by a factor of 10^3. This is, of course, due to the remarkable machinery of photosynthesis, itself a product of life (see discussion in section 2.5). A question then naturally arises: What was the prebiological organic synthesis that led to the spontaneous generation of life? In a series of classic experiments, Miller and Urey demonstrated the synthesis of complex organic compounds from mixtures of simple gases of reducing composition (such as CH_4 and NH_3) using electric discharges. Subsequent observations showed that, indeed, organic synthesis is fairly common without the presence of life in the molecular clouds and in the atmospheres of the planets in the outer solar system. In section 4.4.1 we discuss the organic chemistry of molecular clouds.

A most spectacular achievement of modern radio astronomy is the discovery of large numbers of complex organic molecules in giant molecular clouds. The richness of the rotational spectra observed at millimeter wavelengths is shown in figure 4.5 from a recent survey of the Orion Molecular Cloud. Table 4.1 lists the molecular species identified in the interstellar medium and circumstellar envelopes as of 1996. These molecules include amino acids, which are the building blocks of proteins. There is no doubt that even more complex organic molecules exist, but these molecules have many quantum states that make their detection difficult. The existence of this rich variety of organic compounds implies that the basic chemicals essential for the origin of life are produced in abundance on a cosmic scale. Of course, there is still the problem of delivering this preprocessed material intact to the solar system. There is evidence that at least some of the organic material present during the formation of the solar system have survived in the meteorites.

4.4.1 Organic Synthesis

The basic chemistry that gives rise to the large number of species in table 4.1 is understood. The molecular clouds have low number densities, in the range of 10^3 to 10^6 molecules/cm^3. The temperatures are also low, about 10–50 K. Under these

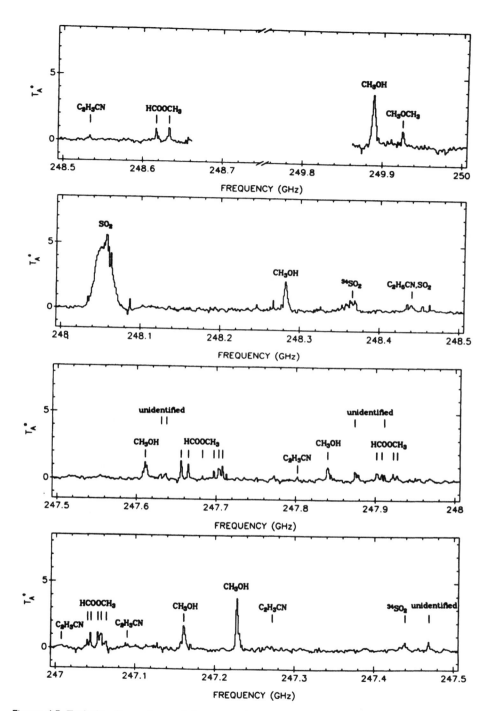

Figure 4.5 Emission lines observed from gas phase molecules in the Orion Molecular Cloud during a spectral scan in 1986. Over 800 resolved spectral lines are present and have been assigned to some 29 molecular species. From Black, G. A., et al., 1986, "The Rotational Emission-Line Spectrum of Orion-A between 247-GHz and 263-GHz." *Astrophys. J. Suppl. Ser.* **60**, 357.

Table 4.1 Molecules identified in the interstellar medium

Species	Name	Species	Name	Species	Name
TWO ATOMS		NH_2	nitrogen dihydride	SiH_4	silane
AlF	Aluminum monofluoride	N_2H^+	protonated nitrogen		
AlCl	Aluminum monochloride	N_2O	nitrous oxide	SIX ATOMS	
C_2	diatomic carbon	NaCN	sodium cyanide	C_5H	pentynylidyne
CH	methylidyne	OCS	carbonyl sulfide	C_5O	1,2-cyclobutadiene
CH^+	methylidyne ion	SO_2	sulfur dioxide	C_2H_4	ethylene
CN	cyanogen	c-SiC_2	silicon dicarbide	CH_3CN	methyl cyanide
CO	carbon monoxide			CH_3NC	methyl isocyanide
CO^+	carbon monoxide ion	FOUR ATOMS		CH_3OH	methanol
CP	carbon phosphide	c-C_3H	cyclopropenylidyne	CH_3SH	methyl mercaptan
CS	carbon monosulfide	l-C_3H	propenylidyne	HC_3NH^+	protonated cyanoacetylene
CSi	silicon carbide	C_3N	cyanoethynyl	HC_3HO	propynal
HCl	hydrogen chloride	C_3O	tricarbon monoxide	$HCOCH_2$	vinoxy
H_2	molecular hydrogen	C_3S	tricarbon monoxide	$HCONH_2$	formamide
KCl	potassium chloride	C_2H_2	acetylene	l-H_2C_4	butatrienylidene
NH	nitrogen hydride	CH_2D^+	methylium		
NO	nitric oxide	HCCN	cyano-methylene	SEVEN ATOMS	
NS	nitrogen sulfide	$HCNH^+$	protonated hydrogen cyanide	C_6H	hexatriynyl
NaCl	sodium chloride			CH_2CHCN	vinyl cyanide
OH	hydroxyl	HNCO	isocyanic acid	CH_3C_2H	methylacetylene
PN	phosphorus nitride	HNCS	isothiocyanic acid	HC_5N	cyanodiacetylene
SO	sulfur monoxide	$HOCO^+$	protonated carbon dioxide	$HCOCH_3$	acetaldehyde
SO^+	sulfoxide ion			NH_2CH_3	methylamine
SiN	silicon nitride	H_2CO	formaldehyde		
SiO	silicon monoxide	H_2CN	methylene-aminylium	EIGHT ATOMS	
SiS	silicon sulfide	H_2CS	thioformaldehyde	CH_3C_3N	methylcyanoacetylene
		H_3O^+	hydronium ion	$HCOOCH_3$	methyl formate
THREE ATOMS		NH_3	ammonia		
C_3	triatomic carbon			NINE ATOMS	
C_2H	ethynyl radical	FIVE ATOMS		CH_3C_4H	methyldiacetylene
C_2O	dicarbon monoxide	C_5	pentatomic carbon	CH_3CH_2CN	ethyl cyanide
C_2S	dicarbon sulfide	C_4H	butadiynyl	$(CH_3)_2O$	dimethyl ether
CH_2	methylene	C_4Si	silicon tetracarbide	CH_3CH_2OH	ethanol
HCN	hydrogen cyanide	l-C_3H_2	propenylidene	HC_7N	cyanohexatriyne
HCO	formyl	c-C_3H_2	cyclopropenylidyne		
HCO^+	formyl ion	CH_2CN	cyanoacetylene	TEN ATOMS	
HCS^+	thioformyl ion	CH_4	methane	CH_3C_5N	methylcyanotetradiyne
HOC^+	isoformyl ion	HC_3N	propadienylidene	$(CH_3)_2CO$	acetone
H_2O	water	HC_2NC	isocyanoacetylene		
H_2S	hydrogen sulfide	HCOOH	formic acid	ELEVEN ATOMS	
HNC	hydrogen isocyanide	H_2CHN	methanimine	HC_9N	cyano-octa-tetra-yne
HNO	nitroxyl	H_2C_2O	ketene		
MgCN	magnesium cyanide	H_2NCN	cyanamide	THIRTEEN ATOMS	
MgNC	magnesium isocyanide	HNC_3	cyanoacetylene	$HC_{11}N$	cyano-deca-penta-yne

From Irvine, W. M., and Knacke, R. F. (1989).

conditions normal neutral chemistry would be too slow. The primary driving forces of disequilibrium chemistry are ionization by cosmic rays and photoionization by starlight. A detailed discussion of the chemistry of the interstellar medium is beyond the scope of this chapter. We show the essential aspects of this chemistry by a few examples. A schematic diagram showing the formation of H_2O, CO, and CO_2 is given in figure 4.6. The bulk of the gas in the interstellar medium is H_2. Cosmic ray or photon ionization leads to the production of H_2^+, which is removed primarily by

$$H_2^+ + H_2 \rightarrow H_3^+ + H \tag{4.1}$$

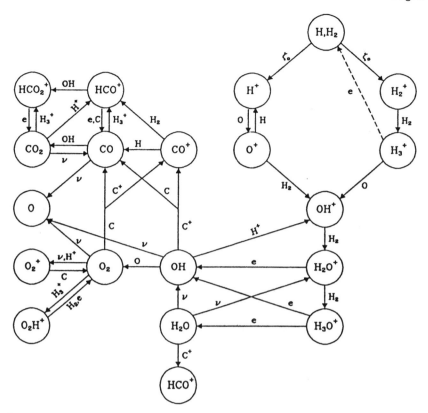

Figure 4.6 The most important chemical pathways involving oxygen-bearing species in the interstellar medium. After van Dishoeck, E. F., Black, G. A., and Lunine, J. I., 1993, "The Chemical Evolution of Protostellar and Protoplanetary Matter," in Levy and Lunine (1993; cited in section 4.1), pp. 163–241.

As shown in figure 4.6, H_3^+ can now react with O:

$$H_3^+ + O \rightarrow OH^+ + H_2 \tag{4.2}$$

This is followed by

$$OH^+ + H_2 \rightarrow H_2O^+ + H \tag{4.3}$$

$$H_2O^+ + H_2 \rightarrow H_3O^+ + H \tag{4.4}$$

$$H_3O^+ + e \rightarrow H_2O + H \tag{4.5}$$

The final product is H_2O, synthesized from H_2 and O. Similar reaction pathways, as illustrated in figure 4.6, produce CO and CO_2.

The synthesis of hydrocarbons may be carried out using similar ion chemistry. The precursor species C and C^+ can react with hydrogen by

$$C^+ + H_2 \rightarrow CH_2^+ + h\nu \tag{4.6}$$

$$C + H_3^+ \rightarrow CH^+ + H_2 \tag{4.7}$$

followed by

$$CH^+ + H_2 \rightarrow CH_2^+ + H \tag{4.8}$$

$$CH_2^+ + H_2 \rightarrow CH_3^+ + H \tag{4.9}$$

$$CH_3^+ + H_2 \rightarrow CH_5^+ + h\nu \tag{4.10}$$

$$CH_5^+ + e \rightarrow CH_4 + H \tag{4.11}$$

The net result is the formation of CH_4. Higher hydrocarbons containing two carbon atoms may be formed in the following reactions:

$$C^+ + CH_4 \rightarrow C_2H_3^+ + H \tag{4.12}$$

$$C_2H_3^+ + e \rightarrow C_2H_2 + H \tag{4.13}$$

$$CH_3^+ + CH_4 \rightarrow C_2H_5^+ + H \tag{4.14}$$

$$C_2H_5^+ + e \rightarrow C_2H_4 + H \tag{4.15}$$

These reactions ultimately produce acetylene and ethylene.

Simple nitrogen compounds may be produced from N^+ as follows:

$$N^+ + H_2 \rightarrow NH^+ + H \tag{4.16}$$

followed by

$$NH_n^+ + H_2 \rightarrow NH_{n+1}^+ + H \tag{4.17}$$

where $n = 1, 2,$ or 3. Ammonia and HCN can then be formed by

$$NH_4^+ + e \rightarrow NH_3 + H \tag{4.18}$$

$$C^+ + NH_3 \rightarrow H_2CN^+ + H \tag{4.19}$$

$$H_2CN^+ + e \rightarrow HCN + H \tag{4.20}$$

We should point out the great efficiency of organic synthesis in the interstellar medium. First, the chemistry is dispersed over a large volume of space. The size of a giant molecular cloud is of the order of 100 pc (1 pc = 3×10^{18} cm = 2×10^5 AU). This increases the collection efficiency of cosmic rays and ultraviolet photons. Second, the chemical synthesis is primarily via ion-molecule reactions that have little or no activation energy, and these are extremely fast even at low temperatures. Third, the ambient radiation field is not intense enough to destroy the synthesized products. These conditions are difficult to duplicate in planetary atmospheres. The only atmosphere in which organic synthesis approaches the richness and complexity of the interstellar medium is that of Titan and presumably the ancient atmospheres of Earth and Mars.

Table 4.2 Deuterium fractionation in interstellar clouds

Molecule	Observed ratio
DCN/HCN	0.002–0.02
DCO^+/HCO^+	0.004–0.02
N_2D^+/N_2H^+	~0.01
DNC/HNC	0.01–0.04
NH_2D/NH_3	0.003–0.14
$HDCO/H_2CO$	~0.01
C_2D/C_2H	0.01–0.05
DC_3N/HC_3N	~0.02
DC_5N/HC_5N	0.02
C_3HD/C_3H_2	0.03–0.15

From Irvine, W. M., and Knacke, R. F. (1989).

4.4.2 Isotopic Fractionation

A distinctive signature of interstellar chemistry is a high degree of isotopic fractionation. Table 4.2 summarizes the deuterium fractionation for hydrogen-bearing species observed in molecular clouds. These values are orders of magnitude higher than the cosmic ratio $D/H = 1.6 \times 10^{-5}$. It is a triumph of ion-molecule reaction studies to provide a satisfactory account of these large fractionations. We consider two examples. The isotope exchange reactions

$$H_3^+ + HD \rightarrow H_2D^+ + H_2 \qquad E_1 \qquad (4.21)$$

$$CH_3^+ + HD \rightarrow CH_2D^+ + H_2 \qquad E_2 \qquad (4.22)$$

are exothermic by $E_1 = 0.46$ and $E_2 = 0.74$ kcal/mole, respectively. The amount of fractionation may be estimated using the formulas developed in section 3.2:

$$\frac{[H_2D^+]}{[H_3^+]} = \frac{[HD]}{[H_2]} e^{E_1/RT} = 2fe^{E_1/RT} \qquad (4.23)$$

$$\frac{[CH_2D^+]}{[CH_3^+]} = \frac{[HD]}{[H_2]} e^{E_2/RT} = 2fe^{E_2/RT} \qquad (4.24)$$

where $f = [D]/[H]$ is the cosmic abundance. Equations (4.23) and (4.24) are good approximations because H_2 and HD are the major reservoirs for hydrogen and deuterium, respectively, in the interstellar medium, and the exchange reactions (4.21) and (4.22) are sufficiently fast that isotopic equilibration is rapidly established. The fractionation may be expressed as

$$\frac{[H_2D^+]}{[H_3^+]} = 3fR_1 \qquad (4.25)$$

$$\frac{[CH_2D^+]}{[CH_3^+]} = 3fR_2 \qquad (4.26)$$

where the fractionation factors are given by

$$R_1 = \frac{2}{3}e^{E_1/RT} \qquad (4.27)$$

$$R_2 = \frac{2}{3}e^{E_2/RT} \qquad (4.28)$$

Note that R_1 and R_2 would be unity if the composition were given by cosmic abundances. At low temperatures (10–50 K), the fractionation factors can be very large. For example, at 30 K, the values of R_1 and R_2 are 1.4×10^3 and 1.5×10^5, respectively. Reactions of the type

$$XH^+ + HD \rightarrow XD^+ + H_2 \qquad E \qquad (4.29)$$

where X is a neutral molecule such as OH, CH_3, or NH_2 favor the right side, and E is positive. XD^+ is usually more tightly bound due to a reduced zero point energy.

We can now appreciate the beauty of a fractionation process such as (4.24). The fractionation factor is very large due to the large energy difference E compared with RT at low temperatures. At the same time reactions such as (4.22) are ion-exchange reactions that can readily proceed without activation energy, as discussed in section 3.4.4. Thus, the observed isotopic fractionations in table 4.2 are simply explained.

4.4.3 Grain Chemistry and Radiative Association

In addition to ion chemistry, catalysis on the surface of grains is believed to play a major role in interstellar chemistry. The most important reaction that is believed to be catalyzed on the surface is the recombination of hydrogen atoms:

$$H + \text{grain} \rightarrow H(\text{ads}) \qquad (4.30)$$

$$H + H(\text{ads}) \rightarrow H_2 \qquad (4.31)$$

Note that the first atom is adsorbed on the grain and waits for the second atom for the reaction to be completed. The efficiency of the recombination depends on the rate of collision between H atoms and grains and on the accommodation coefficient of H on the surface. The latter is highly uncertain and may depend on the number of available sites for chemi-adsorption. Organic synthesis may also be carried out on the surface of grains.

Organic synthesis is initiated by reactions such as (4.1) involving H_2. The sequence of reactions (4.1)–(4.5) does not result in a net increase of the number of chemical bonds, but an exchange of chemical bonds. Therefore, it is extremely important to first form the H—H bond. One may ask how H_2 was formed in the primordial universe without heavy elements and dust grains. The following are alternative processes:

$$H + H \rightarrow H_2 + h\nu \qquad (4.32)$$

$$H + e \rightarrow H^- + h\nu \qquad (4.33)$$

$$H + H^- \rightarrow H_2 + e \qquad (4.34)$$

These processes involving radiative recombination are very inefficient compared with grain catalysis, but in the absence of grains they may have been important during the early development of the universe.

Origins 91

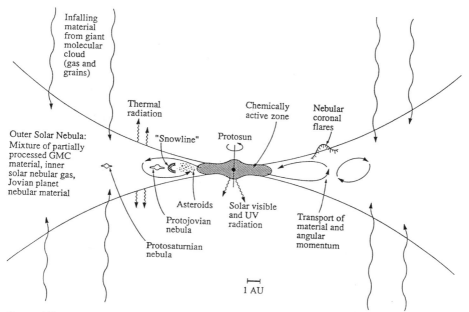

Figure 4.7 Schematic illustration of the physical and chemical processes in the solar nebula. GMC = giant molecular nebula. After van Dishoeck, E. F., Black, G. A., and Lunine, J. I., 1993, "The Chemical Evolution of Protostellar and Protoplanetary Matter," in Levy and Lunine (1993; cited in section 4.1), pp. 163–241.

4.5 The Solar Nebula

The nebula hypothesis for a common origin of the sun and the planets can be traced back to Kant and Laplace in the eighteenth century. The formation of the solar system from a molecular cloud was most probably triggered by a neighboring supernova. The shock from this explosion initiated an instability that led to the collapse of the molecular cloud and the formation of a slowly rotating disklike object known as the solar nebula. A schematic diagram illustrating the physical and chemical processes in the solar nebula is given in figure 4.7. The physical extent of the nebula is about 100 AU, roughly the size of the present solar system. The total amount of material is uncertain. Estimates range from barely 1 solar mass to 2–4 solar masses. The bulk of the mass is concentrated at the center, the protosun. The "flared" geometry of the nebula is due to the weakening of the component of gravity that is perpendicular to the ecliptic plane in the outer part of the nebula. Both gas and solid are present in the nebula. The inner nebula (within 5 AU from the center) is relatively warm, and the solid material is mostly refractory. At larger distances from the center the nebula is cooler, and ices can form. As shown in figure 4.7, the nebula is hardly a "static" system. Without going into the details, we mention three dynamical processes that might have been important for chemistry in the nebula: (a) large-scale convection that mixes material between the inner and outer nebula, (b) viscous shear motion that transfers most of the mass of the nebula to the protosun but leaves most of the

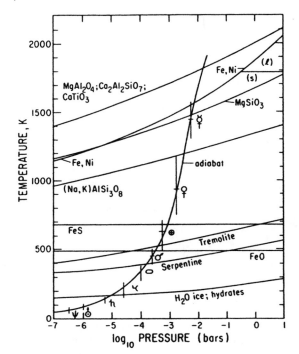

Figure 4.8 Equilibrium condensation sequence for a system of solar elemental composition. After Lewis, J. S., and Prinn, R. G., 1984, *Planets and Their Atmospheres* (New York: Academic Press).

angular momentum to Jupiter, and (c) supersonic shock as the free-falling material enters the "boundary" of the nebula. The effect of (a) and (c) on chemistry is obvious. Process (b) may require hydromagnetic coupling, which in turn requires a high state of ionization of the nebula material. Under these unproven assumptions, ion chemistry may be important.

The physical and chemical conditions of the solar nebula are no longer observable today. Models of the solar nebula can be constructed using information based on the composition of the planets, meteorites, and comets (see sections 4.6–4.8). An example of this class of model is given in figure 4.8. The pressure and temperature conditions in the solar nebula are approximately given by the adiabat. The positions that correspond to the locations for the formation of the planets are indicated in figure 4.8. For example, at 1 AU where Earth is formed, the temperature is about 600 K and the pressure is 10^{-3} bar. In the region close to the sun, the pressure and temperature are higher and only the most refractory oxides of elements such as Al, Ca, Mg, and the rare earths condense, followed by Fe and Ni. The minerals of the alkali metals condense next in the sequence. FeS condenses at about 700 K, at a location between Venus and the Earth. Beyond 1 AU hydrated minerals such as tremolite and serpentine will condense. Finally, as we move out to the outer solar system, volatile material such as H_2O, NH_3, and CH_4 condense as ice and hydrates, $NH_3 \cdot H_2O$ and $CH_4 \cdot 7H_2O$. A major success of this simple model is its ability to account for the bulk density and composition of the planetary bodies in the solar system.

What type of chemistry prevails in the solar nebula depends on the energy source available. Prinn and Fegley (1989) define a ratio that measures the usable energy flux available for driving chemical reactions relative to the thermal flux:

$$R = \frac{\phi(\text{usable flux})}{\phi(\text{thermal})} \quad (4.35)$$

where $\phi(\text{thermal}) = \sigma T^4$ is the thermal flux at temperature T. At 1 AU the estimated ratios for various chemical processes are as follows:

$$R(\text{thermochemistry}) = 4 \times 10^{-5} \quad (4.36)$$

$$R(\text{lightning}) = 4 \times 10^{-6} \quad (4.37)$$

$$R(\text{sunlight}) = 0 \quad (4.38)$$

$$R(\text{starlight}) = 10^{-9} \quad (4.39)$$

$$R(\text{radioactivity}) = 3 \times 10^{-12} \quad (4.40)$$

These estimates are made under the assumptions that the activation energies for chemical reactions are about 10 kcal/mole, the ambient temperature is about 600 K, and the nebula is largely opaque to visible and ultraviolet radiation. The source of energy due to lightning is highly uncertain. Hence, the only proven energy source is thermal energy, derived from the gravitational accretion energy of the nebula itself. Other forms of energy may have existed but need to be verified by observational evidence. We should note that in (4.38) $R(\text{sunlight}) = 0$ not because of the lack of ultraviolet radiation from the sun but because of the enormous opacity due to absorption by dust. As discussed in section 4.7, there may be an indirect source of ultraviolet radiation arising from a hot corona above the nebula.

An example of a thermochemical calculation for the partitioning of the biogeochemical elements C, H, O, and N is shown in figure 4.9. The calculations are based on equilibrium chemistry along the P, T adiabat in the solar nebula. H_2, the dominant species, has a mixing ratio close to unity and is not plotted on this figure 4.9. The most important part of the equilibrium chemistry concerns the following partitioning:

$$CO + 3H_2 \leftrightarrow CH_4 + H_2O \quad (4.41)$$

$$N_2 + 3H_2 \leftrightarrow 2NH_3 \quad (4.42)$$

$$CO + H_2O \leftrightarrow CO_2 + H_2 \quad (4.43)$$

The equilibrium chemistry tends to favor CH_4 and NH_3 in the outer solar nebula and CO and CO_2 in the inner solar nebula. Independent information is required to estimate the temperature at which these reactions are "quenched." As shown in figure 4.9, the quenching temperatures for $N_2 \rightarrow NH_3$, $CO \rightarrow CH_4$, and $CO \rightarrow CO_2$ are 1600, 1470, and 830 K, respectively. Thus, for example, the predicted conversion of CO to CH_4 at the orbit of Jupiter or beyond (T < 300 K) can never be realized using homogeneous equilibrium chemistry alone, because the time required would be too long compared with the age of the nebula.

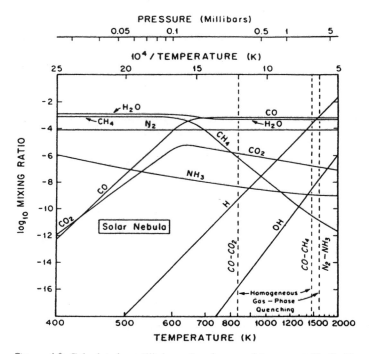

Figure 4.9 Calculated equilibrium abundances of important H, C, N, and O gases along the solar nebula adiabat. After Prinn, R. G., and Fegley, B., Jr., 1989, "Solar Nebula Chemistry: Origin of Planetary, Satellite and Cometary Volatiles," in Atreya et al. (1989; cited in section 4.1), pp. 78–136.

There are obviously other processes that would modify the equilibrium chemistry discussed in the preceding section. Catalysis by transition metals on the surface of grains may accelerate the rate of these reactions. For example, Fe-catalyzed conversion of N_2 to NH_3 and of CO to CH_4 may be important at temperatures as low as 500 K. Additional processes that would modify the simple chemistry outlined here include mixing in the nebula, and similar but distinctive chemistry that can take place in the planetary subnebulae.

4.6 Meteorites

Meteorites are rocks that are of extraterrestrial origin. They are compositionally similar to the sun (in the heavy elements). Radiometric dating places their age at 4.55 Gyr, about the same age as the Moon and Earth. There is strong evidence that meteorites are samples of solid material left over from the solar nebula. The classification of meteorites and their fall frequencies are summarized in the appendix 4.2. Most of the meteorites are chondrites, primitive clastic rocks that formed in the early history of the solar nebula. Although most meteorites consist of unprocessed material, that is not true of iron meteorites. These must have been formed in planetesimals that had melted and were subsequently broke up. The Shergottites, Nakhlites, and Chassignites are most likely from Mars.

4.6.1 Meteorite Chronology

Meteorites contain radioactive nuclei that are extremely useful for dating important events that took place in the solar nebula. The principle of the classical method of dating using radionuclides may be illustrated as follows: The time-dependent concentration of a radioactive element $N(t)$ is

$$N(t) = N(0)e^{-\lambda t} \qquad (4.44)$$

where $N(0)$ is the initial concentration and λ^{-1} is the mean life. In deriving this result we have assumed that there is no new source of the radioactive nuclei after $t = 0$. Consider the example of the two most abundant uranium isotopes. The mean lifetimes of ^{235}U and ^{238}U are 1 and 6.5 Gyr, respectively. The initial ratio $N(^{235}U)/N(^{238}U)$ is 1/3; the current ratio is 1/137.8. From this and (4.44), we may infer the sample is 4.6 Gyr old. The ages of the oldest terrestrial and lunar rocks are determined by this or variants of this method.

There is a class of short-lived nuclei that are extremely important for understanding the chronology of the early solar system. These nuclei have mean lives between 0.1 and 100 Myr, long-lived enough for them to survive from nucleosynthesis (in stars) to incorporation in the meteorites but not long-lived enough to survive to the present. For this reason they are known as extinct nuclei. Since these nuclei are by definition extinct, their abundances can be inferred only from their decay products. An example is ^{26}Al, which has a mean life of 1.1 Myr, decaying to ^{26}Mg. Figure 4.10 shows the variation of the ratio ^{26}Mg/^{24}Mg in an inclusion in the Allende meteorite. The range of variation is of the order of 10%, and there is good correlation with the ratio ^{27}Al/^{24}Mg. Since ^{26}Al decays to ^{26}Mg, it is a good hypothesis that the excess ^{26}Mg is all derived from the now extinct ^{26}Al. The amount needed is ^{26}Al/^{27}Al $= 5 \times 10^{-5}$. Given the short life of ^{26}Al, we conclude that the meteorite must have formed within a few lifetimes of ^{26}Al, or a few million years. Since this pioneering work, the existence of six extinct nuclei in meteorites has been confirmed, as summarized in table 4.3. The results from the other nuclei generally confirm the conclusion based on ^{26}Al. In addition, the large number of nuclei allows us to infer their stellar sources. As shown in table 4.3, for example, the nuclei ^{26}Al, ^{60}Fe, ^{107}Pd, ^{135}Cs, and ^{182}Hf are derived from a low-mass AGB star 2 Myr before the solar system (BSS). At 25 Myr BSS, an OB association is responsible for ^{53}Mn.

4.6.2 Environment of Formation

The meteorites are good probes of the denser and warmer parts of the solar nebula. Based on the fractionation of refractory elements, the maximum temperature of the location where chondrites were formed must be about 2000 K. Calcium- and aluminium-rich inclusions (CAIs) were most likely formed at temperatures between 1800 and 2000 K. In fact, the minerals in CAIs were found to have been formed

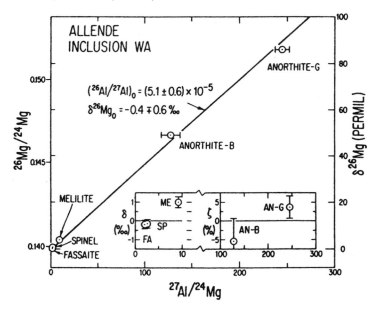

Figure 4.10 Correlation of ^{26}Mg/^{24}Mg with ^{27}Al/^{24}Mg in Allende inclusion. After Lee, T., Papanastassiou, D. A., and Wasserburg, G. J., 1977, "Aluminum-26 in the early solar system: fossil or fuel?" *Astrophys. J. Lett.* **211**, L107. WA is a coarse-grained chondrule containing anorthite, pyroxene, spinel, and melilite.

in exactly the condensation sequence predicted using the thermodynamic equilibrium chemistry described in section 4.5.

The meteorites are not all primitive material. There is evidence for thermal and shock processing that has fundamentally altered the composition and texture of some of the meteoritic material. Theoretically, we would expect a planetesimal exceeding the size of 10 km to undergo differentiation, with ^{26}Al as a possible heat source. The net result of this process is shown in figure 4.11. On breaking up due to collisions, some of the differentiated material is released. This would account for the origin of the highly processed material such as that found in the stony-iron and iron meteorites (see appendix 4.2).

The existence of chondrules in meteorites provides another constraint on the physical conditions in the nebula. These are millimeter-sized igneous-textured spheroids composed mostly of silicate minerals. The spheroid shape suggests that they were once either fully or partially molten droplets. A pulse of intense heating is required for their formation; the temperature must exceed 1800 K for not more than a few minutes. The energetics of chondrule formation are not completely understood and pose a fundamental challenge to the sources of disequilibrium energy discussed in section 4.5. Current ideas include shock during infall of material to the central plane (see figure 4.7), melting during impact, and lightning.

Table 4.3 Extinct nuclides

Radionuclide	Mean life (Myr)	Daughter	Abundance	Comments
Confirmed				
^{26}Al	1.1	^{26}Mg	^{26}Al/^{27}Al = 5 × 10^{-5}	Shortest mean life; possible heat source
^{53}Mn	5.5	^{53}Cr	^{53}Mn/^{55}Mn = 4 × 10^{-5}	Most recent discovery
^{107}Pd	9.4	^{107}Ag	^{107}Pd/^{108}Pd = 2 × 10^{-5}	Only found in metal
^{129}I	23	^{129}Xe	^{129}I/^{127}I = 1 × 10^{-4}	First discovered; found in >75 meteorites
^{146}Sm	149	^{142}Nd	^{146}Sm/^{144}Sm = 0.005 − 0.015	Uncertain abundance
^{244}Pu	120	Fission Xe	^{244}Pu/^{238}U = 0.004 − 0.007	Also α-decays, leaving damage tracks
Strong evidence				
^{92}Nb	231	^{92}Zr	^{92}Nb/^{93}Nb = (2 ± 1) × 10^{-5}	Based on single Nb-rich grain
Hints				
^{60}Fe	1	^{60}Ni	^{60}Fe/^{56}Fe ≤ 2 × 10^{-6}	Excess ^{60}Ni observed; could be nucleosynthetic effect
^{135}Cs	3.3	^{135}Ba	^{135}Cs/^{133}Cs ≤ 2 × 10^{-4}	Deficit in ^{135}Ba observed; could be nucleosynthetic
^{182}Hf	13	^{182}W	^{182}Hf = 1 − 10 × 10^{-5}	Deficit in ^{182}W observed; in Hf-depleted metal
Interesting upper limits				
^{41}Ca	0.19	^{41}K	^{41}Ca/^{40}Ca < 1 × 10^{-8}	
^{247}Cm	22.5	^{235}U	^{247}Cm/^{235}U < 0.004	Must be cogenetic with ^{244}Pu

From Swindle, T. D. (1993).

undifferentiated asteroid (ordinary chondrites)

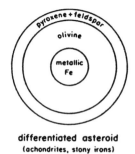
differentiated asteroid (achondrites, stony irons)

Figure 4.11 Schematic cross section of two meteorite parent bodies: undifferentiated and differentiated. After Wetherill, G. W., and Chapman, C. R., 1988, "Asteroids and Meteorites," in Kerridge and Matthews (1988; cited in section 4.1), pp. 35–67.

Table 4.4 Concentrations and molecular characteristics of soluble organic compounds of meteorites

Class	Concentration (ppm)	Compounds identified	Chain length[a]
Amino acids	60	74	C_2–C_7
Aliphatic hydrocarbons	>35	140	C_1–$C_{\geq 23}$
Aromatic hydrocarbons	15–28	87	C_6–C_{20}
Carboxylic acids	>300	20	C_2–C_{12}
Dicarboxylic acids	>30	17	C_2–C_9
Hydroxycarboxylic acids	15	7	C_2–C_6
Purines and pyrimidines	1.3	5	NA
Basic N-heterocycles	7	32	NA
Amines	8	10	C_1–C_4
Amides	55–70	>2	NA
Alcohols	11	8	C_1–C_4
Aldehydes and ketones	27	9	C_1–C_5
Total	≥ 560	411	

[a]From Cronin, J. R. et al. (1988).
NA = not applicable.

Natural remnant magnetization (NRM) has been observed in meteorites. Estimates of paleointensities range from 100 μT in Allende to 18 μT for Murchison. The observed NRM is consistent with the presence of a magnetic field of about 30 μT in the solar nebula that is inherited from the interstellar cloud.

4.6.3 Organic Material

Organic compounds as complex as amino acids have been found in meteorites. Table 4.4 lists important organic species identified in meteorites. There is no doubt as to their abiotic and extraterrestrial origin. The amino acids in meteorites have equal amounts of levo and dextro chirality and are clearly distinguished from those of biological origin with levo chirality only. The organic compounds in CM chondrites include aliphatic and aromatic hydrocarbons, alcohols, carbonyl compounds, amines, amides, and amino acids. There are three possible origins of these organic compounds: (a) molecular cloud, (b) Urey-Miller synthesis, and (c) Fischer-Tropsch process. Note that the first source is pre-solar nebula. The second is in situ gas phase chemistry driven by a disequilibrium energy source. The third is equilibrium chemistry catalyzed on the surface of dust grains in the solar nebula. These sources are not mutually exclusive.

The evidence that at least some of the organic material in the meteorites is from the molecular cloud is its D/H ratio. Some of the organic material is enriched in D by factors of 10–100 relative to the cosmic abundance. Such values are reminiscent of the chemistry of the molecular clouds (see section 4.4.2). There is no other known chemistry that can have caused such large fractionations of deuterium in the solar nebula.

In a series of classic experiments started in 1953, Miller and Urey demonstrated prebiotic synthesis of organic compounds from mixtures of simple gases of reducing composition using electric discharges. The detailed chemical pathways are not understood, but the energy from the discharge breaks up the stable bonds of the reactants. The subsequent chemistry may be a combination of thermal, ion, and radical chemistry. It is a weakness of this type of experiment that only the bulk compositions of the reactants and the final products are quantitatively analyzed. The detailed chemical kinetics and reaction schemes are not known. Nevertheless, qualitative information on the nature of the products and the order of magnitude of the yields may be derived from this type of experiment. Since the early work, there have been two innovations. Similar experiments have been performed with ultraviolet radiation as a source of energy, with high yields of organic compounds. Even when the starting gas mixture is of only mildly reducing composition such as CO and H_2O, organic products such as formaldehyde, alcohol, and formic acids are readily formed. The main uncertainty of the application of these laboratory results to the solar nebula is the lack of a quantitative estimate of the disequilibrium energy source.

It is known from laboratory experiments that the synthesis of organic compounds from simple molecules such as CO and H_2 can readily occur on the surface of grains containing Fe as a catalyst. The process is known as Fischer-Tropsch synthesis. The following are examples of this type of chemistry that might have been important for the production of alkanes, alkenes, and alcohols in the solar nebula:

$$(2n+1)H_2 + nCO \rightarrow C_nH_{2n+2} + nH_2O \qquad (4.45)$$

$$(n+1)H_2 + 2nCO \rightarrow C_nH_{2n+2} + nCO_2 \qquad (4.46)$$

$$2nH_2 + nCO \rightarrow C_nH_{2n} + nH_2O \qquad (4.47)$$

$$nH_2 + 2nCO \rightarrow C_nH_{2n} + nCO_2 \qquad (4.48)$$

$$2nH_2 + nCO \rightarrow C_nH_{2n+1}OH + (n-1)CO_2 \qquad (4.49)$$

$$(n+1)H_2 + (2n-1)CO \rightarrow C_nH_{2n+1}OH + (n-1)CO_2 \qquad (4.50)$$

Similar reactions may lead to the formation of alkynes and aromatic compounds.

4.7 Comets

Comets are the most primitive bodies in the solar system. Formed in the outer parts of the solar nebula and composed primarily of ices, they preserve as a record of the physical and chemical environment in which they were formed. We now know that there are two populations of comets. The long-period comets have periods greater than 200 yr. Their orbits are randomly oriented on the celestial sphere. They have been shown to come from the Oort cloud, a vast spherical cloud of comets surrounding the solar system and extending halfway to the nearest star (about 10^5 AU). The origin of the Oort cloud is still debated. The crucial question concerns whether the comets in the Oort were formed in situ or formed in the accretion disk followed by dynamical ejection to large distances. The short-period comets have periods less than 200 yr, with most of the periods between 5 and 20 yr. Their orbits are within 30° of the ecliptic plane. They are most likely from the Kuiper Belt, a ring of comets beyond the orbit

Table 4.5 Comet composition

Molecule	Relative abundance (by number)	Comments
Comet P/Halley		
H_2O	100	Remote and in situ detections
CO	~7	Direct (native) source
	~8	Distributed source
H_2CO	0–5	Variable
CO_2	3	Infrared (Vega 1 IKS)
CH_4	<0.2–1.2	Ground-based infrared
	0–2	Giotto IMS
NH_3	0.1–0.3	Variable; based on NH_2
	1–2	Giotto IMS
HCN	0.1	Variable; ground-based ratio
	<0.02	Giotto IMS
N_2	~0.02	Ground-based N_2^+ emission
SO_2	<0.002	Ultraviolet (IUE)
H_2S	—	Giotto IMS
CH_3OH	~1	Giotto NMS and IMS
Other Comets		
CO	20	West (1976 VI)
	2	Bradfield (1979 X)
	1–3	Austin (1990 V)
CH_4	<0.2	Levy (1990 XX)
	1.5–4.5	Wilson (1987 VII)
CH_3OH	1–5	Variable
H_2CO	0.1–0.04	If a parent species
HCN	0.03–0.2	Several comets
H_2S	0.2	Austin (1990 V); Levy (1990 XX)
S_2	0.025	IRAS-Araki-Alcock (1983 VII)

Data from Mumma et al. (1993).

of Neptune (30 AU). Occasional perturbation of their orbits by the giant planets sends them to the inner solar system.

From ground-based observations and the recent spacecraft missions to Comet Halley, we now understand that comets were most likely formed at 25–60 K. The bulk of material comprises low-temperature condensates and ices. Solid materials including silicates and organics were found in Halley. Table 4.5 summarizes the chemical composition of the volatile species in recent comets. There is a rich variety of both oxidized and reduced compounds of C, N, and S. It is important to distinguish the photochemical products from the parent molecules. For example, S_2, C_2, and C_3 detected in comets are probably not primordial and may be derived from the fragmentation of parent molecules. Even some of the CO may not be primordial (see discussion in section 4.7.2). We briefly discuss the significance of the observed species shown in table 4.5 for our understanding of the solar system. For comparison, we also show, in table 4.6, the abundances of these species in the interstellar medium. For simplicity, we restrict this discussion to water, carbon, and nitrogen.

Table 4.6 Comet/ISM comparison

Species	Temperature (K)	Distance (AU)	Comets	ISM-gas	ISM-ice
H_2O	152	3.4–5.4	100	≤100	100
CH_3OH	99		1–5	0.01–0.1	7–40
HCN	95		0.02–0.1	0.01–0.1	4
SO_2	83		<0.002 <0.01	0.01–0.1	?
NH_3	78		0.1–0.3	0.2–2	<5
CO_2	72	13.3–14.6	3	<10	?
H_2CO	64		0–5 0.1–0.04	0.1–0.3	<0.2
H_2S	57		0.2	<0.01	0.3
(CO)	(50)	22–25	(if codeposited with water)		
CH_4	31	60–50	0.2–1.2 <0.2	1	<1
CO	25	84–63	~7 0–20	500–2000	0–5
N_2	22	112–79	0.02	100	?
S_2	20		0.025	?	?

Data from Mumma et al. (1993).

4.7.1 Water

Water is believed to be the most abundant volatile in a comet. Indeed, it is generally accepted that a comet is an icy conglomerate containing dust and other frozen volatiles. H_2O molecules have been detected in Comet Halley using infrared spectroscopy. In addition, the measured ortho/para ratio of the hydrogen atoms in water suggests that water was trapped in the comet nucleus at very low temperatures.

4.7.2 Carbon

Carbon compounds of varying degree of oxidation from CH_4 to CO_2 have been detected in comets. CO appears to be the dominant carbon species. Observations at Halley indicate that there are two sources of CO: a direct source, and a delayed source that can be modeled as photolytic fragmentation of polymeric formaldehyde. The CH_4/CO ratio in comets provides a key test of whether the carbon species are derived from the interstellar medium or chemically produced in the solar nebula. The ratio CH_4/CO is of the order of 0.1 in comets but less than 10^{-3} in the interstellar medium. CO_2 is within an order of magnitude of CO in comets, but its upper limit in the interstellar medium is several orders of magnitude below the concentration of CO. Referring to table 4.6, we note that CO is the dominant carbon species both for comets and the interstellar medium. However, the other carbon species such as CH_4, CH_3OH, and CO_2 are far more abundant relative to CO in the comets than in the interstellar medium. This is circumstantial evidence that the materials in the comets are not simply condensed from the interstellar medium but have been processed chemically in the solar nebula. However, the time needed for conversion from CO to CH_4 by equilibrium chemistry appears to be longer than the lifetime of the nebula. Processing

4.7.3 Nitrogen

The principal nitrogen species such as NH_3 and N_2 have never been detected in comets by direct measurements but are inferred from observations of molecules such as NH_2 and N_2^+. There is much more NH_3 than N_2 in comets. The relative abundances of NH_3 and N_2 are reversed in the interstellar medium. Therefore, the partitioning of nitrogen species in the solar nebula was not the same as the interstellar medium. Local chemistry had altered the speciation. Similar conclusions can be drawn from the ratios of HCN/N_2 and HCN/NH_3. These ratios are not characteristic of the interstellar medium. Nonequilibrium processes in the nebula might have been important for the production of CN compounds.

The preceding discussion points to the possibility of disequilibrium chemistry in the nebula. What was the energy source? Solar ultraviolet radiation is the major source for disequilibrium chemistry today. According to section 4.5, the direct solar radiation in the nebula is negligible due to grain opacity. However, recently Shu and colleagues (1993) showed that in the active T-Tauri phase of the sun, absorption of extreme ultraviolet by H atoms can lead to the formation of a corona as hot as 10^4 K. Radiation from this corona due to the radiative recombination of H^+ and e may provide a large source of diffuse ultraviolet radiation to the nebula. It remains to be shown that this source of energy is capable of driving photochemistry in the nebula and thereby producing the chemical species observed in comets.

4.8 Formation of Planets

There are two pathways for matter in the solar nebula to organize. The bulk of matter formed the protosun via a gravitational instability that occurred in a rarefied gaseous medium. The planets formed by a gradual process of accretion that started from dust and icy grains. As a consequence of these two very different mechanisms, we now understand why there are no planetary bodies of a size between the mass of the sun (2×10^{33} g) and that of Jupiter (1.9×10^{30} g) but there is a hierarchy of solar system objects with masses between Jupiter and a comet, such as asteroids, satellites, and the non-Jovian planets. We are primarily concerned with planetary formation.

The essential data on the size of planetary cores and gaseous envelopes of the giant planets are summarized in table 4.7. The giant planets all have similar cores of 10–30 M_E (M_E = 1 Earth mass). Jupiter has a massive envelope of the order of 300 M_E, while Saturn's gaseous envelope is somewhat smaller, about 95 M_E. The gaseous envelopes of Uranus and Neptune are much smaller, in the range of 1–3 M_E. The terrestrial planets have small cores of size 0.1–1 M_E and no massive gaseous envelopes. The small atmospheres on these planetary bodies are not captured from the solar nebula but are of secondary origin (see discussion in section 4.9).

The first step in the formation of planets is the formation of solid material by condensation. In the inner solar system solids consist of high-temperature condensates, "rocky" material such as that in the meteorites. In the outer solar system the temperature is lower and solids are composed of "icy" material such as that preserved in

Table 4.7 Bulk properties of the giant planets[a]

Planet	Total mass	Low Z mass[b]	High Z mass
Jupiter	318.1	254.1–292.1	26–64
Saturn	95.1	72.1–79.1	16–23
Uranus	14.6	1.3–3.6	11–13.3
Neptune	17.2	0.7–3.2	14–16.5

From Pollack, J. B., and Bodenheimer, P. (1989).
[a] In units of Earth masses.
[b] Z is the atomic number.

comets. The initial growth of dust grains from atomic size to planetesimals (1 km) is rapid and likely is complete within 10^4 yr. The subsequent growth of the small bodies into planets is believed to involve five steps:

1. accretion of planetesimals to the size of a planetary embryo,
2. runaway growth of embryos by "feeding" on the available planetesimals,
3. a slower growth and agglomeration of embryos into the cores of the present planets,
4. growth of cores to critical size (10–20 M_E) that is massive enough to accrete a gaseous envelope, and
5. termination of the growth process due to tidal interactions between the massive planet and the nebula.

Not all the planets participate in the full five stages of growth. The terrestrial planets complete only the first two steps. There is not enough time to grow bigger than the size of about 1 M_E. Step 4 explains why all the giant planets have core masses in the range of 10–30 M_E as shown in table 4.7. Once this critical size is reached it is easier to accumulate gas than solids. Thus, the gaseous envelope grows but the core remains roughly the same size. Step 5 places an ultimate limit on the size of a planet. When a planet reaches the size of Jupiter, its tidal force is large enough to open a gap around the planet. This inhibits further growth of the planet. Of all the planets, only Jupiter has completed part of step 5. Uranus and Neptune do not have enough time or material to accrete large gaseous envelopes. Saturn is an intermediate case between Jupiter, and Uranus and Neptune.

4.9 Atmospheres of Terrestrial Planets and Satellites

As stated earlier, the terrestrial planets and satellites did not accrete cores large enough to accumulate large gaseous envelopes from the gas in the nebula. The atmospheres of these planetary bodies are acquired later from degassing of solid material. In sections 4.9.1–4.9.3 we briefly review the important processes and supporting evidence.

4.9.1 Noble Gases

As early as 1949 Suess and Brown independently recognized the secondary origin of the terrestrial atmosphere based on an analysis of the concentrations of noble gases in the atmospheres of the sun and Earth. Today their arguments are even more compelling on the basis of an expanded set of data on noble gases. The concentrations of noble gases in the atmospheres of the terrestrial planets and type I carbonaceous chondrites are shown in figure 4.12. The patterns of the terrestrial planets and meteorites are similar but are distinctly different from that of the sun. The depletion factors for the noble gases in Earth relative to the sun are shown in figure 4.13. The depletion is largest for the lightest noble gas, He, and the factor is about 10^{14}. The heavier noble gases are depleted by smaller factors, about 10^6. The sun's noble gas pattern is close to the cosmic abundances and is characteristic of the gas in the solar nebula. Based on the great difference in the concentrations and distributions of noble gases, we conclude that the atmospheres of the terrestrial planets were not captured from the gas in the nebula but were derived from devolatization from solid material in the solar nebula.

4.9.2 Proto-atmospheres

From figure 4.13 we notice that the terrestrial planets are massively depleted in the noble gases relative to the sun, but the depletion is less for volatiles that can be chemically bound. There is little depletion in the refractory material such as the metals. Therefore, to first order, the terrestrial planets are made of the same material as the meteorites. The volatiles are derived from the meteorites by impact volatization. Recent experiments by Ahrens and colleagues (1989) have determined the amount of shock devolatization of serpentine, brucite, calcite, and Murchison meteorite in high-energy impacts. The results indicate that impact devolatization occurred when Earth and Venus reached about 0.12 of their present radius. Complete devolatization occurs when planetesimals collide with the protoplanets that have reached 0.3 of their final radii. Thus it is relatively easy for proto-planets to acquire protoatmospheres composed primarily of CO_2 and H_2O. These are good greenhouse gases and can efficiently trap the heat into which the gravitational energy of accretion has been converted. The surface temperature of the proto-planet rises rapidly to above 1500 K, creating what is known as a magma ocean on the surface. The hot atmosphere and magma ocean will further facilitate the dehydration of the infalling material, thereby maintaining their existence. As shown in a model calculation in figure 4.14a, the magma ocean and the massive proto-atmosphere on Earth were created when the planet was about 0.3 times its final radius and persisted to the end of the accretional epoch. The corresponding growth history of water is shown in figure 4.14b. Note that the total quantity of water at the end of accretion is about 1 Earth ocean (1.5×10^{24} g). Ultimately the hot proto-atmosphere collapsed because of the termination of accretion and the associated impact energy.

On a smaller planet like Mars, loss of part of the atmosphere due to impacts is also possible. The process, known as atmospheric cratering, is capable of eroding away the growing atmosphere. What survives is the net result of a balance between the acquisition and ejection of gases by impacts. Thus, Mars might have lost 1–100 times the total water inventory that accreted onto the planet. But once a protoplanet grows beyond the size of Mars, loss by atmospheric cratering becomes inefficient.

4.9.3 Loss of Proto-atmospheres

Loss of proto-atmospheres by thermal evaporation (Jeans escape) is inefficient except for the lightest gases. The process likely to be more important is hydrodynamic escape

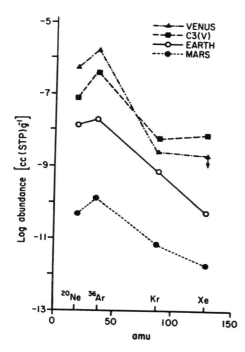

Figure 4.12 Atmospheric abundances (per unit of whole-planet mass) of the noble gases for Venus, Earth, Mars, and a C3(V) meteorite. After Cogley, J. G., and Henderson-Sellers, A., 1984, "The Origin and the Earliest State of the Earth's Hydrosphere." *Rev. Geophys. Space Phys.* **22**, 131.

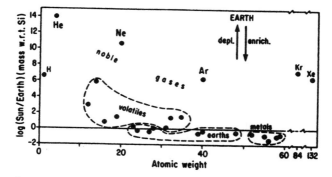

Figure 4.13 Depletion factors for major elements in Earth with respect to the sun. After Walker, J. C. G., 1977, *Evolution of the Atmosphere* (New York: Macmillan).

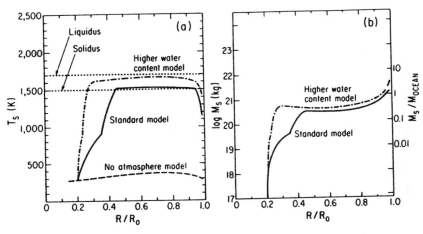

Figure 4.14 (a) The evolution of surface temperature during accretion for standard Earth (solid line) and higher water (dot-dashed line) models. The dashed line traces the surface temperature of the model without an impact-generated atmosphere. R_0 is the final radius. (b) The total mass of the impact-generated H_2O atmosphere is plotted against the normalized radius. After Ahrens, T. J., O'Keefe, J. D., and Lange, M. A., 1989, "Formation of Atmospheres during Accretion of the Terrestrial Planets," in Atreya et al. (1989; cited in section 4.1), pp. 328–385.

fueled perhaps by the production of large amounts of H_2 in the iron-water reaction

$$MgSiO_3 + Fe + H_2O \rightarrow \frac{1}{2}Mg_2SiO_4 + \frac{1}{2}Fe_2SiO_4 + H_2 \quad (4.51)$$

The presence of large amounts of a light gas like H_2 in a terrestrial planet is gravitationally unstable and results in a hydrodynamic flow. An important consequence of this mechanism of escape is that heavier species will be "dragged off" with a characteristic mass fractionation described as follows. Consider the simple case of an atmosphere with two components with masses m_1 and m_2 and escape fluxes F_1 and F_2, respectively. Let X_1 and X_2 be the mole fractions of the species such that $X_1 + X_2 = 1$. The fluxes F_1 and F_2 are related by

$$F_2 = \frac{X_2}{X_1}F_1 - \alpha X_2(m_2 - m_1) \quad (4.52)$$

where α is a parameter that is independent of m, X, and F, m_1 is the lighter mass (of H_2), and F_1 is the flux of H_2. Expression (4.52) suggests that to first order the flux of the heavier gas is proportional to that of the lighter gas plus a mass-dependent correction term. This provides a beautifully simple explanation of the fractionation pattern of the isotopes of heavier noble gases such as Xe in planetary atmospheres relative to meteorites, as shown in figure 4.15. Note the approximately linear dependence of the Xe isotopes on the mass of the species. A similar fractionation pattern is also observed for Kr. The hydrodynamic escape may be further facilitated by the enhanced energy output from a more active phase (T-Tauri) of the protosun.

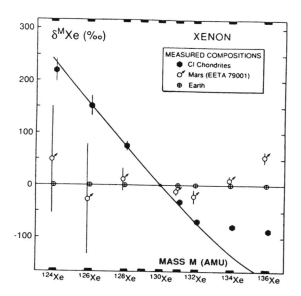

Figure 4.15 Isotopic composition of xenon in meteorites and in planetary atmospheres, plotted relative to terrestrial atmospheric xenon. Units are permil deviation from the terrestrial composition. After Pepin, R. O., 1989, "Atmospheric Composition: Key Similarities and Differences," in Atreya et al. (1989; cited in section 4.1), pp. 291–305.

4.10 Miscellaneous Topics

4.10.1 Deuterium in the Solar System

As discussed in section 4.3.1, deuterium was synthesized in the Big Bang during the first minutes of the origin of the universe. Subsequent nuclear reactions may have destroyed deuterium by converting it to heavier elements but would not be a significant source of fresh deuterium. This fact, in addition to deuterium's chemical reactivity and larger mass relative to hydrogen, makes deuterium one of the most useful tracers for studying the origin and evolution of solar system objects. The overall picture of the D/H ratio in the solar system and the interstellar medium is summarized in figure 4.16. For the bulk of hydrogen in the universe, the principal reservoir is H or H_2, and the D/H ratio is about 10^{-5}. The interstellar medium (bulk), the protosun, and the giant planets Jupiter and Saturn all share the same reservoir of hydrogen and this D/H value.

In solar system objects smaller than the giant planets, the dominant hydrogen-containing species is not H or H_2 but the condensible molecules such as CH_4, H_2O, and NH_3. Earth's reservoir of hydrogen is ocean water. Its D/H ratio is known as SMOW (standard mean ocean water) and has the value $1.5576 \pm 0.0005 \times 10^{-4}$. The small bodies in the solar system, meteorites, Comet Halley, and Titan, have D/H ratios close to SMOW. The meteorites also contain organic compounds that have much higher D/H ratio (labeled meteorite organics in figure 4.16). These organic compounds are believed to have originated from the interstellar medium, where organic synthesis via ion-molecule reactions has a distinct D/H signature (see section 4.4.2). It is clear that the hydrogen in the small bodies is not derived directly from the bulk of the hydrogen reservoir, but is formed in ices at low temperature in the outer part of the

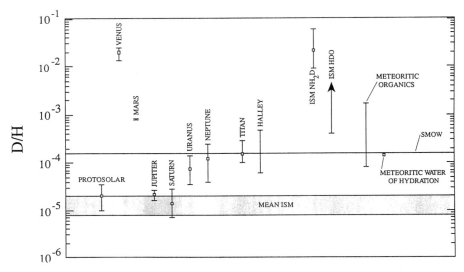

Figure 4.16 D/H ratios in the solar system and interstellar molecular clouds. The uncertainties in the measurements are enclosed by vertical bars. SMOW = standard mean ocean water, ISM = interstellar medium. After Yung, Y. L., and Dissly, R. W., 1992, "Deuterium in the Solar System," in *Isotope Effects in Gas-Phase Chemistry*, J. A. Kaye, editor (Washington, D.C.: American Chemical Society), pp. 369–389.

solar nebula. The question arises as to what determines the D/H ratio in the ices at the time of formation.

Since equilibrium chemistry is believed to dominate in the solar nebula, there is an obvious possibility of exchange between the condensible species (CH_4, H_2O, and NH_3) and bulk hydrogen, as in the following reactions:

$$CH_4 + HD \leftrightarrow CH_3D + H_2 \quad (4.53)$$

$$H_2O + HD \leftrightarrow HDO + H_2 \quad (4.54)$$

$$NH_3 + HD \leftrightarrow NH_2D + H_2 \quad (4.55)$$

At low temperatures ($T < 500$ K) the equilibrium favors the right side of reactions (4.53)–(4.55) and deuterium tends to become more concentrated in CH_3D, HDO, and NH_2D with respect to HD. As shown in figure 4.17, the fractionation is as much as a factor of 10 and becomes even larger at lower temperatures. Thus, if all ices in the solar nebula formed at 200 K or colder, exchange reactions (4.53)–(4.55) would explain the observed large enrichment in deuterium. But this beautifully simple mechanism has an intrinsic self-limitation. Reactions (4.53)–(4.55) all have high activation energies. To achieve high enrichment factors the temperature must be sufficiently low, but at low temperatures the kinetic rates of the reactions are too slow. The time constants for the reactions to reach equilibrium are excessively long compared with the duration of the solar nebula. The efficiency of the reactions may be enhanced by catalysis on the surface of grains, but there may not be enough metallic grains in the solar nebula

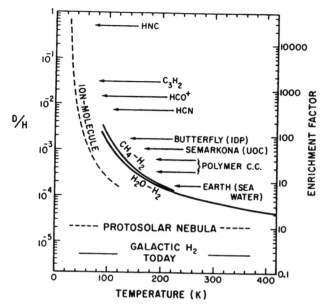

Figure 4.17 Deuterium abundances galactic reservoirs, the protosolar nebula, Earth, and meteorites. The curves give the D/H ratios expected from isotopic exchange between neutral species and ion-molecule reactions as a function of temperature. IDP = Interplanetary dust particles; UOC = Unequilibrated ordinary chondrites; C.C. = Carbonaceous chondrites. After Zinner, E., 1988, "Interstellar Cloud Material in Meteorites," in Kerridge and Matthews (1988; cited in section 4.1), pp. 956–983.

for this purpose (Grinspoon and Lewis, 1987). The failure of this theory to account for deuterium fractionation in the solar system serves to enhance our appreciation of the beauty of the ion-exchange mechanism that successfully accounts for the huge deuterium fractionation in the molecular clouds. Exchange reactions (4.21) and (4.22) are similar to (4.53)–(4.55), except that, being ion-molecule reactions, the former have no activation energy and can proceed rapidly at low temperatures where the fractionation is greatest. Were there ion-molecule reactions or other disequilibrium reactions in the solar nebula? In view of the characteristic speciation in comets discussed in section 4.7, we believe that this is possible. However, no quantitative modeling has been performed.

If the hydrogen-bearing ices in the nebula are enriched in deuterium by some mechanism that remains to be identified, we can understand the enhanced values of D/H in all the small bodies and Earth, whose volatiles are derived from planetesimals. The cores of giant planets presumably contain ices of similar origin, but due to the enormous amounts of H_2 in these planets, the enrichment in the ices is greatly diluted. The lesser giant planets, Uranus and Neptune, have relatively larger cores and smaller gaseous envelopes compared with Jupiter and Saturn; the dilution effect is less. As a consequence, the D/H ratios of Uranus and Neptune are intermediate between the interstellar medium value and the SMOW value (see figure 4.16).

Now to explain the enrichment factors in deuterium for Mars and Venus. Relative to SMOW, D/H ratios on Mars and Venus are enriched by factors of 5 and 150, respectively. This is most likely due to atmospheric evolution. Hydrogen is known to escape from the terrestrial planets today. The efficiency of escape for deuterium is less than that for hydrogen. Hence, over the age of the solar system D/H values tend to increase with time. This enrichment is greatly diluted on Earth by the large reservoir of hydrogen in the terrestrial ocean. But Mars and Venus do not have oceans, and the enrichment is preserved in the atmospheric water vapor. These topics are pursued in greater detail in chapters 7 and 8.

A summary of our current understanding of the origin and fractionation of deuterium is presented in figure 4.18. The vertical arrow indicates increasing values of the D/H ratio, with the cosmic abundance of D to H (in HD in molecular clouds) as the lower limit, and chemically processed material in molecular clouds as the upper limit. Thus, the precursor of the solar nebula, the molecular cloud, is the source of both unfractionated and highly fractionated deuterated material. The horizontal arrow indicates the direction of time. The D/H values in the present solar system are the end-product of cumulative chemical, physical, and evolutionary processes in the solar nebula and planetary atmospheres.

4.10.2 Oxygen Isotopes in the Solar System

Oxygen has three isotopes, ^{16}O, ^{17}O, and ^{18}O, with abundances roughly in the proportions $1 : 4 \times 10^{-4} : 2 \times 10^{-3}$. The standard isotopic composition of oxygen is that of SMOW. The variations in the isotopic composition in solar system material are small, and it is convenient to define a fractional variation relative to SMOW:

$$\delta Q = \left[\frac{(Q/O)_{\text{sample}}}{(Q/O)_{\text{SMOW}}} - 1 \right] \times 1000 \quad (4.56)$$

where $O = {}^{16}O$ and $Q = {}^{17}O$ or ^{18}O. It is often useful to plot the variations of $\delta^{17}O$ with that of $\delta^{18}O$ and study the anomalies. Figure 4.19 summarizes the $\delta^{17}O$ and $\delta^{18}O$ data for solar system objects. Note that the bulk of isotopic variations falls on the terrestrial line. The slope of this line is close to 0.5, consistent with equilibrium fractionation. Most material in the solar system, for example, meteorites and lunar and SNC meteorites (from Mars), also fall on a line with this slope. However, the oxygen isotopes in the calcium-aluminum inclusions (CAIs) in the Allende carbonaceous chondrite are fundamentally different. The Allende mixing line has a slope of 1 instead of 0.5. Discovered in 1971 by Clayton and colleagues, there is no generally accepted explanation for this anomaly. We should point out that only oxygen appears to be unusual in this respect. A similar study for the three isotopes of silicon, ^{28}Si, ^{29}Si, and ^{30}Si, in Allende chondrules reveals the expected 0.5 slope between the variations of ^{29}Si and ^{30}Si relative to ^{28}Si. There is no anomaly for silicon.

There are at least three hypotheses proposed to explain the oxygen anomaly. One proposal is that the solar nebula was subject to an injection of pure ^{16}O from a nearby supernova explosion. The net addition of ^{16}O would obviously cause the $\delta^{17}O$ and $\delta^{18}O$ values to decrease in a mass-independent way consistent with the Allende observations.

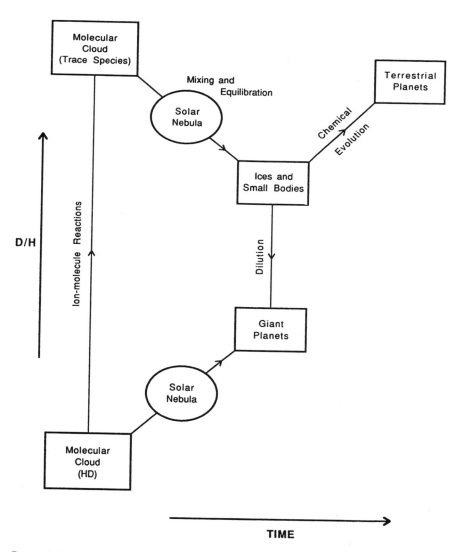

Figure 4.18 Simplified schematic showing the origin of deuterium and important fractionation processes. The D/H values of the rectangular boxes near the top are higher than those below. Time evolution is from left to right. After Yung, Y. L., and Dissly, R. W., 1992, "Deuterium in the Solar System," in *Isotope Effects in Gas-Phase Chemistry*, J. A. Kaye, editor (Washington, D.C.: American Chemical Society), pp. 369–389.

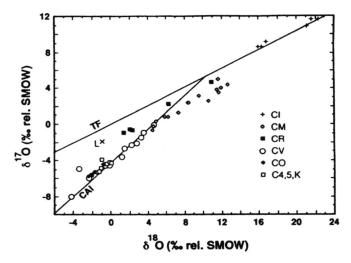

Figure 4.19 Oxygen isotopic compositions of carbonaceous chondrites. Terrestrial fractionation (TF) and refractory inclusion (CAI) lines are included for reference. After Clayton, R. N., 1993, "Oxygen Isotopes in Meteorites." *Ann. Rev. Earth Planet. Sci.* **21**, 112–149.

Note that in this case the ultimate source of the oxygen anomaly is extra-solar system. Another class of explanation is based on the assumption that the origin of the anomaly is within the solar system. The mechanisms fall into two categories: nuclear and chemical. An explanation of the first type invokes a large flux of energetic particles from the protosun, resulting in the preferential production of ^{16}O and the destruction of ^{17}O and ^{18}O. Quantitative modeling shows that an enormous proton flux and a special distribution of matter in the solar nebula are needed. The chemical explanation is based on laboratory experiments on the mass-independent isotopic fractionations between O_2 and O_3 when the former is subject to dissociation in an electric discharge experiment. However, the laboratory results have not been used to model the observed effects in the meteorites. We conclude by pointing out that there is a mass-independent isotopic anomaly in atmospheric ozone in the terrestrial stratosphere.

Appendix

Appendix 4.1 Cosmic abundances of the elements

Z	Element		A	Atoms (%)	Process[a]	Abundance[b] at./10^6Si
1	Hydrogen	H	1	~100		2.72×10^{10}
			2	0.002	U	5.40×10^5
2	Helium	He	3	0.0142	U, H	3.10×10^5
			4	~100	U, H	2.18×10^9

(continued)

Appendix 4.1 (continued)

Z	Element		A	Atoms (%)	Process[a]	Abundance[b] at./10^6Si
3	Lithium	Li	6	7.5	X	4.48
			7	92.5	X, H, U	55.52×10^1
4	Beryllium	Be	9	100	X	7.80×10^{-1}
5	Boron	B	10	19.8	X	4.80
			11	80.2	X	1.92×10^1
6	Carbon	C	12	98.89	He	1.20×10^7
			13	1.11	H	1.34×10^5
7	Nitrogen	N	14	99.634	H	2.47×10^6
			15	0.366	H	9.08×10^3
8	Oxygen	O	16	99.76	He	2.01×10^7
			17	0.038	H	7.64×10^3
			18	0.204	He, N	4.10×10^4
9	Fluorine	F	19	100	N	8.43×10^2
10	Neon	Ne	20	92.99	Ex	3.25×10^6
			21	0.226	He, N	7.91×10^3
			22	6.79	He, N	2.38×10^5
11	Sodium	Na	23	100	Ex	5.70×10^4
12	Magnesium	Mg	24	78.99	Ex	8.49×10^5
			25	10.00	Ex	1.07×10^5
			26	11.01	Ex	1.18×10^5
13	Aluminum	Al	27	100	Ex	8.49×10^4
14	Silicon	Si	28	92.23	Ex	9.22×10^5
			29	4.67	Ex	4.67×10^4
			30	3.10	Ex	3.10×10^4
15	Phosphorus	P	31	100	Ex	1.04×10^4
16	Sulfur	S	32	95.02	Ex	4.89×10^5
			33	0.75	Ex	3.86×10^3
			34	4.21	Ex	2.17×10^4
			36	0.017	Ex	8.80×10^1
17	Chlorine	Cl	35	75.77	Ex	3.97×10^3
			37	24.23	Ex	1.27×10^3
18	Argon	Ar	36	84.2	Ex	8.76×10^4
			38	15.8	Ex	1.64×10^4
			40	0.00061	Ex	5.50×10^{-1}
			40			2.00×10^{-2}
19	Potassium	K	39	93.258	Ex	3.516×10^3
			40	0.01167	Ex	4.40×10^{-1}
			40			5.48
			41	6.730	Ex	2.537×10^2
20	Calcium	Ca	40	96.94	Ex	5.92×10^4
			42	0.647	Ex, HeS	3.95×10^2
			43	0.135	Ex, HeS	8.25×10^1
			44	2.09	Ex, HeS	1.277×10^3
			46	0.0035	Ex	2.14
			48	0.187	Ex	1.44×10^2
21	Scandium	Sc	45	100	Ex, E	3.38×10^1
22	Titanium	Ti	46	8.2	E	1.97×10^2
			47	7.4	E	1.78×10^2
			48	73.7	E	1.769×10^3
			49	5.4	E	1.3×10^2
			50	5.2	E, Ex	1.25×10^2

(continued)

Appendix 4.1 (*continued*)

Z	Element		A	Atoms (%)	Process[a]	Abundance[b] at./10^6Si
23	Vanadium	V	50	0.25	E	7.4×10^{-1}
			51	99.75	E	2.94×10^2
24	Chromium	Cr	50	4.35	E	5.83×10^2
			52	83.79	E	1.12×10^4
			53	9.50	E	1.28×10^3
			54	2.36	E	3.16×10^2
25	Manganese	Mn	55	100	E	9.51×10^3
26	Iron	Fe	54	5.8	E	5.22×10^4
			56	91.8	E	8.26×10^5
			57	2.15	E	1.94×10^4
			58	0.29	E	2.61×10^3
27	Cobalt	Co	59	100	E	2.25×10^3
28	Nickel	Ni	58	68.3	E	3.37×10^4
			60	26.1	E	1.29×10^4
			61	1.13	E	5.57×10^2
			62	3.59	E	1.77×10^3
			64	0.91	E	4.49×10^2
29	Copper	Cu	63	69.1	E	3.56×10^2
			65	30.8	E	1.58×10^2
30	Zinc	Zn	64	48.6	E	6.12×10^2
			66	27.9	E	3.52×10^2
			67	4.10	E, HeS	5.17×10^1
			68	18.8	E, HeS	2.34×10^2
			70	0.62	E, HeS	7.81
31	Gallium	Ga	69	60.1	E, HeS	2.27×10^1
			71	39.9	E, HeS	1.51×10^1
32	Germanium	Ge	70	20.5	E, HeS	2.42×10^1
			72	27.4	E, HeS	3.23×10^1
			73	7.8	E, HeS	9.20
			74	36.5	E, HeS	4.31×10^1
			76	7.8	E, HeS	9.20
33	Arsenic	As	75	100	HeS, ?	6.79
34	Selenium	Se	74	0.87	P	5.4×10^{-1}
			76	9.0	HeS	5.59
			77	7.6	HeS, ?	4.72
			78	23.5	HeS, ?	1.46×10^1
			80	49.8	HeS, ?	3.09×10^1
			82	9.2	?	5.71
35	Bromine	Br	79	50.69	HeS, ?	5.98
			81	49.31	HeS, ?	5.82
36	Krypton	Kr	78	0.339	P	1.54×10^{-1}
			80	2.22	HeS, P	1.01
			82	11.45	HeS	5.19
			83	11.47	HeS, ?	5.20
			84	57.11	HeS, ?	2.59×10^1
			86	17.42	R	7.89
37	Rubidium	Rb	85	72.17	S, R	5.12
			87	27.83	R	1.97
			87			2.10

Appendix 4.1 (continued)

Z	Element		A	Atoms (%)	Process[a]	Abundance[b] at./10^6Si
38	Strontium	Sr	84	0.56	P	1.32×10^{-1}
			86	9.82	S	2.34
			87	7.41	S	1.76
			87			1.63
			88	82.22	S, R	1.957×10^1
39	Yttrium	Y	89	100	S, R	4.64
40	Zirconium	Zr	90	51.5	S, R	5.51
			91	11.2	S, R	1.20
			92	17.1	S, R	1.83
			94	17.4	S, R	1.86
			96	2.80	R	3.00×10^{-1}
41	Niobium	Nb	93	100	S, R	7.1×10^{-1}
42	Molybdenum	Mo	92	14.8	P	3.73×10^{-1}
			94	9.3	P	2.34×10^{-1}
			95	15.9	S, R	4.01×10^{-1}
			96	16.7	S	4.21×10^{-1}
			97	9.6	S, R	2.42×10^{-1}
			98	24.1	S, R	6.07×10^{-1}
			100	9.6	R	2.42×10^{-1}
44	Ruthenium	Ru	96	5.5	P	1.02×10^{-1}
			98	1.86	P	3.46×10^{-2}
			99	12.7	S, R	2.36×10^{-1}
			100	12.6	S	2.34×10^{-1}
			101	17.0	S, R	3.16×10^{-1}
			102	31.6	S, R	5.88×10^{-1}
			104	18.7	R	3.48×10^{-1}
45	Rhodium	Rh	103	100	S, R	3.44×10^{-1}
46	Palladium	Pd	102	1.0	P	1.39×10^{-2}
			104	11.0	S	1.53×10^{-1}
			105	22.2	S, R	3.09×10^{-1}
			106	27.3	S, R	3.79×10^{-1}
			108	26.7	S, R	3.71×10^{-1}
			110	11.8	R	1.64×10^{-1}
47	Silver	Ag	107	51.83	S, R	2.74×10^{-1}
			109	48.17	S, R	2.55×10^{-1}
48	Cadmium	Cd	106	1.25	P	1.99×10^{-2}
			108	0.89	P	1.42×10^{-2}
			110	12.5	S	1.99×10^{-1}
			111	12.8	S, R	2.04×10^{-1}
			112	24.1	S, R	3.83×10^{-1}
			113	12.2	S, R	1.94×10^{-1}
			114	28.7	S, R	4.56×10^{-1}
			116	7.5	R	1.19×10^{-1}
49	Indium	In	113	4.3	P, S, R	7.9×10^{-3}
			115	95.7	S, R	1.76×10^{-1}
50	Tin	Sn	112	1.01	P	3.86×10^{-2}
			114	0.67	P	2.56×10^{-2}
			115	0.38	P, S, R	1.45×10^{-2}
			116	14.8	S	5.65×10^{-1}
			117	7.75	S, R	2.96×10^{-1}
			118	24.3	S, R	9.29×10^{-1}

(continued)

Appendix 4.1 (*continued*)

Z	Element		A	Atoms (%)	Process[a]	Abundance[b] at./10^6Si
			119	8.6	S, R	3.29×10^{-1}
			120	32.4	S, R	1.24
			122	4.56	R	1.74×10^{-1}
			124	5.64	R	2.15×10^{-1}
51	Antimony	Sb	121	57.3	S, R	2.02×10^{-1}
			123	42.7	S, R	1.50×10^{-1}
52	Tellurium	Te	120	0.091	P	4.5×10^{-3}
			122	2.5	S	1.23×10^{-1}
			123	0.89	S	4.4×10^{-2}
			124	4.6	S	2.26×10^{-1}
			125	7.0	S, R	3.44×10^{-1}
			126	18.7	S, R	9.18×10^{-1}
			128	31.7	R	1.56
			130	34.5	R	1.69
53	Iodine	I	127	100	S, R	9.0×10^{-1}
54	Xenon	Xe	124	0.114	P	4.96×10^{-3}
			126	0.111	P	4.83×10^{-3}
			128	2.16	S	9.39×10^{-2}
			129	27.60	S, R	1.20
			130	4.34	S	1.89×10^{-1}
			131	21.64	S, R	9.4×10^{-1}
			132	26.53	S, R	1.15
			134	9.69	R	4.21×10^{-1}
			136	7.82	R	3.4×10^{-1}
55	Cesium	Cs	133	100	S, R	3.72×10^{-1}
56	Barium	Ba	130	0.106	P	4.62×10^{-3}
			132	0.101	P	4.40×10^{-3}
			134	2.42	S	1.06×10^{-1}
			135	6.59	S, R	2.87×10^{-1}
			136	7.85	S	3.42×10^{-1}
			137	11.2	S, R	4.88×10^{-1}
			138	71.7	S, R	3.13
57	Lanthanum	La	138	0.089	P	4.0×10^{-4}
			139	99.911	S, R	4.48×10^{-1}
58	Cerium	Ce	136	0.190	P	2.2×10^{-3}
			138	0.254	P	2.9×10^{-3}
			140	88.5	S, R	1.026
			142	11.1	R	1.29×10^{-1}
59	Praseodymium	Pr	141	100	S, R	1.74×10^{-1}
60	Neodymium	Nd	142	27.2	S	2.27×10^{-1}
			143	12.2	S, R	1.02×10^{-1}
			143			1.01×10^{-1}
			144	23.8	S, R	1.99×10^{-1}
			145	8.3	S, R	6.94×10^{-2}
			146	17.2	S, R	1.44×10^{-1}
			148	5.7	R	4.77×10^{-2}
			150	5.6	R	4.68×10^{-2}
62	Samarium	Sm	144	3.1	P	8.1×10^{-3}
			147	15.1	S, R	3.94×10^{-2}
			147			4.06×10^{-2}
			148	11.3	S	2.95×10^{-2}
			149	13.9	S, R	3.63×10^{-2}

Appendix 4.1 (*continued*)

Z	Element		A	Atoms (%)	Process[a]	Abundance[b] at./10^6Si
			150	7.4	S	1.93×10^{-2}
			152	26.6	R	6.94×10^{-2}
			154	22.6	R	5.89×10^{-2}
63	Europium	Eu	151	47.9	S, R	4.66×10^{-2}
			153	52.1	S, R	5.06×10^{-2}
64	Gadolinium	Gd	152	0.20	P	6.6×10^{-4}
			154	2.1	S	6.95×10^{-3}
			155	14.8	S, R	4.90×10^{-2}
			156	20.6	S, R	6.82×10^{-2}
			157	15.7	S, R	5.20×10^{-2}
			158	24.8	S, R	8.21×10^{-2}
			160	21.8	R	7.22×10^{-2}
65	Terbium	Tb	159	100	S, R	5.89×10^{-2}
66	Dysprosium	Dy	156	0.057	P	2.27×10^{-4}
			158	0.100	P	3.98×10^{-4}
			160	2.3	S	9.15×10^{-3}
			161	19.0	S, R	7.56×10^{-2}
			162	25.5	S, R	1.01×10^{-1}
			163	24.9	S, R	9.91×10^{-2}
			164	28.18	S, R	1.12×10^{-1}
67	Holmium	Ho	165	100	S, R	8.75×10^{-2}
68	Erbium	Er	162	0.14	P	3.54×10^{-4}
			164	1.56	P, S	3.95×10^{-3}
			166	33.4	S, R	8.45×10^{-2}
			167	22.9	S, R	5.79×10^{-2}
			168	27.1	S, R	6.86×10^{-2}
			170	14.9	R	3.77×10^{-2}
69	Thulium	Tm	169	100	S, R	3.86×10^{-2}
70	Ytterbium	^{70}Yb	168	0.135	P	3.28×10^{-4}
			170	3.1	S	7.53×10^{-3}
			171	14.4	S, R	3.50×10^{-2}
			172	21.9	S, R	5.32×10^{-2}
			173	16.2	S, R	3.94×10^{-2}
			174	31.6	S, R	7.68×10^{-2}
			176	12.6	R	3.06×10^{-2}
71	Lutetium	Lu	175	97.39	S, R	3.59×10^{-2}
			176	2.61	S	9.64×10^{-4}
			176			1.06×10^{-3}
72	Hafnium	Hf	174	0.16	P	2.8×10^{-4}
			176	5.2	S	9.2×10^{-3}
			176			9.02×10^{-3}
			177	18.6	S, R	3.27×10^{-2}
			178	27.1	S, R	4.77×10^{-2}
			179	13.7	S, R	2.41×10^{-2}
			180	35.2	S, R	6.20×10^{-2}
73	Tantalum	Ta	180	0.0123	P, S, R	2.78×10^{-6}
			181	99.9877	S, R	2.26×10^{-2}
74	Tungsten	W	180	0.13	P	1.78×10^{-4}
			182	26.3	S, R	3.60×10^{-2}

(*continued*)

Appendix 4.1 (continued)

Z	Element		A	Atoms (%)	Process[a]	Abundance[b] at./10^6Si
			183	14.3	S, R	1.96×10^{-2}
			184	30.7	S, R	4.21×10^{-2}
			186	28.6	R	3.92×10^{-2}
75	Rhenium	Re	185	37.40	S, R	1.90×10^{-2}
			187	62.60	S, R	3.17×10^{-2}
			187			*3.43×10^{-2}*
76	Osmium	Os	184	0.018	P	1.29×10^{-4}
			186	1.60	S	1.15×10^{-2}
			187	1.60	S	1.15×10^{-2}
			187			*8.9×10^{-3}*
			188	13.3	S, R	9.54×10^{-2}
			189	16.1	S, R	1.15×10^{-1}
			190	26.4	S, R	1.89×10^{-1}
			192	41.0	R	2.94×10^{-1}
77	Iridium	Ir	191	37.3	S, R	2.46×10^{-1}
			193	62.7	S, R	4.14×10^{-1}
78	Platinum	Pt	190	0.013	P	1.78×10^{-4}
			192	0.78	S	1.07×10^{-2}
			194	32.9	S, R	4.51×10^{-1}
			195	33.8	S, R	4.63×10^{-1}
			196	25.3	S, R	3.47×10^{-1}
			198	7.2	R	9.86×10^{-2}
79	Gold	Au	197	100	S, R	1.86×10^{-1}
80	Mercury	Hg	196	0.15	P	7.8×10^{-4}
			198	10.0	S	5.2×10^{-2}
			199	16.8	S, R	8.74×10^{-2}
			200	23.1	S, R	1.20×10^{-1}
			201	13.2	S, R	6.86×10^{-2}
			202	29.8	S, R	1.55×10^{-1}
			204	6.9	R	3.59×10^{-2}
81	Thallium	Ti	203	29.5	S, R	5.42×10^{-2}
			205	70.5	S, R	1.30×10^{-1}
82	Lead	Pb	204	1.94	S	6.12×10^{-2}
			206	19.12	S, R	6.03×10^{-1}
			206			*5.94×10^{-1}*
			207	20.62	S, R	6.50×10^{-1}
			207			*6.44×10^{-1}*
			208	58.31	S, R	1.838
			208			*1.830*
83	Bismuth	Bi	209	100	S, R	1.44×10^{-1}
90	Thorium	Th	232	100	RA	3.35×10^{-2}
			232			*4.20×10^{-2}*
92	Uranium	U	235	0.720	RA	6.49×10^{-5}
			235			*5.73×10^{-3}*
			238	99.275	RA	8.94×10^{-3}
			238			*1.81×10^{-2}*

Source: Anders, E., and Ebihara, M., 1982, "Solar-System Abundances of the Elements." *Geochim. Cosmochim. Acta* **46**, 2363–2380.

[a] U = cosmological nucleosynthesis; H = hydrogen burning; N = hot hydrogen burning; He = helium burning; Ex = explosive nucleosynthesis; E = nuclear statistical equilibrium; S = s process; HeS = helium-burning s process; R = r process; RA = r process producing actinides; P = p process; X = cosmic ray spallation.
[b] Italicized values refer to abundances 4.55 Gyr ago.

Appendix 4.2 Meteorite classes and numbers

Class	Falls	Fall frequency (%)	Finds non-Antarctic	Finds Antarctic
Chondrites				
CI	5	0.60	0	0
CM	18	2.2	5	34
CO	5	0.60	2	6
CV	7	0.84	4	5
H	276	33.2	347	671
L	319	38.3	286	224
LL	66	7.9	21	42
EH	7	0.84	3	6
EL	6	0.72	4	1
Other	3	0.36	3	3
Achondrites				
Eucrites	25	3.0	8	13
Howardites	18	2.2	3	4
Diogenites	9	1.1	0	9
Ureilites	4	0.48	6	9
Aubrites	9	1.1	1	17
Shergottites	2	0.24	0	2
Nakhlites	1	0.12	2	0
Chassignites	1	0.12	0	0
Anorthositic breccias	0	0	0	1
Stony irons				
Mesosiderites	6	0.72	22	2
Pallasites	3	0.36	34	1
Irons				
IAB	6	0.73	97	4
IC	0	0.08	11	0
IIAB	5	0.45	60	6
IIC	0	0.05	7	0
IID	3	0.09	12	0
IIE	1	0.10	13	0
IIF	1	0.03	4	0
IIIAB	8	1.42	189	0
IIICD	2	0.14	19	0
IIIE	0	0.10	13	0
IIIF	0	0.05	6	0
IVA	3	0.39	52	1
IVB	0	0.09	12	0
Other irons	13	1.32	175	0

Data from Sears, D. W. G., and Dodd, R. T. (1988).

5

Jovian Planets

5.1 Introduction

The four giant planets in the outer solar system, Jupiter, Saturn, Uranus, and Neptune, are a distinct group by themselves. The essential astronomical and atmospheric aspects of these planets are summarized in table 5.1. The significance of this group in the chemistry of the solar system is briefly pointed out in chapter 4. These planets are composed primarily of the lightest elements, hydrogen and helium, which were captured from the solar nebula during formation. The planets have rocky cores made of heavier elements. In the case of Jupiter and Saturn the mass of the gas greatly exceeds that of the core, whereas for Uranus and Neptune the masses of gas and core are comparable. Due to the enormous gravity of the giant planets, little mass has escaped from their atmospheres. Hence, the bulk composition of these planets provides a good measure of the initial composition of the solar nebula from which they were derived. Of all planetary bodies in the solar system, the constituents of giant planets are the closest to the cosmic abundances of the elements.

The chemistry of the atmospheres of the giant planets is interesting for the following reasons:

1. chemistry in a dominantly reducing atmosphere
2. interplay between photochemistry and equilibrium chemistry
3. ion chemistry in polar auroral regions
4. heterogeneous chemistry of aerosols
5. chemistry of meteoritic debris
6. lack of a planetary "surface"

Table 5.1 Astronomical and atmospheric data

Characteristic	Jupiter	Saturn	Uranus	Neptune
Radius (equatorial) (km)	71492	60268	25559	24764
Mass (kg)	1.899×10^{27}	5.688×10^{26}	8.68×10^{25}	1.027×10^{26}
Mean density (g cm^{-3})	1.33	0.69	1.29	1.64
Gravity (surface, equator) (m s^{-2})	23.12	8.96	8.69	11.00
Semimajor axis (AU)	5.2028	9.5388	19.1914	30.0611
Obliquity (relative to orbital plane) (deg)	3.08	26.73	97.92	28.80
Eccentricity of orbit	0.0483	0.0556	0.0461	0.0086
Period of revolution (Earth days)	4332.71	10759.5	30684	60190
Orbital velocity (km s^{-1})	13.06	9.64	6.81	5.43
Period of rotation (h)	9.841	10.233	17.9	19.21
Mass of atmospheric column (1 bar) (kg cm^{-2})	0.447	1.15	1.19	0.94
Total atmospheric mass[a] (1 bar) (kg)	2.87×10^{20}	5.26×10^{20}	9.76×10^{19}	7.24×10^{19}
Equilibrium temperature (K)	128	98	56	57
Temperature at 1 bar[b] (K)	165	134	76	71.5
Annual variation in solar insolation[c]	1.21	1.25	1.21	1.04
Atmospheric scale height (km)	27.0	59.5	27.7	19.1
Atmospheric lapse rate (K km^{-1})	1.8	6.9	6.7	8.5
Escape velocity (km s^{-1})	59.5	35.6	21.22	23.3

[a]Total mass of the atmosphere above 1 bar pressure level.
[b]Data from Lindal et al. (1992).
[c]Based on the ratio of aphelion to perihelion distance squared.

We briefly comment on these reasons in this section. Each topic will receive a more detailed treatment in later sections. First of all, the atmospheres of the Jovian planets are more than 90% hydrogen and helium. Since helium is inert, the atmospheric chemistry is dominated by hydrogen. Therefore, we would expect the most stable compounds of carbon, oxygen, nitrogen, and phosphorus to be CH_4, H_2O, NH_3, and PH_3. This is in fact confirmed by the available observed composition of the bulk atmospheres of these planets. However, in the upper atmospheres of these planets, the composition is controlled by photochemistry. For example, one consequence of the dissociation of CH_4 following the absorption of ultraviolet radiation is the production of C_2H_6, as summarized schematically by

$$2CH_4 \rightarrow C_2H_6 + 2H \tag{5.1}$$

(A discussion of the detailed chemical pathways is given in section 5.3.1.) The fate of the H atoms produced in (5.1) is to recombine to form H_2. Thus, photochemistry is capable of driving the net endothermic reaction

$$2CH_4 \rightarrow C_2H_6 + H_2 \tag{5.2}$$

Here we point out a fundamental difference between the giant planets and other planetary bodies in the solar system. If the H and H_2 produced in (5.1) and (5.2) escape from planetary atmospheres, this results in irreversible change in the chemistry of these atmospheres. As discussed in chapter 1, H and H_2 do not escape from giant planets. All the products of photochemistry are preserved in the atmosphere. Since

C_2H_6 is stable in the atmosphere once it is removed from the photochemically active region (mesosphere), its ultimate sink is in the deep atmosphere, where thermodynamic equilibrium chemistry favors the reverse of (5.2):

$$C_2H_6 + H_2 \rightarrow 2CH_4 \tag{5.3}$$

Reactions such as (5.3) that tend to restore the atmospheric composition to thermodynamic equilibrium occur at pressures greater than 1 kbar and at temperatures above 1000 K. From this discussion we may infer that the chemistry of the giant planets consists of two distinct regimes: photochemistry in the upper atmosphere driven by solar ultraviolet light, and equilibrium chemistry in the interior of the planet driven by thermal energy. Mixing processes deliver the photochemical products from the upper atmosphere to the interior, and the equilibrium products are recycled back to the upper atmosphere. This is a closed loop because there is no loss of even the lightest molecules.

All giant planets have extensive magnetic fields, giving rise to complex interactions with the solar wind. The deposition of energetic particles in the polar regions gives rise to aurorae that have been observed from both spacecraft and Earth. The dominant chemistry that can occur in the auroral zones is ion chemistry, reminiscent of what is known to be important in the molecular clouds (section 4.4). Organic synthesis is believed to be important.

The net result of the photochemical destruction of CH_4 is its conversion to higher hydrocarbons. Ultimately the product species become heavy enough to condense, giving rise to a ubiquitous aerosol layer in the atmospheres of the giant planets known as Axel-Danielson dust. The presence of this aerosol layer has the most profound consequence for the thermal budget of the upper atmosphere. Indeed, aerosol heating is a primary cause of the thermal inversion in the atmospheres of the giant planets above the tropopause (see figure 1.2). The aerosols may also present a surface for possible heterogeneous chemical reactions.

The giant planets, with their strong gravitational fields, tend to focus meteoritic materials into their atmospheres. In addition, extensive ring systems are also sources of extraplanetary material as the orbits of the ring particles decay due to internal collisions, tidal interactions, or atmospheric drag. The nature of this influx of material is unknown, but we can assume that it is chemically very different from that of the ambient atmosphere. This material is certainly rich in the heavy elements such as O, C, N, S, Fe, Na, K, Mg, Al, and Si. The fate of these elements in a reducing atmosphere is poorly understood and may give rise to species that we would not expect from an atmosphere in thermodynamic equilibrium.

The Jovian planets are characterized by the lack of planetary surfaces. The atmosphere merges continuously with the bulk of the planet. "Geochemistry" as defined on the surface of smaller planetary bodies is absent. The visible atmosphere interacts directly with the deep interior with time constants that are orders of magnitude smaller than "geological" time. As a consequence, the interior has a direct impact on the composition of the upper atmosphere.

In this chapter the composition of the atmospheres of the Jovian planets is critically examined and discussed in terms of the physical and chemical processes that are known or remain to be identified. In the general discussions the four planets will be taken as a group. In presenting quantitative results of model calculations we first report them for Jupiter, followed by similar results for the other giant planets.

5.1.1 Voyager Observations

Our knowledge of the atmospheres of the giant planets has advanced by a quantum jump from the two spectacularly successful missions Voyager 1 and 2 to the outer solar system. We show examples of the Voyager observations to convey the sense of excitement of these discoveries. The Voyager spacecraft each carried 11 instruments, of which the most important for atmospheric studies are the ultraviolet spectrometer (UVS), the radio science (RSS), and the infrared radiometer, interferometer, and spectrometer (IRIS).

The ultraviolet Spectrometer probes the uppermost regions of the atmosphere, the thermosphere and the mesosphere where the sun's extreme ultraviolet (EUV) energy is absorbed and scattered. The thermosphere is also the region where a new source of ultraviolet radiation, known as the electroglow, may be produced by an unknown mechanism. The Voyager UVS is a grating spectrometer with 128 channels covering a range from 512 to 1690 Å. The spectral resolution is 33 Å. The instrument can operate in the airglow mode, by looking at the radiation emanating from the disk of the planet, or in the occultation mode, by looking at the sun or a bright star as it is occulted by the planet. Figure 5.1 shows the airglow spectrum of Jupiter (and Saturn) obtained by Voyager 1. The single brightest feature in the spectrum is Lyman α at 1216 Å, with disk-center intensity of 15 kR (1 kilo-rayleigh $= 1 \times 10^9$ photons cm^{-2} s^{-1}), arising from the resonance scattering of the solar Lyman α line by H atoms in the upper atmosphere of Jupiter. To first order we may regard the H atoms as a layer of white scatterers lying above a deep layer of perfect absorbers—CH_4. The column density of H atoms required to reflect this much Lyman α is of the order of 10^{17} cm^{-2}. The strong emission features between Lyman α and 800 Å are mostly due to H_2 Lyman and Werner bands. The integrated intensity is about 3 kR and cannot be readily accounted for by the scattering of solar EUV radiation. In view of this difficulty, this intense EUV radiation from Jupiter (and other Jovian planets) has been named "electroglow." An obvious energy source is the precipitating electrons from the magnetosphere, but then why is this emission not observed on the dark side of the planets? The upper limit of H_2 bands on the night side is about 70 times below the observed brightness on the day side. Thus, the discovery of the intense EUV emissions raised a fundamental question as deep as the discovery itself. In addition to the electroglow, Jupiter also has intense auroral emissions caused by energetic particles from the magnetosphere. The latter phenomenon has a terrestrial analog and is better understood.

Figure 5.2 gives the spectrum of the sun during occultation as the tangent ray descends deeper and deeper into the atmosphere of Jupiter. At the altitude of 920 km above an arbitrary reference level, there is virtually no attenuation of the solar beam. Attenuation increases as the altitude decreases. Most of the absorption is by H_2 under 1000 Å, but CH_4 and other hydrocarbons become important at longer wavelengths. The exact wavelengths where absorption by various molecules become important have been discussed in chapter 2 (see figures 2.8 and 2.10) At -80 km, nearly all the solar flux has been extinguished by the atmosphere. The nature of the absorption spectrum yields information on the atmospheric composition. By modeling the rate of change

124 Photochemistry of Planetary Atmospheres

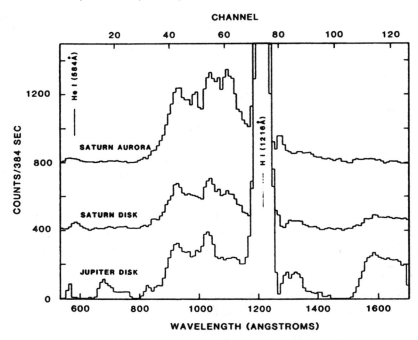

Figure 5.1 Extreme ultraviolet airglow observations of Jupiter and Saturn from Voyager 1 encounters in July 1979 and November 1980, respectively. After Broadfoot, L. et al., 1981, "Extreme Ultraviolet Observations from Voyager 1 Encounter with Saturn." *Science* **212**, 206.

of atmospheric opacity with altitude, valuable information concerning the temperature (scale height) and distribution of the absorbing species may be extracted. One of the important results of this analysis is the confirmation of a hot thermosphere of Jupiter.

The RSS observations on Voyager were carried out in the occultation mode using the S-band ($\lambda = 13$ cm, $\nu = 2.3$ GHz) and the X-band ($\lambda = 3.5$ cm, $\nu = 8.6$ GHz) channels of the spacecraft's telemetry as the signal, which were received on Earth. As the spacecraft went behind the planet, bending of the tangent ray occurred due to the difference between the index of refraction of the atmosphere and the vacuum. Three regions of the atmosphere that can be most sensitively probed by this technique are the ionosphere, the stratosphere, and the troposphere. Unfortunately, the mesosphere does not have enough electrons or neutral molecules to produce a detectable signal in this measurement. An illustration of the ionospheric measurements by RSS is given in figure 5.3 for Saturn. Vertical profiles of the density of electrons are derived from the radio occultation experiments on Voyager 1 and 2 (labeled V_1 and V_2 in the figure). For comparison, the results from the earlier Pioneer mission (labeled P/S in the figure) are included. This figure shows that Saturn has an extensive ionosphere that exhibits considerable spatial and temporal variations (the spacecrafts' encounters were separated by a number of years). The topside scale height for the ionosphere is in the range of 2000–3000 km. This determines the value T_p/m_i, where T_p is the

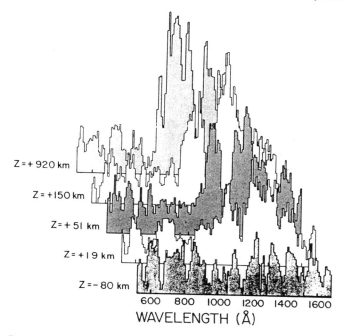

Figure 5.2 Extreme ultraviolet solar occultation observations of Jupiter. The altitude scale is arbitrary. Large Z values refer to high atmosphere. Lower Z values refer to deeper regions of atmosphere where solar UV radiation is progressively absorbed. After Festou, M. C. et al., 1981, "Composition and Thermal Structure Profiles of the Jovian Upper Atmosphere Determined by the Voyager Ultraviolet Stellar Occultation Experiment." *J. Geophys. Res.* **86**, 5715.

sum of the electron (T_e) and the ion (T_i) temperature and m_i is the mean mass of the ion. The most likely ion at this high altitude is a proton. Therefore, we may conclude that the temperature of the ionosphere (assuming $T_e = T_i$) of Saturn is in the range of 300–600 K. This is much greater than that expected on the basis of solar EUV heating. The new energy source has not been identified to date. A similar problem exists with the thermospheric temperatures of all other Jovian planets. The lower part of the electron density profile reveals the existence of sharp layers. These features have been observed in all radio occultations of the giant planets. One explanation is that they are layers of metal ions derived from meteoritic debris.

Figure 5.4 shows an example of what can be deduced from RSS measurements in the stratosphere and the troposphere. The RSS signal yields T/m as a function of altitude, where T = mean temperature and m = mean molecular mass. From an independent knowledge of atmospheric composition (e.g., from UVS and IRIS), the thermal structure of the atmosphere may be determined. Figure 5.4 shows that the RSS-derived profiles exhibit considerably more vertical structure than those from IRIS measurements and model computations. The reason is that the latter give mean

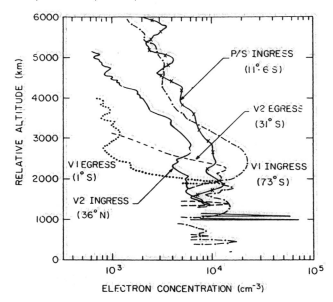

Figure 5.3 Radio science radio occultation measurements of electron concentrations in the ionosphere of Saturn by Voyagers 1 and 2 (labeled V1 and V2). For comparison, the Pioneer (labeled P/S) results are also shown. After Kliore, A. J. et al., 1980, "Vertical Structure of the Ionosphere and Upper Neutral Atmosphere of Saturn from Pioneer Radio Occultation." *Science* **207**, 446; and Lindal, G. F., Sweetnam, D. N., and Eshleman, V. R., 1985, "The Atmosphere of Saturn: An Analysis of the Voyager Radio Occultation Measurements." *Astron. J.* **90**, 1136.

properties of the atmosphere averaged over a scale height. The wave features in the RSS profiles are authentic atmospheric waves.

The IRIS instrument on Voyager consists of a Michelson interferometer that records the spectrum from 180 to 2500 cm^{-1} (4–56 μm) with a resolution of 4.3 cm^{-1}, plus a single-channel radiometer. At this resolution it is possible to observe the detailed vibrational and rotational spectra of simple molecules present in the atmosphere. Figure 5.5 shows three spectra that are typical of the observations of Jupiter and Saturn. The broad feature between 250 and 700 cm^{-1} due to pressure-induced transitions in H_2 is clearly present on both Jupiter and Saturn. The NH_3 feature near 200 cm^{-1} is prominent in Jupiter but is absent from Saturn. The strong ν_4 CH_4 band in the 1200–1400 region is present in all spectra. C_2H_6 and C_2H_2 appear in these spectra as strong emission features. This is caused by a combination of the thermal inversion and the high concentrations of C_2H_6 and C_2H_2 in the stratosphere. Figure 5.6 shows the IRIS spectrum for the 5 μm window region of Jupiter. In this region it is possible to see deep into the troposphere to the level of several bars. Minor constituents such as H_2O, GeH_4, CH_3D, and PH_3 can be measured from such observations.

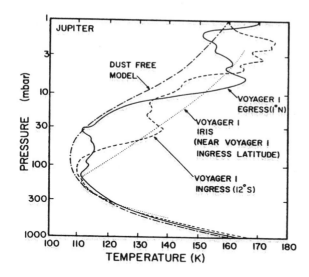

Figure 5.4 Radio science radio occultation measurements of the temperature profile in the troposphere and the middle atmosphere to 1 mbar. After Lindal, G. F. et al., 1981, "The Atmosphere of Jupiter: An Analysis of the Voyager Radio Occultation Measurements." *J. Geophys. Res.* **86**, 8721.

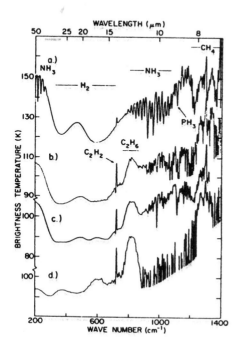

Figure 5.5 IRIS observations of the infrared spectra of Jupiter's atmosphere. After Hanel, R. et al., 1981, "Infrared Observations of the Saturnian System from Voyager 1." *Science* **212**, 192.

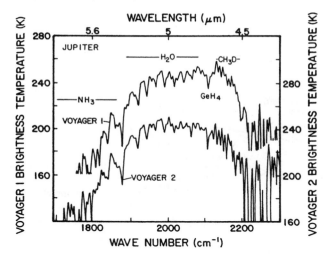

Figure 5.6 IRIS observations of the infrared spectra of Jupiter's atmosphere in the 5 μm "window" region. After Hanel, R. et al., 1979, "Infrared Observations of the Jovian System from Voyager 2." *Science* **206**, 952.

It is fair to say that the Voyager results are the single greatest contribution to our knowledge of the atmospheres of the outer solar system. Of course, these results do not stand alone. Observations from ground-based facilities and Earth-orbiting satellites also contribute to greatly advance understanding of the planetary atmospheres before and after the Voyager encounters. We have chosen the Voyager observations here to illustrate how a concentrated technological initiative can revolutionize a field.

5.1.2 Chemical Composition

The composition of the atmosphere of Jupiter as deduced from Voyager and other measurements is summarized in table 5.2. Less is known about Saturn, Uranus, and Neptune, and tables 5.3 and 5.4 provide summaries of our knowledge of these planets. It is convenient to classify the species in these tables according to the following categories:

(a) low Z elements,
(b) high Z elements,
(c) gases that can condense,
(d) gases from equilibrium chemistry,
(e) photochemical products,
(f) products of ion chemistry, and
(g) isotopomers,

where H and He are the low Z elements, and high Z elements include all elements heavier than He. The above classification is not intended to be exclusive. A species may belong to more than one category. For example, HCN can be derived as a photochemical product of NH_3 and CH_4 in the stratosphere or as an equilibrium product from

Table 5.2 Composition of the atmosphere of Jupiter

Species	Abundance	Remarks
H_2	0.898 ± 0.02	primordial, electroglow observed
He	0.102 ± 0.02	primordial, resonance line observed
CH_4	$3.0 \pm 1.0 \times 10^{-3}$	photolyzed in mesosphere
NH_3	$2.6 \pm 0.4 \times 10^{-3}$	condensation in upper troposphere, photolyzed in troposphere and stratophere
PH_3	$7.0 \pm 1.0 \times 10^{-7}$	photolyzed in troposphere and stratophere
H_2O	$3.0 \pm 2.0 \times 10^{-5}$	at 6 bar level
	$4.0 \pm 1.0 \times 10^{-6}$	at 2–4 bar level
CO	$1.6 \pm 0.3 \times 10^{-9}$	mixed from deep atmosphere
GeH_4	$7.0^{+4}_{-2} \times 10^{-10}$	mixed from deep atmosphere
AsH_3	$2.2 \pm 1.1 \times 10^{-10}$	mixed from deep atmosphere
C_2H_6	$5.8 \pm 1.5 \times 10^{-6}$	photochemical, stratosphere
C_2H_4	$7.0 \pm 3.0 \times 10^{-9}$	photochemical, stratosphere
C_2H_2	$1.1 \pm 0.3 \times 10^{-7}$	photochemical, stratosphere
C_4H_2	$3.0 \pm 2.0 \times 10^{-10}$	photochemical, stratosphere
CH_3C_2H	$2.5^{+2}_{-1} \times 10^{-9}$	polar auroral region
C_6H_6	$2.0^{+2}_{-1} \times 10^{-9}$	polar auroral region
H	Dayglow = 15 kR	nonbulge, disk-centered, solar maximum
e	maximum density = $3 \times 10^5 cm^{-3}$	ionosphere
H_3^+	detected	polar auroral region
HD	$2.8^{+3.0}_{-0.8} \times 10^{-5}$	gives D/H $\sim 1.6 \times 10^{-5}$
CH_3D	$2.0 \pm 4.0 \times 10^{-7}$	gives D/H $\sim 1.7 \times 10^{-5}$
$^{13}CH_4$	3.3×10^{-5}	gives $^{12}C/^{13}C \sim 90$
$^{13}CCH_6$	5.8×10^{-8}	gives $^{12}C/^{13}C \sim 94$
$^{13}CCH_2$	1.0×10^{-8}	gives $C_2H_2/^{13}CCH_2 \sim 10$
$^{15}NH_3$	2.0×10^{-6}	gives $^{14}N/^{15}N \sim 125$
C_3H_8	$< 6.0 \times 10^{-7}$	polar auroral region
H_2S	$< 3.3 \times 10^{-8}$	troposphere
	$< 2.0 \times 10^{-9}$	stratosphere

From Fegley, B. Jr. (1995).

the planetary interior. In addition, HCN can condense at the tropopause of Neptune's atmosphere. Therefore, HCN belongs simultaneously to (b)–(e). We briefly discuss the species in tables 5.2–5.4 in terms of the preceding classification. The photochemical products are special topics that are elaborated in sections 5.3–5.6.

(a) Low Z Elements

H_2 and He are obviously derived from the gaseous component of the solar nebula. But the abundance of He relative to H_2 is not constant among the giant planets. Table 5.5 summarizes the He/H ratios for the giant planets, along with internal heat fluxes. The first notable difference is between Jupiter and Saturn. The He/H ratio in Jupiter is 71% of the cosmic abundance value but only 21% for Saturn. It is difficult to imagine that He was nonuniform in the solar nebula. The explanation is found in the internal heat of the giant planets, as summarized in table 5.5. It is known that some giant planets radiate more heat than what is absorbed from the sun. The ratio of the total heat flux to the absorbed solar flux provides a measure of the excess heat generated

Table 5.3 Composition of the atmosphere of Saturn

Species	Abundance	Remarks
H_2	0.963 ± 0.024	
He	0.0325 ± 0.024	
CH_4	$4.5^{+2.4}_{-1.9} \times 10^{-3}$	
NH_3	$0.5 - 2.0 \times 10^{-4}$	
PH_3	$1.4 \pm 0.8 \times 10^{-6}$	
CO	$1.0 \pm 0.3 \times 10^{-9}$	
GeH_4	$4.0 \pm 4.0 \times 10^{-10}$	
AsH_3	$3.0 \pm 1.0 \times 10^{-9}$	
C_2H_6	$7.0 \pm 1.5 \times 10^{-6}$	
C_2H_2	$3.0 \pm 1.0 \times 10^{-7}$	
H	Dayglow = 3.3 kR	disk-centered, solar maximum
e	maximum density = $2.3 \times 10^4 cm^{-3}$	south polar latitudes
HD	$1.10 \pm 0.58 \times 10^{-4}$	
CH_3D	$3.9 \pm 2.5 \times 10^{-7}$	
$^{13}CH_4$	5.1×10^{-5}	gives $^{13}C/^{12}C \sim 89$
CH_3C_2H	tentatively detected	
C_3H_8	tentatively detected	
H_2S	$< 2.0 \times 10^{-7}$	
H_2O	$< 2.0 \times 10^{-8}$	
HCN	$< 4.0 \times 10^{-9}$	
SiH_4	$< 4.0 \times 10^{-9}$	

From Fegley, B. Jr. (1995).

Table 5.4 Composition of the atmospheres of Uranus and Neptune

Species	Uranus	Neptune	Remarks
H_2	0.825 ± 0.033	0.80 ± 0.03	
He	0.152 ± 0.033	0.19 ± 0.032	
CH_4	2.3×10^{-2}	1.0–2.0×10^{-2}	troposphere
	2.0×10^{-5}	6.0–50.0×10^{-4}	stratosphere
C_2H_6	1.0–20.0×10^{-9}	$1.5^{+2.5}_{-0.5} \times 10^{-6}$	
C_2H_2	1.0×10^{-8}	$6.0^{+1.4}_{-4.0} \times 10^{-8}$	
CO	$< 3.0 \times 10^{-8}$	1.2×10^{-6}	
HCN	$< 1.0 \times 10^{-10}$	1.0×10^{-9}	
H	—	—	
e	—	—	
HD	1.48×10^{-4}	1.92×10^{-4}	Gives D/H = 9.0×10^{-5} for Uranus, 1.2×10^{-4} for Neptune
CH_3D	8.3×10^{-6}	1.2×10^{-5}	Gives D/H = 9.0×10^{-5} for Uranus, 1.2×10^{-4} for Neptune
H_2S	$< 8.0 \times 10^{-7}$	$< 3.0 \times 10^{-6}$	
NH_3	$< 1.0 \times 10^{-7}$	$< 6.0 \times 10^{-7}$	

From Fegley, B., Jr. (1995).

Table 5.5 Helium to hydrogen ratio and internal heat flux for giant planets

	Jupiter	Saturn	Uranus	Neptune
He/H	0.0568	0.0169	0.0921	0.119
(He/H)/(He/H)$_\odot$	0.709	0.211	1.15	1.48
Total heat flux/absorbed solar flux	1.67 ± 0.09	1.78 ± 0.09	1.06 ± 0.08	2.61 ± 0.28
Excess heat density (10^{-13} W/g)	1.76 ± 0.14	1.52 ± 0.11	< 0.16	0.34 ± 0.11

in the planet. A value of unity for this ratio implies the absence of an internal heat source. The ratios for Jupiter and Saturn are 1.67 and 1.78, respectively, implying that these planets radiate 67% and 78% more energy than what they receive from the sun. A more useful quantity is the excess heat density (in units of 10^{-13} W g^{-1}) derived from the excess heating rate divided by the mass of the planet. This quantity is 1.76 and 1.52 for Jupiter and Saturn, respectively. There are two major sources of heating for the giant planets. The first is heat derived from gravitational contraction of the planets. According to models of the origin and evolution of these planets, there was a period of rapid contraction at the beginning (10^5 yr), followed by a longer period of slow contraction over the next few billion years. This is the principal source of internal heat for Jupiter, but this source would be insufficient for a smaller planet like Saturn. The second major source of internal heat derives from the separation of helium from hydrogen. At pressures greater than 3 Mbar, which occurs in the interior of Jupiter and Saturn, hydrogen becomes metallic. If the temperature is sufficiently high (> 10^4 K), helium forms an immiscible mixture with hydrogen. However, at lower temperatures, He and H$_2$ are only partially miscible. The heavier element tends to sink to the interior, much as differentiation occurs in the interior of the terrestrial planets. In Saturn this process is shown to be important, and as the helium-rich gas sinks to the core, its viscous interactions with the surrounding air convert the gravitational energy into heat. Thus, the depletion of He in the gaseous envelope of Saturn and the internal heating of the planet are beautifully and simply related. Since He is also somewhat depleted on Jupiter, this process may also be occurring on that planet, albeit at a reduced level. Pressure at the centers of Uranus and Neptune does not exceed 200 kbar, and there is no separation of helium from hydrogen by this mechanism.

The only planet in table 5.5 with He/H ratio equal to the cosmic abundance ratio is Uranus. The question then arises as to why Neptune is enriched in helium by about 30% relative to Uranus. There is no simple explanation. There is, however, an alternative interpretation of the Voyager data on which this helium abundance is based. The Voyager data could be explained with a model containing the cosmic abundance of He and 0.3% N$_2$. As we shall argue, the latter hypothesis may be needed to account for the HCN observed in Neptune. The ultimate difference between Uranus and Neptune is attributed to the absence and presence, respectively, of an internal heat source, which affects the interior convection of these planets.

(b) High Z Elements

From tables 5.2–5.4 we may note that the giant planets are enriched in the high Z elements relative to the cosmic abundances. Table 5.6 expresses the same information

Table 5.6 High Z elements and bulk properties of the giant planets

Property	Jupiter	Saturn	Uranus	Neptune
$C/C_\odot$	2.3 ± 0.2	5.1 ± 2.3	35.0 ± 15.0	40.0 ± 20.0
$N/N_\odot$	2.0	3.0 ± 1.0	—	33.0 ?
$P/P_\odot$	1.4 ± 0.4	2.8 ± 1.6	—	—
$O/O_\odot$	0.02	—	—	—
$S/S_\odot$	$< 3.7 \times 10^{-3}$	—	—	—
Total mass[a]	318.1	95.1	14.6	17.2
Low Z mass[b]	254.1–292.1	72.1–79.1	1.3–3.6	0.7–3.2
High Z mass (core)	26.0–64.0	16.0–23.0	11.0–13.3	14.0–16.5

[a] All masses are in units of Earth masses.
[b] Low Z elements = H + He.
[c] High Z elements = all elements heavier than He.

relative to the cosmic standard. Thus, carbon (primarily as CH_4) is enriched by factors of 2–3 in the atmospheres of Jupiter and Saturn, and by about 30–40 times in the atmospheres of Uranus and Neptune. The reason may become clear if we examine the bulk properties of the giant planets inferred from independent sources. As summarized in table 5.6 all giant planets have substantial cores, which may contribute significantly to the budget of high Z elements. The cores of Jupiter and Saturn are much smaller than the gaseous envelopes of low Z elements, but the situation is reversed for Uranus and Neptune, which have large cores relative to gas. Nitrogen and phosphorus appear to be enriched also, but observing NH_3 and PH_3 in the atmospheres of Uranus and Neptune is difficult due to condensation. The 33 times enrichment tentatively assigned to Neptune is based on a plausible but unverified interpretation of the origin of HCN in Neptune. Oxygen (in H_2O) has been positively identified only in Jupiter and appears to be depleted relative to the cosmic abundance by a factor of 50. This may not be real because water may be sequestered in the clouds. The upper limit for sulfur (in H_2S) established for Jupiter suggests that Jupiter is depleted in sulfur by a factor greater than 300. Again, in this case the problem may lie with the sequestering of sulfur in NH_4SH clouds. This topic is discussed in section 5.1.2(c).

(c) Gases That Can Condense

At the low temperatures that are typical of the visible troposphere and the tropopause region in the giant planets a large number of gases can condense. The vapor pressures for a selected number of simple molecules over the temperature range of interest are given in figure 5.7. Of all the molecules compared in this figure, H_2O condenses most easily, followed by NH_3, PH_3, C_2 hydrocarbons, and CH_4. This has an important consequence on the interpretation of the observed composition of the atmospheres of these planets. Figure 5.8 shows a model of Jupiter with regions of the atmosphere where clouds of NH_3, NH_4SH, and H_2O are expected to form. The concentrations of the condensing species are solar. For water, the dashed lines show the location of H_2O ice formation if its abundance is only 10^{-3} times solar. For an understanding of atmospheric composition, there are two crucial regions of the atmosphere where an estimate of the saturation vapor pressure of molecular species is useful. For the

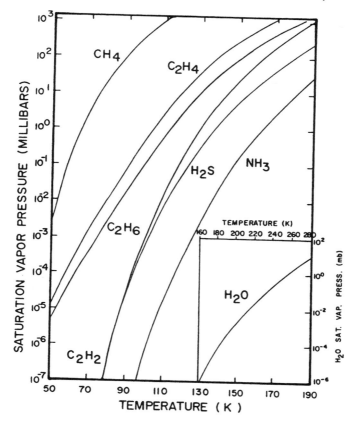

Figure 5.7 Saturation vapor pressures of selected molecules of importance in the atmosphere of the outer solar system. After Atreya, S., 1986, *Atmospheres and Ionospheres of the Outer Planets and their Satellites* (New York: Springer-Verlag).

visible atmosphere part we shall choose the pressure level of 1 bar, with approximate temperatures of 165, 135, and 75 K, respectively, for Jupiter, Saturn, and Uranus. For the purpose of this discussion we take the thermal structures of Uranus and Neptune to be the same. The tropopauses of the giant planets are between 100 and 200 mbar; we arbitrarily adopt a pressure level of 150 mbar. The minimum temperatures are 110, 85, and 50 K, respectively for Jupiter, Uranus, and Saturn. Using this information it is possible to compute the saturation vapor pressures of simple molecules at 1 bar and the tropopauses. The results are summarized in table 5.7. Comparing the results of this table with those in figure 5.7 and tables 5.3 and 5.4, we can understand the nondetection of H_2O, H_2S, and HCN in Saturn and the apparent depletion of NH_3, PH_3, and some of the higher hydrocarbons in Uranus and Neptune. These species may all be present in the atmosphere but may all be in the frozen solids at the ambient atmospheric temperatures. There is a major surprise when we compare table 5.4 with

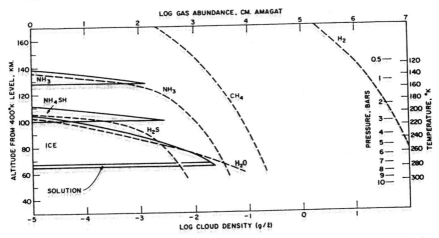

Figure 5.8 A model of the clouds of Jupiter based on cosmic abundances of the elements. The main cloud layers, composed of aqueous NH_3 solution plus H_2O ice, solid NH_4SH, and solid NH_3, are formed by condensation of H_2O, NH_3, and H_2S, which are stable in the lower atmosphere. After Weidenschilling, S. J., and Lewis, J. S., 1973, "Atmospheric and Cloud Structures of the Jovian Planets." *Icarus* **20**, 465.

table 5.7. The CH_4 abundance in the stratosphere of Neptune is in excess of its saturation value by factors of 10–100. The tropopause cold trap must have failed to remove CH_4 from air that is entering the stratosphere from the troposphere. A special convective injection mechanism may be operating on Neptune but not on Uranus, whose CH_4 abundance in the stratosphere is consistent with that given by the cold trap. The ultimate difference between Uranus and Neptune may be in the absence and presence, respectively, of excess heat flux that drives convection in the troposphere (see table 5.5).

Table 5.7 Saturation mole fraction of simple molecules in the troposphere (at the 1 bar level) and the tropopause of giant planets.

	CH_4	C_2H_2	H_2O
Troposphere (1 bar)			
Jupiter	NS	NS	2.3×10^{-9}
Saturn	NS	2.2×10^{-2}	—
Uranus/Neptune	7.7×10^{-3}	—	—
Tropopause			
Jupiter	NS	4.8×10^{-3}	S
Saturn	NS	1.3×10^{-5}	S
Uranus/Neptune	1.9×10^{-5}	—	S

Notes: The temperatures at 1 bar are 165, 135, and 75 K for Jupiter, Saturn, and Uranus/Neptune, respectively. The corresponding adopted temperatures (not the real temperatures) at 150 mbar are 110, 85, and 50 K. NS denotes mixing ratios > 0.1 and S denotes values $< 1.0 \times 10^{-10}$.

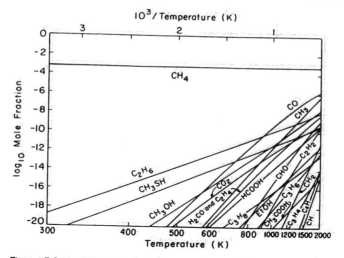

Figure 5.9 Partitioning of carbon species in the deep atmosphere of Jupiter. Equilibrium chemistry with cosmic abundances of elements is assumed in the model. After Prinn, R. G., and Barshay, S. S., 1977, "Carbon Monoxide on Jupiter and Implications for Atmospheric Convection." *Science* **198**, 1031.

(d) Gases from Equilibrium Chemistry

As pointed out in section 5.1, in the interior of the giant planets at pressures and temperatures in excess of 1 kbar and 1000 K, respectively, equilibrium chemistry is important. The partitioning of chemical species is carried out according to the laws of thermodynamics, as described in section 3.2. An example of this type of chemistry for carbon is given in figure 5.9 for Jupiter. Assuming cosmic abundances of the elements, the model predicts appreciable amounts of species such as CO and C_2H_6. Vigorous mixing from the level of about 1100 K can bring up parcels of air with mixing ratios of CO and C_2H_6 equal to 10^{-9} and 10^{-10}, respectively. This source can account for the observed concentration of CO in Jupiter but is too small to explain the observation of C_2H_6. This process has been invoked to account for the observed HCN, GeH_4, and AsH_3 in the atmosphere of Jupiter. The observed species CO, GeH_4, and AsH_3 in Saturn and CO in the atmosphere of Neptune may have a similar origin in the planetary interior. However, the observed HCN in Neptune cannot be accounted for by this mechanism, because the stratospheric abundance far exceeds that of the saturation vapor at the tropopause. As discussed in section 5.4.2, HCN in Neptune must be locally synthesized in the upper atmosphere. A key uncertainty concerning the validity of equilibrium chemistry for the giant planets is the time constant required for equilibrium to be reached. A crucial quenching temperature must be estimated so that the chemical time equals that for atmospheric transport. We note that Uranus is a very quiescent planet compared with Neptune. Hence, we expect both CO and HCN to be significantly depleted in Uranus relative to Neptune. This is borne out by the observations summarized in table 5.4.

(e) Photochemical Products

Free electrons, protons, H_3^+ and H atoms are readily produced in the thermospheres of the giant planets. EUV flux from the sun and energetic particles precipitated from the magnetosphere are the principal energy sources. The C_2 hydrocarbons, C_2H_6, C_2H_4, and C_2H_2, are obviously disequilibrium products generated by the photochemistry of CH_4 in the upper atmosphere (mesosphere) of the giant planets. More complex hydrocarbons are also produced, especially in the polar region with energy derived from magnetospheric particles. Since this topic constitutes the principal theme of this chapter, we defer the discussion to sections 5.2–5.6.

(f) Products of Ion Chemistry

The polar regions of Jupiter contain enhanced concentrations of H_3^+ and complex hydrocarbons such as methyl acetylene (CH_3C_2H) and benzene (C_6H_6). This may be the result of an enhanced energy source associated with the observed aurorae. This topic will be deferred to section 5.2.3.

(g) Isotopomers

A large number of isotopomers have been measured in the atmosphere of the giant planets. The most interesting case is the trend in D/H as deduced from HD and CH_3D observations, as discussed in section 4.10.1. The inferred isotopic ratios $^{12}C/^{13}C$ and $^{15}N/^{14}N$ are consistent with cosmic abundances. The only exception is the ratio $C_2H_2/^{13}C_2H_2$ with value equal to about 10 for Jupiter, which defies explanation. There may be large errors associated with this work.

The above brief survey provides an overview of the observed composition of the atmospheres of the giant planets as summarized in tables 5.2–5.4. The purpose is to provide a broader framework in which photochemistry plays a part. As we shall see, photochemistry is but one link in the overall chemical cycles of the elements. Having cast the problem in its proper context, we now discuss the quantitative aspects of modeling.

5.1.3 The One-Dimensional Model

The fundamental quantity in atmospheric chemistry is the number density, n_i (molecules/cm^3), of species i. In general n_i is a function of space (in three dimensions) and time. It is convenient to define a dimensionless quantity, the mole fraction (or mixing ratio), of species i:

$$f_i = \frac{n_i}{\sum_i n_i} \tag{5.4}$$

The equation that governs the time rate of change of the number density of species i is the continuity equation:

$$\frac{\partial n_i}{\partial t} + \nabla \cdot \vec{\phi}_i = P_i - L_i \tag{5.5}$$

where $\vec{\phi}_i$ is the flux, P_i is the chemical production rate, and L_i is the chemical loss rate. The units of flux are molecules cm^{-2} s^{-1}; P_i and L_i have units of molecules

$cm^{-3} s^{-1}$. All the chemistry discussed in chapters 2 and 3 goes into the computation of P_i and L_i. If we neglect the flux term, there is no transport. The number densities are then determined locally only by P_i and L_i. This is the limit of a pure photochemical system. In general, transport between different parts of the atmosphere is important. The flux ϕ_i is given by

$$\phi_i = n_i v_i \tag{5.6}$$

where v_i is the three-dimensional velocity field in the atmosphere, independently determined from dynamical modeling. The computation of wind fields from the primitive equations of fluid dynamics requires detailed input that is usually not available in planetary atmospheres. A class of simplified models in one dimension has been developed. Transport is restricted to the vertical (z), and the flux, instead of (5.6), consists of diffusive processes given by

$$\phi_i = -D_i \left(\frac{\partial n_i}{\partial z} + \frac{n_i}{H_i} + \frac{1+\alpha_i}{T} \frac{\partial T}{\partial z} n \right)$$
$$- K \left(\frac{\partial n_i}{\partial z} + \frac{n_i}{H_a} + \frac{1}{T} \frac{\partial T}{\partial z} n \right) \tag{5.7}$$

where D_i is the coefficient of molecular diffusion, T is the temperature, α_i is the thermal diffusivity factor, and H_i and H_a are, respectively, the scale height of species i and the mean scale height of the ambient atmosphere. K is the coefficient of eddy diffusion, an empirical quantity with the same units as D_i ($cm^2 s^{-1}$). Note that the flux in (5.7) consists of two parts. One part is the well-known molecular diffusion that can be rigorously derived from the molecular theory of ideal gases. The second part is based on an approximate theory of eddy transport. The values of the eddy diffusion coefficient K must be determined empirically from atmospheric observations.

A fundamental difference between the molecular diffusion and the eddy diffusion is in the scale height term (the thermal gradient term is usually unimportant). Molecular diffusion tends to drive the species toward "diffusive equilibrium" with n_i given by

$$n_i(z) = n_i(0) e^{-z/H_i} \tag{5.8}$$

in the isothermal case. That is, each species will follow its own scale height in the atmosphere. Eddy diffusion tends to drive the species to a well-mixed state given by

$$n_i(z) = n_i(0) e^{-z/H_a} \tag{5.9}$$

in the isothermal case. In this case each species has the same scale height as the bulk atmosphere. Molecular diffusion dominates in the thermosphere, but eddy diffusion becomes more important at lower altitudes. The place where $D_i = K$ is known as the homopause. Below the homopause the atmosphere is homogeneously mixed, but above it the atmospheric species are diffusively separated according to their own scale heights. These two distinct regions of the atmosphere are known as the homosphere and heterosphere, respectively.

In the simplified one-dimension model, the equation of continuity becomes

$$\frac{\partial n_i}{\partial t} + \frac{\partial \phi_i}{\partial z} = P_i - L_i \tag{5.10}$$

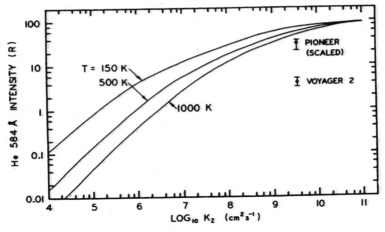

Figure 5.10 Concentration of He in the atmosphere of Jupiter computed for various values of the vertical eddy diffusion coefficient. After McConnell, J. C., Sandel, B. R., and Broadfoot, L., 1981, "Voyager UV Spectrometer Observations of He 584 Å Dayglow at Jupiter." *Planet. Space Sci.* **29**, 283.

where all relevant quantities are functions of t and z, and ϕ_i is given by (5.7). Equations (5.7) and (5.10) together constitute a system of nonlinear partial differential equations that are first order in time (t) and second order in space (z). The nonlinear terms usually appear in the chemistry (P_i and L_i). These equations can be solved by finite difference in z and time marching in t. To accelerate the rate of convergence in the presence of the nonlinear chemistry, Newton's method is used. We are mainly interested in the asymptotic solution as $t \to \infty$, or the steady state. The quantities P_i and L_i are appropriate diurnal averages in this case.

We briefly comment on the derivation of the eddy diffusion K defined in (5.7). In general it is a function of z, and $K(z)$ is known as the eddy diffusivity profile. Since $K(z)$ is independent of species, it can be based on any species. We use the atmosphere of Jupiter for an illustrative purpose. Helium has no chemistry; its distribution is determined by the balance between molecular and eddy diffusion. For small values of K, there is little He in the thermosphere. For higher values of K, He is readily mixed to the upper atmosphere. By comparing the model and observations of He emissions from the Voyager UVS experiment, one can infer the magnitude of K in the homopause region, K_H (see figure 5.10). Once the eddy diffusion coefficient is derived in this manner, we must use the same value for all other species; for example, CH$_4$. The functional dependence of $K(z)$ is usually parameterized as

$$K(z) = K_H \left[\frac{n_H}{n(z)}\right]^\gamma \qquad P \leq P_T \qquad (5.11a)$$

$$K(z) = K_T \qquad P > P_T \qquad (5.11b)$$

where $n(z)$ is the total number density, n_H is $n(z)$ at the homopause, P_T is the pressure at the tropopause, and γ is a coefficient around 0.5. The value of γ would be exactly 0.5 if the principal mechanism for eddy mixing in the atmosphere were the dissipation

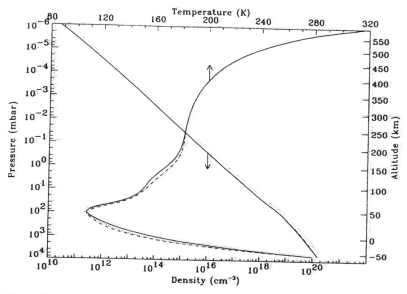

Figure 5.11 Total density and temperature profiles as a function of pressure for the standard NEB (solid line), NTeZ (dotted line), and POL (dashed line) model atmospheres. Altitudes marked on the right are measured from the 1 bar level and apply to the NEB model only. From Gladstone et al. 1996; cited in section 5.3.

of upward-propagating gravity waves generated in the troposphere. For Jupiter the best choices of the parameters are: $K_H = 1.4 \times 10^6$ cm^2 s^{-1}, $K_T = 1.0 \times 10^3$, $n_H = 1.4 \times 10^{13}$, and $\gamma = 0.45$. This formulation allows for a discontinuity in $K(z)$ across the tropopause. Since $n = 6.6 \times 10^{18}$ cm^{-3} at the tropopause, $K = 3.9 \times 10^3$ cm^2 s^{-1} just above the tropopause but is equal to 1×10^3 cm^2 s^{-1} just below the tropopause. The eddy diffusion time constants at the homopause and the tropopause are of the order of 3 months and 130 yr, respectively.

A model of the atmosphere of Jupiter that is consistent with recent observations is given in figure 5.11. The model is based on Voyager IRIS data in the lower atmosphere and stellar occultation experiments on Voyager UVS. It is in hydrostatic equilibrium and appropriate for the north equatorial belt region. In the thermosphere the temperature asymptotically approaches 1100 K.

5.2 Thermosphere

The thermosphere of the giant planets is where the EUV solar radiation is absorbed. This region includes the ionosphere produced by the ionizing solar photons. A number of outstanding issues are currently unresolved. These include the source of energy for maintaining the high temperature of the thermosphere, the mechanism for exciting the electroglow, and the layered structure of the lower ionosphere.

5.2.1 Energetics

The temperature in the thermospheres of planets reflects primarily a balance between energy input from the absorption of solar EUV energy and energy loss by conduction to the lower parts of the atmosphere where radiative cooling at infrared wavelengths becomes important. Early calculations based on an energy balance model predicted that the temperatures in thermospheres of the giant planets should not exceed 200 K. Therefore, it came as a great surprise that the observed temperatures are so high: 1000 K, 420 K, 800 K, and 750 K for Jupiter, Saturn, Uranus, and Neptune, respectively. The equation that governs the globally averaged vertical thermal structure in the thermosphere of a planet is

$$\rho C_p \frac{\partial T}{\partial t} = \frac{\partial}{\partial z}\left(K \frac{\partial T}{\partial z}\right) + Q_H - Q_{IR} \qquad (5.12)$$

where T is the temperature, ρ = density of ambient atmosphere, C_p = specific heat at constant pressure, K = thermal conductivity, Q_H = heating rate, and Q_{IR} = infrared cooling rate. The thermal conductivity may be parameterized by

$$K = AT^s \qquad (5.13)$$

where $A = 252$ and $s = 0.751$ for an H_2 atmosphere. K is in units of erg cm^{-1} s^{-1} K^{-1}. Since H and H_2 are inefficient radiators in the infrared, Q_{IR} is negligible in the thermosphere and becomes important only near the homopause, where radiative cooling by CH_4 and C_2H_2 becomes important. The time constant is sufficiently long that we can take a diurnally averaged temperature and drop the time-dependent term in (5.12). With these simplifying assumptions (5.12) becomes

$$\frac{d}{dz}\left(K \frac{dT}{dz}\right) = -Q_H \qquad (5.14)$$

If we assume that the thermosphere is asymptotically isothermal with temperature T_∞ and $dT/dz = 0$ and that the homopause temperature is T_h, then (5.14) and (5.13) can be integrated to yield the relation

$$\frac{A}{s+1} \frac{T_\infty^{s+1} - T_h}{h} = \phi_0 \qquad (5.15)$$

$$\phi_0 = \int_{z_h}^{\infty} Q_H dz \qquad (5.16)$$

where h is the distance between the thermosphere and the homopause, and z_h is the altitude of the homopause. The right side of (5.15), ϕ_0, is the column-integrated rate of energy absorption. Equations (5.15) and (5.16) give a simple relation between the temperature of the thermosphere and the homopause, and the integral of the energy flux absorbed. Using the observed temperatures, we can use (5.15) and (5.16) to estimate the energy fluxes needed to account for the hot thermospheres. The results for ϕ_0 are shown in table 5.8. These values are significantly higher than those of the incident solar EUV radiation at solar maximum. What is the source of energy? Is it an internal source common to all planets in the solar system? Is it an internal source

Table 5.8 Thermospheric temperatures of Jupiter, Saturn, Uranus, and Neptune and estimated energy fluxes required to maintain these temperatures.

	Jupiter	Saturn	Uranus	Neptune
Temperature (K)	1000[a]	420[b]	800[c]	750[d]
Mean energy flux required to maintain observed temperature ϕ_0 (erg cm^{-2}s^{-1})	0.45[c,l]	0.05[f,l]	0.05[c,e,l]	0.10[d,l]
Total power P_0 (W)[g]	2.8×10^{13}[g]	2.3×10^{12}	3.8×10^{11}	7.9×10^{11}
Global mean incident solar EUV radiation flux below 1100Å ϕ_s (erg cm^{-2}s^{-1})	6.2×10^{-2}[h]	1.8×10^{-2}	4.6×10^{-3}	1.9×10^{-3}
Total solar power P_s (W)	4.0×10^{12}	8.2×10^{11}	3.4×10^{10}	1.5×10^{10}
Total aurora power input P_A (W)	1.7×10^{14}[i]	2.0×10^{11}[j]	10^{11}[k]	10^{9}[d]

[a] Atreya et al. (1981), McConnell et al. (1982).
[b] Smith et al. (1983).
[c] Herbert et al. (1987).
[d] Broadfoot et al. (1989).
[e] Waite et al. (1983).
[f] Sandel et al. (1982).
[g] $P_0 = 4\pi R_p^2 \phi_0$, where R_p = planetary radius.
[h] $\phi_s = \frac{1}{4}\pi F/d^2$, where πF is the incident solar energy flux below 1100 Å at 1 AU at solar maximum, and d is the distance of the planet from the sun in AU.
[i] Deduced from observed aurora intensity (f), area of emissions and efficiency of excitation by precipitating particles ($\sim 1/30$).
[j] Broadfoot et al. (1981).
[k] Broadfoot et al. (1986).
[l] Strobel et al. (1990); we did not take the alternate interpretation that the mean energy flux for Saturn ϕ_0 is 0.4 erg cm^{-2}s^{-1}.

Note: Note that the estimated energy fluxes are model dependent and must be regarded as order of magnitude estimates.

that is specific to the H$_2$-dominated atmospheres of the giant planets? Is it related to to the electroglow? The answers to these questions are unknown at present. A more detailed analysis than that given here for the thermosphere of Uranus gives the profile of heating, Q_H, that is required. As shown in figure 5.12, the heating must occur very high in the thermosphere. This rules out, for example, precipitation of high-energy particles from the magnetosphere as the energy source, because they will penetrate too deeply into the atmosphere. The best hypothesis is given by Dessler and Hunten, who postulated the existence of a population of soft electrons, with energy of the order of a few to tens of volts, that can get accelerated in the magnetosphere and deposit their energy in the thermosphere.

The hot thermosphere of Jupiter was first inferred from Pioneer 10 radio occultation studies in 1973. Voyager confirmed and extended the results to the other giant planets. But a definitive solution to the energetics problem has not been found in more than 20 years. This is a challenge for fundamentally new ideas.

5.2.2 Electroglow

As discussed in section 5.1.1, Voyager discovered extensive EUV emissions from the giant planets. An example of the observed spectrum is given in figure 5.1. While

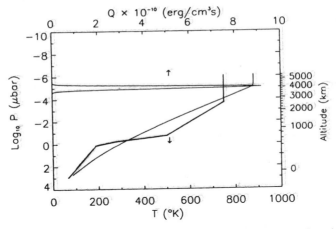

Figure 5.12 Heating rates in the thermosphere of Uranus inferred from Voyager 2 UVS observations. The two temperature profiles represent possible solutions to the Voyager 2 Ultraviolet Spectrometer observations. After Stevens, M. H., Strobel, D. F., and Herbert, F., 1993, "An Analysis of the Voyager-2 Ultraviolet Spectrometer Occultation Data at Uranus—Inferring Heat-Sources and Model Atmospheres." *Icarus* **101**, 45.

there is no doubt about the identity of the emissions, there is a controversy about the energy source because the solar EUV input appears to be insufficient to supply the radiated power. Solar EUV can excite the H_2 emissions either by direct absorption of the solar photons, followed by fluorescence, or indirectly via the production of photoelectrons, followed by electron impact excitation of H_2. A major difficulty in discriminating between the excitation mechanisms is due to the poor spectral resolution (33 Å) of the Voyager UVS instrument. Post-Voyager observations of Jupiter by the Hopkins Ultraviolet Telescope at much higher spectral resolution (3 Å) show that solar fluorescence can only account for about 20% of the H_2 band emissions. The rest of the emission cannot be due to photoelectrons for two reasons. First, there are not enough photoelectrons to make up the deficit. Second, the detailed excitation mechanism derived from the high-resolution observed spectrum is inconsistent with photoelectron excitation. It is possible that there is a local acceleration mechanism that can energize the ambient electrons to energies that are above the threshold of the Lyman and Werner bands but below that of the photoelectrons. It is also possible that the source of energy for the electroglow is related to that of the heating of the thermospheres of the giant planets. This is one of the outstanding unresolved issues of the thermospheres of the giant planets.

5.2.3 Ionosphere

Solar EUV radiation is readily absorbed in the thermosphere, resulting in photoionization:

$$H_2 + h\nu \rightarrow H_2^+ + e \quad (5.17a)$$
$$\rightarrow H^+ + H + e \quad (5.17b)$$
$$He + h\nu \rightarrow He^+ + e \quad (5.18)$$
$$H + h\nu \rightarrow H^+ + e \quad (5.19)$$

The thresholds for (5.17a), (5.18) and (5.19) are 804, 504, and 912 Å, respectively. The electrons produced in these reactions can have excess energy, and these are known as photoelectrons. In addition to direct photoionization, reactions (5.17)–(5.19) may also be driven by photoelectrons, as for example in

$$H_2 + e \rightarrow H_2^+ + 2e \quad (5.20a)$$
$$\rightarrow H^+ + H + 2e \quad (5.20b)$$

These ions will undergo exchange reactions in the atmosphere:

$$He^+ + H_2 \rightarrow H_2^+ + He \quad (5.21)$$
$$H_2^+ + H_2 \rightarrow H_3^+ + H \quad (5.22)$$

The fate of the molecular ions is rapid recombination with electrons:

$$H_2^+ + e \rightarrow H + H \quad (5.23)$$
$$H_3^+ + e \rightarrow H_2 + H \quad (5.24a)$$
$$\rightarrow 3H \quad (5.24b)$$

The loss of the atomic ion H^+ by radiative recombination,

$$H^+ + e \rightarrow H + h\nu \quad (5.25)$$

is inefficient (see section 3.7.3). Since the ionization potential of H (13.59 eV) is lower than that of H_2 (15.41 eV), charge transfer from H^+ to H_2 is endothermic. However, the reaction is exothermic if the H_2 molecule is in a vibrationally excited state:

$$H^+ + H_2^* \rightarrow H_2^+ + H \quad (5.26)$$

where H_2^* is in a state with $v' \geq 4$. Thus, H^+ is long-lived or short-lived according to the population of H_2^* in the thermosphere.

H^+ ions that are not chemically removed in the upper region of the ionosphere will be transported to the lower ionosphere, where they can transfer charge to the hydrocarbon ions:

$$H^+ + CH_4 \rightarrow CH_4^+ + H \quad (5.27a)$$
$$\rightarrow CH_3^+ + H_2 \quad (5.27b)$$

The hydrocarbon ions, being molecular ions, are rapidly removed by dissociative recombination. In the lower part of the thermosphere, there are other molecules or atoms of lower ionization potential than the species we have discussed. Reactions

$$CH_3 + h\nu \rightarrow CH_3^+ + e \qquad (5.28)$$

$$Si + h\nu \rightarrow Si^+ + e \qquad (5.29)$$

$$Na + h\nu \rightarrow Na^+ + e \qquad (5.30)$$

have ionization thresholds at 1262 Å (9.82 eV), 1520 Å (8.15 eV), and 2410 Å (5.14 eV), respectively. The source of CH_3 radicals is CH_4 photochemistry (see section 5.3). The Si and Na atoms may be derived from the burning up of micrometeoroids in the upper atmosphere. The atomic ions, such as the ones produced in (5.29) and (5.30), have low ionization potentials and cannot be easily removed by charge transfer. Radiative recombination is very slow. They may be a candidate for the multiple layers of ions observed in the lower ionosphere of the giant planets.

5.3 Hydrocarbon Chemistry

The thermodynamically stable form of carbon in the giant planets is methane. However, in the mesosphere region of the giant planets the molecule is not stable against photolysis. The destruction of CH_4 leads to the production of higher hydrocarbons, some of which have been detected by Voyager and Earth-based observations. The dominant chemistry of the mesosphere and the stratosphere is that of CH_4 and its photochemical products. An unusual property of carbon is to form complex organic compounds with multiple bonds. A list of organic compounds up to C_4, including polyynes up to C_8H_2, is given in table 5.9. Compounds beyond C_4 are probably important but have not been detected. A set of reactions that are important for the interconversion of carbon species is listed in tables 5.10 and 5.11.

There are at least two major problems with the basic photochemical data. The first is that there is a lack of detailed knowledge of the branching ratios of photodissociation as a function of wavelength. In general, it is difficult to measure all of the dissociation products, some of which may be short-lived radicals. This lack of knowledge is so serious that even for CH_4 photolysis at Lyman α there is no agreement over the branching ratios (see discussion in section 5.3.1). The second problem is the lack of information of the chemical rate coefficients at the temperatures appropriate for the atmospheres of the outer solar system, 50–200 K. The bulk of kinetic rate coefficients are measured at room temperature or higher (for combustion studies), and it is a professional hazard to extrapolate these measurements to such low temperatures.

In the following sections we divide the discussion of photochemistry and chemical kinetics into several parts:

a. C_1 and C_2 compounds
b. photosensitized dissociation
c. C_3 compounds
d. polyynes
e. H atoms
f. C_4 compounds and C_6H_6

Table 5.9 Model species and heats of formation

Formula	Name	Alternate names	$\Delta_f H°$ (kJ mol^{-1})[a]
He	helium		0
H	atomic hydrogen		218.0
H$_2$	molecular hydrogen		0
CH	methylidyne	methine, methenyl	595.8
^{3}CH$_2$	methylene (ground state)		386.4
^{1}CH$_2$	methylene (excited)		417.5
CH$_3$	methyl		145.7
CH$_4$	methane		−74.5
C$_2$	diatomic carbon		837.7
C$_2$H	ethynyl		477
C$_2$H$_2$	acetylene	ethyne	226.7
C$_2$H$_3$	vinyl	ethenyl	265.3
C$_2$H$_4$	ethylene	ethene	52.5
C$_2$H$_5$	ethyl		107.5
C$_2$H$_6$	ethane		−83.9
C$_3$H$_2$	propargylene	propynylidene	
	allenylcarbene	propadienylidene	195
C$_3$H$_3$	propargyl	propynyl	343
	allenyl	propadienyl	
CH$_3$C$_2$H	methylacetylene	propyne, allylene	186.6
CH$_2$CCH$_2$	allene	propadiene	191.3
C$_3$H$_5$	propenyl		161
	allyl		
	propylidene		
C$_3$H$_6$	propylene	propene	20.2
C$_3$H$_7$	propyl		100.5
C$_3$H$_8$	propane		−104.7
C$_4$H	butadiynyl		
C$_4$H$_2$	diacetylene	1,3-butadiyne	472.8
C$_4$H$_3$	butenynyl		294
	butatrienyl		
C$_4$H$_4$	vinylacetylene	1-butene-3-yne	289.5
	butatriene		349
C$_4$H$_5$	1,3-butadienyl		
	2-butenylidyne		
	butynyl		305
1-C$_4$H$_6$	ethylacetylene	1-butyne	165.2
1,2-C$_4$H$_6$	methylallene	1,2-butadiene	162.3
1,3-C$_4$H$_6$	bivinyl	1,3-butadiene	108.8
C$_4$H$_8$	1-butene		−0.5
	cis-2-butene		−7.6
	trans-2-butene		−11.0
	isobutylene	2-methylpropene	−17.9
C$_4$H$_9$	butyl		74
C$_4$H$_{10}$	butane	n-butane	−125.7
	isobutane	2-methylpropane	−134.5
C$_6$H	hexatriynyl		
C$_6$H$_2$	hexatriyne		652
C$_6$H$_3$	hexadiynyliumylidene		
C$_6$H$_6$	benzene		82.9
C$_8$H$_2$	octatetrayne		864

[a] Heat of formation at $T = 298$ K and $P = 1$ bar in the ideal gas state. Data from Lias et al. (1988,1994), Domalski and Hearing (1993), Chase et al. (1985), and Wagman et al. (1982).

Table 5.10 Photodissociation reactions

	Reaction	Rate coefficient	Reference
R1	$H_2 + h\nu \rightarrow 2H$	$0, 8.1 \times 10^{-10}$ $690 \leq \lambda \leq 1133$ Å	1; see text
R2	$^3CH_2 + h\nu \rightarrow CH + H$	$1.5 \times 10^{-6}, 1.8 \times 10^{-6}$ $990 \leq \lambda \leq 1975$ Å	2; 3
R3	$CH_3 + h\nu \rightarrow CH + H_2$	$2.2 \times 10^{-27}, 2.7 \times 10^{-8}$ $1475 \leq \lambda \leq 1525$ Å	4; 5; 6; 7; 8
R4	$\rightarrow {}^1CH_2 + H$	$5.0 \times 10^{-6}, 5.1 \times 10^{-6}$ $1475 \leq \lambda \leq 2225$ Å	
R5	$CH_4 + h\nu \rightarrow {}^1CH_2 + H_2$	$2.6 \times 10^{-15}, 1.3 \times 10^{-7}$ $750 \leq \lambda \leq 1625$ Å	9; 10; 11; 12; 13; 14; 15; 16; 17; 18
R6	$\rightarrow {}^3CH_2 + 2H$	$0, 1.2 \times 10^{-7}$ $750 \leq \lambda \leq 1375$ Å	
R7	$\rightarrow CH + H + H_2$	$0, 2.6 \times 10^{-8}$ $790 \leq \lambda \leq 1375$ Å	
R8	$C_2H_2 + h\nu \rightarrow C_2H + H$	$4.1 \times 10^{-9}, 1.3 \times 10^{-7}$ $670 \leq \lambda \leq 2175$ Å	19; 20; 21; 22; 23
R9	$\rightarrow C_2 + H_2$	$2.1 \times 10^{-9}, 5.1 \times 10^{-8}$ $690 \leq \lambda \leq 1975$ Å	24; 25; 26 see text
R10	$C_2H_3 + h\nu \rightarrow C_2H_2 + H$	$7.8 \times 10^{-6}, 7.8 \times 10^{-6}$ $4150 \leq \lambda \leq 4250$ Å	27
R11	$C_2H_4 + h\nu \rightarrow C_2H_2 + H_2$	$4.5 \times 10^{-8}, 3.9 \times 10^{-7}$ $930 \leq \lambda \leq 2025$ Å	28; 29
R12	$\rightarrow C_2H_2 + 2H$	$5.2 \times 10^{-8}, 6.1 \times 10^{-7}$ $930 \leq \lambda \leq 2025$ Å	
R13	$\rightarrow C_2H_3 + H$	$3.5 \times 10^{-9}, 3.6 \times 10^{-8}$ $1425 \leq \lambda \leq 2025$ Å	
R14	$C_2H_5 + h\nu \rightarrow CH_3 + {}^1CH_2$	$4.6 \times 10^{-6}, 4.6 \times 10^{-6}$ $2325 \leq \lambda \leq 2563$ Å	30; 31
R15	$C_2H_6 + h\nu \rightarrow C_2H_4 + H_2$	$1.8 \times 10^{-12}, 1.4 \times 10^{-8}$ $930 \leq \lambda \leq 1625$ Å	13; 32; 33; 34
R16	$\rightarrow C_2H_4 + 2H$	$8.2 \times 10^{-13}, 1.0 \times 10^{-7}$ $930 \leq \lambda \leq 1625$ Å	
R17	$\rightarrow C_2H_2 + 2H_2$	$4.2 \times 10^{-12}, 1.0 \times 10^{-7}$ $930 \leq \lambda \leq 1625$ Å	
R18	$\rightarrow CH_4 + {}^1CH_2$	$1.3 \times 10^{-13}, 6.8 \times 10^{-8}$ $930 \leq \lambda \leq 1575$ Å	
R19	$\rightarrow 2CH_3$	$6.7 \times 10^{-14}, 2.3 \times 10^{-8}$ $930 \leq \lambda \leq 1575$ Å	
R20	$C_3H_3 + h\nu \rightarrow C_3H_2 + H$	$4.2 \times 10^{-7}, 4.2 \times 10^{-7}$ $2475 \leq \lambda \leq 3050$ Å	30; 35
R21	$CH_3C_2H + h\nu \rightarrow C_3H_3 + H$	$4.6 \times 10^{-8}, 4.3 \times 10^{-7}$ $1048 \leq \lambda \leq 2225$ Å	36; 22; 37; 38
R22	$\rightarrow C_3H_2 + H_2$	$1.7 \times 10^{-8}, 1.6 \times 10^{-7}$ $1048 \leq \lambda \leq 2225$ Å	
R23	$CH_2CCH_2 + h\nu \rightarrow C_3H_3 + H$	$5.0 \times 10^{-7}, 1.5 \times 10^{-6}$ $1200 \leq \lambda \leq 2225$ Å	39; 40; see text
R24	$\rightarrow C_3H_2 + H_2$	$1.9 \times 10^{-7}, 5.7 \times 10^{-7}$ $1200 \leq \lambda \leq 2225$ Å	
R25	$C_3H_5 + h\nu \rightarrow CH_3C_2H + H$	$1.7 \times 10^{-5}, 1.8 \times 10^{-5}$ $1975 \leq \lambda \leq 2563$ Å	41; 42

Table 5.10 (*continued*)

	Reaction	Rate coefficient	Reference
R26	$\rightarrow CH_2CCH_2 + H$	7.2×10^{-5}, 7.3×10^{-5}	
		$1975 \leq \lambda \leq 2563$ Å	
R27	$\rightarrow C_2H_2 + CH_3$	7.8×10^{-6}, 7.8×10^{-6}	
		$1975 \leq \lambda \leq 2563$ Å	
R28	$C_3H_6 + h\nu \rightarrow C_3H_5 + H$	3.6×10^{-7}, 9.8×10^{-7}	43; 44
		$1048 \leq \lambda \leq 2025$ Å	
R29	$\rightarrow CH_3C_2H + H_2$	7.5×10^{-9}, 1.1×10^{-7}	
		$1048 \leq \lambda \leq 2025$ Å	
R30	$\rightarrow CH_2CCH_2 + H_2$	8.3×10^{-9}, 1.9×10^{-7}	
		$1048 \leq \lambda \leq 2025$ Å	
R31	$\rightarrow C_2H_4 + {}^1CH_2$	1.4×10^{-8}, 7.7×10^{-8}	
		$1048 \leq \lambda \leq 2025$ Å	
R32	$\rightarrow C_2H_3 + CH_3$	2.4×10^{-7}, 7.8×10^{-7}	
		$1048 \leq \lambda \leq 2025$ Å	
R33	$\rightarrow C_2H_2 + CH_4$	2.7×10^{-8}, 1.1×10^{-7}	
		$1048 \leq \lambda \leq 2025$ Å	
R34	$C_3H_8 + h\nu \rightarrow C_3H_6 + H_2$	1.9×10^{-10}, 1.0×10^{-7}	32; 45
		$1200 \leq \lambda \leq 1675$ Å	
R35	$\rightarrow C_2H_6 + {}^1CH_2$	6.4×10^{-12}, 3.6×10^{-8}	
		$1200 \leq \lambda \leq 1575$ Å	
R36	$\rightarrow C_2H_5 + CH_3$	2.9×10^{-11}, 1.7×10^{-7}	
		$1200 \leq \lambda \leq 1575$ Å	
R37	$\rightarrow C_2H_4 + CH_4$	2.2×10^{-11}, 1.1×10^{-7}	
		$1200 \leq \lambda \leq 1675$ Å	
R38	$C_4H_2 + h\nu \rightarrow C_4H + H$	3.2×10^{-7}, 6.3×10^{-7}	24; 46
		$1200 \leq \lambda \leq 2325$ Å	
R39	$\rightarrow C_2H_2 + C_2$	2.0×10^{-8}, 1.8×10^{-7}	
		$1200 \leq \lambda \leq 2075$ Å	
R40	$\rightarrow 2C_2H$	5.4×10^{-9}, 6.0×10^{-8}	
		$1200 \leq \lambda \leq 2025$ Å	
R41	$C_4H_4 + h\nu \rightarrow C_4H_2 + H_2$	3.4×10^{-5}, 3.5×10^{-5}	47; 48
		$1675 \leq \lambda \leq 2325$ Å	
R42	$\rightarrow 2C_2H_2$	8.5×10^{-6}, 8.7×10^{-6}	
		$1675 \leq \lambda \leq 2325$ Å	
R43	$1{-}C_4H_6 + h\nu \rightarrow C_4H_4 + 2H$	2.9×10^{-8}, 4.3×10^{-7}	36; 22; 49; 50
		$1060 \leq \lambda \leq 2075$ Å	
R44	$\rightarrow C_3H_3 + CH_3$	1.0×10^{-7}, 3.5×10^{-7}	
		$1060 \leq \lambda \leq 2225$ Å	
R45	$\rightarrow C_2H_5 + C_2H$	2.1×10^{-8}, 1.5×10^{-7}	
		$1060 \leq \lambda \leq 2225$ Å	
R46	$\rightarrow C_2H_4 + C_2H + H$	7.6×10^{-9}, 1.3×10^{-7}	
		$1060 \leq \lambda \leq 1875$ Å	
R47	$\rightarrow C_2H_3 + C_2H + H_2$	1.9×10^{-9}, 2.4×10^{-7}	
		$1060 \leq \lambda \leq 1625$ Å	
R48	$\rightarrow 2C_2H_2 + H_2$	4.7×10^{-10}, 9.2×10^{-8}	
		$1060 \leq \lambda \leq 1625$ Å	
R49	$1, 2{-}C_4H_6 + h\nu \rightarrow C_4H_5 + H$	1.5×10^{-7}, 2.8×10^{-7}	32; 51; 52; 53
		$1700 \leq \lambda \leq 2325$ Å	

(*continued*)

Table 5.10 (continued)

	Reaction	Rate coefficient	Reference
R50	$\rightarrow C_4H_4 + 2H$	$5.4 \times 10^{-7}, 1.2 \times 10^{-6}$ $1700 \leq \lambda \leq 2025$ Å	
R51	$\rightarrow C_3H_3 + CH_3$	$9.0 \times 10^{-7}, 1.5 \times 10^{-6}$ $1700 \leq \lambda \leq 2325$ Å	
R52	$\rightarrow C_2H_4 + C_2H_2$	$5.0 \times 10^{-8}, 8.2 \times 10^{-8}$ $1700 \leq \lambda \leq 2325$ Å	
R53	$\rightarrow C_2H_3 + C_2H_2 + H$	$6.6 \times 10^{-8}, 1.2 \times 10^{-7}$ $1700 \leq \lambda \leq 2125$ Å	
R54	$\rightarrow C_2H_3 + C_2H + H_2$	$1.3 \times 10^{-8}, 4.3 \times 10^{-8}$ $1700 \leq \lambda \leq 1875$ Å	
R55	$\rightarrow 2C_2H_2 + H_2$	$9.2 \times 10^{-8}, 1.7 \times 10^{-7}$ $1700 \leq \lambda \leq 2325$ Å	
R56	$1,3-C_4H_6 + h\nu \rightarrow C_4H_5 + H$	$2.0 \times 10^{-5}, 2.1 \times 10^{-5}$ $1675 \leq \lambda \leq 2325$ Å	32; 54; 55
R57	$\rightarrow C_4H_4 + H_2$	$3.5 \times 10^{-6}, 3.7 \times 10^{-6}$ $1675 \leq \lambda \leq 2325$ Å	
R58	$\rightarrow C_3H_3 + CH_3$	$2.8 \times 10^{-5}, 2.9 \times 10^{-5}$ $1675 \leq \lambda \leq 2325$ Å	
R59	$\rightarrow C_2H_4 + C_2H_2$	$1.2 \times 10^{-5}, 1.2 \times 10^{-5}$ $1675 \leq \lambda \leq 2325$ Å	
R60	$\rightarrow 2C_2H_3$	$7.1 \times 10^{-6}, 7.3 \times 10^{-6}$ $1675 \leq \lambda \leq 2325$ Å	
R61	$C_4H_8 + h\nu \rightarrow 1,3-C_4H_6 + 2H$	$1.4 \times 10^{-7}, 6.3 \times 10^{-7}$ $1048 \leq \lambda \leq 2025$ Å	44; 32; 56; 57; 58
R62	$\rightarrow C_3H_5 + CH_3$	$5.8 \times 10^{-7}, 1.5 \times 10^{-6}$ $1048 \leq \lambda \leq 2025$ Å	
R63	$\rightarrow CH_3C_2H + CH_4$	$3.0 \times 10^{-8}, 6.7 \times 10^{-8}$ $1048 \leq \lambda \leq 2025$ Å	
R64	$\rightarrow CH_2CCH_2 + CH_4$	$6.1 \times 10^{-9}, 1.7 \times 10^{-7}$ $1048 \leq \lambda \leq 1725$ Å	
R65	$\rightarrow C_2H_5 + C_2H_3$	$2.1 \times 10^{-8}, 3.2 \times 10^{-7}$ $1048 \leq \lambda \leq 1825$ Å	
R66	$\rightarrow 2C_2H_4$	$6.6 \times 10^{-8}, 1.6 \times 10^{-7}$ $1048 \leq \lambda \leq 2025$ Å	
R67	$\rightarrow C_2H_2 + 2CH_3$	$1.5 \times 10^{-8}, 8.0 \times 10^{-8}$ $1048 \leq \lambda \leq 1825$ Å	
R68	$C_4H_{10} + h\nu \rightarrow C_4H_8 + H_2$	$3.5 \times 10^{-10}, 3.3 \times 10^{-7}$ $1200 \leq \lambda \leq 1675$ Å	32; 59; 60
R69	$\rightarrow C_3H_8 + {}^1CH_2$	$0, 1.6 \times 10^{-8}$ $1200 \leq \lambda \leq 1425$ Å	
R70	$\rightarrow C_3H_6 + CH_4$	$5.0 \times 10^{-12}, 3.3 \times 10^{-8}$ $1200 \leq \lambda \leq 1675$ Å	
R71	$\rightarrow C_3H_6 + CH_3 + H$	$2.0 \times 10^{-11}, 7.5 \times 10^{-8}$ $1200 \leq \lambda \leq 1675$ Å	
R72	$\rightarrow C_2H_6 + C_2H_4$	$6.0 \times 10^{-11}, 1.7 \times 10^{-7}$ $1200 \leq \lambda \leq 1675$ Å	
R73	$\rightarrow 2C_2H_5$	$5.0 \times 10^{-11}, 1.2 \times 10^{-7}$ $1200 \leq \lambda \leq 1675$ Å	

Table 5.10 (continued)

	Reaction	Rate coefficient	Reference
R74	$\rightarrow C_2H_4 + 2CH_3$	1.5×10^{-11}, 8.3×10^{-8}	
		$1200 \leq \lambda \leq 1675$ Å	
R75	$C_6H_2 + h\nu \rightarrow C_6H + H$	$= J_{38}$	Estimate
R76	$\rightarrow C_4H + C_2H$	$= J_{40}$	Estimate
R77	$C_8H_2 + h\nu \rightarrow C_6H + C_2H$	$= J_{40}$	Estimate
R78	$\rightarrow 2C_4H$	$= J_{40}$	Estimate

References: (1) Mentall and Gentieu 1970; (2) van Dishoeck 1989; (3) Pilling et al. 1971; (4) Okabe 1978; (5) Parkes et al. 1976; (6) Arthur 1986a; (7) Yu et al. 1984; (8) Ye et al. 1988; (9) Samson et al. 1989; (10) Backx et al. 1975; (11) Lee and Chiang 1983; (12) Mount et al. 1977; (13) Mount and Moos 1978; (14) Ditchburn 1955; (15) Brion and Thomson 1984; (16) Rebbert and Ausloos 1972/1973; (17) Slanger and Black 1982; (18) Laufer and McNesby 1968; (19) Han et al. 1989; (20) Suto and Lee 1984; (21) Wu 1990; (22) Hamai and Hirayama 1979; (23) Cooper et al. 1988; (24) Okabe 1981a,b, 1983a,b; (25) Fahr and Laufer 1986; (26) Wodtke and Lee 1985; (27) Hunziker et al. 1983; (28) Zelikoff and Watanabe 1953; (29) Hara and Tanaka 1973; (30) Adachi et al. 1979; (31) Blomberg and Liu 1985; (32) Calvert and Pitts 1966; (33) Lias et al. 1970; (34) Akimoto et al. 1965, 1973; (35) Jacox and Milligan 1974; (36) Nakayama and Watanabe 1964; (37) Person and Nicole 1970; (38) Stief et al. 1971; (39) Rabalais et al. 1971; (40) Fuke and Schnepp 1979; (41) Shimo et al. 1986; (42) Gierczak et al. 1988; (43) Samson et al. 1962; (44) Collin 1988; (45) Johnston et al. 1978; (46) Glicker and Okabe 1987; (47) Braude 1945; (48) Yung et al. 1984; (49) Hill and Doepker 1972; (50) Deslauriers et al. 1980; (51) Doepker and Hill 1969; (52) Collin and Deslauriers 1986; (53) Diaz and Doepker 1977; (54) Doepker 1968; (55) Bergmann and Demtröder 1968 (56) Collin and Więckowski 1978; (57) Collin 1973; (58) Niedzielski et al. 1978, 1979; (59) Jackson and Lias 1974; (60) Okabe and Becker 1963.

[a] Photodissociation rate constants for reaction i (denoted J_i) in units of s^{-1}. Values are for 5 mbar and 1 nbar, respectively, at 10° latitude and are calculated utilizing diurnally averaged optical depths and the solar maximum irradiances of Mount and Rottman (1981) and Torr and Torr (1985). The adopted heliocentric distance of Jupiter is 5.2 AU and the subsolar latitude is 0°. Also indicated is the wavelength range (above 360 Å) in which the cross sections are nonzero.

5.3.1 Photochemistry and Chemical Kinetics

(a) C_1 and C_2 Compounds

Although the bond energy of methane (CH_3-H) is only 4.55 eV (105 kcal mole^{-1}), corresponding to a wavelength threshold of 2725 Å, absorption of ultraviolet radiation by the molecule is negligible above 1600 Å. This accounts for the remarkable stability of CH_4 in the atmospheres of the outer solar system. In fact, the fractional abundance of CH_4 remains roughly constant in the stratospheres of the Jovian planets and starts to fall off only around the homopause.

Absorption of a Lyman α photon by methane results in dissociation by the following possible paths:

$$CH_4 + h\nu \rightarrow {}^1CH_2 + H_2 \qquad q_1 = 0.41 \qquad (5.31a)$$
$$\rightarrow CH_2 + 2H \qquad q_2 = 0.51 \qquad (5.31b)$$
$$\rightarrow CH + H + H_2 \qquad q_3 = 0.08 \qquad (5.31c)$$
$$\rightarrow CH_3 + H \qquad q_4 = 0.00 \qquad (5.31d)$$

where $^1CH_2$ and CH_2 are, respectively, the first excited (singlet) state and ground (triplet) state of the methylene radical, and CH and CH_3 are, respectively, the methylidyne and methyl radical (a complete listing of organic species in the model is given in table 5.9). The quantum yields (q) are uncertain. In the early laboratory studies

Table 5.11 Chemical reactions

	Reaction	Rate coefficient	Reference
R79	$2H + M \rightarrow H_2 + M$	$1.5 \times 10^{-29} T^{-1.3}$	1
R80	$H + {}^3CH_2 \rightarrow CH + H_2$	$4.7 \times 10^{-10} e^{-370/T}$	2
R81	$H + {}^3CH_2 + M \rightarrow CH_3 + M$	$= k_{119}$	Estimate
R82	$H + CH_3 + M \rightarrow CH_4 + M$	$k_0 = 3.8 \times 10^{-28} e^{-20/T}$ $(T < 200$ K$)$	3
		$= 5.8 \times 10^{-29} e^{355/T}$ $(T > 200$ K$)$	
		$k_\infty = 1 \times 10^{-9} T^{-0.4}$	
R83	$H + CH_4 \rightarrow CH_3 + H_2$	$3.73 \times 10^{-20} T^3 e^{-4406/T}$	1
R84	$H + C_2H_2 + M \rightarrow C_2H_3 + M$	$k_0 = 6.4 \times 10^{-25} T^{-2} e^{-1200/T}$	4
		$k_\infty = 3.8 \times 10^{-11} e^{-1374/T}$	5
R85	$H + C_2H_3 \rightarrow C_2H_2 + H_2$	6.0×10^{-12}	3
R86	$H + C_2H_4 + M \rightarrow C_2H_5 + M$	$k_0 = 2.15 \times 10^{-29} e^{-349/T}$	6
		$k_\infty = 4.39 \times 10^{-11} e^{-1087/T}$	3
R87	$H + C_2H_5 \rightarrow 2CH_3$	$7.95 \times 10^{-11} e^{-127/T}$	7
R88	$\rightarrow C_2H_4 + H_2$	3.0×10^{-12}	1
R89	$H + C_2H_5 + M \rightarrow C_2H_6 + M$	$k_0 = 5.5 \times 10^{-23} T^{-2} e^{-1040/T}$	8
	$k_\infty = 1.5 \times 10^{-10}$		9
R90	$H + C_2H_6 \rightarrow C_2H_5 + H_2$	$9.2 \times 10^{-22} T^{3.5} e^{-2600/T}$	1
R91	$H + C_3H_2 + M \rightarrow C_3H_3 + M$	$k_0 = 10 \times k_{0,82}$	10
	$k_\infty = 1.1 \times 10^{-11}$		11
R92	$H + C_3H_3 + M \rightarrow CH_3C_2H$	$k_0 = 10 \times k_{0,82}$	10
		$k_\infty = 1.15 \times 10^{-10} e^{-276/T}$	12
R93	$\rightarrow CH_2CCH_2 + M$	$= k_{92}$	Estimate
R94	$H + CH_3C_2H + M \rightarrow CH_3 + C_2H_2 + M$		
R95	$\rightarrow C_3H_5 + M$		
R96	$H + CH_2CCH_2 + M \rightarrow CH_3 + C_2H_2 + M$	$k_0 = 8 \times 10^{-24} T^{-2} e^{-1225/T}$	4
		$k_\infty = 1.4 \times 10^{-11} e^{-1006/T}$	13
R97	$\rightarrow C_3H_5$	$k_0 = 8 \times 10^{-24} T^{-2} e^{-1225/T}$	4
		$k_\infty = 6.6 \times 10^{-12} e^{-1359/T}$	13
R98	$H + CH_2CCH_2 \rightarrow CH_3C_2H + H$	$1 \times 10^{-11} e^{-1000/T}$	4
R99	$H + C_3H_5 \rightarrow CH_3C_2H + H_2$	$= k_{85}$	Estimate
R100	$\rightarrow CH_2CCH_2 + H_2$	$= k_{85}$	Estimate
R101	$\rightarrow CH_4 + C_2H_2$	$= k_{85}$	Estimate
R102	$H + C_3H_5 + M \rightarrow C_3H_6 + M$	$k_0 = k_{0,95}$	Estimate
		$k_\infty = 1.66 \times 10^{-10}$	14
R103	$H + C_3H_6 \rightarrow C_3H_5 + H_2$	$1.66 \times 10^{-10} e^{-1761/T}$	14
R104	$H + C_3H_6 + M \rightarrow C_3H_7 + M$	$k_0 = 10 \times k_{0,86}$	
		$k_\infty = 2.6 \times 10^{-11} e^{-798/T}$	
R105	$H + C_3H_7 \rightarrow C_3H_6 + H_2$	5×10^{-11}	14
R106	$\rightarrow C_2H_5 + CH_3$	6×10^{-11}	15
R107	$H + C_3H_7 + M \rightarrow C_3H_8 + M$	$k_0 = k_{0,140}$	Estimate
		$k_\infty = 1.66 \times 10^{-10}$	14
R108	$H + C_3H_8 + M \rightarrow C_3H_7 + H_2 + M$	$2.16 \times 10^{-18} T^{2.4} e^{2250/T}$	15
R109	$H + C_4H_2 + M \rightarrow C_4H_3 + M$	$k_0 = 1 \times 10^{-28}$	4
		$k_\infty = 1.39 \times 10^{-10} e^{-1184/T}$	16
R110	$H + C_4H_3 \rightarrow 2C_2H_2$	3.3×10^{-12}	17
R111	$\rightarrow C_4H_2 + H_2$	$= k_{85}$	Estimate
R112	$H + C_4H_5 \rightarrow C_4H_4 + H_2$	$= k_{85}$	Estimate
R113	$H + C_4H_5 + M \rightarrow 1-C_4H_6 + M$	$k_0 = 10 \times k_{0,135}$	
		$k_\infty = 1 \times 10^{-10}$	
R114	$H + C_4H_9 \rightarrow C_4H_8 + H_2$	1×10^{-10}	Estimate
R115	$H + C_6H_2 + M \rightarrow C_6H_3 + M$	$= k_{109}$	Estimate
R116	$H + C_6H_3 \rightarrow C_2H_2 + C_4H_2$	$= k_{110}$	Estimate
R117	$\rightarrow C_6H_2 + H_2$	$= k_{111}$	Estimate

Table 5.11 (continued)

	Reaction	Rate coefficient	Reference
R118	$CH + H_2 \to {}^3CH_2 + H$	$2.38 \times 10^{-10} e^{-1760/T}$	2
R119	$CH + H_2 + M \to CH_3 + M$	$k_0 = 0.1 \times k_{0,82}$	10
		$k_\infty = 2.37 \times 10^{-12} e^{523/T}$	18
R120	$CH + CH_4 \to C_2H_4 + H$	$3 \times 10^{-11} e^{200/T}$	19, 20
R121	$CH + C_2H_2 \to C_3H_2 + H$	$3.49 \times 10^{-10} e^{61/T}$	21
R122	$CH + C_2H_4 \to CH_2CCH_2 + H$	$2.23 \times 10^{-10} e^{173/T}$	21
R123	$CH + C_2H_6 \to C_3H_6 + H$	$1.8 \times 10^{-10} e^{132/T}$	19
R124	${}^1CH_2 + H_2 \to {}^3CH_2 + H_2$	1.26×10^{-11}	22, 23
R125	$\to CH_3 + H$	9.24×10^{-11}	22, 23
R126	${}^1CH_2 + CH_4 \to {}^3CH_2 + CH_4$	1.2×10^{-11}	24
R127	$\to 2CH_3$	5.9×10^{-11}	24
R128	$2\,{}^3CH_2 \to C_2H_2 + 2H$	$2.1 \times 10^{-10} e^{-408/T}$	1, 25
R129	${}^3CH_2 + CH_3 \to C_2H_4 + H$	7×10^{-11}	26
R130	${}^3CH_2 + CH_4 \to 2CH_3$	$7.1 \times 10^{-12} e^{-5051/T}$	24
R131	${}^3CH_2 + C_2H_2 \to C_3H_2 + H_2$	$1 \times 10^{-11} e^{-3332/T}$	24
R132	$\to C_3H_3 + H$	$1 \times 10^{-11} e^{-3332/T}$	27
R133	${}^3CH_2 + C_2H_5 \to CH_3 + C_2H_4$	3×10^{-11}	1
R134	$CH_3 + H_2 \to CH_4 + H$	$3.31 \times 10^{-11} e^{-7200/T}$	28
R135	$2CH_3 + M \to C_2H_6$	$k_0 = 8.76 \times 10^{-7} T^{-7.03} e^{-1390/T}$	29
		$k_\infty = 1.5 \times 10^{-7} T^{-1.18} e^{-329/T}$	
R136	$CH_3 + C_2H_3 \to CH_4 + C_2H_2$	6.5×10^{-13}	1
R137	$\to C_3H_5 + H$		
R138	$CH_3 + C_2H_3 + M \to C_3H_6$		
R139	$CH_3 + C_2H_5 \to CH_4 + C_2H_4$	$3.3 \times 10^{-11} T^{-0.5}$	1
R140	$CH_3 + C_2H_5 + M \to C_3H_8$	$k_0 = 1.01 \times 10^{-22} e^{341/T}\ (T < 200\ K)$	10
		$= 2.22 \times 10^{-26} e^{2026/T}\ (T > 200\ K)$	1
		$k_\infty = 8.1 \times 10^{-10} T^{-0.5}$	
R141	$CH_3 + C_3H_3 + M \to 1,2\text{-}C_4H_6$	$k_0 = 10 \times k_{0,135}$	10
		$k_\infty = 5 \times 10^{-11}$	4
R142	$+ M \to 1\text{-}C_4H_6$	$k_0 = 10 \times k_{0,135}$	10
		$k_\infty = 5 \times 10^{-11}$	4
R143	$CH_3 + C_3H_5 \to CH_4 + CH_3C_2H$	$= k_{136}$	Estimate
R144	$\to CH_4 + CH_2CCH_2$	$= k_{136}$	Estimate
R145	$CH_3 + C_3H_5 + M \to C_4H_8$	$k_0 = k_{0,118}$	Estimate; 14
		$k_\infty = 3.3 \times 10^{-11}$	
R146	$CH_3 + C_3H_6 + M \to C_4H_9$	$k_0 = 10 \times k_{0,86}$	Estimate; 14
		$k_\infty = 5.3 \times 10^{-13} e^{-3724/T}$	
R147	$CH_3 + C_3H_7 \to CH_4 + C_3H_6$	$1.85 \times 10^{-12} e^{339/T}$	14, 30
R148	$CH_3 + C_3H_7 + M \to C_4H_{10}$	$k_0 = 7.12 \times 10^{-22} e^{715/T}\ (T < 200\ K)$	10
		$= 4.57 \times 10^{-24} e^{2184/T}\ (T > 200\ K)$	30
		$k_\infty = 3.47 \times 10^{-11}$	
R149	$CH_3 + C_3H_8 \to CH_4 + C_3H_7$	$1.3 \times 10^{-24} T^4 e^{-4175/T}$	31
R150	$CH_3 + C_4H_5 \to CH_4 + C_4H_4$	$= k_{136}$	Estimate
R151	$CH_3 + C_4H_5 + M \to$ products	$= k_{148}$	Estimate
R152	$C_2 + H_2 \to C_2H + H$	$1.77 \times 10^{-10} e^{-1469/T}$	32
R153	$C_2 + CH_4 \to C_2H + CH_3$	$5.05 \times 10^{-11} e^{-297/T}$	32
R154	$C_2H + H_2 \to C_2H_2 + H$	$5.6 \times 10^{-11} e^{-1443/T}$	3
R155	$C_2H + CH_4 \to C_2H_2 + CH_3$	$9.0 \times 10^{-12} e^{-250/T}$	3
R156	$C_2H + C_2H_2 \to C_4H_2 + H$	1.5×10^{-10}	33

(continued)

Table 5.11 (continued)

	Reaction	Rate coefficient	Reference
R157	$C_2H + C_2H_4 \to C_4H_4 + H$	2×10^{-11}	1
R158	$C_2H + C_2H_6 \to C_2H_2 + C_2H_5$	2.1×10^{-11}	3
R159	$C_2H + C_3H_8 \to C_2H_2 + C_3H_7$	1.4×10^{-11}	34
R160	$C_2H + C_4H_2 \to C_6H_2 + H$	$= k_{156}$	35
R161	$C_2H + C_4H_{10} \to C_2H_2 + C_4H_9$	$= k_{159}$	Estimate
R162	$C_2H + C_6H_2 \to C_8H_2 + H$	$= 0.025 \times k_{156}$	35
R163	$C_2H + C_8H_2 \to$ products	$= 0.025 \times k_{156}$	Estimate
R164	$C_2H_3 + H_2 \to C_2H_4 + H$	$2.6 \times 10^{-13} e^{-2646/T}$	3
R165	$C_2H_3 + C_2H_2 + M \to C_4H_5 + M$	$k_0 = 10 \times k_{84}$ $k_\infty = 3.1 \times 10^{-12} e^{-2495/T}$	Estimate, 36
R166	$2C_2H_3 \to C_2H_4 + C_2H_2$	1.8×10^{-11}	37
R167	$2C_2H_3 + M \to 1,3-C_4H_6 + M$	$k_0 = 10 \times k_{0,135}$ $k_\infty = 8.2 \times 10^{-11}$	10, 37
R168	$C_2H_3 + C_2H_5 \to 2C_2H_4$	8×10^{-13}	1
R169	$\to C_2H_6 + C_2H_2$	8×10^{-13}	1
R170	$\to CH_3 + C_3H_5$		
R171	$C_2H_3 + C_2H_5 + M \to C_4H_8 + M$		
R172	$C_2H_5 + H_2 \to C_2H_6 + H$	$5.1 \times 10^{-24} T^{3.6} e^{-4253/T}$	1
R173	$2C_2H_5 \to C_2H_6 + C_2H_4$	$1.2 \times 10^{-11} e^{-540/T}$	38, 39
R174	$2C_2H_5 + M \to C_4H_{10} + M$	$k_0 = 1.55 \times 10^{-22} e^{586/T}\ (T < 200\ K)$ $= 5.52 \times 10^{-24} e^{1253/T}\ (T > 200\ K)$ $k_\infty = 1.4 \times 10^{-11} e^{35/T}$	10 38 39
R175	$C_3H_2 + C_2H_2 \to C_4H_2 + {}^3CH_2$	5×10^{-13}	11
R176	$C_3H_3 + C_2H_2 \to C_4H_2 + CH_3$	2×10^{-13}	11
R177	$C_3H_5 + H_2 \to C_3H_6 + H$	$5.25 \times 10^{-11} e^{-9913/T}$	14
R178	$C_3H_7 + H_2 \to C_3H_8 + H$	$3 \times 10^{-21} T^{2.84} e^{-4600/T}$	15
R179	$C_4H + H_2 \to C_4H_2 + H$	$6.3 \times 10^{-12} e^{-1450/T}$	40
R180	$C_4H + CH_4 \to C_4H_2 + CH_3$	$= k_{155}$	Estimate
R181	$C_4H + C_2H_2 \to C_6H_2 + H$	$= k_{156}$	35
R182	$C_4H + C_2H_6 \to C_4H_2 + C_2H_5$	$= k_{158}$	Estimate
R183	$C_4H + C_4H_2 \to C_8H_2 + H$	$= 0.025 \times k_{156}$	Estimate
R184	$C_4H + C_6H_2 \to$ products	$= 0.025 \times k_{156}$	Estimate
R185	$C_4H + C_8H_2 \to$ products	$= 0.025 \times k_{156}$	Estimate
R186	$C_4H_5 + H_2 \to 1-C_4H_6 + H$	$1.8 \times 10^{-11} e^{-1699/T}$	36
R187	$C_4H_5 + C_2H_2 \to C_6H_6 + H$	$3.0 \times 10^{-12} e^{-1861/T}$	41
R188	$C_6H + H_2 \to C_6H_2 + H$	$= k_{179}$	Estimate
R189	$C_6H + CH_4 \to C_6H_2 + CH_3$	$= k_{155}$	Estimate
R190	$C_6H + C_2H_2 \to C_8H_2 + H$	$= 0.025 \times k_{156}$	Estimate
R191	$C_6H + C_2H_6 \to C_6H_2 + C_2H_5$	$= k_{158}$	Estimate
R192	$C_6H + C_4H_2 \to$ products	$= 0.025 \times k_{156}$	Estimate
R193	$C_6H + C_6H_2 \to$ products	$= 0.025 \times k_{156}$	Estimate
R194	$C_6H + C_8H_2 \to$ products	$= 0.025 \times k_{156}$	Estimate

Note: Units are s^{-1} for photolysis reactions, $cm^3 s^{-1}$ for two-body reactions, and $cm^6 s^{-1}$ for three-body reactions. k_0 and k_∞ are the low and high pressure rate coefficients, respectively, for three body reactions.

References: (1) Tsang and Hampson 1986; (2) Zabarnick, Fleming, and Lin 1986 estimate; (3) Allen, Yung, and Gladstone 1992; (4) Yung, Allen, and Pinto 1984; (5) Sugawara, Okazaki, and Sato 1981; (6) Lightfoot and Pilling 1987; (7) Pratt and Wood 1984; (8) Teng and Jones 1972; (9) Munk et al. 1986; (10) Laufer et al. 1983; (11) Homann and Schweinfurth 1981; (12) Homann and Wellmann 1983; (13) Wagner and Zellner 1972b; (14) Allara and Shaw 1980; (15) Tsang 1988; (16) Nava, Mitchell, and Stief 1986; (17) Schwanebeck and Warnatz 1975; (18) Berman and Lin 1984; (19) Berman and Lin 1983; (20) Anderson, Freedman, and Kolb 1987; (21) Berman et al. 1982; (22) Braun, Bass, and Pilling 1970; (23) Langford, Petek, and Moore 1983; (24) Bohland, Temps, and Wagner 1985; (25) Frank, Bhaskaran, and Just 1986; (26) Laufer 1981; (27) Bohland, Temps, and Wagner 1986; (28) Moller, Mozzhukhin, and Wagner 1986; (29) Slagle et al. 1988; (30) Anastasi and Arthur 1987; (31) Hautman et al. 1981; (32) Pitts, Pasternack, and McDonald 1982; (33) Stephens et al. 1987; (34) Okabe 1983a,b; (35) Tanzawa and Gardiner 1980; (36) Callear and Smith 1986; (37) Fahr and Laufer 1990; (38) Pacey and Wimalasena 1984; (39) Arthur 1986b; (40) Laufer and Bass 1979; (41) Cole et al. 1984.

the principal branches (5.31a)–(5.31c) were identified. There was an upper limit of 10% for (5.31d). In a more recent study, (5.31d) was shown to account for as much as 50% of of the total dissociation. The difference in the model predictions resulting from adopting the different branching ratios is not large, partly because the radicals can interconvert, as in the following reactions:

$$^1CH_2 + H_2 \rightarrow CH_3 + H \tag{5.32}$$

$$CH + H_2 \rightarrow CH_2 + H \tag{5.33}$$

$$CH_2 + H \rightarrow CH + H_2 \tag{5.34}$$

$$CH_3 + h\nu \rightarrow CH + H_2 \tag{5.35a}$$

$$\rightarrow {}^1CH_2 + H \tag{5.35b}$$

The discussion of CH_4 photochemistry in this chapter is based on the earlier laboratory studies. The radicals produced in (5.31a)-(5.31c) and interconverted in (5.32)–(5.35) are ultimately removed by the formation of C_2 hydrocarbons in reactions such as

$$CH + CH_4 \rightarrow C_2H_4 + H \tag{5.36}$$

$$CH_2 + CH_2 \rightarrow C_2H_2 + H_2 \tag{5.37}$$

$$CH_2 + CH_3 \rightarrow C_2H_4 + H \tag{5.38}$$

$$CH_3 + CH_3 + M \rightarrow C_2H_6 + M \tag{5.39}$$

where M is a third body in the ternary reaction (5.39). Radicals can also be removed by the formation of C_3 and more complex hydrocarbons, which is discussed in section 5.3.1. Note that in each of reactions (5.36)–(5.39) a new C_2 bond is formed. The net result of the photolysis of CH_4 may be summarized in the following overall schemes for forming C_2H_6, C_2H_4, and C_2H_2:

$$\begin{array}{lr} 2(CH_4 + h\nu \rightarrow {}^1CH_2 + H_2) & (5.31a) \\ 2({}^1CH_2 + H_2 \rightarrow CH_3 + H) & (5.32) \\ \underline{CH_3 + CH_3 + M \rightarrow C_2H_6 + M} & (5.39) \\ net \quad 2CH_4 \rightarrow C_2H_6 + 2H & (I) \end{array}$$

$$\begin{array}{lr} CH_4 + h\nu \rightarrow {}^1CH_2 + H_2 & (5.31a) \\ CH_4 + h\nu \rightarrow CH_2 + 2H & (5.31b) \\ {}^1CH_2 + H_2 \rightarrow CH_3 + H & (5.32) \\ \underline{CH_2 + CH_3 \rightarrow C_2H_4 + H} & (5.38) \\ net \quad 2CH_4 \rightarrow C_2H_4 + 4H & (II) \end{array}$$

$$\begin{array}{lr} CH_4 + h\nu \rightarrow CH_2 + 2H & (5.31b) \\ CH_2 + H \rightarrow CH + H_2 & (5.34) \\ \underline{CH + CH_4 \rightarrow C_2H_4 + H} & (5.36) \\ net \quad 2CH_4 \rightarrow C_2H_4 + 2H + H_2 & (III) \end{array}$$

$$\begin{array}{lr} CH_4 + h\nu \rightarrow CH + H + H_2 & (5.31c) \\ \underline{CH + CH_4 \rightarrow C_2H_4 + H} & (5.36) \\ net \quad 2CH_4 \rightarrow C_2H_4 + 2H + H_2 & (IV) \end{array}$$

$$2(CH_4 + h\nu \rightarrow CH_2 + 2H) \quad (5.31b)$$
$$\underline{CH_2 + CH_2 \rightarrow C_2H_2 + H_2} \quad (5.37)$$
$$net \quad 2CH_4 \rightarrow C_2H_2 + 4H + H_2 \quad (V)$$

The above are some illustrative examples of simple chemical schemes that convert CH_4 into C_2 hydrocarbons. This not a complete listing of the important schemes for the synthesis of C_2 compounds.

Once formed, C_2H_6 and C_2H_4 may be photolyzed by hydrogen elimination or fragmentation. The principal pathways are as follows:

$$C_2H_6 + h\nu \rightarrow C_2H_4 + H_2 \quad (5.40a)$$
$$\rightarrow C_2H_4 + 2H \quad (5.40b)$$
$$\rightarrow C_2H_2 + 2H_2 \quad (5.40c)$$
$$\rightarrow CH_4 + {}^1CH_2 \quad (5.40d)$$
$$C_2H_4 + h\nu \rightarrow C_2H_2 + H_2 \quad (5.41a)$$
$$\rightarrow C_2H_2 + 2H \quad (5.41b)$$

Most higher alkanes, including ethane, are well shielded by CH_4. Photolysis is only a minor loss for C_2H_6. However, there is little shielding of C_2H_4 by CH_4 or C_2H_6. Photolysis of C_2H_4 is a major loss process, resulting in the production of acetylene. The photolysis of C_2H_2 has more complicated implications and is discussed in section 5.3.1(b).

(b) Photosensitized Dissociation

The photolysis of C_2H_2 leads to the production of reactive fragments:

$$C_2H_2 + h\nu \rightarrow C_2H + H \quad q_1 \quad (5.42a)$$
$$\rightarrow C_2 + H_2 \quad q_2 \quad (5.42b)$$

with quantum yield $q_1 = 0.3$ and 0.06 at 1470 Å and 1849 Å, respectively, and $q_2 = 0.1$ at both wavelengths. Both the ethynyl radical (C_2H) and diatomic carbon (C_2) can react with H_2, CH_4, and C_2H_6:

$$C_2H + H_2 \rightarrow C_2H_2 + H \quad (5.43)$$
$$C_2 + H_2 \rightarrow C_2H + H \quad (5.44)$$
$$C_2H + CH_4 \rightarrow C_2H_2 + CH_3 \quad (5.45)$$
$$C_2 + CH_4 \rightarrow C_2H + CH_3 \quad (5.46)$$
$$C_2H + C_2H_6 \rightarrow C_2H_2 + C_2H_5 \quad (5.47)$$
$$C_2 + C_2H_6 \rightarrow C_2H + C_2H_5 \quad (5.48)$$

Table 5.12 Maximum dissociation rates in the atmosphere of Jupiter due to interaction with ultraviolet sunlight and cosmic rays in the atmosphere of Jupiter

Energy source	Potential absorber	Maximum dissociation rate[a] ($cm^{-2}s^{-1}$)
Solar flux		
below 900Å	H_2, He	8.8×10^8
Lyα(1216Å)	CH_4	5.2×10^9
1250–1600Å	CH_4	5.5×10^9
1600–2000Å	C_2H_2	1.0×10^{11}
2000–2300Å	C_4H_2	1.1×10^{12}
Cosmic rays[b]	all species	4.0×10^7

[a]Computed using the formula $F \times f_1 \times f_2$, where F = incident solar flux, f_1 = attenuation due to distance from the sun = 0.037, and f_2 = global average factor = 0.25. The values used for F are for solar maximum.
[b]Capone et al. (1980).

Note that the net result of (5.42a) and (5.43) is the photosensitized dissociation of H_2

$$C_2H_2 + h\nu \rightarrow C_2H + H \quad (5.42a)$$
$$\underline{C_2H + H_2 \rightarrow C_2H_2 + H} \quad (5.43)$$
$$net \quad H_2 \rightarrow 2H \quad (VI)$$

Similarly, the net result of (5.40a), (5.44), and (5.46) is the photosensitized dissociation of CH_4 and C_2H_6.

$$C_2H_2 + h\nu \rightarrow C_2H + H \quad (5.42a)$$
$$\underline{C_2H + CH_4 \rightarrow C_2H_2 + CH_3} \quad (5.45)$$
$$net \quad CH_4 \rightarrow CH_3 + H \quad (VII)$$

$$C_2H_2 + h\nu \rightarrow C_2H + H \quad (5.42a)$$
$$\underline{C_2H + C_2H_6 \rightarrow C_2H_2 + C_2H_5} \quad (5.47)$$
$$net \quad C_2H_6 \rightarrow C_2H_5 + H \quad (VIII)$$

Photosensitized dissociation can also be effected using C_2 produced in (5.42b) instead of C_2H. The details are the same as those of cycles (VII) and (VIII), and will not be repeated. Note that in these cycles, C_2H_2 is used as a catalyst, a collector of ultraviolet photons, with the net result that H_2, CH_4, and C_2H_6 are dissociated at much longer wavelengths than can happen via direct absorption of ultraviolet radiation. The potential impact of photosensitized dissociation may be appreciated from table 5.12, which summarizes the solar fluxes in relevant wavelength intervals. H_2 is dissociated by UV radiation below 1000 Å, and CH_4 by ultraviolet radiation below 1600 Å, including Lyman α. C_2H_2 absorbs photons below 2300 Å. Since the solar flux increases rapidly between 1000 and 2300 Å, the photosensitized dissociation plays a fundamental role in the atmospheres of the outer solar system.

Photosensitized dissociation provides an efficient path for the synthesis of higher alkanes. Chemical schemes (VII) and (VIII) produce alkyl radicals deep in the atmosphere, where ternary reactions are efficient. In addition to methyl radical recombination (5.39), we can now have alkyl recombinations such as

$$CH_3 + C_2H_5 + M \rightarrow C_3H_8 + M \quad (5.49)$$

$$C_2H_5 + C_2H_5 + M \rightarrow C_4H_{10} + M \quad (5.50)$$

We give two examples of simple chemical schemes for synthesizing propane (C_3H_8) and butane (C_4H_{10}) from the simpler alkanes CH_4 and C_2H_6:

$$
\begin{array}{lr}
2(C_2H_2 + h\nu \rightarrow C_2H + H) & (5.42a) \\
C_2H + CH_4 \rightarrow C_2H_2 + CH_3 & (5.45) \\
C_2H + C_2H_6 \rightarrow C_2H_2 + C_2H_5 & (5.47) \\
CH_3 + C_2H_5 + M \rightarrow C_3H_8 + M & (5.49) \\
\text{net} \quad CH_4 + C_2H_6 \rightarrow C_3H_8 + 2H & (IX)
\end{array}
$$

$$
\begin{array}{lr}
2(C_2H_2 + h\nu \rightarrow C_2H + H) & (5.42a) \\
2(C_2H + C_2H_6 \rightarrow C_2H_2 + C_2H_5) & (5.47) \\
C_2H_5 + C_2H_5 + M \rightarrow C_4H_{10} + M & (5.50) \\
\text{net} \quad 2C_2H_6 \rightarrow C_4H_{10} + 2H & (X)
\end{array}
$$

Note the difference between direct photolysis and photosensitized dissociation. In photolysis, the alkane usually loses an even number of H atoms to become an alkene or alkyne. However, in the photosensitized dissociation only one H atom is abstracted. The resulting radical can then react to form higher alkanes. It is straightforward to generalize schemes (IX) and (X) to produce even more complex alkanes, but there is no motivation to go beyond C_4 compounds due to the lack of observations of these species. In a variant of schemes (IX) and (X), we may obtain the ethyl radical from the direct recombination of C_2H_4 and H instead of the photosensitized dissociation of C_2H_6.

(c) C_3 Compounds

Our knowledge of hydrocarbons that are more complex than C_2 is limited. Many of the rate coefficients are not known, and we have to estimate their magnitude based on analogies with similar reactions. Another problem is the large number of isomers associated with the higher hydrocarbons. The list compiled for table 5.9 does not include all the isomers. Few of these more complex hydrocarbons computed in the model have been observed.

The principal C_3 compounds considered in the model are the two isomers of C_3H_4, methyl acetylene (CH_3C_2H) and allene (CH_2CCH_2), and propene (C_3H_6) and propane (C_3H_8). C_3 compounds can be formed by insertion of a C_1 radical into a C_2 compound, as in the following examples:

$$CH + C_2H_2 \rightarrow C_3H_2 + H \quad (5.51)$$

$$CH + C_2H_4 \rightarrow CH_2CCH_2 + H \quad (5.52)$$

$$CH + C_2H_6 \rightarrow C_3H_6 + H \quad (5.53)$$

where the methylidyne radical, CH, is derived from either CH_4 photolysis or the reaction $CH_2 + H$ (5.34). The radical C_3H_2 can undergo successive hydrogenation to form methyl acetylene and allene:

$$C_3H_2 + H + M \rightarrow C_3H_3 + M \tag{5.54}$$

$$C_3H_3 + H + M \rightarrow CH_3C_2H + M \tag{5.55a}$$

$$\rightarrow CH_2CCH_2 + M \tag{5.55b}$$

The allene formed in (5.52) and (5.55b) isomerizes rapidly through the exchange reaction

$$H + CH_2CCH_2 \rightarrow CH_3C_2H + H \tag{5.56}$$

This is known as a telescopic reaction: The H atom is attached to one end of allene, followed by the ejection of the H atom at the other end of the molecule. The net result is the conversion of a pair of double bonds into a triple and a single bond. The reaction is exothermic to the right by about one kcal mole^{-1}.

The simplest paths to form methyl acetylene and allene would be the insertion reactions

$$CH_2 + C_2H_2 + M \rightarrow CH_3C_2H + M \tag{5.57a}$$

$$\rightarrow CH_2CCH_2 + M \tag{5.57b}$$

Early laboratory experiments showed that the reactions could be fast. However, a more recent study showed that the reactions have high activation energies and cannot play a significant role in the atmospheres of the outer solar system.

Recombination of radicals provides alternative pathways of forming C_3 compounds. Examples include the recombination of alkyl radicals mentioned earlier (5.49) and the following reaction forming propene:

$$CH_3 + C_2H_3 + M \rightarrow C_3H_6 + M \tag{5.58}$$

where the vinyl radical (C_2H_3) is produced by reaction (5.72) between H and C_2H_2 [see section 5.3.1(e)].

Photolysis provides a major pathway for the destruction of C_3 compounds:

$$C_3H_4 + h\nu \rightarrow C_3H_3 + H \tag{5.59a}$$

$$\rightarrow C_3H_2 + H_2 \tag{5.59b}$$

$$C_3H_6 + h\nu \rightarrow C_3H_5 + H \tag{5.60a}$$

$$\rightarrow CH_3C_2H + H_2 \tag{5.60b}$$

$$\rightarrow CH_2CCH_2 + H_2 \tag{5.60c}$$

$$\rightarrow C_2H_3 + CH_3 \tag{5.60d}$$

$$C_3H_8 + h\nu \rightarrow C_3H_6 + H_2 \tag{5.61a}$$

$$\rightarrow C_2H_5 + CH_3 \tag{5.61b}$$

$$\rightarrow C_2H_4 + CH_4 \tag{5.61c}$$

where C_3H_4 is either CH_3C_2H or CH_2CCH_2, and we have listed only the most important channels of dissociation (for a complete listing, see table 5.10). Of the above reactions, C_3H_8 dissociation (5.61) is partially shielded by CH_4 and is not as fast as those of C_3H_6 and C_3H_4. Most of the C_3 dissociations result in the elimination of hydrogen. There is no net loss of C_3 compounds. Only reactions such as (5.60d), (5.61b), and (5.61c) lead to fragmentation of C_3 into species of lower carbon number. In addition to loss by photolysis, we show in section 5.3.1(e) that the unsaturated compounds such as C_3H_4 and C_3H_6 may be destroyed by cracking by hydrogen.

(d) Polyynes

The polyynes are long-chain acetylenelike compounds. The simplest example is diacetylene (C_4H_2) formed by

$$C_2H_2 + h\nu \rightarrow C_2H + H \quad (5.42a)$$
$$C_2H + C_2H_2 \rightarrow C_4H_2 + H \quad (5.62)$$
$$\text{net} \quad 2C_2H_2 \rightarrow C_4H_2 + 2H \quad \text{(XI)}$$

This process can be repeated with the ethynyl radical inserting into the C_4H_2 and higher polyynes:

$$C_2H_2 + h\nu \rightarrow C_2H + H \quad (5.42a)$$
$$C_2H + C_{2n}H_2 \rightarrow C_{2n+2}H_2 + H \quad (5.63)$$
$$\text{net} \quad C_2H_2 + C_{2n}H_2 \rightarrow C_{2n+2}H_2 + 2H \quad \text{(XIIa)}$$

where, for $n = 1$, chemical scheme (XII) is the same as scheme (XI). For $n = 2$, 3, 4, we have the production of C_6H_2, C_8H_2, and $C_{10}H_2$, respectively. In the model we do not go beyond C_8H_2. All polyynes with more than eight carbon atoms are labeled "products." They are expected to condense in the form of polymers and will not further participate in chemical reactions. The polyynes absorb ultraviolet radiation, resulting in the production of radicals that further participate in the synthesis of higher polyynes. The following example illustrates this mechanism:

$$C_{2n}H_2 + h\nu \rightarrow C_{2n}H + H \quad (5.64)$$
$$C_{2n}H + C_2H_2 \rightarrow C_{2n+2}H_2 + H \quad (5.65)$$
$$\text{net} \quad C_2H_2 + C_{2n}H_2 \rightarrow C_{2n+2}H_2 + 2H \quad \text{(XIIb)}$$

There are alternative paths to synthesize the most important of the polyynes, diacetylene:

$$C_3H_2 + C_2H_2 \rightarrow C_4H_2 + CH_2 \quad (5.66)$$
$$C_3H_3 + C_2H_2 \rightarrow C_4H_2 + CH_3 \quad (5.67)$$

where the C_3H_2 and C_3H_3 radicals are derived from the insertion reaction of CH (5.51) and photolysis of C_3H_4 (5.59). The chemical schemes are

$$CH + C_2H_2 \rightarrow C_3H_2 + H \quad (5.51)$$
$$C_3H_2 + C_2H_2 \rightarrow C_4H_2 + CH_2 \quad (5.66)$$
$$CH_2 + H \rightarrow CH + H_2 \quad (5.34)$$
$$\text{net} \quad 2C_2H_2 \rightarrow C_4H_2 + H_2 \quad \text{(XIII)}$$

$$C_3H_4 + h\nu \rightarrow C_3H_3 + H \quad (5.59a)$$
$$C_3H_3 + C_2H_2 \rightarrow C_4H_2 + CH_3 \quad (5.66)$$
$$\underline{CH_3 + H + M \rightarrow CH_4 + M} \quad (5.68)$$
$$net \quad C_3H_4 + C_2H_2 \rightarrow C_4H_2 + CH_4 \quad (XIV)$$

where C_3H_4 is either CH_3C_2H or CH_2CCH_2; the recombination of CH_3 and H (5.68) is discussed in section 5.3.1(e).

By analogy with acetylene, photolysis of polyynes can also result in the photosensitized dissociation of H_2, CH_4, and higher alkanes. The primary reaction is the production of the $C_{2n}H$ radical, where $n = 2, 3, 4$, in reaction (5.64). The $C_{2n}H$ radical readily attacks H_2 and alkanes:

$$C_{2n}H + H_2 \rightarrow C_{2n}H_2 + H \quad (5.69)$$
$$C_{2n}H + CH_4 \rightarrow C_{2n}H_2 + CH_3 \quad (5.70)$$
$$C_{2n}H + C_2H_6 \rightarrow C_{2n}H_2 + C_2H_5 \quad (5.71)$$

The net results may be summarized by chemical schemes similar to (VI)–(VIII) with the radical $C_{2n}H$ taking the place of C_2H. Since the polyynes absorb ultraviolet radiation at even longer wavelengths than C_2H_2, the contribution of higher polyynes to photosensitized dissociation is appreciable.

(e) H Atoms

Direct photolysis of H_2 by the absorption of EUV radiation is a minor source of H atoms in the upper atmosphere of Jupiter. The most important sources are photolysis of CH_4 and C_2H_4, photosensitized dissociation by schemes (VI), (XIIa), and (XIIb) and abstraction reactions such as (5.32) and (5.33). In addition, impact dissociation by energetic electrons in the aurorae at the polar atmosphere of the planet provides a net source of H atoms.

The presence of large quantities of H atoms has a serious impact on the chemical composition of the atmosphere. In section 5.3.1(d) we discussed the formation of higher hydrocarbons from CH_4, a molecule with the highest H/C ratio. Since the higher hydrocarbons have a lower H/C ratio, this organic synthesis is always accompanied by the production of hydrogen, as either H_2 or H. If the by-product is H_2, there is no further reaction due to the great stability of H_2. However, if the by-product is H, this can lead to further reactions. First, H atoms can scavenge CH_3 radicals via reaction (5.68). The net result is to restore the dissociation products back to CH_4. Indeed, a substantial fraction of the dissociation (direct photolysis or photosensitized dissociation) of CH_4 does not result in the synthesis of higher hydrocarbons but returns to CH_4 via (5.68).

The efficiency at which the higher hydrocarbons can be synthesized depends to a large extent on how rapidly H atoms can be removed. The direct recombination of H atoms to form a simple molecule like H_2 is too slow (see section 3.5). There is no effective sink for H atoms above the homopause. This explains the high abundance of H in the thermosphere, where its maximum mixing ratio becomes as high as 1%. The primary loss of H is via transport to below the homopause, where catalytic recombination can take place. The most important scheme for removing H atoms is

160   Photochemistry of Planetary Atmospheres

$$C_2H_2 + H + M \to C_2H_3 + M \quad (5.72)$$
$$\underline{C_2H_3 + H \to C_2H_2 + H_2} \quad (5.73)$$
$$net \quad H + H \to H_2 \quad (XV)$$

where the rate-limiting step is the formation of the vinyl radical (C_2H_3). There are similar schemes involving diacetylene and higher polyynes, and these schemes are important for scavenging H atoms.

Hydrogen atoms that are not catalytically recombined can drive a number of interesting reactions, including hydrogenation, as shown by

$$C_2H_2 + H + M \to C_2H_3 + M \quad (5.72)$$
$$\underline{C_2H_3 + H_2 \to C_2H_4 + H} \quad (5.74)$$
$$net \quad C_2H_2 + H_2 \to C_2H_4 \quad (XVI)$$

where abstraction of a hydrogen atom from H_2 by the vinyl radical is a slow reaction with a large activation energy. It is important in the atmosphere of Jupiter but is expected to be less important in the colder atmospheres of the other giant planets. In the presence of H, further hydrogenation of C_2H_4 is possible:

$$C_2H_4 + H + M \to C_2H_5 + M \quad (5.75)$$
$$C_2H_5 + H \to 2CH_3 \quad (5.76)$$
$$\underline{CH_3 + CH_3 + M \to C_2H_6 + M} \quad (5.39)$$
$$net \quad C_2H_4 + 2H \to C_2H_6 \quad (XVII)$$

The general tendency is that unsaturated hydrocarbons can be hydrogenated in the presence of H atoms to become more saturated compounds. At the same time, H atoms can be recombined using the unsaturated hydrocarbons as catalysts.

A variant of scheme (XVII) may result in hydrogenation of C_2H_4 all the way back to CH_4:

$$C_2H_4 + H + M \to C_2H_5 + M \quad (5.75)$$
$$C_2H_5 + H \to 2CH_3 \quad (5.76)$$
$$\underline{2(CH_3 + H + M \to CH_4 + M)} \quad (5.68)$$
$$net \quad C_2H_4 + 4H \to 2CH_4 \quad (XVII)$$

In general, unsaturated compounds may be "cracked" by H atoms, as in the following examples:

$$CH_3C_2H + H \to CH_3 + C_2H_2 \quad (5.77)$$
$$\underline{CH_3 + H + M \to CH_4 + M} \quad (5.68)$$
$$net \quad CH_3C_2H + 2H \to CH_4 + C_2H_2 \quad (XVIII)$$

$$C_4H_2 + H + M \to C_4H_3 + M \quad (5.78)$$
$$\underline{C_4H_3 + H \to 2C_2H_2} \quad (5.79)$$
$$net \quad C_4H_2 + 2H \to 2C_2H_2 \quad (XIX)$$

Note that the unsaturated species can scavenge H atoms as in catalytic cycle (XV) and its analogs, but the unsaturated compounds may in turn be cracked by H atoms. There is thus a complicated interaction between unsaturated species and H atoms. They mutually regulate each others' abundances.

(f) C_4 Compounds and C_6H_6

The important C_4 compounds are vinyl acetylene (C_4H_4), 1-butyne (1-C_4H_6), 1,2-butadiene (1,2-C_4H_6), 1,3-butadiene (1,3-C_4H_6), 1-butene (C_4H_8), diacetylene (C_4H_2), and butane (C_4H_{10}). The last two molecules are discussed in sections 5.3.1(b) and 5.3.1(d) and are not further discussed here. With the exception of C_4H_2, none of the above molecules has been detected in the atmospheres of the giant planets.

The principal reactions that form the C_4 bond are the following radical-radical reactions:

$$C_2H_3 + C_2H_3 + M \rightarrow 1,3-C_4H_6 + M \qquad (5.80)$$

$$CH_3 + C_3H_3 + M \rightarrow 1,2-C_4H_6 + M \qquad (5.81)$$

$$C_2H_3 + C_2H_5 + M \rightarrow C_4H_8 + M \qquad (5.82)$$

$$CH_3 + C_3H_5 + M \rightarrow C_4H_8 + M \qquad (5.83)$$

where the origin of the alkyl radicals explained in section 5.3.1(b); the vinyl radical (C_2H_3) is from the reaction H + C_2H_2 (5.72), and the propargyl radical (C_3H_3) is from the photolysis of C_3H_4 (5.59a). The C_3H_5 radical is produced by

$$C_3H_4 + H + M \rightarrow C_3H_5 + M \qquad (5.84)$$

$$C_3H_6 + h\nu \rightarrow C_3H_5 + H \qquad (5.85)$$

where C_3H_4 includes both CH_3C_2H and CH_2CCH_2. Photolysis transforms one C_4 compound into another C_4 compound:

$$1-C_4H_6 + h\nu \rightarrow C_4H_4 + 2H \qquad (5.86a)$$

$$1,2-C_4H_6 + h\nu \rightarrow C_4H_4 + 2H \qquad (5.87a)$$

$$C_4H_8 + h\nu \rightarrow 1,3-C_4H_6 + 2H \qquad (5.88a)$$

Loss of C_4 compounds occurs when the photolysis leads to the fragmentation of the C_4 species, as in the following examples:

$$C_4H_4 + h\nu \rightarrow 2C_2H_2 \qquad (5.89)$$

$$1-C_4H_6 + h\nu \rightarrow C_3H_3 + CH_3 \qquad (5.86b)$$

$$1,2-C_4H_6 + h\nu \rightarrow C_3H_3 + CH_3 \qquad (5.87b)$$

$$1,3-C_4H_6 + h\nu \rightarrow C_3H_3 + CH_3 \qquad (5.90)$$

$$C_4H_8 + h\nu \rightarrow C_3H_5 + CH_3 \qquad (5.88b)$$

where we have shown only the more important branches (for a complete listing of dissociation branches, see table 5.10). The dissociative losses are sufficiently fast that most C_4 species except butane are destroyed in the stratosphere before they can be transported to the lower atmosphere.

Besides the polyynes, the only species of complex hydrocarbon containing more than four carbon atoms considered in the model is benzene (C_6H_6), formed by

$$C_4H_5 + C_2H_2 \rightarrow C_6H_6 + H \qquad (5.91)$$

The primary source of the C_4H_5 radical is C_2H_2:

$$C_2H_2 + C_2H_3 + M \rightarrow C_4H_5 + M \tag{5.92}$$

The chemical scheme for synthesizing benzene from acetylene is

$$\begin{aligned}
C_2H_2 + H + M &\rightarrow C_2H_3 + M &(5.72)\\
C_2H_2 + C_2H_3 + M &\rightarrow C_4H_5 + M &(5.92)\\
\underline{C_4H_5 + C_2H_2} &\underline{\rightarrow C_6H_6 + H} &(5.91)\\
net \quad 3C_2H_2 &\rightarrow C_6H_6 &(XX)
\end{aligned}$$

The efficiency for forming benzene is reduced by the reaction with H_2:

$$C_4H_5 + H_2 \rightarrow 1-C_4H_6 + H \tag{5.93}$$

This reaction competes with (5.91) for the C_4H_5 radical. It is conceivable that synthesis reaction (5.91) could be accelerated if C_2H_2 were in an excited state or by ion reactions in auroral regions of the atmosphere. Benzene is a very stable molecule. Once formed, it cannot be chemically destroyed in the upper atmosphere. The ultimate sink is pyrolysis in the interior of the planet.

5.3.2 Modeling Results

The results of a model that includes the photochemistry outlined in section 5.3.1 are briefly described. The standard model is appropriate for the north equatorial belt (at a latitude of 10° N) of Jupiter. There are a total of 39 species and 194 reactions in the model, as listed in tables 5.9, 5.10, and 5.11, respectively. The set of continuity (5.10) and diffusion (5.7) equations are solved for all species. The eddy diffusivity profile is given in figure 5.11. The molecular diffusion coefficients are evaluated using standard kinetic theory. The lower boundary is at 6 bar, well below the tropopause region (0.1 bar), where most of the long-lived species and photochemical aerosols tend to accumulate. At the lower boundary the mixing ratios of H_2, He, and CH_4 are fixed at 0.89, 0.11, and 2.2×10^{-3}, respectively. All other species (i) are assumed to be lost from the lower boundary at the maximum deposition velocity w_i given by

$$w_i = -K \left(\frac{d\log n_i}{dz} + \frac{1}{H_a} \right) \tag{5.94}$$

where the number density (n_i), eddy diffusion coefficient (K), and mean scale height (H_a) all refer to 6 bar. The upper boundary is placed at 10^{-9} bar, well above the homopause region (10^{-6} bar) where most of the primary photochemistry occurs. We impose a downward flux of hydrogen atoms of 4×10^9 cm^{-2} s^{-1} from the dissociation of H_2 in the thermosphere. Zero flux condition is imposed for all other species.

(a) Photodissociation

Table 5.10 lists 78 photodissociation reactions, along with their rate coefficients at the upper boundary (1 nbar) and in the lower stratosphere (8.9 mbar). The solar flux used corresponds to solar maximum at a distance of 5.3 AU, conditions that are appropriate for the Voyager encounters.

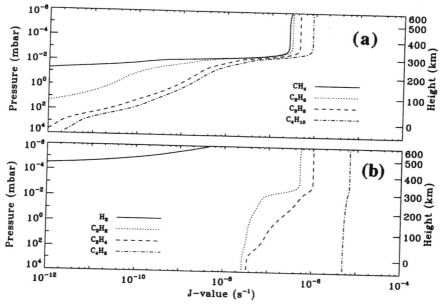

Figure 5.13 Photoabsorption rate coefficients for the major gases as a function of pressure and altitude in the standard NEB model atmosphere. (a) J values for the alkanes, CH_4, C_2H_6, C_3H_8, and C_4H_{10}. (b) J values for the species H_2, C_2H_2, C_2H_4, and C_4H_2. From Gladstone et al. 1996; cited in section 5.3.

Figure 5.13a shows the photodissociation coefficients (J) of the alkanes in the model. The J values are of the order of 10^{-6} s^{-1} above the homopause but decrease rapidly below the homopause, due to self-shielding in the case of CH_4 and due to partial shielding by CH_4 for the higher alkanes. As we show in section 5.3.2(b), when the photolytic process becomes inefficient, other processes such as transport would determine the distribution of the species. Figure 5.13b shows the J values for H_2, C_2H_2, C_2H_4, and C_4H_2. J_{H_2} drops rapidly near the upper boundary of the model due to self shielding. $J_{C_2H_2}$ is high above the homopause but decreases by an order of magnitude in the stratosphere due to shielding by CH_4. C_4H_2 absorbs at longer wavelengths, and the lack of shielding makes the J value nearly constant throughout the middle atmosphere. C_2H_4 is partially shielded by CH_4 and C_2H_2.

The vertical profiles of the J values of the higher hydrocarbons are not plotted. As can be seen from table 5.10, their values are generally high, in the range of 10^{-6} to 10^{-8}, with little attenuation between the mesosphere and the stratosphere. This accounts for the instability of C_3 and C_4 species in the model.

(b) C_1 and C_2 Species, and H

A schematic diagram showing the principal pathways by which the C_2 species are formed from the dissociation products of CH_4 is given in figure 5.14a. The mixing ratios of the stable C_1 and C_2 hydrocarbons are shown in figure 5.14b. The C_1 and C_2 radical species are given in figures 5.14c and d, respectively. CH_4 is the par-

164 Photochemistry of Planetary Atmospheres

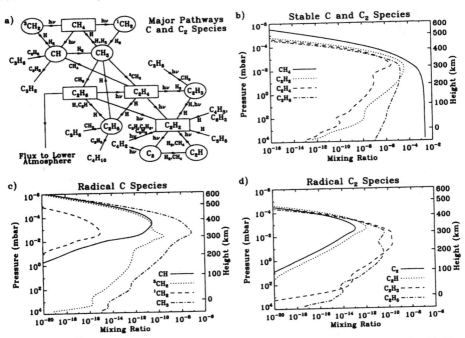

Figure 5.14 (a) A scheme for the major reactions among the C and C_2 compounds. (b) The abundances of stable C and C_2 species in the standard NEB model atmosphere as a function of pressure and altitude. (c) The abundances of radical C species. (d) The abundances of radical C_2 species. From Gladstone et al. 1996; cited in section 5.3.

ent molecule of all hydrocarbons in the model. Its mixing ratio is roughly constant throughout the middle atmosphere. Rapid destruction occurs around and above the homopause; the rate coefficient is given in figure 5.13a. The destruction of CH_4 gives rise to the production of radical species, shown in figure 5.14c. The peak concentrations of the radicals are located near the homopause, where maximum destruction of CH_4 occurs. As discussed in section 5.3.1, reactions between radicals and stable molecules and between radicals and radicals give rise to C_2 species C_2H_6, C_2H_4, and C_2H_2, as shown in figure 5.14b. The peak mixing ratios of the C_2 hydrocarbons occur near the homopause and decrease away from this region due to destruction. At the peak (5–10 μbar) the C_2H_2 mixing ratio of 40 ppm is the largest of all disequilibrium species in the model. In the stratosphere C_2H_6 is most abundant, followed by C_2H_2 and C_2H_4. The abundances correlate inversely with the rate coefficients for photolytic destruction (figure 5.13b). Photolysis of the C_2 hydrocarbons gives rise to C_2 radicals, shown in figure 5.14d.

As shown earlier, the initiation of all hydrocarbon chemistry is the breakup of the CH_4 molecule. The rates of the major reactions that break up CH_4 are given in figure 5.15a. The principal reactions include direct photolysis, R5–R7 (see table 5.10), photosensitized dissociation (sum of R153, R155, R180, and R189; see table 5.11), and reactions R120: $CH + CH_4$ and R83: $H + CH_4$. Figure 5.15a shows that the

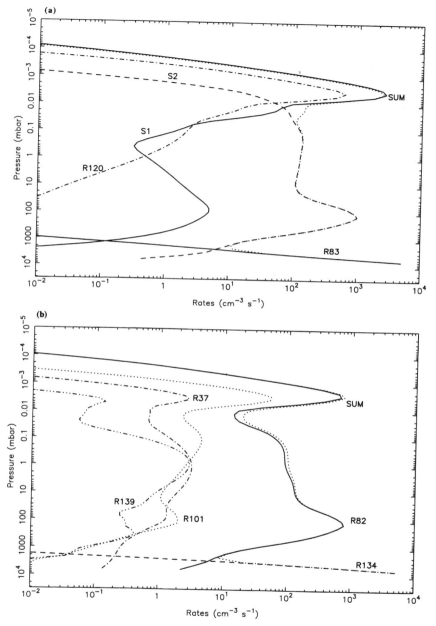

Figure 5.15 (a) Rates for major reactions destroying CH_4 in the model. S1 = R5 + R6 + R7: $CH_4 + h\nu$, R120: $CH + CH_4$, S2 = R153 + R155 + R180 + R189 = photosensitized dissociation (see section 5.3.1), R83: $H + CH_4$, SUM = summation of all reactions that destroy CH_4. (b) Rates for major reactions producing CH_4 in the model. R82: $CH_3 + H$, R37: $C_3H_8 + h\nu$, R101: $H + C_3H_5$, R139: $CH_3 + C_2H_5$, R134: $CH_3 + H_2$, SUM = summation of all reactions that produce CH_4. From Gladstone et al. 1996; cited in section 5.3.

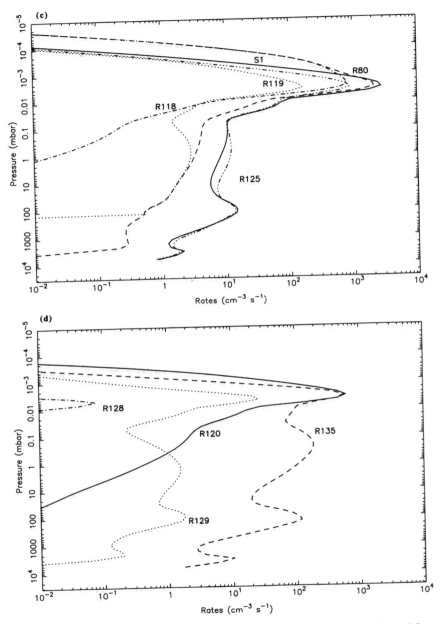

Figure 5.15 (c) Rates for major reactions that recycle C1 compounds. S1 = R5 + R6 + R7, R118: CH + H_2, R80: CH_2 + H (solid line), R119: CH + H_2, and R125: $^1CH_2$ + H_2. (d) Rates for major reactions that form the C_2 bond. R120: CH + CH_4, R135: CH_3 + CH_3, R128: $^3CH_2$ + $^3CH_2$, R129: $^3CH_2$ + CH_3. From Gladstone et al. 1996; cited in section 5.3.

Table 5.13 Column-integrated production rates, loss rates, column abundances, and fluxes of major hydrocarbon species in the model.

Species	Production rate ($cm^{-2} s^{-1}$)	Loss rate ($cm^{-2} s^{-1}$)	Column abundance (cm^{-2})	Flux ($cm^{-2} s^{-1}$)
H	5.3×10^{10}	5.7×10^{10}	9.3×10^{16}	-4.0×10^9
CH_4	1.1×10^{10}	1.8×10^{10}	1.5×10^{24}	7.2×10^9
C_2H_2	2.1×10^{10}	2.1×10^{10}	1.0×10^{18}	-9.7×10^3
C_2H_4	7.6×10^9	7.6×10^9	2.3×10^{16}	-3.3×10^3
C_2H_6	4.3×10^9	1.4×10^9	1.7×10^{20}	-2.9×10^9
C_3H_8	1.2×10^9	8.2×10^8	2.5×10^{19}	-4.3×10^8
C_4H_{10}	8.9×10^7	6.6×10^7	1.4×10^8	-2.3×10^7

From Gladstone et al. (1996), cited in section 5.3.
All fluxes refer to the lower boundary, except for the flux of H, which refers to the upper boundary.

primary photolysis of CH_4 is only a small fraction (20%) of the total destruction. Photosensitized dissociation is more important, especially in the stratosphere. (See table 5.13 for a listing of the column-integrated rates.) Not all the destruction of CH_4 produces higher hydrocarbons; part of the dissociation products is recycled back to CH_4. The major reactions that produce CH_4 in the model, R82: $CH_3 + H$ and R134: $CH_3 + H_2$, are shown in figure 5.15b. For comparison, we also show the sum of all reactions that destroy CH_4 as well as the sum of all reactions that restore CH_4. A detailed breakdown of the destruction and restoration rates of CH_4 is summarized in table 5.13. Only about 70% of the CH_4 destruction results in the synthesis of higher hydrocarbons; the rest is restored to CH_4. As can be seen in figure 5.15c, the C_1 radical species CH, $^1CH_2$, CH_2, and CH_3 play a fundamental role in the synthesis of hydrocarbons. The rates of fast recycling between the C_1 radicals are shown in figure 5.15c. The most important reactions included are CH to CH_2 via R118: $CH + H_2$, return of CH_2 to CH via R80: $CH_2 + H$, CH to CH_3 via R119: $CH + H_2$, and $^1CH_2$ to CH_3 via R125: $^1CH_2 + H_2$. Note that the production rate of CH by R80 is more than an order of magnitude greater than that of direct photolysis, R7: $CH_4 + h\nu$. Nearly all the $^1CH_2$ produced in R5: $CH_4 + h\nu$ is converted to CH_3 even though there is little or no production of CH_3 from direct photolysis. The production rates of C_2 compounds are determined by the rate-limiting reactions R120: $CH + CH_4$ and R135: $CH_3 + CH_3$, as shown in figure 5.15d. The principal reactions of destruction of the C_2 compounds are R87: $C_2H_5 + H$ and R122: $CH + C_2H_4$, as shown in figure 5.15d. The last reaction results in the production of a C_3 compound. The detailed production and destruction rates of the C_2 hydrocarbons are summarized in table 5.13.

The Lyman α brightness of Jupiter indicates that H atoms are abundant in the upper atmosphere of the planet. Radiative models that can simulate the observations suggest that the column abundance of H atoms is about 10^{17} cm^{-2}, consistent with the prediction of the present model. As shown later (in figure 5.17), the maximum mixing ratio above the homopause (10^{-3} mbar) approaches 1%, making it the most abundant radical in the model. A large number of reactions produce H atoms in the model; the rates of a few more important reactions are given in figure 5.16a, along with the total rate of production. As can be seen from figure 5.16a, the most important source is scheme (VI), driven by the photosensitized dissociation of H_2 by C_2H_2. The total

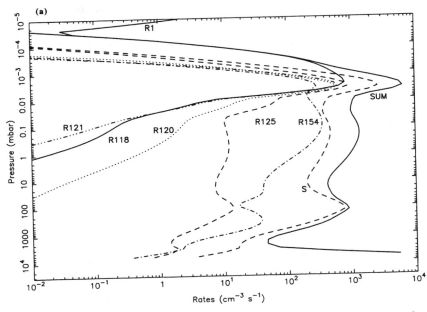

Figure 5.16 (a) Rates for major reactions producing H in the model. R1: $H_2 + h\nu$, R118: $CH + H_2$, R125: $^1CH_2 + H_2$, R154: $C_2H + H_2$, R120: $CH + CH_4$, R121: $CH + C_2H_2$, S = R4 + 2R6 + R7 + R8 + 2R12 + R13 + R16 + R38 + R75 (R4: $CH_3 + h\nu$, R6 + R7 + R8: $CH_4 + h\nu$, R12 + R13: $C_2H_4 + h\nu$, R16: $C_2H_6 + h\nu$, R38: $C_4H_2 + h\nu$, R75: $C_6H_2 + h\nu$), SUM = summation of all reactions that produce H. From Gladstone et al. 1996; cited in section 5.3.

column-integrated rate of H production is 5.8×10^{10} cm^{-2} s^{-1}, which is significantly larger than the downward flux of H atoms from the thermosphere, 4×10^9 cm^{-2} s^{-1}, and the direct photolysis of CH_4 (R4), 5.4×10^8 cm^{-2} s^{-1}. The rates of the most important reactions for removing H atoms in the model are given in figure 5.16b, along with the total removal rate. The removal of hydrogen is dominated by the catalytic cycles involving C_2H_2 and polyynes in scheme (XV) and its analogs. The detailed budget of the rates of H atom production and destruction is summarized in table 5.13.

According to the equations that determine the distribution of chemical species in the model, (5.7) and (5.10), the relative importance of chemistry and transport depends on the time constants. A convenient measure of the time constant due to molecular diffusion is H_i^2/D_i. The corresponding time constant of eddy diffusion is taken to be H_a^2/K. The quantities H_i, D_i, H_a, and K are as defined in (5.7). The chemical time constant is derived from the rate of chemical destruction of a species according to the reactions listed in table 5.11. The chemical loss time scales for selected species and the transport time scales for molecular diffusion and eddy diffusion are presented in figure 5.16c. Above the homopause atomic H is controlled by transport; all other species are chemically controlled. Below the homopause, H atoms are efficiently scavenged and the chemical lifetime becomes shorter than that of transport. The alkanes are long-lived with respect to transport throughout most of the middle atmosphere.

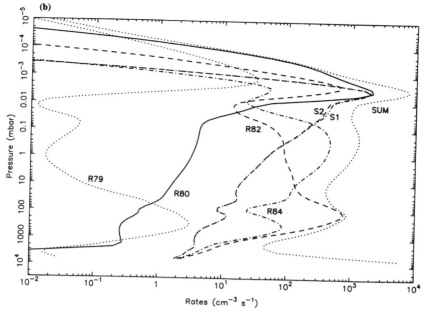

Figure 5.16 (b) Rates for major reactions removing H in the model. R79: H + H, R80: H + 3CH$_2$, R82: H + CH$_3$, R84: H + C$_2$H$_2$, S1 = R109 + R115 (R109: H + C$_4$H$_2$, R115: H + C$_6$H$_2$), S2 = R110 + R111 + R117 (R110: H + C$_4$H$_3$, R111 + R117: H + C$_6$H$_3$), SUM = summation of all reactions that remove H. From Gladstone et al. 1996; cited in section 5.3.

The chemical lifetimes of C$_3$H$_8$ and C$_4$H$_{10}$ are comparable to that of C$_2$H$_6$ and are not shown in figure 5.16c. The unsaturated species, C$_2$H$_4$ and C$_2$H$_2$, are short-lived with respect to transport. However, the destruction of acetylene is usually not terminal but results in recycling C$_2$H$_2$, as in chemical cycles (VI)–(VIII) and (XV). Hence, if we consider the extended family of C$_2$H$_2$ (C$_2$, C$_2$H, C$_2$H$_2$, C$_2$H$_3$) as a single species labeled "C$_2$H$_2$" in figure 5.16c, then the chemical lifetime is much longer than the transport, except in the lower stratosphere. The chemical lifetimes of other unsaturated species such as polyynes, CH$_3$C$_2$H, and C$_4$ species are all much shorter than the transport time in the stratosphere.

For the long-lived species transport is important. As shown in figure 5.16d, the flux of H atoms is from the lower thermosphere to the homopause region, where they are catalytically recombined. In the middle atmosphere, a major loss of the long-lived alkanes is by transport to the troposphere and then to the planetary interior. Figure 5.16d shows the fluxes of C$_2$H$_6$ and C$_3$H$_8$. More than half of the alkanes produced in the upper atmosphere are lost by transport to the lower atmosphere. The flux of C$_2$H$_2$, an unsaturated species, is much less, amounting to a few percent of the total production. Most of the unsaturated species are chemically destroyed in the upper atmosphere, and their downward fluxes across the tropopause are negligible.

The most abundant disequilibrium species computed in the model, H, C$_2$H$_2$, C$_2$H$_4$, C$_2$H$_6$, C$_3$H$_8$, and C$_4$H$_{10}$, are compared with observations in figure 5.17, which gives

170 Photochemistry of Planetary Atmospheres

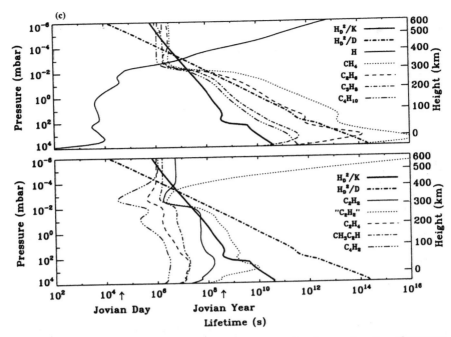

Figure 5.16 (c) Chemical loss time constants of the major species as a function of pressure and altitude in the model. Also shown for comparison are the time-constant profiles for transport by eddy diffusion (H_D^2/K) and by molecular diffusion for CH$_4$ (H_D^2/D). From Gladstone et al. 1996; cited in section 5.3.

the number densities of H atoms and the mixing ratios of the other species, along with available observations. To compare the model H atoms and the observations of Lyman α brightness, we need a radiative transfer calculation, described in section 5.3.2(c). Comparison between the model and observed C$_2$H$_2$, C$_2$H$_4$, and C$_2$H$_6$ species indicates good agreement. The model predictions are consistently higher than the observations in the upper atmosphere near the homopause.

(c) H and He Airglow

The observed airglow at the resonance line of hydrogen at Lyman α (1216 Å) and at the resonance line of helium at 584 Å imposes constraints on the abundances of H atoms and helium in the upper atmosphere. The Lyman α emission is extremely intense in the aurorae at the poles of the planet. In the equatorial region, there is a bulge in the emission that is locked to the rotation of the magnetic field. The nonaurora and nonbulge emission at the subsolar point observed by International Ultraviolet Explorer (IUE) telescope at the time of the Voyager encounter in 1979 was 11–13 kR. The disk-center He 584 Å brightness observed by Voyager UVS was about 4 R.

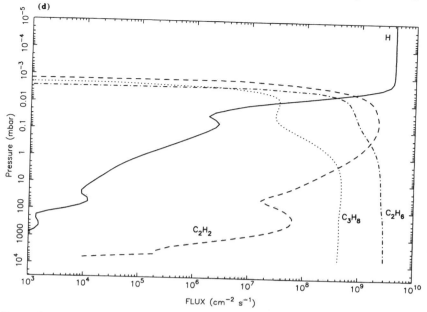

Figure 5.16 (d) Flux of H, and long-lived hydrocarbons: C_2H_6, C_2H_2, and C_3H_8. From Gladstone et al. 1996; cited in section 5.3.

The total column abundance of H atoms in the model is 6×10^{16} cm^{-2}. Only the H atoms that lie above the $\tau = 1$ level of CH_4 effectively scatter Lyman α photons; this column is 5.9×10^{16} cm^{-2}. Radiative transfer calculation yields a disk-center brightness of 8.6 kR, for an assumed incident solar Lyman α flux of 5.1×10^{11} photons cm^{-2} s^{-1} Å^{-1}, appropriate for the time of the observations. The corresponding calculation for He 584 Å emission yields a sub-solar brightness of 4.4 R, assuming a solar flux of 4.0×10^9 photons cm^{-2} s^{-1} at 1 AU, and a gaussian line with full-width half-maximum of 120 mÅ. Not all the Lyman α emission should be attributed to resonance scattering of sunlight. At least part of the emission may be attributed to the electroglow. We conclude that the model H and He concentrations in the standard north equatorial belt (NEB) model provide a satisfactory explanation of the airglow emissions.

(d) C_3 Species

A schematic diagram illustrating the principal sources, sinks, and transformations between the C_3 species is given in figure 5.18a. Figure 5.18b presents altitude profiles of the major C_3 species in the model, CH_3C_2H, CH_2CCH_2, C_3H_6, and C_3H_8. The most stable of the C_3 species is propane, with mixing ratio exceeding 10^{-8} throughout the stratosphere. The unsaturated species have high concentrations near the region of production but, because of their short lifetimes, are rapidly destroyed as they are

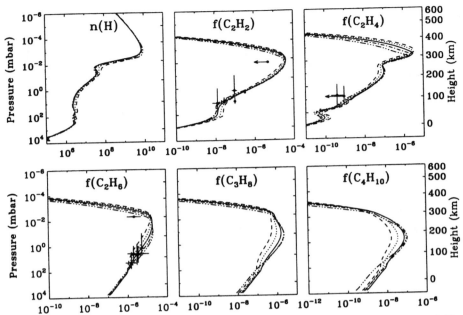

Figure 5.17 Comparison between observations and model values of H density and C_2H_2, C_2H_4, C_2H_6, C_3H_8, and C_4H_{10} mixing ratios. The standard model, represented by solid lines, uses values of $K_H = 1.4 \times 10^6$ cm^2 s^{-1} and $K_T = 3.9 \times 10^3$ cm^2 s^{-1}, with the eddy diffusion profile given by $K(z) = K_H[n_H/n(z)]^\gamma$, with $\gamma = 0.45$ and $n_H = 1.4 \times 10^{13}$ cm^{-3}. The dashed, dotted, dash-dot, and dash-dot-dot-dot lines show the profiles obtained with $\gamma = 0.35, 0.40, 0.50$, and 0.55, respectively. The values of K_H and K_T were held constant at the standard NEB values for these runs. Representative Voyager and ground-based measurements are indicated for C_2H_2, C_2H_4, and C_2H_6 (see section 5.x). From Gladstone et al. 1996; cited in section 5.3.

transported to the lower atmosphere. Between the homopause and the tropopause the drop in mixing ratios exceeds six orders of magnitude. The altitude profiles of C_3 radical species, C_3H_2, C_3H_3, C_3H_5, and C_3H_7, are presented in figure 5.18c. The maximum concentrations occur near the homopause and decrease rapidly away from this region.

The production rates of the major reactions that lead to the formation of the C_3 bond are shown in figure 5.19a, R121: $CH + C_2H_2$, R123: $CH + C_2H_6$, R138: $CH_3 + C_2H_3$, and R140: $CH_3 + C_2H_5$. The production is divided into two regimes: a peak production region near the homopause, driven by the primary photolysis of CH_4, and an extended production region throughout the bulk of the stratosphere, driven by the photosensitized dissociation of CH_4 catalyzed by polyynes. Figure 5.19b shows the rates of the main reactions that result in a net loss of C_3 species (i.e., to C_1, C_2, or C_4), R175: $C_3H_2 + C_2H_2$, R176: $C_3H_3 + C_2H_2$, R94: $CH_3C_2H + H$, and R31 + R32 + R33: $C_3H_6 + h\nu$. The reactions whose net results are the interconversion of C_3 species are not shown.

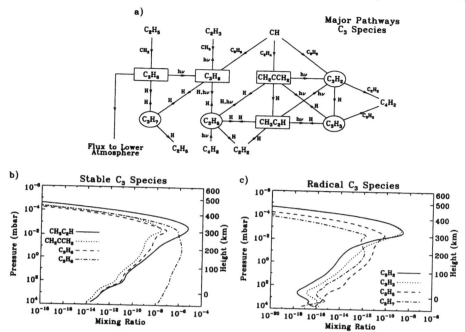

Figure 5.18 (a) A scheme for the major reactions among the C_3 compounds. (b) The abundances of stable C_3 species in the standard North Equatorial Belt (NEB) model atmosphere as a function of pressure and altitude. (c) The abundances of radical C_3 species. From Gladstone et al. 1996; cited in section 5.3.

(e) Polyynes

The chemistry of C_2H_2 has been discussed in section on 5.3.2(b). The polyynes are derived from the chemistry of acetylene. A schematic of the important reactions producing and destroying polyynes is shown in figure 5.20a. The mixing ratios of the major polyynes, C_4H_2, C_6H_2, and C_8H_2, are presented in figure 5.20b, which also includes C_2H_2 for comparison. The most abundant of the polyynes is diacetylene, with maximum mixing ratio of about 1 ppm at 4 μbar. The mixing ratios of the polyyne radical species, C_4H, C_4H_3, C_6H, and C_6H_3, are presented in figure 5.20c. All polyynes and their associated radicals are short-lived species. They are heavily concentrated near the homopause, the region of active photochemistry, but decrease rapidly in the lower part of the stratosphere. The rates of the important reactions that produce polyynes are shown in figure 5.21, R175: $C_3H_2 + C_2H_2$, R156: $C_2H + C_2H_2$, R181: $C_4H + C_2H_2$, and R190: $C_6H + C_2H_2$. Note that the most important source of diacetylene is via the C_3H_2 radical in schemes (XIII) and (XIV). The loss of a polyyne is primarily through photolysis and cracking by hydrogen. We show only the rates of destruction of C_4H_2 as an illustration of the loss mechanisms of a polyyne. Figure 5.21 shows the rates of the important reactions for removing C_4H_2 (by destroying the C_4 bond), R39 + R40: $C_4H_2 + h\nu$ and R110: $C_4H_3 + H$. These rates are an order of magnitude less than the cycling rates of C_4H_2, R109: $C_4H_2 + H$, as shown in

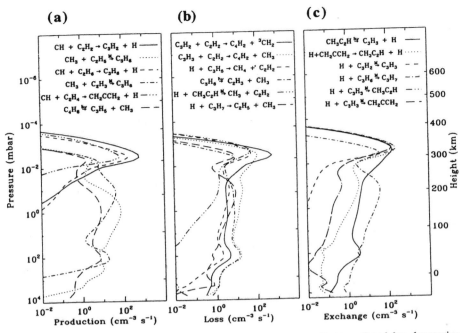

Figure 5.19 The rate profiles for the six most important reactions, ordered by decreasing column reaction rates, of the entire set of reactions leading to the production (a), loss (b), and exchange (c), of C_3 compounds in the standard NEB model. From Gladstone et al. 1996; cited in section 5.3.

figure 5.16b. Note the fundamental difference between the rates of destruction of C_4H_2 in figure 5.21, in which a C_4 bond is destroyed, and the apparent loss rate of C_4H_2 in R109, which, together with R110: $C_4H_3 + H$, is part of the catalytic cycle for removing H atoms.

(f) C_4 Species

A schematic diagram showing the principal sources, sinks, and transformations between the C_4 species is given in figure 5.22a. Figure 5.22b presents the altitude profiles of the major C_4 species in the model: C_4H_4, 1-C_4H_6, 1,2-C_4H_6, 1,3-C_4H_6, C_4H_8, and C_4H_{10}. Altitude profiles of the radicals C_4H_5 and C_4H_9 are presented in figure 5.22c. Butane is the only long-lived C_4 species, with mixing ratios between 10^{-7} and 10^{-8} in the stratosphere. All other C_4 species and radicals are short-lived, and their concentrations fall rapidly away from the region of active photochemistry. The rates of the principal reactions leading to the formation of the C_4 bond are shown in figure 5.21, R174: $C_2H_5 + C_2H_5$, R148: $CH_3 + C_2H_7$, R141 + R142: $CH_3 + C_2H_3$, R167: $C_2H_3 + C_2H_3$, R145: $C_3H_5 + CH_3$, and R171: $C_2H_5 + C_2H_3$. The major reactions that result in the destruction of the C_4 bond are photodissociations. The rates of the principal reactions that fragment the C_4 species are given in figure 5.21, R69–R74: $C_4H_{10} + h\nu$, R42: $C_4H_4 + h\nu$, R44–R48: 1-$C_4H_6 + h\nu$, and R62–R67: $C_4H_8 + h\nu$,

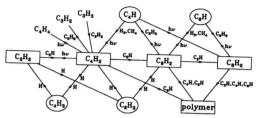
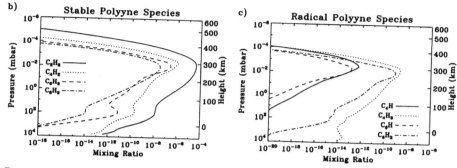

Figure 5.20 (a) A scheme for the major reactions among the polyyne compounds. (b) The abundances of stable polyyne species in the standard NEB model as a function of pressure and altitude. (c) The abundances of radical polyyne species. From Gladstone et al. 1996; cited in section 5.3.

where all the branches that destroy a C_4 molecule have been added. For C_4H_6, the dissociation of one isomer, 1-C_4H_6, is given in figure 5.21. The rates for the breakup of the other isomers are not shown.

(g) Column-Integrated Abundances and Fluxes

For all hydrocarbon species computed in the model, it is convenient to define column-integrated abundances of all computed species above the tropopause. The results are summarized in table 5.13. Since all higher hydrocarbons are derived from CH_4 and are ultimately returned to the lower atmosphere, the fluxes across the tropopause provide a check on the fate of the upward-flowing CH_4 and the downward-flowing hydrocarbons. These fluxes are summarized in table 5.13.

The total downward flux of all nonmethane hydrocarbons is 7.2×10^9 C-atoms cm^{-2} s^{-1} (see table 5.13 for relative contribution of various species). This must be balanced by an equivalent upward flux of CH_4. For comparison, we note that the total rate of destruction of CH_4 is 1.8×10^{10} cm^{-2} s^{-1}. Therefore, only about 40% of the CH_4 destruction results in the irreversible conversion to higher hydrocarbons in the upper atmosphere of Jupiter. The downward flux of higher hydrocarbons across the tropopause is a measure of the "carbon cycle" of Jupiter. The total planetary flux is about 2.5 Gt-C yr^{-1} (1 Gt = 1 gigaton = 10^{15} g).

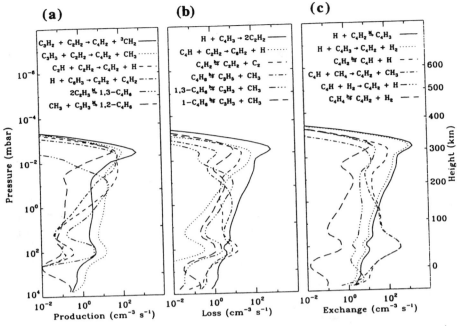

Figure 5.21 The rate profiles for the six most important reactions, ordered by decreasing column reaction rates, of the entire set of reactions leading to the production (a), loss (b), and exchange (c), of C_4 compounds in the standard NEB model. From Gladstone et al. 1996; cited in section 5.3.

5.3.3 Comparison of Giant Planets

Methane, the parent molecule of all hydrocarbons, and the most abundant photochemical products of methane, ethane and acetylene, have all been detected in the other giant planets, Saturn, Uranus, and Neptune. A comparison of the abundances of these species is given in table 5.14. For future reference, we have also included the concentrations of hydrocarbons in the atmosphere of Titan (see chapter 6). There is considerable range in the fractional abundances of CH_4 in these atmospheres, as a consequence of the capture efficiency of the low Z elements from the solar nebula at the time of planetary formation (see section 5.1.2). The concentrations of CH_4 for Uranus and Neptune shown in table 5.14 correspond to stratospheric values, which are lower than the tropospheric values due to condensation of CH_4 at the cold tropopause of these planets. The difference in methane in the upper atmosphere between Uranus and Neptune is attributed to convective penetration in Neptune (see section 5.1.2).

The question arises as to the causes of the variations in the observed abundances of C_2H_6 and C_2H_2, and in the ratio C_2H_6/C_2H_2. No systematic study of the hydrocarbon chemistry of the giant planets using the same set of chemical reactions has been carried out. After Jupiter, the only other atmosphere of the giant planets that has been studied to date using the new chemistry shown in table 5.11 is Neptune. One main difference between Neptune and Jupiter is the lower temperature in Neptune. Condensation of

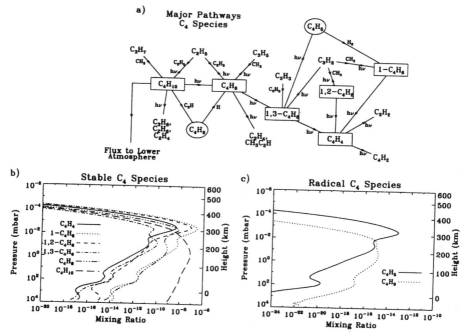

Figure 5.22 (a) A scheme for the major reactions among the C_4 compounds. (b) The abundances of stable C_4 species in the standard NEB model atmosphere as a function of pressure and altitude. (c) The abundances of radical C_4 species. From Gladstone et al. 1996; cited in section 5.3.

C_2H_6 and C_2H_2 plays a major role in Neptune, and this accounts for their lower concentrations. Based on our experience in the detailed modeling of the atmosphere of Jupiter, we have identified the following factors as important for controlling the concentrations of the hydrocarbons in the model: (a) temperature, (b) eddy diffusivity profile, (c) solar flux, (d) downward flux of H atoms at the upper boundary, and (e) condensation of hydrocarbons.

Photochemistry of hydrocarbons may be a source of the dark Axel-Danielson dust in the stratosphere of the giant planets and Titan. The detailed mechanism of polymerization is not known—it is likely that the pathway of formation is heterogeneous

Table 5.14 Comparison of major hydrocarbon species in the giant planets and their geometric albedos

Planet	CH_4	C_2H_6	C_2H_2	C_2H_6/C_2H_2	A
Jupiter	3.0×10^{-3}	5.8×10^{-6}	1.1×10^{-7}	53	0.25
Saturn	4.5×10^{-3}	7.0×10^{-6}	3.0×10^{-7}	23	0.3
Uranus	2.0×10^{-5}	$1-20 \times 10^{-9}$	1.0×10^{-8}	0.1-2	0.5
Neptune	$6-50 \times 10^{-4}$	1.5×10^{-6}	6.0×10^{-8}	25	0.5
Titan	$1-3 \times 10^{-2}$	2.0×10^{-5}	2.0×10^{-6}	10	0.054

chemistry of C_2H_2. The geometric albedos of the giant planets and Titan at 3075 Å listed in table 5.14 may provide a hint of this connection. A conservative purely gaseous atmosphere (scattering according to the Rayleigh phase function) would have a geometric albedo of 0.80. The observed ultraviolet albedos of the planetary bodies listed in table 5.14 are all much less than that of the nonabsorbing scattering atmosphere limit, due to the presence of the Axel-Danielson dust. Note the qualitative anticorrelation between the abundance of C_2H_2 and the geometric albedo of the planet, suggesting that C_2H_2 may be the precursor of the dark aerosols.

5.4 Nitrogen Chemistry

The only nitrogen species that have been detected in Jupiter are NH_3 and HCN. Ammonia is the thermodynamically stable form of nitrogen in the interior of the planet. Most of the active photochemistry of NH_3 takes place in the upper troposphere of Jupiter, resulting in the formation of hydrazine aerosols. Molecular nitrogen is a minor by-product. The origin of HCN is still debated. Thermodynamic equilibrium chemistry in the planetary interior is capable of producing the observed abundance of HCN. A photochemical source requires the coupled chemistry of NH_2 and hydrocarbon radicals. None of the photochemical products of ammonia, with the exception of HCN (which may or may not be derived from NH_3), has been detected. It is also not known whether the hydrazine aerosols have the same ultraviolet absorption properties as the Axel-Danielson dust.

5.4.1 Ammonia

The bulk of ultraviolet radiation below 1600 Å is absorbed by H_2, CH_4, C_2H_6 in the upper atmosphere. Only the radiation at longer wavelengths can reach the troposphere. Photolysis of ammonia at long wavelength yields

$$NH_3 + h\nu \rightarrow NH_2 + H \tag{5.95a}$$

$$\rightarrow NH(a^1\Delta) + H_2 \tag{5.95b}$$

The most likely fate of $NH(a^1\Delta)$ is to react with H_2:

$$NH(a^1\Delta) + H_2 \rightarrow NH_2 + H \tag{5.96}$$

The amine radical (NH_2) can react with H or with itself:

$$NH_2 + H + M \rightarrow NH_3 + M \tag{5.97}$$

$$NH_2 + NH_2 + M \rightarrow N_2H_4 + M \tag{5.98}$$

The first pathway results in the restoration of NH_3. The second pathway results in the formation of hydrazine (N_2H_4). At the low temperature in the upper troposphere of Jupiter, N_2H_4 can readily condense and may be a major constituent of the Axel-Danielson dust. The gaseous N_2H_4 may be photolyzed or attacked by H atoms:

$$N_2H_4 + h\nu \rightarrow N_2H_3 + H \tag{5.99}$$

$$N_2H_4 + H \rightarrow N_2H_3 + H_2 \tag{5.100}$$

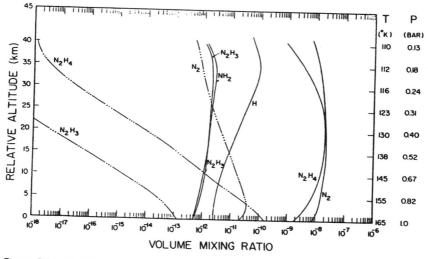

Figure 5.23 Distributions of the photochemical products of NH$_3$. Pressures and temperatures corresponding to the altitudes are shown on the right ordinate. N$_2$ distributions are shown for the case where N$_2$H$_4$ is enormously supersaturated (solid lines) and where the N$_2$H$_4$ mixing ratio is limited to its saturated value (dash-dot lines). After Atreya, S. K., Donahue, T. M., and Kuhn, W. R., 1977, "The Distribution of Ammonia and Its Photochemical Products on Jupiter." *Icarus* **31**, 348.

N$_2$H$_3$ can react further with H or self-react:

$$N_2H_3 + H \rightarrow N_2H_2 + H_2 \quad (5.101a)$$

$$\rightarrow 2NH_2 \quad (5.101b)$$

$$N_2H_3 + N_2H_3 \rightarrow N_2H_4 + N_2H_2 \quad (5.102a)$$

$$\rightarrow 2NH_3 + N_2 \quad (5.102b)$$

where (5.101a) and (5.102a) are the preferred channels. N$_2$H$_2$ is unstable and dissociates into N$_2$ and H$_2$:

$$N_2H_2 + h\nu \rightarrow N_2 + H_2 \quad (5.103)$$

Therefore, the ultimate fate of NH$_3$ in the atmosphere is to form aerosols of hydrazine, or N$_2$. Figure 5.23 shows the mixing ratios of the major nitrogen species in the atmosphere of Jupiter: NH$_3$, NH$_2$, N$_2$H$_4$, N$_2$H$_3$, N$_2$H$_2$, and N$_2$. This may not be the principal source of N$_2$ in Jupiter. The photochemical theory predicts a concentration of the order of 10^{-8}, and this is small compared with the value of the order of 10^{-6} predicted by equilibrium chemistry in the interior of Jupiter.

5.4.2 HCN

The detection of HCN in the atmosphere of Jupiter is not confirmed. An amount of HCN of the order of that reported for Jupiter could be produced from equilibrium chemistry in the planet's interior. Upward mixing would then bring it to the visible part of the atmosphere. There is an alternative way to produce HCN, by photochemistry. The obvious sources of carbon and nitrogen in the Jovian atmosphere are CH_4 and NH_3, respectively. However, as pointed out in sections 5.3 and 5.4.1, the primary photolysis of CH_4 and NH_3 occurs in very different regions of the atmosphere, with the former in the mesosphere and the latter in the troposphere. Since the photolysis of NH_3 is a source of hot H atoms, this is one way to break the CH_4 molecule in the lower atmosphere, as shown in the following:

$$NH_3 + h\nu \rightarrow NH_2 + H^* \quad (5.95a)$$

$$H^* + M \rightarrow H + M \quad (5.104)$$

$$H^* + CH_4 \rightarrow CH_3 + H_2 \quad (5.105)$$

where H^* is a hot hydrogen atom. Most of the 4H^* produced in (5.95a) is quenched by the ambient atmosphere (5.104), but enough survives to react with CH_4 (5.105). Photosensitized dissociation may also contribute to the production of CH_3 radicals (chemical scheme VII). The CH_3 radicals readily recombine with NH_2 radicals produced in (5.95a) and (5.96) to form methylamine:

$$CH_3 + NH_2 + M \rightarrow CH_3NH_2 + M \quad (5.106)$$

A minor branch (8%) of the dissociation of CH_3NH_2 yields HCN:

$$CH_3NH_2 + h\nu \rightarrow HCN + H_2 + 2H \quad (5.107)$$

An alternative photochemical scheme involves the interaction between the radicals NH_2 and C_2H_3:

$$NH_3 + h\nu \rightarrow NH_2 + H \quad (5.95a)$$

$$C_2H_2 + H + M \rightarrow C_2H_3 + M \quad (5.72)$$

$$NH_2 + C_2H_3 + M \rightarrow C_2H_5N + M \quad (5.108)$$

$$C_2H_5N + h\nu \rightarrow HCN + CH_3 + H \quad (5.109)$$

where there are four isomers of C_2H_5N: cyclic aziridine, vinylamine ($CH_2 = CH-NH_2$), ethylideneimine ($CH_3-CH = NH$), and N-methyleneimine ($CH_2 = N-CH_3$). The best known of the isomers is the first one, also known as ethyleneimine, with the following structure:

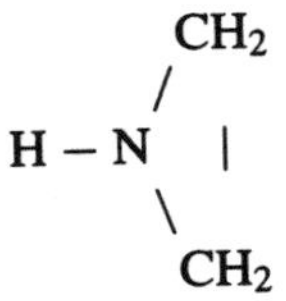

$$H-N \begin{array}{c} CH_2 \\ | \\ CH_2 \end{array}$$

Its photolysis is known to produce HCN. The results of a simple model that includes the above chemistry are shown in figure 5.24. It is clear that the photochemical model is capable of generating more than 1 ppb of HCN in the lower atmosphere.

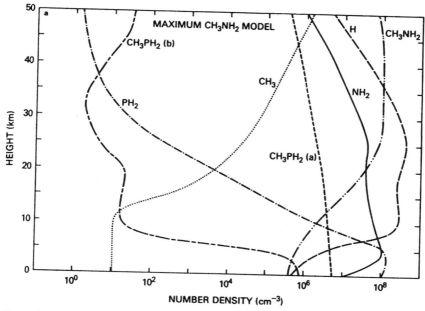

Figure 5.24 Number density of the indicated species as a function of altitude in the catalytic chemical model. For CH_3PH_2 two profiles are shown, for model a (no CH_3PH_2 loss) and for model b (loss of CH_3PH_2 by photolysis) maximum CH_3NH_2 model. After Kaye, J. A., and Strobel, D. F., 1983, "Formation amd Photochemistry of Methylamine in Jupiter's Atmosphere." *Icarus* **55**, 399.

5.4.3 Comparison of Giant Planets

Ammonia has been detected in the atmosphere of Saturn. The abundance is comparable to that in Jupiter. Ammonia has not been detected in the atmospheres of Uranus and Neptune, due to the condensation.

HCN has been detected in the atmosphere of Neptune but not in the atmosphere of any other planet. It is not entirely clear whether the origin of HCN is the interior of the planet or photochemical production in the stratosphere. Both theories have fundamental difficulties. The difficulty with the first theory is to find a mechanism for transporting HCN through the cold trap at the tropopause, where the allowed saturated vapor pressure of HCN is several orders of magnitude less than the observed concentration. The difficulty with the photochemical theory is to identify a sufficient source of nitrogen for synthesizing HCN. N_2 is an obvious candidate, but the theory would require an N_2 mole fraction of the order of 0.1%, making N_2 the dominant form of nitrogen in Neptune. Although this is in conflict with the predictions of equilibrium chemistry, the amount is about right to provide an alternative explanation of the unusually high He abundance observed in Neptune (table 5.4).

5.5 Phosphorus Chemistry

The photochemistry of phosphine is similar to that of ammonia. The absorption of ultraviolet radiation is followed by dissociation:

$$PH_3 + h\nu \rightarrow PH_2 + H \tag{5.110}$$

PH_3 can be further attacked by H atoms generated from PH_3 or NH_3 photolysis:

$$PH_3 + H \rightarrow PH_2 + H_2 \tag{5.111}$$

The PH_2 radicals recombine to form diphosphine (P_2H_4) and its subsequent chemistry follows a path similar to that for hydrazine:

$$PH_2 + PH_2 + M \rightarrow P_2H_4 + M \tag{5.112}$$

$$P_2H_4 + H \rightarrow P_2H_3 + H \tag{5.113}$$

$$P_2H_4 + PH_2 \rightarrow P_2H_3 + PH_3 \tag{5.114}$$

$$P_2H_3 + PH_2 \rightarrow P_2H_2 + PH_3 \tag{5.115}$$

$$P_2H_3 + P_2H_3 \rightarrow P_2H_2 + P_2H_4 \tag{5.116}$$

$$P_2H_2 + h\nu \rightarrow P_2 + H_2 \tag{5.117}$$

The production of P_2 as a result of the photochemistry of PH_3 gives rise to the possibility of making P_4 by recombination:

$$P_2 + P_2 + M \rightarrow P_4 + M \tag{5.118}$$

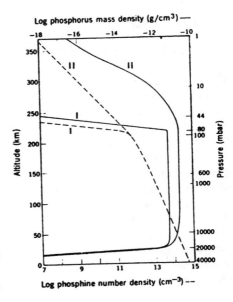

Figure 5.25 Mass density of red phosphorus particles and molecular number density of PH_3 gas as functions of altitude and pressure on Jupiter. Run I simulates average planetary conditions, and run II simulates the Great Red Spot, which we argue is a region of strong upward mixing. After Prinn, R. G., and Lewis, J. S., 1975, "Phosphine on Jupiter and Implications for the Great Red Spot." *Science* **190**, 294.

Since P_4 is a chromophore (red), it is a candidate for the material in the Great Red Spot of Jupiter. Figure 5.25 shows a model calculation of the important species associated with the photochemistry of PH_3 in the Jovian atmosphere.

Phosphine has been observed in the atmosphere of Saturn, with an abundance somewhat higher than that in the atmosphere of Jupiter. It has not been observed in the atmospheres of Uranus and Neptune, due to its condensation in these atmospheres.

5.6 Oxygen Chemistry

5.6.1 H_2O

Based on thermodynamic stability, H_2O is expected to be the most abundant oxygen-bearing molecule. Since the cosmic abundance of oxygen is 3.7×10^{-4} relative to H_2, this is the expected mole fraction of H_2O in Jupiter. The observed abundance suggests that water is depleted by a factor of 50 with respect to the cosmic abundance, a surprising result in view of the enhancements of the other heavy elements, C and N, over their cosmic abundances (see table 5.6). The current explanation is that H_2O is sequestered in water clouds in the deep atmosphere and is thus not detectable spectroscopically. To date, there is no observational proof of this interesting idea.

H_2O has not been detected in the other giant planets, due to condensation at the low temperatures.

5.6.2 CO in Jupiter

The detection of CO in the atmosphere of Jupiter initiated a lively debate on its origin. The atmosphere is dominated by H_2, a reducing gas. The presence of an oxygen-bearing molecule, even at the level of 1 ppb, is an indication of an unusual nonequilibrium activity in the planet. One possibility is an extraplanetary source, with the oxygen atom being derived from the ablation of micrometeoroids or the torus of Io. Subsequent photochemistry would convert the oxygen to CO. Another possibility is based on the equilibrium chemistry between H_2O and CH_4 in the interior of the planet, where the physical environment favors the production of CO. Vigorous mixing would then pump the interior CO to the upper atmosphere. Based on the present information, the interior source is considered a more plausible explanation of the observed CO in Jupiter. The photochemical theory of CO is not discussed here, but is deferred to chapter 6, where we develop this theory to explain the presence of CO in Titan's atmosphere.

The equilibrium theory is based on the reaction between H_2O and CH_4 in the planetary interior at high temperature and pressure:

$$H_2O + CH_4 \leftrightarrow CO + 3H_2 \tag{5.119}$$

A complete model including all the major hydrogen, carbon, and oxygen species for the interior of Jupiter is shown in figure 5.9. The model shows that for atmospheric composition that corresponds to cosmic abundances, the mixing ratio of CO could be as high as 10^{-6} in the planet's interior at 2000 K and a few kilobars. However, near the visible part of the planet (around 1 bar), the abundance of CO becomes negligible, falling to below 10^{-20}. Therefore, equilibrium chemistry alone could not account for the observed high abundance of CO. However, if we postulate that there is vigorous mixing between the deep atmosphere and the troposphere, then it is possible to bring the CO-rich air to where it is observable.

To "freeze" the composition of an air parcel as it is transported upward, it is important that the time constant for chemical removal be long compared with the transport time constant. For CO, the principal removal reactions are believed to be

$$CO + H_2 + M \rightarrow H_2CO + M \tag{5.120}$$

$$H_2CO + H_2 \rightarrow OH + CH_3 \tag{5.121}$$

$$H_2CO + H \rightarrow O + CH_3 \tag{5.122}$$

The laboratory and theoretical knowledge for these reactions is poor. A more likely path for the reduction of CO to CH_4 involves the formation of the methoxy radical (CH_3O):

$$CO + H + M \rightarrow HCO + M \tag{5.123}$$

$$HCO + HCO \rightarrow H_2CO + CO \tag{5.124}$$

$$H_2CO + H + M \rightarrow CH_3O + M \tag{5.125}$$

$$CH_3O + H \rightarrow CH_3 + OH \tag{5.126}$$

Once the strong C–O bond is broken, the oxygen that is released either as O or OH will readily react with H_2 to form H_2O. An estimate of the chemical time constant of CO may be made using reactions (5.123)–(5.126). The result is strongly dependent on the ambient temperature, and it is sometimes convenient to associate the time constant with a "quenching temperature." For CO in Jupiter the quenching temperature is around 1100 K.

The transport time constant may be parameterized as H^2/K, where K is the eddy diffusivity coefficient and H is the scale height. We do not know the correct value of K in the deep atmosphere. The theory of free convection suggests a value of 10^7 to 10^9 cm^2 s^{-1}. The mixing ratio of CO in the troposphere for various values of the mixing coefficient K is presented in figure 5.26. For very high values of K, in the range of 10^{15} cm^2 s^{-1} (clearly unrealistic), quenching occurs at pressures much greater than 1 kbar. The predicted abundance of CO exceeds 10^{-7}. Conversely, for very small values of K, of the order of 100 cm^2 s^{-1} (also clearly unrealistic), the expected CO abundance is less than 10^{-12}. K is most likely of the same magnitude as that predicted by free convection. In this case this simple and beautiful combination of thermodynamics and transport can provide a satisfactory account of the observed CO in Jupiter.

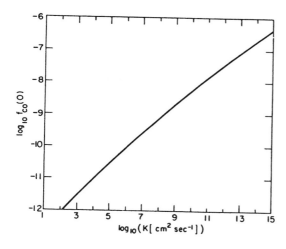

Figure 5.26 Values of the eddy diffusion coefficient K in the deep Jovian atmosphere required to give various values for the well-mixed CO mixing ratio in the observable portion of the atmosphere. After Prinn, R. G., and Barshay, S. S., 1977, "Carbon Monoxide on Jupiter and Implications for Atmospheric Convection." *Science* **198**, 1031.

5.6.3 CO in the Outer Solar System

CO has been identified in the atmospheres of Saturn and Neptune but not in the atmosphere of Uranus. The abundance of CO in Saturn is about the same as that in Jupiter and is most probably transported from the interior of the planet. However, there may be an extraplanetary contribution arising from the oxygen atoms derived from the erosion of ices in the Saturnian rings.

The observed mixing ratio of CO in Neptune is 1.2 ppm, about three orders of magnitude higher than those in Jupiter and Saturn. This large abundance rules out an extraplanetary explanation. An interior source also has its own difficulties. It requires an unusually large eddy diffusion coefficient or the assumption that the conversion of CO to CH_4 is less efficient than that given by the reactions (5.120)–(5.122); that is, the quenching temperature computed from these reactions is incorrect. A satisfactory resolution of this puzzle awaits in the future.

It remains a puzzle why the upper limit of CO in Uranus is 40 times less than the observed abundance of CO in Neptune. One main reason must be the fact that Uranus does not have an internal heat source, and hence there is less dynamical activity (see section 5.1.2 and table 5.5). A lower value of the eddy diffusion coefficient implies a lower quenching temperature and hence a lower abundance of CO in the upwelling air parcel.

CO is a ubiquitous molecule in the universe. It is abundant in the interstellar medium and is present in the atmosphere of the sun. In the solar system, the molecule has been detected in most planetary atmospheres, including the atmospheres of comets. CO is known to be present in all three terrestrial planets, Mars, Venus, and Earth. In the small bodies in the outer solar system CO has been detected in Titan, but not in Io, Triton, and Pluto, though CO ice has been identified on the surface of Triton. The reason for the ubiquity of CO is the great stability of the molecule and the abundance of the elements C and O.

5.7 Miscellaneous Topics

5.7.1 AsH$_3$ and GeH$_4$

Small concentrations (less than 1 ppb) of arsine (AsH$_3$) and germane (GeH$_4$) have been detected in the atmosphere of Jupiter but not in any other giant planet. The observed abundances are in agreement with thermodynamic models based on cosmic composition of the elements and upward mixing from the deep atmosphere.

5.7.2 Sulfur Species

Conspicuous by its absence is the detection of any sulfur species in the giant planets. A serious search was made for H$_2$S in the atmosphere of Jupiter, yielding an upper limit that is about 300 times less than the cosmic abundance. There is no reason to expect such a large depletion of sulfur in the giant planets during formation. We believe that most of the H$_2$S is sequestered in NH$_4$SH clouds.

5.7.3 Isotopomers

Deuterated hydrogen (HD) and deuterated methane (CH$_3$D) have been detected in all the giant planets. This provides valuable information on the D/H ratios in these planets. The D/H ratios in the atmospheres of Jupiter and Saturn are close to the cosmic abundance value of 1.6×10^{-5}. The D/H ratios in Uranus and Neptune show a significant enhancement over the cosmic abundance value. This could be explained as follows. There were two distinct reservoirs of deuterium in the solar nebula: gaseous HD and ices such as HDO or CH$_3$D. The former is characterized by a D/H ratio given by cosmic abundance. The latter has a higher D/H ratio, caused by a hitherto unidentified mechanism. The hydrogen and deuterium in the giant planets are derived from a combination of these two reservoirs. In Jupiter and Saturn the gaseous contribution dominates; in Uranus and Neptune the ices are an important component of the total budget. See section 4.10.1 for comparison of D/H values in the solar system and a discussion of the mechanisms of isotopic fractionation.

The ^{13}C and ^{15}N isotopic species have been detected in the atmosphere of Jupiter. The derived isotopic ratios, ^{13}C/^{12}C and ^{15}N/^{14}N, appear to be close to the cosmic abundance values, except for that inferred from ^{13}C$_2$H$_2$. This last anomaly may be due to the exceedingly low concentration of C$_2$H$_2$ in the atmosphere of Jupiter. In the other giant planets, only ^{13}CH$_4$ has been detected in the atmosphere of Saturn, and the inferred ^{13}C/^{12}C appears to be close to cosmic abundance.

5.7.4 Unsolved Problems

The vitality of a field is measured not only by the number of solved problems but also by the number of new questions that are raised in the investigations and the problems that remain to be solved. The following list of unsolved problems is given as a challenge to future investigators. Unless otherwise stated, the problems refer to all four giant planets.

(a) Energetics and Dynamics

1. What is the energy source for maintaining the high temperatures in the thermospheres?
2. What is the excitation mechanism for the excess ultraviolet emissions that cannot be accounted for by resonance fluorescence of sunlight?
3. Why are the internal heat fluxes of Uranus and Neptune so different for two planets that are so similar?
4. Why is the upper atmosphere of Uranus so quiescent, whereas the upper atmosphere of Neptune is vigorously mixed?

(b) Atmospheric Composition

5. What is the mechanism for producing the multiple layering in the lower ionosphere?
6. What is the reason for an apparent excess of He over the cosmic abundance value in Neptune?
7. Why are oxygen and sulfur compounds apparently deficient?
8. What are the precursor molecules of the Axel-Danielson dust in the stratosphere?
9. What is the chemical nature of the chromophores in the atmosphere?
10. Does ion chemistry driven by magnetospheric particles affect the global chemical composition?
11. What is the cause of the Lyman α bulge in Jupiter?

(c) Photochemistry and Kinetics

These two areas desperately need new input. In photochemistry we need measurements of the branching ratios and dissociation products as a function of wavelength. The measurements are more important near the thresholds of dissociation, where the cross sections are smaller but the solar flux is larger. In chemical kinetics a whole new class of experiments needs to be done at low pressure (1 mtorr to 1 torr) and low temperature (50–200 K).

12. What are the photodissociation products of CH_4 at Lyman α (1216 Å) and at longer wavelengths (1450 Å)?
13. What are the quantum yields and branching ratios of the photodissociation of C_2H_2 from Lyman α to threshold (around 2200 Å)?
14. What is the fate of excited C_2H_2 in the atmosphere? Does it take part in the polymerization of C_2H_2?
15. What are the cross sections and branching ratios of photodissociation of C_4H_2 and higher polyynes?
16. What are the rate coefficients of the reactions forming higher polyynes at low temperature? The reactions include R181: $C_4H + C_2H_2$ and R190: $C_6H + C_2H_2$ (see table 5.11).
17. What are the low-temperature rate coefficients for the CH insertion reactions: R120: $CH + CH_4$, R121: $CH + C_2H_2$, R122: $CH + C_2H_4$, R123: $CH + C_2H_6$?
18. What are the photodissociation cross sections, branching ratios, and dissociation products for methyl acetylene (CH_3C_2H), allene (CH_2CCH_2), and propene (C_3H_6) from Lyman α to threshold?

19. What are the low-temperature rate coefficients for the following reactions that are important in the synthesis of higher hydrocarbons: R135: $CH_3 + CH_3$, R138: $CH_3 + C_2H_3$, R141: $CH_3 + C_3H_3$, R143: $CH_3 + C_3H_5$, R167: $C_2H_3 + C_2H_3$, R171: $C_2H_3 + C_2H_5$?
20. What are the low-temperature rate coefficients for the following reactions that scavenge H atoms: R84: $H + C_2H_2$, R109: $H + C_4H_2$, R164: $C_2H_3 + H_2$?
21. What are the low-temperature rate coefficients for cracking of higher hydrocarbons by H atoms: R94: $H + CH_3C_2H$, R110: $C_4H_3 + H$?
22. What is the role of heterogeneous chemistry on the surface of the Axel-Danielson dust?
23. What are the optical properties of hydrazine after exposure to ultraviolet radiation?
24. What is the combined chemistry of CP and PN, which are chemical analogs of CN?
25. What are the photochemical products of AsH_3 and GeH_4 photochemistry?

6

Satellites and Pluto

6.1 Introduction

The presence of an atmosphere on a small planetary body the size of the Moon is surprising. Loss of material by escape would have depleted the atmosphere over the age of the solar system. Since these objects are not large enough to possess, or to sustain for long, a molten core, continued outgassing from the interior is not expected. However, it is now known that four small bodies in the outer solar system possess substantial atmospheres: Io, Titan, Triton, and Pluto. These atmospheres range from the very tenuous on Io (of the order of a nanobar) to the very massive on Titan (of the order of a bar). The atmospheric pressures on Triton and Pluto are of the order of 10 μbar. Perhaps the most interesting questions about these atmospheres concern their unusual origin and their chemical evolution.

Io is the innermost of the four Galilean satellites of Jupiter, the other three being Ganymede, Europa, and Callisto. All the Galilean moons are comparable in size, but there is no appreciable atmosphere on the other moons. The first indications that Io possesses an atmosphere came in 1974 with the discovery of sodium atoms surrounding the satellite and the detection of a well-developed ionosphere from the Pioneer 10 radio occultation experiment. The Voyager encounter in 1979 established the existence of active volcanoes as well as SO_2 gas. These are the only extraterrestrial active volcanoes discovered to date, and they owe their existence to a curious tidal heating mechanism associated with the 2:1 resonance between the orbits of Io and Europa. This tidal heating generates a total power of 10^{13} to 10^{14} W, a value that may be compared to the total geothermal heat flux of Earth, 3×10^{13} W. Io's heat is released over a surface area that is about 12 times smaller than that of the Earth. Hence, Io's internal heating per unit area is an order of magnitude larger than that of Earth.

The atmosphere of Io is continuously being eroded by bombardment by the energetic particles in the Jovian magnetosphere, but the atmosphere is resupplied by material of volcanic origin. The sputtered products escape into the magnetosphere of Jupiter and create an extended cloud of neutrals around Io and a torus of heavy ions in the orbit of Io. The major ions are oxygen and sulfur ions derived from the dissociation products of SO_2. In this chapter, we examine the bound atmosphere of Io: SO_2 photochemistry, the ionosphere, atmospheric sputtering, and the torus.

Titan, the largest satellite of Saturn, is known to have an atmosphere, since CH_4 lines were spectroscopically identified in its spectrum in 1944. Very little was known about Titan's atmosphere until the Voyager encounters, which revealed that Titan has a massive N_2 atmosphere, with surface pressure equal to 1.5 bar. A rich variety of organic molecules and extensive aerosol layers were found to be present in the atmosphere. Titan is believed to have formed in the Saturnian subnebula at the time of the formation of Saturn. Due to the lower temperatures in this region of the solar nebula, ices were common, and Titan could have accreted material that is rich in ices. As the atmospheric constituents are photochemically processed and converted into condensible material, the ices on the surface or outgassing from the interior must maintain the supply of gas to the atmosphere. The emphasis of this chapter is on the organic chemistry in Titan's atmosphere and its implications for evolution. The hydrocarbon chemistry on Titan provides an interesting comparison with that of the giant planets. Organic synthesis is greatly facilitated in this atmosphere due to the lower amount of the reducing gas H_2. This, together with the smaller size of the planetary body, results in a higher rate of chemical evolution.

In bulk composition the atmosphere of Titan is mildly reducing, with an oxidation state intermediate between that of the giant planets and the terrestrial planets. This chemical environment may be similar to that of Earth at the time of formation, an environment that is conducive to the synthesis of complex organic compounds that may lead to the spontaneous generation of life. Since there are no preserved records of the early chemical environment of the Earth, Titan offers an exciting analog of the prebiological Earth.

Triton, the largest satellite of Neptune, is believed to have formed from the icy debris in the outer part of the solar nebula. It was subsequently captured by Neptune, and this accounts for its unusual retrograde orbit. The bulk composition of Triton is interesting because it provides a sampling of the condensible species in the solar nebula such as CH_4, CO, CO_2, NH_3, N_2, and H_2O. These species are also present in molecular clouds, precursors of the solar nebula. Any major difference in the inventory of the major volatiles between the solar nebula and the molecular clouds would yield valuable clues on physical and chemical processing that must have occurred during the formation of the solar system. The atmosphere of Triton consists primarily of N_2, with trace concentrations of CH_4. CO and CO_2 have been detected in the ice on the surface. There is an upper limit of 1% for CO in the atmosphere based on a combination of observation and modeling. The vapor pressure of CO_2 is sufficiently low that the gas is negligible in the atmosphere. The photochemistry of N_2 and CH_4 in the atmosphere of Triton is similar to that of Titan, except that the temperature is much lower and the surface pressure of Triton corresponds to that of the mesosphere of Titan.

Pluto, the ninth planet of the solar system, has little in common with the other eight planets but seems to closely resemble Triton. Both are believed to be icy planetesimals formed between 30 and 50 AU in the solar nebula. We know much less about Pluto than Triton. The bulk atmosphere is believed to be composed of N_2, in vapor equilibrium with N_2 ice on the surface. CH_4 and CO ices have been identified in the reflection spectrum of Pluto, and we expect trace amounts of these gases to be present in the atmosphere. The photochemistry of the atmosphere of Pluto is expected to be similar to that of Triton, given our current lack of knowledge of Pluto.

An important aspect of the atmospheres of the small bodies is their physical extent. In our experience with the terrestrial planets and the giant planets, the altitude of the exobase of the atmosphere is usually small compared with the planetary radius. Thus, the plane parallel approximation is valid. However, in the case of the small bodies, this approximation is invalid. The extension of the atmosphere is comparable to the radius of the solid body. The equations of continuity must be formulated in spherical coordinates.

6.2 Io

6.2.1 Neutral Atmosphere

Pre-Voyager ground-based infrared reflectance measurements have shown the presence of SO_2 frost on the surface of Io. Gaseous SO_2 has been definitively identified on Io by the Voyager infrared radiometer, interferometer and spectrometer (IRIS) instrument. The original report of these data interpreted the observed gas as vapor in equilibrium with SO_2 frost on the surface at 130 K. However, the Voyager observations may be associated with volcanic plumes. Therefore, exactly how much SO_2 the atmosphere contains is an unresolved issue. The problem can best be appreciated by referring to figure 6.1a, which shows the pressure dependence of SO_2 on temperature. In regions of active volcanic activity the temperature may exceed 300 K; the corresponding SO_2 vapor is exceedingly high. At the subsolar point, the temperature could be as high as 130 K and the vapor pressure of SO_2 exceeds 10^{-7} bar. However, in the polar regions and on the night side, the temperatures drop below 90 K and the vapor pressure of SO_2 is less than 10^{-12} bar, which is close to the lower limit of a collisional atmosphere.

The simplest picture is that Io's atmosphere may be of a transient nature: thick and dense atmosphere near the volcanic plumes and at the subsolar point, gradually decaying away toward the poles and the nightside. The best observational evidence in support of this picture is the microwave observation. The experiment reported detection of $4-35 \times 10^{-9}$ bar of SO_2 covering 3–15% of the surface of Io, a result consistent with SO_2 being in equilibrium with the surface frost. There is circumstantial evidence for the presence of about 20×10^{-9} bar of a noncondensible gas, such as O_2 or SO, in the atmosphere. O_2 has been postulated to impede the lateral flow of SO_2 away from the subsolar point and to account for the warm exosphere of Io. There is no direct observation to support this hypothesis. Two models of Io's atmosphere, a high-density and a low-density model, are shown in figure 6.1b.

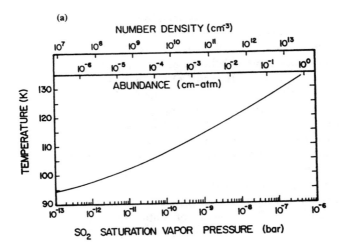
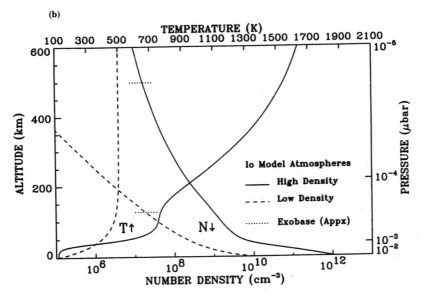

Figure 6.1 (a) SO$_2$ saturation vapor pressure as a function of temperature. The corresponding surface number density and column number density are indicated on the upper abscissae. After Kumar, S., and Hunten, D. M., 1982, *The Satellites of Jupiter* (Tucson: University Arizona Press), p. 782. (b) Model atmosphere of Io, showing the SO$_2$ number density (N) and temperature (T) for both the high-density and low-density cases. The approximate altitude location of the exobase is shown by short horizontal dotted lines. The pressure scale is for the high-density model only. After Summers, M. E., and Strobel, D. F., 1996, "Photochemistry and Vertical Transport in Io's Atmosphere and Ionosphere." *Icarus* **120**, 290–316.

The photochemistry of SO_2 in the atmosphere is initiated by the photolysis of SO_2:

$$SO_2 + h\nu \rightarrow SO + O \qquad (6.1a)$$
$$\rightarrow S + O_2 \qquad (6.1b)$$

where the thresholds of reactions (6.1a) and (6.1b) are 2170 and 2084 Å, respectively. The products can undergo further dissociation:

$$SO + h\nu \rightarrow S + O \qquad (6.2)$$
$$O_2 + h\nu \rightarrow O + O \qquad (6.3)$$

Some of these atoms can diffuse to the exosphere, where they are lost due to sputtering by the magnetospheric energetic particles. Recombination of the dissociation products can also occur:

$$SO + O + M \rightarrow SO_2 + M \qquad (6.4)$$
$$O + O + M \rightarrow O_2 + M \qquad (6.5)$$

where the third body (M) is either the ambient atmosphere of SO_2 or the surface. Formation of more complex compounds can readily occur, as in the following examples:

$$SO_2 + O + M \rightarrow SO_3 + M \qquad (6.6)$$
$$O_2 + O + M \rightarrow O_3 + M \qquad (6.7)$$

But these compounds are unstable in the atmosphere of Io and are readily removed by reactions such as

$$SO_3 + SO \rightarrow 2SO_2 \qquad (6.8)$$
$$O_3 + h\nu \rightarrow O_2 + O \qquad (6.9)$$

A simple reaction set describing the photochemistry of SO_2 is given in table 6.1. The main uncertainties are the chemical kinetics at low temperature near the surface and the fate of the dissociation products at the surface. We adopt the "reasonable" assumption that S atoms stick to the surface but O atoms recombine to form O_2 and are released back to the atmosphere. The results of a representative model are given in figure 6.2a for neutral species. SO_2 is the most abundant molecule in the atmosphere near the lower boundary. At higher altitudes O and S atoms become the dominant species. The photochemical products SO and O_2 are next in abundance. The abundance of O_2 is small compared with the other dissociation products primarily because of the reaction that removes O_2 rapidly via

$$O_2 + S \rightarrow SO + O \qquad (6.10)$$

where the S atoms are derived from (6.1b) and (6.2).

Since the atmosphere is so thin, ternary reactions that require collisional stabilization are too slow in the gas phase. However, the surface may serve as a good "sponge layer" for the atoms to react to form stable molecules. The photochemical model predicts a high abundance of SO, formed mainly from SO_2 photolysis and by recombination of S and O atoms on the surface of Io. Since SO_2 condenses on the

Table 6.1 List of essential reactions for the neutral atmosphere of Io with their preferred rate coefficients

	Reaction		Rate coefficient[a]	Reference
R1a	$SO_2 + h\nu$	$\rightarrow SO + O$	1.0×10^{-5}	Okabe (1971), Welge (1984)
b		$\rightarrow S + O_2$	6.3×10^{-7}	Driscoll & Warneck (1968)
R2	$SO + h\nu$	$\rightarrow S + O$	1.8×10^{-5}	Phillips (1981)
R3a	$O_2 + h\nu$	$\rightarrow O + O$	3.9×10^{-10}	Hudson (1971)
b		$\rightarrow O(^1D) + O$	9.4×10^{-8}	
R4	$S_2 + h\nu$	$\rightarrow S + S$	9.2×10^{-5}	Brewer & Brabson (1966)
R5	$Na_2 + h\nu$	$\rightarrow Na + Na$	3.0×10^{-4}	
R6a	$NaO_2 + h\nu$	$\rightarrow NaO + O$	9.0×10^{-5}	Plane (1989)
b		$\rightarrow Na + O_2$	1.0×10^{-5}	0.1 branching assumed
R7a	$NaS_2 + h\nu$	$\rightarrow NaS + S$	9.0×10^{-5}	same as $NaO_2 + h\nu$
b		$\rightarrow Na + S_2$	1.0×10^{-5}	0.1 branching assumed
R8	$Na_2O + h\nu$	$\rightarrow NaO + Na$	1.0×10^{-5}	assumed
R9	$Na_2S + h\nu$	$\rightarrow NaS + Na$	1.0×10^{-5}	assumed
R10	$S + O_2$	$\rightarrow SO + O$	2.3×10^{-12}	JPL92
R11	$SO + SO$	$\rightarrow SO_2 + S$	$5.8 \times 10^{-12} e^{-1760/T}$	HH92
R12	$SO + O_2$	$\rightarrow SO_2 + O$	$2.6 \times 10^{-13} e^{-2400/T}$	JPL92
R13	$O + S_2$	$\rightarrow SO + S$	$2.2 \times 10^{-11} e^{-84/T}$	Yung & DeMore (1982)
R14	$SO + SO_3$	$\rightarrow 2SO_2$	2.5×10^{-15}	Yung & DeMore (1982)
R15	$O + O + M$	$\rightarrow O_2 + M$	$1.0 \times 10^{-26} T^{-2.9}$	NIST (1994)
R16	$SO + O + M$	$\rightarrow SO_2 + M$	7.7×10^{-31}	Yung & DeMore (1982)
R17	$SO_2 + O + M$	$\rightarrow SO_3 + M$	$3.4 \times 10^{-32} e^{-1120/T}$	Yung & DeMore (1982)
R18	$S + S + M$	$\rightarrow S_2 + M$	$1.0 \times 10^{-26} T^{-2.9}$	Baulch & Drsydale (1973)
R19	$NaO + O$	$\rightarrow Na + O_2$	3.7×10^{-10}	JPL92
R20	$NaO_2 + O$	$\rightarrow NaO + O_2$	5.0×10^{-13}	Plane (1989)
R21a	$Na_2O + O$	$\rightarrow 2NaO$	1.0×10^{-12}	assumed
b		$\rightarrow Na_2 + O_2$	$\phi_{Na_2} k_{19a}$	
R22a	$Na_2O + S$	$\rightarrow NaO + NaS$	1.0×10^{-12}	assumed
b		$\rightarrow Na_2 + SO$	$\phi_{Na_2} k_{20a}$	
R23	$Na + O + M$	$\rightarrow NaO + M$	1.0×10^{-33}	assumed
R24	$Na + O_2 + M$	$\rightarrow NaO_2 + M$	$2.2 \times 10^{-27} T^{-1.2}$	JPL92
R25	$NaO + O_2 + M$	$\rightarrow NaO_3 + M$	$3.1 \times 10^{-25} T^{-2.0}$	JPL92
R26	$NaS + O$	$\rightarrow Na + SO$	3.7×10^{-10}	same as R19
R27	$NaS_2 + O$	$\rightarrow NaS + SO$	5.0×10^{-13}	same as R20
R28a	$Na_2S + O$	$\rightarrow NaS + NaO$	1.0×10^{-12}	assumed
b		$\rightarrow Na_2 + SO$	$\phi_{Na_2} k_{26a}$	
R29a	$Na_2S + S$	$\rightarrow 2NaS$	1.0×10^{-12}	assumed
b		$\rightarrow Na_2 + S_2$	$\phi_{Na_2} k_{27a}$	
R30	$Na_2 + O$	$\rightarrow NaO + Na$	5.0×10^{-10}	assumed
R31	$Na_2 + S$	$\rightarrow NaS + Na$	5.0×10^{-10}	

[a] Photodissociation coefficients for zero optical depth, hemispheric average, with units of s^{-1}. Bimolecular and termolecular rate coefficients have units of $cm^3 s^{-1}$ and $cm^6 s^{-1}$, respectively.

night side and the polar regions of Io and SO does not, it is possible that SO forms a residual atmosphere on Io away from the sunlight side and in the polar regions. The predicted pressure is of the order of a nanobar and may serve to buffer the atmosphere against supersonic winds of SO_2 that would otherwise be blowing from the dayside to the nightside.

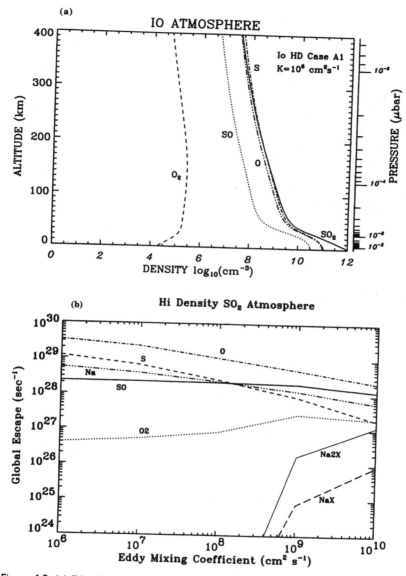

Figure 6.2 (a) Distribution of major neutral constituents in the atmosphere of Io in the high-density model. The units are molecules/second. (b) Escape rates for major species from the atmosphere of Io in the high-density model. After Summers, M. E., and Strobel, D. F., 1996, "Photochemistry and Vertical Transport in Io's Atmosphere and Ionosphere." *Icarus* **120**, 290–316.

Oxygen atoms are the most abundant species at high altitudes, and their concentration ultimately determines the level of the exobase. Oxygen and sulfur atoms as well as atoms of alkali metal readily escape from Io. The neutral atoms get ionized and form a torus around Jupiter in the orbit of Io. The chemical composition and the dynamics of the torus are discussed in section 6.2.3.

Escape rates for major species from the atmosphere of Io are summarized in figure 6.2b. One consequence of the higher rate of escape of O relative to S is the deposition of S atoms on the surface. This depositional rate of sulfur is of the order of 10^{11} atoms cm^{-2} s^{-1}. This is equivalent to 10^{-5} to 10^{-4} cm yr^{-1}, or 0.5–5 km of sulfur over the age of the solar system. The total rate of mass loss from Io is of the order of 3×10^{28} atoms s^{-1}, or 1000 kg s^{-1}. This is an interesting example of the profound impact that atmospheric photochemistry can have on the evolution of the surface of a planetary body. This result may be compared with the overall resurfacing rate of 10^{-3} to 10 cm yr^{-1} deduced from the absence of craters on the surface of Io. The loss of material to the torus is thus a minor fraction of the average volcanic output over geologic time.

6.2.2 Ionosphere

A well-developed ionosphere of Io was detected by the radio occultation experiment on Pioneer 10 in 1974 at a local time of 5:24 p.m. The corresponding solar zenith angle equals 81°. The most likely ions that can be formed in the atmosphere of Io are SO_2^+, SO^+, O^+, and S^+:

$$SO_2 + h\nu \rightarrow SO_2^+ + e \qquad (6.11)$$

$$SO + h\nu \rightarrow SO^+ + e \qquad (6.12)$$

$$O + h\nu \rightarrow O^+ + e \qquad (6.13)$$

$$S + h\nu \rightarrow S^+ + e \qquad (6.14)$$

Of these ions, SO^+ has the lowest ionization potential, 10.2 eV, which corresponds to a photon with wavelength 1215 Å. Hence, SO_2^+, O^+, and S^+ produced in reactions (6.11), (6.13), and (6.14) will all rapidly undergo charge transfer to form SO^+:

$$SO_2^+ + O \rightarrow SO^+ + O_2 \qquad (6.15)$$

$$SO_2^+ + SO \rightarrow SO^+ + SO_2 \qquad (6.16)$$

$$O^+ + SO_2 \rightarrow SO^+ + O_2 \qquad (6.17a)$$

$$\rightarrow O_2^+ + SO \qquad (6.17b)$$

$$O_2^+ + S \rightarrow SO^+ + O \qquad (6.18)$$

$$S^+ + O_2 \rightarrow SO^+ + O \qquad (6.19)$$

$$S^+ + SO \rightarrow SO^+ + S \qquad (6.20)$$

The ultimate fate of SO^+ in the ionosphere of Io is dissociative recombination,

$$SO^+ + e \rightarrow S + O \qquad (6.21)$$

Table 6.2 List of essential reactions for the ionosphere of Io with their preferred rate coefficients.

	Reaction			Rate coefficient[a]	Reference[b]
R32	$SO_2 + h\nu$	$\rightarrow$	$SO_2^+ + e$	4.2×10^{-8}	Wu & Judge (1981)
R33	$O_2 + h\nu$	$\rightarrow$	$O_2^+ + e$	1.8×10^{-8}	
R34	$SO + h\nu$	$\rightarrow$	$SO^+ + e$	1.8×10^{-8}	
R35	$O + h\nu$	$\rightarrow$	$O^+ + e$	7.6×10^{-9}	McGuire (1968)
R36	$S + h\nu$	$\rightarrow$	$S^+ + e$	7.6×10^{-9}	McGuire (1968)
R37	$Na + h\nu$	$\rightarrow$	$Na^+ + e$	3.6×10^{-7}	McGuire (1968)
R38	$O^+ + SO_2$	$\rightarrow$	$O_2^+ + SO$	8.0×10^{-10}	AH86
R39	$O^+ + SO$	$\rightarrow$	$SO^+ + O$	5.0×10^{-10}	assumed
R40	$O^+ + O_2$	$\rightarrow$	$O_2^+ + O$	1.1×10^{-10}	A93
R41	$O^+ + S$	$\rightarrow$	$S^+ + O$	5.0×10^{-10}	assumed
R42	$O^+ + Na$	$\rightarrow$	$Na^+ + O$	5.0×10^{-10}	assumed
R43	$S^+ + SO$	$\rightarrow$	$SO^+ + S$	1.0×10^{-9}	assumed
R44	$S^+ + O_2$	$\rightarrow$	$SO^+ + O$	2.3×10^{-11}	AH86
R45	$S^+ + Na$	$\rightarrow$	$Na^+ + S$	5.0×10^{-10}	assumed
R46	$SO^+ + SO$	$\rightarrow$	$S^+ + SO_2$	1.0×10^{-11}	assumed
R47	$SO^+ + Na$	$\rightarrow$	$Na^+ + SO$	5.0×10^{-10}	assumed
R48	$O_2^+ + S$	$\rightarrow$	$S^+ + O_2$	5.0×10^{-10}	assumed
R49	$O_2^+ + SO$	$\rightarrow$	$SO^+ + O_2$	1.0×10^{-10}	assumed
R50a	$O_2^+ + Na$	$\rightarrow$	$Na^+ + O_2$	6.3×10^{-10}	AH86
b		$\rightarrow$	$NaO^+ + O$	7.1×10^{-11}	
R51a	$SO_2^+ + S$	$\rightarrow$	$SO^+ + O_2$	5.0×10^{-10}	assumed
b		$\rightarrow$	$S^+ + SO_2$	1.0×10^{-10}	
R52	$SO_2^+ + SO$	$\rightarrow$	$SO^+ + SO_2$	5.0×10^{-10}	assumed
R53	$SO_2^+ + O$	$\rightarrow$	$SO^+ + O_2$	1.0×10^{-10}	assumed
R54	$SO_2^+ + O_2$	$\rightarrow$	$O_2^+ + SO_2$	2.5×10^{-10}	AH86
R55a	$SO_2^+ + Na$	$\rightarrow$	$Na^+ + SO_2$	5.0×10^{-10}	assumed
b		$\rightarrow$	$NaO^+ + SO$	5.0×10^{-10}	assumed
R56	$Na^+ + Na_2$	$\rightarrow$	$Na_2^+ + Na$	5.0×10^{-10}	assumed
R57	$Na^+ + Na_2O$	$\rightarrow$	$Na_2O^+ + Na$	1.0×10^{-9}	assumed
R58	$Na^+ + Na_2S$	$\rightarrow$	$Na_2S^+ + Na$	1.0×10^{-9}	assumed
R59	$Na^+ + e$	$\rightarrow$	Na	$2.7 \times 10^{-12} (\frac{300}{T})^{.69}$	PH80
R60	$NaO^+ + e$	$\rightarrow$	$Na + O$	$2.0 \times 10^{-7} (\frac{300}{T})^{.50}$	assumed
R61	$S^+ + e$	$\rightarrow$	S	$3.9 \times 10^{-12} (\frac{300}{T})^{.63}$	PH80
R62	$O^+ + e$	$\rightarrow$	O	$3.9 \times 10^{-12} (\frac{300}{T})^{.63}$	assumed
R63	$O_2^+ + e$	$\rightarrow$	$O + O$	$2.0 \times 10^{-7} (\frac{300}{T})^{.50}$	assumed
R64	$SO^+ + e$	$\rightarrow$	$S + O$	$2.0 \times 10^{-7} (\frac{300}{T})^{.50}$	assumed
R65	$SO_2^+ + e$	$\rightarrow$	$SO + O$	$3.0 \times 10^{-7} (\frac{300}{T})^{.50}$	assumed
R66	$Na_2^+ + e$	$\rightarrow$	$Na + Na$	$3.0 \times 10^{-7} (\frac{300}{T})^{.50}$	assumed
R67	$Na_2O^+ + e$	$\rightarrow$	$NaO + Na$	$2.0 \times 10^{-7} (\frac{300}{T})^{.50}$	assumed
R68	$Na_2S^+ + e$	$\rightarrow$	$NaS + Na$	$2.0 \times 10^{-7} (\frac{300}{T})^{.50}$	assumed

[a] Photoionization coefficients for zero optical depth, hemispheric average, with units of s^{-1}. Ion-molecule rate coefficients have units of $cm^3 \ s^{-1}$.
[b] AH86 is Anicich and Huntress (1986), PH80 is Prasad and Huntress (1980), A93 is Anicich (1993).

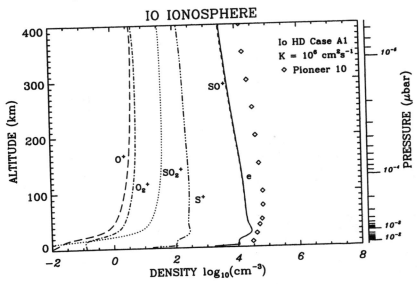

Figure 6.3 Major ions and electron number density. The data (diamonds) were taken by Pioneer 10 radio occultation on the dayside. After Summers, M. E., and Strobel, D. F., 1996, "Photochemistry and Vertical Transport in Io's Atmosphere and Ionosphere." *Icarus* **120**, 290–316.

or charge transfer to an alkali metal,

$$SO^+ + Na \rightarrow Na^+ + SO \qquad (6.22a)$$
$$\rightarrow NaO^+ + S \qquad (6.22b)$$

A summary of the important ion reactions in the atmosphere of Io is given in table 6.2. The results of a simple photochemical model incorporating these reactions are summarized in figure 6.3. SO^+ is the major ion with peak concentrations of the order of 5×10^4 cm^{-3}, followed by smaller amounts of S^+ and SO_2^+.

6.2.3 Torus

Even before the Voyager encounters, optical emissions of Na and K were discovered in a tenuous cloud surrounding Io, and S^+ ions were observed in a torus in the orbit of Io around Jupiter. The Voyager spacecraft discovered a spectacular torus of heavy ions that radiate profusely in the ultraviolet. Figure 6.4 presents the ultraviolet spectrum obtained by the Voyager ultraviolet spectrometer (UVS). The strongest ultraviolet features have been identified and are due to doubly and triply ionized atoms of O and S. The total radiation power of the torus is $3-6 \times 10^{12}$ W.

The theory of the torus is simple in concept, though the details are rather intricate. The escape velocity for a particle to leave from the surface Io is 2.65 km s^{-1}, but the velocity is even smaller if the escape is from the exosphere, which may be located at

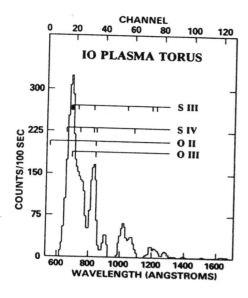

Figure 6.4 Extreme ultraviolet spectrum of the plasma torus of Jupiter observed during the Voyager 1 encounter. Multiplets of identified species are indicated. After Broadfoot, A. L. et al., 1979, *Science* **204**, 979–982.

a fraction of Io's radius above the surface. The atoms that escape from Io will orbit around Jupiter in a torus because the velocity to escape from Jupiter at the orbit of Io is 24 km s^{-1}. These atoms eventually get ionized and corotate with the magnetosphere of Jupiter, forming a plasma of heavy ions. Voyager observations of the shape of the ultraviolet torus are shown in figure 6.5a.

It is now known that the torus of Io consists of three components. First there is the neutral cloud of Na surrounding Io. This cloud also contains other neutral atoms such as K, O, and S. Due to the short lifetime (of the order of hours) of the neutral atoms against ionization, the neutral cloud does not extend significantly beyond Io. The ionized torus consists of a cold plasma confined to the orbit of Io and a hot plasma that extends beyond the orbit of Io. The mean temperatures of the cold and hot plasma are a few eV (10^4 K) and 80 eV (10^6 K), respectively. The total number density of electrons (equal to the number of ions) in a cross section of the torus is shown in figure 6.5b. The "width" of the torus is of the order of a Jovian radius, R_J. The maximum electron density is 3000 cm^{-3} at 5.7 R_J, the orbit of Io. The composition of the torus of Io inferred from optical and ultraviolet emissions and Voyager measurements is summarized in table 6.3. This composition is consistent with the material derived from the photochemistry of SO_2 described in section 6.2.1.

The energetics of the plasma torus are not completely understood. When the ions of oxygen and sulfur are picked up by the magnetic field, each gains 260 and 520 eV of gyro energy, respectively. Coulomb scattering and plasma wave interactions transfer the initial ion energy to the thermal plasma in the torus. The ultimate fate of the particles is to diffuse inward toward Jupiter or outward away from Jupiter with time constants of the order of 1 yr and 10–100 days, respectively. The enormous difference in the rates of diffusion is revealed in the spacing of the contours for charge density inside and and outside of 5.7 R_J (see figure 6.5b).

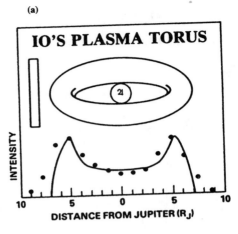

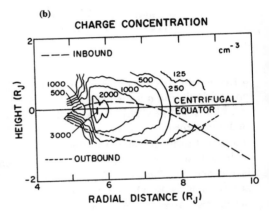

Figure 6.5 (a) Measured intensity (points) of the 685 Å feature of the EUV missions as a function of distance from Jupiter in the orbital plane of the satellites. A model torus used to fit the data is shown to scale above the data; the intensity predicted by the model is shown by the solid line. The bar on the left gives the field of view of the observations. After Broadfoot, A. L. et al., 1979, *Science* **204**, 979–982. (b) A contour map of charge concentrations in the vicinity of Io constructed from measurements made along the inbound trajectory of Voyager 1. The units are electrons/cm^3. R_J = Jupiter radius. After Belcher, J. W., 1983, *Physics of the Jovian Magnetosphere* (Cambridge: Cambridge University Press), p. 68.

The plasma in the torus of Io interacts strongly with Io. This Jovian wind can impinge on Io and result in efficient sputtering of its atmosphere. The sputtered material can produce even more ions, which in turn can contribute to atmospheric sputtering. This unstable nonlinear interaction (leading to the catastrophic erosion of the atmosphere and surface of Io) has never been observed. In fact, the opposite of this—the stability of the torus—has been established by more than a decade of observations. This is remarkable in view of the large variations in the rate of volcanic output and the solar cycle variations in ultraviolet flux and solar wind. It is now believed that the interaction is self-regulating. A greatly enhanced Jovian wind would set up a current loop in the ionosphere of Io, which would generate a local magnetic field that can divert the incident plasma of heavy ions at a greater distance from Io. Alternatively, the mass loading of the torus affects the dynamics and the energetics of the magnetosphere in such a way as to remove the ions at a faster rate. It is beyond the scope of this chapter to discuss the merits of the competing theories of Io's interaction with the Jovian magnetosphere.

Table 6.3 Composition of Io's environment

Constituent	Concentration (cm^{-3})	Reference
Ionosphere		Kliore et al. (1974)
electrons, e^- (dawn, 5.5h)	9×10^3	
electrons, e^- (dusk, 17.5h)	6×10^4	
Plasma torus		Bagenal et al. (1994)
Singly ionized sulfur, SII	1000	
Doubly ionized sulfur, SIII	500	
Triply ionized sulfur, SIV	40	
Singly ionized oxygen, OII	1000	
Doubly ionized oxygen, OIII	30	
Total electrons, e^-	2000–4000	
Neutral cloud		Brown (1974)
Sodium, Na	10	

6.3 Titan

Titan possesses a mildly reducing atmosphere in the outer solar system. It has a rather massive atmosphere that is different from the atmospheres of the giant planets in at least two aspects. First, there is little H_2 in the atmosphere of Titan relative to the major gas N_2. The gravity is low enough that light gases like H and H_2 readily escape from the satellite. In contrast, the composition of the giant planets is dominated by H_2. Second, unlike the giant planets, Titan has a cold surface. Organic compounds that are synthesized in the atmosphere are deposited on the surface, resulting in their permanent sequestration. This implies an irreversible chemical evolution of the atmosphere and the surface. There is no recycling of organic species as in the giant planets. The atmosphere is being gradually destroyed and must be resupplied by primordial ice on the surface or outgassing from the interior.

Most of our knowledge of Titan was obtained from the two Voyager flyby missions to Saturn. Table 6.4 summarizes the physical parameters and composition measurements of Titan. Most of the information was derived from Voyager UVS, IRIS, and RSS (radio science). It is illuminating to display the compositional information as a comparison between Titan and Jupiter in figure 6.6. For future reference, we display similar compositional information for Earth's atmosphere in this figure. As pointed out earlier, the great contrast between Titan and Jupiter is in that N_2 dominates in Titan whereas H_2 dominates in Jupiter. Methane is a minor species in both atmospheres and is the parent molecule for the higher hydrocarbons, the C_2, C_3, and C_4 species observed in these atmospheres. The detailed photochemistry of CH_4 in Titan's atmosphere and its comparison with that of Jupiter is discussed in sections 6.3.1 and 6.3.2. Methane is also the source of the small amount of H_2 present in the atmosphere of Titan. H_2 is easily lost from Titan by thermal escape, and its abundance must be maintained by a constant photochemical source.

The dominance of nitrogen on Titan gives rise to the rich coupled chemistry between nitrogen and carbon. The variety of nitrile species on Titan appears to be unique in the solar system. The detailed chemistry is discussed in section 6.3.3. The existence

Table 6.4 Physical parameters and chemical composition measurements for Titan

Physical or chemical data

At the surface (altitude $z = 0$)
  r_0 = distance to center = 2575 km
  g_0 = gravity = 135 cm s^{-2}
  P_0 = total pressure = 1.5 bar
  T_0 = temperature = 94 K
  n_0 = number density = 1.2×10^{20} cm^{-3}
Composition of the troposphere (volume mixing ratio)
  $N_2 > 0.97$
  $CH_4 < 0.03$
  $H_2 = 0.002$
At the tropopause ($z = 45$ km)
  $P = 130$ mbar
  $T = 71.4$ K
  $n = 1.1 \times 10^{19}$ cm^{-3}
Composition of the stratosphere (volume mixing ratio)
  $CH_4 = 1\text{–}3 \times 10^{-2}$
  $H_2 = 2 \times 10^{-3}$
  $C_2H_6 = 2 \times 10^{-5}$
  $C_2H_2 = 2 \times 10^{-6}$
  $C_2H_4 = 4 \times 10^{-7}$
  $C_2H_8 = 2\text{–}4 \times 10^{-6}$
  $CH_3C_2H = 3 \times 10^{-8}$
  $C_4H_2 = 10^{-8}\text{–}10^{-7}$
  $HCN = 2 \times 10^{-7}$
  $HC_3N = 10^{-8}\text{–}10^{-7}$
  $C_2N_2 = 10^{-8}\text{–}10^{-7}$
  $CO = 6 \times 10^{-5}$
  $CO_2 = 1.5 \times 10^{-9}$
  $H_2O < 1 \times 10^{-9}$
  $CH_3D = CH_4 \times 6.4 \times 10^{-4}$
Composition of mesosphere and thermosphere
  $N_2 = 2.7 \times 10^8$ cm^{-3} at $z = 1280$ km
  $CH_4 = 1.2 \times 10^8$ cm^{-3} at $z = 1140$ km (mixing ratio = 0.08)
  C_2H_2 mixing ratio = 1–2% at $z = 840$ km
  H atmos: disk-averaged Lyα airglow = 500 R[a]
Haze layers
  Optical haze at $z = 300$ km
  UV haze at $z = 400$ km

[a] The observed airglow on the the dayside is about 1 kR. However, based on our knowledge of the Lyα nightglow and other emissions, we conclude that the dayglow consists of roughly equal contributions from resonance scattering of solar radiation and excitation by electron impact. The value quoted here refers to the component of Lyα airglow arising from resonance scattering. From Yung, Y. L. et al. (1984).

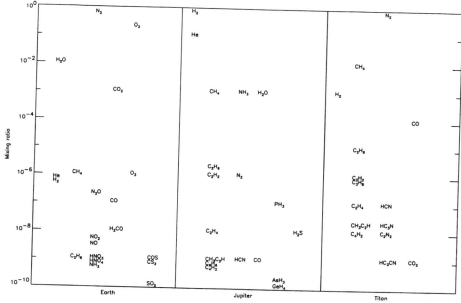

Figure 6.6 Comparison of the approximate speciation of hydrogen, carbon, nitrogen, oxygen, sulfur, and other trace elements in the atmospheres of Earth, Jupiter, and Titan.

of CO_2, a highly oxidized species, was a great surprise. However, it can be readily synthesized from CO if we postulate a source of extraplanetary oxygen in the form of meteoritic H_2O. The question then arises as to the origin of CO. In view of the ubiquitous existence of CO in the outer solar system and the interstellar medium, it is conceivable that Titan's CO is primordial. Conversely, it is also possible to produce CO from the photochemistry of CH_4 and oxygen derived from meteoritic H_2O. The coupled chemistry of carbon and oxygen species is the subject of section 6.3.4.

As summarized in figure 6.6, the greatest contrast between the atmospheres of Earth and Titan is the large abundance of O_2 in the former. The high concentration of O_2 gives rise to the O_3 layer. Like O_2, most of the important parent molecules in the terrestrial atmosphere are of biological origin. These include species such as CH_4, N_2O, NH_3, and COS. The abundance of CO_2 in the atmosphere is biologically regulated. H_2 is not stable in the atmosphere and is readily oxidized to become the more stable form of hydrogen—H_2O. Biogenic CH_4 is the ultimate source of the observed H_2 in the atmosphere. We defer further discussion of Earth's atmosphere to chapters 9 and 10.

The photochemistry of Titan is discussed in some detail in the sections 6.3.1–6.3.5. Figure 6.7a shows a schematic of the thermal structure and various regions of the atmosphere of Titan where different photochemical and physical processes are believed to be important. A model atmosphere of Titan is presented in figure 6.7b. The eddy and molecular diffusivity profiles are given in figure 6.7c. The former is derived on the basis of trial and error and by analogy with the terrestrial atmosphere. The reason for the low eddy mixing near Titan's tropopause is the thermal inversion in this

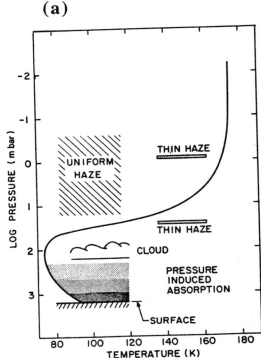

Figure 6.7 (a) Schematic of the vertical temperature profile and various regions of the atmosphere of Titan where different processes are believed to be important. After Samuelson, R. E. et al., 1981, "Mean Molecular Weight and Hydrogen Abundance of Titan's Atmosphere." *Nature* **292**, 688.

part of the atmosphere. At higher altitudes eddy mixing increases due to the breaking of upward-propagating gravity waves. The molecular diffusivity profile is based on the kinetic theory of gases. Absorption of extreme ultraviolet (EUV) wavelengths in the thermosphere of Titan leads to the production of N atoms and ions of nitrogen. These rapidly transfer charge to the hydrocarbons and initiate a complex chain of interactions. In the mesosphere region, photolysis of CH_4 becomes important, leading to the formation of higher hydrocarbons and a high-altitude haze layer. The chemical kinetics peculiar to Titan are briefly reviewed. Nitrogen atoms derived from the thermosphere and oxygen-containing molecules derived from meteoritic H_2O interact with hydrocarbon radicals, resulting in a rich suite of C–N and C–O compounds. The regions where some of the organic species are expected to condense are indicated in figure 6.7a. It is clear that the upper atmosphere must be a source of complex organic compounds. The tropopause serves as a cold trap where the organic compounds are removed, followed by transport to their repository on the surface of Titan. The photochemically generated aerosol layer above the tropopause is sufficiently dense that no ultraviolet photons can penetrate to the troposphere.

6.3.1 Ionosphere

Interaction between solar EUV radiation and the thermosphere of Titan is well documented in the Voyager UVS spectrum of Titan shown in figure 6.8. The most prominent

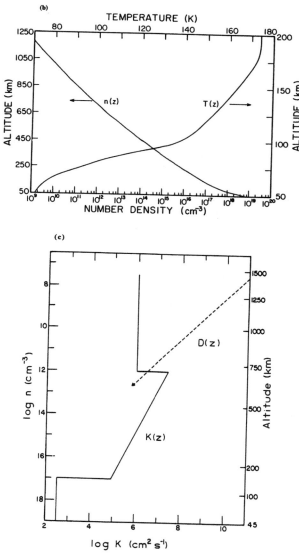

Figure 6.7 (b) Altitude profiles for the total number density $n(z)$ and temperature $T(z)$. $T(z) = 174$ K for $z > 1200$ km. After Yung, Y. L. et al., 1984, "Photochemistry of the Atmosphere of Titan: Comparison between Model and Observations." *Astrophys. J. Suppl.* **55**, 465. (c) Eddy diffusivity profile, $K(z)$, in cm^2 s^{-1}, and molecular diffusivity profile for CH$_4$ in N$_2$. $D(z)$, in cm^2 s^{-1}, is given for comparison, (dashed line). After Yung, Y. L. et al., 1984, "Photochemistry of the Atmosphere of Titan: Comparison between Model and Observations." *Astrophys. J. Suppl.* **55**, 465.

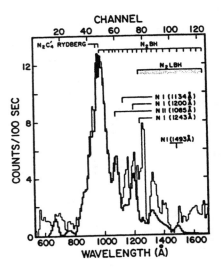

Figure 6.8 The principal N, N$^+$, and N$_2$ emissions from Titan detected by Voyager UVS. The effects of instrumental scattering and Lyman α have been removed from the spectra. A synthetic model based on the electron impact ionization of N$_2$ is shown by heavy lines. After Strobel, D. F., and Shemansky, D. E., 1982, "EUV Emission from Titan's Upper Atmosphere: Voyager I Encounter." *J. Geophys. Res.*, **87**, 1361.

observed features result from emissions by various excited states of N$_2$, N, and N$^+$. The total averaged radiated power from the dayside is 2×10^{10} W. In this region of the atmosphere the primary interaction appears to be between solar EUV and N$_2$ gas. N$_2$ is readily dissociated or ionized by solar EUV:

$$N_2 + h\nu \rightarrow N + N(^2D) \tag{6.23}$$

$$N_2 + h\nu \rightarrow N_2^+ + e \tag{6.24a}$$

$$\rightarrow N^+ + N + e \tag{6.24b}$$

where N(^{2}D) is the first excited state of the N atom. These processes may also be driven by energetic electrons from the Saturnian magnetosphere, augmenting the solar driven reactions by 20%. The primary removal mechanism for N atoms is formation of N$_2$ via

$$N + NH \rightarrow N_2 + H \tag{6.25}$$

where NH is derived from

$$N(^2D) + CH_4 \rightarrow NH + CH_3 \tag{6.26}$$

The N atoms that are not removed flow downward to the mesosphere region, where they react with hydrocarbon radicals to form HCN (see section 6.3.3).

A comprehensive set of ion reactions is referred to the literature. The primary fate of the ions of nitrogen is loss by charge transfer to the hydrocarbon species:

$$N_2^+ + CH_4 \rightarrow N_2 + CH_3^+ + H \tag{6.27a}$$

$$\rightarrow N_2 + CH_2^+ + H_2 \tag{6.27b}$$

$$N^+ + CH_4 \rightarrow N + CH_4^+ \tag{6.28a}$$

$$\rightarrow NH + CH_3^+ \tag{6.28b}$$

$$\rightarrow HCN^+ + H_2 + H \tag{6.28c}$$

More complex hydrocarbon ions can be formed, as in the following examples:

$$CH_2^+ + CH_4 \rightarrow C_2H_5^+ + H \tag{6.29}$$

$$CH_3^+ + CH_4 \rightarrow C_2H_5^+ + H_2 \tag{6.30}$$

$$CH_4^+ + CH_4 \rightarrow CH_5^+ + CH_3 \tag{6.31}$$

$$CH_5^+ + C_2H_4 \rightarrow C_2H_5^+ + CH_4 \tag{6.32}$$

Further reactions lead to the production of the terminal ion H_2CN^+ by

$$C_2H_5^+ + HCN \rightarrow H_2CN^+ + C_2H_4 \tag{6.33}$$

$$HCN^+ + CH_4 \rightarrow H_2CN^+ + CH_3 \tag{6.34}$$

As shown below, H_2CN^+ is the most abundant ion in the ionosphere of Titan. A schematic diagram summarizing the principal pathways that result in the production of H_2CN^+, including most of the above reactions, is given in figure 6.9a.

Complex hydrocarbon ions may be produced as follows:

$$C_2H_5^+ + C_2H_2 \rightarrow C_3H_3^+ + CH_4 \tag{6.35}$$

$$C_2H_5^+ + C_2H_4 \rightarrow C_3H_5^+ + CH_4 \tag{6.36}$$

$$C_3H_3^+ + C_2H_2 \rightarrow C_5H_3^+ + H_2 \tag{6.37}$$

$$C_3H_3^+ + C_2H_4 \rightarrow C_5H_5^+ + H_2 \tag{6.38}$$

$$CH_5^+ + C_2H_2 \rightarrow C_2H_3^+ + CH_4 \tag{6.39}$$

$$C_2H_3^+ + C_2H_2 \rightarrow C_4H_3^+ + H_2 \tag{6.40}$$

A schematic diagram showing the principal pathways for the production of complex hydrocarbon ions is given in figure 6.9b. All ions containing more than three carbon atoms are labeled "$C_nH_m^+$." Note that charge transfer reactions are capable of producing interesting neutral molecules such as NH_3 by

$$N^+ + H_2 \rightarrow NH^+ + H \tag{6.41}$$

$$NH^+ + C_2H_4 \rightarrow C_2H_2^+ + NH_3 \tag{6.42}$$

although the amount that can be produced is insignificant.

The ultimate fate of H_2CN^+ and the hydrocarbon ions is loss by dissociative recombination in reactions such as

$$H_2CN^+ + e \rightarrow HCN + H \tag{6.43}$$

$$C_nH_m^+ + e \rightarrow C_nH_{m-1} + H \tag{6.44a}$$

$$\rightarrow C_pH_q + C_rH_s \tag{6.44b}$$

where $p + r = n$ and $q + s = m$. The number densities of the major ions in the model are presented figure 6.10a. H_2CN^+ is the most abundant ion, with peak concentration in excess of 10^3 cm^{-3} at about 1200 km above the surface. The second most abundant ion is the sum of complex hydrocarbon ions containing more than three carbon atoms ($C_nH_m^+$). N_2^+ is a minor ion even though the production rate of N_2^+ by (6.24a) and

(a)

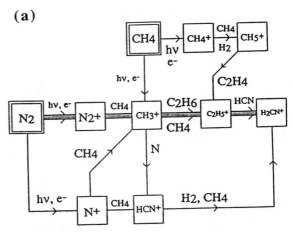

(b)

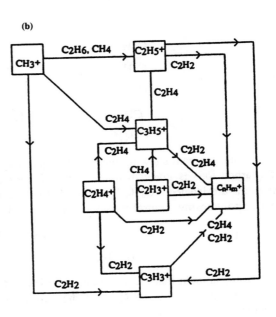

Figure 6.9 (a) Ion-neutral reaction pathways leading to the formation of the major ion, H_2CN^+. (b) Ion-neutral reaction pathways leading to the formation of the higher mass hydrocarbon ions, $C_nH_m^+$. Note that CH_3^+, $C_2H_4^+$, and $C_2H_3^+$ are produced by photoionization and electron impact ionization. $C_2H_5^+$ is removed primarily by reaction with HCN. After Keller, C. N. et al., 1992, "A Model of the Ionosphere of Titan." *J. Geophys. Res.* **97**, 12117.

electron impact is the most important ionization reaction in the model (see figure 6.11, a and b). The reason is the rapid charge exchange reactions such as (6.27a and 6.27b) that remove N_2^+. The total ion density (equal to the electron density) at the peak is about 5×10^3 cm^{-3} and is consistent with the upper limit obtained by the Voyager observations. The concentrations of minor ions are shown in figures 6.10, b and c. In the region of the main peak CH_3^+ and $C_2H_3^+$ are abundant due to their low

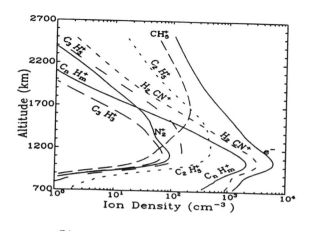
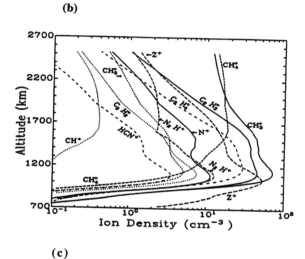
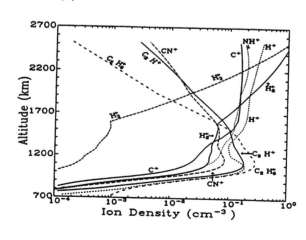

Figure 6.10 (a) Density profiles of the seven major ions. (b) Density profiles of the minor ions in the model. (c) Density profiles of the minor ions (of smaller concentrations) in the model. Ionization from both electron impact ionization and photoionization is included. The solar zenith angle is 60°. After Keller, C. N. et al., 1992, "A Model of the Ionosphere of Titan." *J. Geophys. Res.* **97**, 12117.

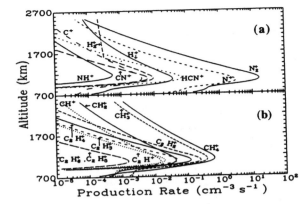

Figure 6.11 Ionization production rate profiles for all species. Purely photoionization rates are shown here, and the solar zenith angle is 60°. The ions are divided into groups as nonhydrocarbons (a) and hydrocarbons (b). After Keller, C. N. et al., 1992, "A Model of the Ionosphere of Titan." *J. Geophys. Res.* **97**, 12117.

ionization potential relative to other ions. In the upper ionosphere simple ions such as H_2^+, NH^+, and H_3^+ and atomic ions such as C^+ and H^+ become important. The rates of production of nonhydrocarbon and hydrocarbon ions in the model are shown in figure 6.11. Only solar EUV ionization is used in the model. The peak production rate of N_2^+ occurs at an altitude of 1050 km, with a rate of 17 ions cm^{-3} s^{-1}. The minor ionization reactions include photoionization of N, CH_4, and CH_3.

6.3.2 Hydrocarbon Chemistry

The essence of hydrocarbon chemistry has been described in chapter 5 and is not repeated here. The important reactions are summarized in tables 6.5, 6.6, and 6.7 and are briefly discussed here. A great difference between the atmosphere of Titan and those of the giant planets is the dominance of N_2 on Titan. The presence of large amounts of N_2 implies that the preferred pathway for $^1CH_2$ is quenching by

$$^1CH_2 + N_2 \rightarrow CH_2 + N_2 \tag{6.45}$$

rather than the reaction with H_2 to form CH_3 by (5.32). However, CH_3 may still be produced by photosensitized dissociation via the $C_{2n}H_2$ cycle (see, e.g., chemical scheme VII in section 5.3.1). In fact, the efficiency of this scheme is enhanced because the competing scheme (VI in section 5.3.1) for the photosensitized dissociation of H_2 is not important on Titan.

Another great difference between Titan and the giant planets is the ability of hydrogen (H and H_2) to escape from Titan. As pointed out in chapter 5, the synthesis of higher hydrocarbons from the parent molecule CH_4 is accompanied by the production of H or H_2 (see, e.g., chemical schemes I–V discussed in section 5.3.1). The gravity of Titan is sufficiently low that H and H_2 can readily escape. With the loss of hydrogen the production of higher hydrocarbons becomes irreversible. In addition, the loss of H results in a much smaller concentration of H atoms in the atmosphere. Consequently, the cracking of unsaturated hydrocarbons by H atoms becomes less important when compared with the giant planets.

Table 6.5 List of molecular dissociation reactions for the atmosphere of Titan

	Reaction[a]			Rate coefficient[b]	Reference
R1	$N_2 + e$	$\rightarrow$	$N + N(^2D) + e$	—	Strobel and Shemansky (1982)
R2	$N_2 + CR$	$\rightarrow$	$N + N(^2D)$	—	Capone et al. (1980)
R3a	$CO + e$	$\rightarrow$	$C + O + e$	—	Lee and McKay (1982)
b		$\rightarrow$	$C + O(^1D) + e$	—	
R4	$CH_3 + h\nu$	$\rightarrow$	$^1CH_2 + H$	1.0×10^{-6}	Parkes et al. (1973)
R5a	$CH_4 + h\nu$	$\rightarrow$	$^1CH_2 + H_2$	2.2×10^{-8}	Watanabe et al. (1953)
b		$\rightarrow$	$CH_2 + 2H$	2.0×10^{-8}	Strobel (1973); Mount
c		$\rightarrow$	$CH + H + H_2$	3.2×10^{-9}	and Moos (1978)
R6a	$C_2H_2 + h\nu$	$\rightarrow$	$C_2H + H$	4.5×10^{-8}	Nakayama and Watanabe (1964)
b		$\rightarrow$	$C_2 + H_2$	2.5×10^{-8}	Okabe (1981, 1983a)
R7a	$C_2H_4 + h\nu$	$\rightarrow$	$C_2H_2 + H_2$	1.4×10^{-7}	Zelikoff and Watanabe (1953)
b		$\rightarrow$	$C_2H_2 + 2H$	1.2×10^{-7}	Back and Griffiths (1967)
R8a	$C_2H_6 + h\nu$	$\rightarrow$	$C_2H_4 + H_2$	1.1×10^{-8}	Akimoto et al. (1965)
b		$\rightarrow$	$C_2H_4 + 2H$	9.1×10^{-9}	Hampson and McNesby (1965)
c		$\rightarrow$	$C_2H_2 + 2H_2$	9.9×10^{-9}	Lisa et al. (1970)
d		$\rightarrow$	$CH_4 + ^1CH_2$	6.1×10^{-9}	Mount and Moos (1978)
e		$\rightarrow$	$2CH_3$	3.5×10^{-9}	
R9	$C_3H_3 + h\nu$	$\rightarrow$	$C_3H_2 + H$	1.2×10^{-7}	Poole and Anderson (1959); Ramsay and Thistlewaithe (1996); Jacox and Milligan (1974);
R10a	$CH_3C_2H + h\nu$	$\rightarrow$	$C_3H_3 + H$	9.0×10^{-8}	Stief et al. (1971)
b		$\rightarrow$	$C_3H_2 + H_2$	2.4×10^{-8}	Hamai and Hirayama (1979)
c		$\rightarrow$	$CH_3 + C_2H$	4.5×10^{-9}	Heller and Milne (1978)
R11a	$CH_2CCH_2 + h\nu$	$\rightarrow$	$C_3H_3 + H$	3.2×10^{-7}	Sutcliffe and Walsh (1952)
b		$\rightarrow$	$C_3H_2 + H_2$	1.2×10^{-7}	Rabalais et al. (1979)
c		$\rightarrow$	$C_2H_2 + CH_2$	4.9×10^{-8}	Heller and Milne (1978)
R12a	$C_3H_6 + h\nu$	$\rightarrow$	$CH_2CCH_2 + H_2$	3.5×10^{-7}	Calvert and Pitts (1966)
b		$\rightarrow$	$C_2H_4 + CH_2$	1.2×10^{-8}	Borrell et al. (1971)
c		$\rightarrow$	$C_2H_2 + CH_3 + H$	2.0×10^{-7}	Collin et al. (1979)
d		$\rightarrow$	$C_2H + CH_4 + H$	3.0×10^{-8}	
R13a	$C_3H_8 + h\nu$	$\rightarrow$	$C_3H_6 + H_2$	4.6×10^{-8}	Calvert and Pitts (1966)
b		$\rightarrow$	$C_2H_6 + ^1CH_2$	7.6×10^{-9}	
c		$\rightarrow$	$C_2H_5 + CH_3$	3.3×10^{-8}	
d		$\rightarrow$	$C_2H_4 + CH_4$	1.8×10^{-8}	
R14a	$C_4H_2 + h\nu$	$\rightarrow$	$C_4H + H$	2.6×10^{-7}	Georgieff and Richard (1958)
b		$\rightarrow$	$2C_2H$	1.7×10^{-7}	Okabe (1981); Heller and Milne (1978)

(continued)

Table 6.8 summarizes the basic assumptions and boundary conditions of the model. The lower boundary of the model is placed at the tropopause (45 km), where the CH_4 mixing ratio is assumed to be 2.0×10^{-2}. The upper boundary of the photochemical model is at 1160 km for all species, except H and H_2, for which the upper boundary is the exobase, at 1425 km where the species can escape at the velocity given by the Jeans escape formula. The special boundary conditions for nitrogen and oxygen species are discussed in sections 6.3.3 and 6.3.4. Unless otherwise stated, for all other species the lower boundary condition is maximum deposition velocity (followed by loss to the troposphere) and the boundary condition at the upper boundary is zero flux.

Table 6.5 (continued)

	Reaction[a]		Rate coefficient[b]	Reference
R15a	$C_6H_2 + h\nu$	$\to C_6H + H$	2.6×10^{-7}	Kloster-Jensen et al. (1974)
b		$\to C_4H + C_2H$	1.7×10^{-7}	estimated
R16a	$C_8H_2 + h\nu$	$\to C_6H + C_2H$	2.6×10^{-7}	Kloster-Jensen et al. (1974)
b		$\to 2C_4H$	1.7×10^{-7}	estimated
R17	$HCN + h\nu$	$\to H + CN$	1.1×10^{-7}	West (1975); Lee (1980)
R18	$HC_3N + h\nu$	$\to C_2H + CN$	4.0×10^{-7}	Connors et al. (1974)
R19	$C_2N_2 + h\nu$	$\to 2CN$	2.7×10^{-7}	Connors et al. (1974) Nuth and Glicker (1982)
R20a	$CO_2 + h\nu$	$\to CO + O$	7.4×10^{-11}	Shemansky (1972)
b		$\to CO + O(^1D)$	5.5×10^{-9}	Demore and Patapoff (1972)
R21a	$H_2O + h\nu$	$\to H + OH$	5.1×10^{-8}	Allen et al. (1981)
b		$\to H_2 + O(^1D)$	2.0×10^{-8}	
R22a	$H_2CO + h\nu$	$\to H_2 + CO$	2.5×10^{-7}	Pinto et al. (1980)
b		$\to H + HCO$	1.8×10^{-7}	
R23	$HCO + h\nu$	$\to H + CO$	1.0×10^{-4}	Pinto et al. (1980)
R24	$CH_2CO + h\nu$	$\to CH_2 + CO$	1.3×10^{-6}	Okabe (1978)

[a] Excited atom or molecule: $N(^2D)$, $O(^1D)$, $^1CH_2 = CH_2(\tilde{a}^1A_1)$; CR = cosmic rays.
[b] The values for the diurnally averaged dissociation coefficient refer to the top of the atmosphere in units of s^{-1}.

(a) C_1 and C_2 Species, and H and H_2

The primary driving force of the photochemistry of hydrocarbons in Titan's atmosphere is photolysis. Figure 6.12 presents the diurnally averaged dissociation coefficients for the important hydrocarbons in the model. The major absorber at short wavelengths (below 1600 Å) is CH_4. Absorption by C_2H_2, C_2H_4, and HCN becomes important at longer wavelengths. Aerosol opacity provides significant attenuation of solar radiation below 240 km.

The mixing ratios of H_2 and the major hydrocarbon species CH_4, C_2H_2, C_2H_4, and C_2H_6 are given in figure 6.13a. The number densities of the important radicals H, CH, $^1CH_2$, CH_2, CH_3, C_2H_3, C_2H_5, C_2H, C_4H, and C_6H are shown in figure 6.13, b and c. The computed mixing ratio at 1140 km is 8.1%, in good agreement with the value 8 ± 3% deduced from Voyager UVS observations. The increase of CH_4 at high altitudes is due to diffusive separation above the homopause. The maximum abundances of all photochemically produced species occur near the homopause region in the mesosphere, where CH_4 dissociation is greatest. The most abundant radical is H, with peak concentration equal to 5×10^8 cm^{-3}. Since H atoms escape from Titan, their abundance is an order of magnitude less than that in the Jovian atmosphere. The consequence of this for photochemistry on Titan is discussed later in this section. The rapid decrease in the concentrations of the higher hydrocarbons near the tropopause is due to condensation. Table 6.9 gives the saturation mixing ratio of selected species in the model. It is clear that most of the heavier hydrocarbons would condense and be removed from the gas phase. The radicals CH_3, C_2H_3, and C_2H_5 all exhibit a secondary maximum at about 200 km. This is driven by the photosensitized dissociation of CH_4 by photons at long wavelength (see discussion in the following paragraph).

Table 6.6 List of essential hydrocarbon reactions for the atmosphere of Titan

	Reaction	Rate coefficient[a]	Reference
R25	$H + C_2H_2 + M \rightarrow C_2H_3 + M$	$k_0 = 6.4 \times 10^{-25} T^{-2} e^{-1200/T}$ $k_\infty = 9.2 \times 10^{-12} e^{-1200/T}$	Payne and Stief (1976)
R26	$H + C_2H_4 + M \rightarrow C_2H_5 + M$	$k_0 = 1.1 \times 10^{-23} T^{-2} e^{-1040/T}$ $k_\infty = 3.7 \times 10^{-11} e^{-1040/T}$	Michael et al. (1973); Lee et al. (1978)
R27	$CH + CH_4 \rightarrow C_2H_4 + H$	1.0×10^{-10}	Butler et al. (1981)
R28	$^1CH_2 + N_2 \rightarrow CH_2 + N_2$	7.9×10^{-12}	Ashfold et al. (1980);
R29a	$^1CH_2 + H_2 \rightarrow CH_2 + H_2$	1.0×10^{-12}	Laufer (1981a)
b	$\rightarrow CH_3 + H$	7.0×10^{-12}	
R30a	$^1CH_2 + CH_4 \rightarrow CH_2 + CH_4$	1.6×10^{-12}	Laufer (1981a)
b	$\rightarrow 2CH_3$	1.9×10^{-12}	
R31a	$CH_2 + C_2H_2 + M \rightarrow CH_3C_2H + M$	$k_0 = 3.8 \times 10^{-25}$ $k_\infty = 2.2 \times 10^{-12}$	Laufer et al. (1983); Laufer (1981a)
b	$\rightarrow CH_2CCH_2 + M$	$k_0 = 3.8 \times 10^{-25}$ $k_\infty = 3.7 \times 10^{-12}$	
R32	$2CH_2 \rightarrow C_2H_2 + H_2$	5.3×10^{-11}	Banyard et al. (1980)
R33	$CH_3 + H + M \rightarrow CH_4 + M$	$k_0 = 1.7 \times 10^{-27}$ $k_\infty = 1.5 \times 10^{-10}$	Laufer et al. (1983); Patrick et al. (1980)
R34	$CH_3 + CH_2 \rightarrow C_2H_4 + H$	7.0×10^{-11}	Laufer et al. (1981a)
R35	$2CH_3 + M \rightarrow C_2H_6 + M$	$k_0 = 1.3 \times 10^{-23}$ $k_\infty = 5.5 \times 10^{-11}$	Laufer et al. (1983); Callear and Metcalfe (1976)
R36	$C_2H_3 + H \rightarrow C_2H_2 + H_2$	1.5×10^{-11}	Keil et al. (1976)
R37a	$C_2H_3 + CH_3 \rightarrow C_2H_2 + CH_4$	9.1×10^{-12}	estimated
b	$(+M) \rightarrow C_3H_6 + M$	$k_0 = 1.3 \times 10^{-22}$ $k_\infty = 9.1 \times 10^{-12}$	Laufer et al. (1983)
R38	$2C_2H_3 \rightarrow C_2H_4 + C_2H_2$	5.3×10^{-12}	MacFadden and Currie (1973)
R39	$C_2H_5 + H \rightarrow 2CH_3$	$1.9 \times 10^{-10} e^{-440/T}$	Teng and Jones (1972)
R40a	$C_2H_5 + CH_3 \rightarrow C_2H_4 + CH_4$	$1.7 \times 10^{-12} e^{-200/T}$	Teng and Jones (1972)
b	$(+ M) \rightarrow C_3H_8 + M$	$k_0 = 5.6 \times 10^{-22}$ $k_\infty = 4.2 \times 10^{-11} e^{-200/T}$	Laufer et al. (1983)
R41a	$C_2H_5 + C_2H_3 \rightarrow C_2H_6 + C_2H_2$	6.0×10^{-12}	Laufer et al. (1983)
b	$\rightarrow 2C_2H_4$	3.0×10^{-12}	
R42a	$2C_2H_5 \rightarrow C_2H_6 + C_2H_2$	$1.7 \times 10^{-12} e^{-90/T}$	Teng and Jones (1972)
b	$(+ M) \rightarrow C_4H_{10} + M$	$k_0 = 2.9 \times 10^{-21}$ $k_\infty = 1.3 \times 10^{-11} e^{-95/T}$	Laufer et al. (1983)
R43	$C_3H_2 + H + M \rightarrow C_3H_3 + M$	$k_0 = 1.7 \times 10^{-26}$ $k_\infty = 1.5 \times 10^{-10}$	Laufer et al. (1983)
R44a	$C_3H_3 + H + M \rightarrow CH_3C_2H + M$	$k_0 = 1.7 \times 10^{-26}$ $k_\infty = 1.5 \times 10^{-10}$	
b	$\rightarrow CH_2CCH_2 + M$	same as branch[a]	
R45a	$CH_3C_2H + H + M \rightarrow (C_3H_5^* + M)$	$k_0 = 8.0 \times 10^{-24} T^{-2} e^{-1225/T}$ $k_\infty = 9.7 \times 10^{-12} e^{-1550/T}$	Whytock et al. (1976) Wagner and Zellner (1972a)
	$\rightarrow CH_3 + C_2H_2 + M$		
b	$\rightarrow C_3H_5 + M$	same as branch[a]	

(continued)

A schematic diagram showing the pathways for the interconversion of major C_1 and C_2 species is given in figure 6.13d.

The destruction of CH_4 is the ultimate source of all higher hydrocarbons. Figure 6.14a presents the rates of the major reactions destroying CH_4, R5: $CH_4 + h\nu$, R55: $CH_4 + C_2H$, R61: $CH_4 + C_4H$, and R65: $CH_4 + C_6H$ (reaction numbers refer to tables 6.5–6.7). Direct photolysis of CH_4 is important in the mesosphere; photosensitized dissociation driven by C_2H_2 and polyynes is more important in the lower stratosphere. The most important reaction for producing CH_4, R33: $CH_3 + H$, is also included in figure 6.14a. Since there are fewer H atoms in the atmosphere of Ti-

Table 6.6 (continued)

Reaction			Rate coefficient[a]	Reference
R46a	$CH_2CCH_2 + H + M$	$\rightarrow (C_3H_5^* + M)$	$k_0 = 8.0 \times 10^{-24} T^{-2} e^{-1225/T}$	Wagner and Zellner (1972b)
		$\rightarrow CH_3 + C_2H_2 + M$	$k_\infty = 9.7 \times 10^{-13} e^{-1550/T}$	
b		$\rightarrow C_3H_5 + M$	$k_0 = 8.0 \times 10^{-24} T^{-2} e^{-1225/T}$	
			$k_\infty = 1.4 \times 10^{-11} e^{-1000/T}$	
R47a	$C_3H_5 + H$	$\rightarrow CH_3C_2H + H_2$	1.5×10^{-11}	Zellner (1972a,b)
b		$\rightarrow CH_2CCH_2 + H_2$	1.5×10^{-11}	
c	$(+M)$	$\rightarrow C_3H_6 + M$	$k_0 = 1.0 \times 10^{-28}$	
			$k_\infty = 1.0 \times 10^{-11}$	
d		$\rightarrow CH_4 + C_2H_2$	1.5×10^{-11}	
R48	$C_3H_6 + H + M$	$\rightarrow C_3H_7 + M$	$k_0 = 1.5 \times 10^{-29}$	Analogy with $C_2H_4 + H$
			$k_\infty = 3.7 \times 10^{-11} e^{-1040/T}$	
R49	$C_3H_7 + H$	$\rightarrow C_2H_5 + CH_3$	$1.9 \times 10^{-10} e^{-440/T}$	Analogy with $C_2H_3 + H$
R50a	$C_3H_5 + CH_3$	$\rightarrow CH_3C_2H + CH_4$	4.5×10^{-12}	Analogy with $C_2H_3 + CH_3$
b		$\rightarrow CH_2CCH_2 + CH_4$	4.5×10^{-12}	
R51a	$C_3H_7 + CH_3$	$\rightarrow C_3H_6 + CH_4$	$2.5 \times 10^{-12} e^{-200/T}$	Analogy with $C_2H_5 + CH_3$
b	$(+M)$	$\rightarrow C_4H_{10} + M$	$k_0 = 2.5 \times 10^{-19}$	Laufer et al. (1983)
			$k_\infty = 4.2 \times 10^{-11} e^{-200/T}$	
R52	$C_2 + H_2$	$\rightarrow C_2H + H$	1.4×10^{-12}	Pasternack et al. (1979)
R53	$C_2 + CH_4$	$\rightarrow C_2H + CH_3$	1.9×10^{-11}	Pasternack et al. (1979)
R54	$C_2H + H_2$	$\rightarrow C_2H_2 + H$	$1.9 \times 10^{-11} e^{-1450/T}$	Brown and Laufer (1981)
R55	$C_2H + CH_4$	$\rightarrow C_2H_2 + CH_3$	$2.8 \times 10^{-12} e^{-250/T}$	Brown and Laufer (1981)
R56	$C_2H + C_2H_6$	$\rightarrow C_2H_2 + C_2H_5$	6.5×10^{-12}	Okabe (1983b)
R57	$C_2H + C_3H_8$	$\rightarrow C_2H_2 + C_3H_7$	1.4×10^{-11}	Brown and Laufer (1981)
R58	$C_2H + C_2H_2$	$\rightarrow C_4H_2 + H$	3.1×10^{-11}	Estimated
R59	$C_2H + C_4H_2$	$\rightarrow C_6H_2 + H$	3.1×10^{-11}	Estimated
R60	$C_2H + C_6H_2$	$\rightarrow C_8H_2 + H$	3.1×10^{-11}	Estimated
R61	$C_4H + CH_4$	$\rightarrow C_4H_2 + CH_3$	$9.3 \times 10^{-13} e^{-250/T}$	Estimated
R62	$C_4H + C_2H_6$	$\rightarrow C_4H_2 + C_2H_5$	2.2×10^{-12}	Estimated
R63	$C_4H + C_2H_2$	$\rightarrow C_6H_2 + H$	1.0×10^{-11}	Estimated
R64	$C_4H + C_4H_2$	$\rightarrow C_8H_2 + H$	1.0×10^{-11}	Estimated
R65	$C_6H + CH_4$	$\rightarrow C_6H_2 + CH_3$	$3.1 \times 10^{-13} e^{-250/T}$	Estimated
R66	$C_6H + C_2H_6$	$\rightarrow C_6H_2 + C_2H_5$	6.5×10^{-12}	Estimated
R67	$C_6H + C_2H_2$	$\rightarrow C_8H_2 + H$	3.3×10^{-12}	Estimated
R68	$H + C_4H_2 + M$	$\rightarrow C_4H_3 + M$	$k_0 = 1.0 \times 10^{-28}$	Schwanebeck et al. (1975)
			$k_\infty = 2.0 \times 10^{-12}$	
R69a	$H + C_4H_3$	$\rightarrow 2C_2H_2$	3.3×10^{-12}	Schwanebeck et al. (1975)
b		$\rightarrow C_4H_2 + H_2$	1.2×10^{-11}	
R70	$H + CH_2CCH_2$	$\rightarrow CH_3C_2H + H$	$1.0 \times 10^{-11} e^{-1000/T}$	

[a] Units for two-body and three-body rate coefficients are $cm^3 s^{-1}$ and $cm^6 s^{-1}$, respectively; the effective two-body rate coefficient for a three-body reaction is given by $k = k_0 k_\infty M/(k_0 + k_\infty M)$, where k_0, k_∞, and M are, respectively, the rate coefficients in the low/pressure and high/pressure limits, and the number density of ambient atmosphere in molecules/cm^3.

tan, this reaction is much less important than it is in the giant planets. As much as 96% of the dissociation of CH_4 results in the production of higher hydrocarbons. The corresponding efficiency for Jupiter is only 66%. The reason, as stated above, is the escape of H from Titan. The net column-integrated rate of destruction of CH_4 is 1.5×10^{10} cm^{-2} s^{-1}, and this must be balanced by an equivalent upward flux at the tropopause. Figure 6.14b gives the rates of the major synthesis reactions, R32: $2CH_2 \rightarrow C_2H_2 + H_2$, R34: $CH_2 + CH_3 \rightarrow C_2H_4 + H$, R35: $2CH_3 \rightarrow C_2H_6$, and R40b: $CH_3 + C_2H_5 \rightarrow C_3H_8$. It is clear that the synthesis of alkenes and alkynes (except polyynes) occurs preferentially in the mesosphere, where the CH_2 radicals

Table 6.7 List of nitrogen and oxygen reactions for the atmosphere of Titan

	Reaction	Rate coefficient	Reference
R71a	$N(^2D) \to N + h\nu$	2.3×10^{-5}	Okabe (1978)
b	$N(^2D) + N_2 \to N + N_2$	$< 6 \times 10^{-15}$	Black et al. (1969)
R72	$N(^2D) + CH_4 \to NH + CH_3$	3.0×10^{-12}	Black et al. (1969)
R73	$N + CH_2 \to HCN + H$	$5.0 \times 10^{-11} e^{-250/T}$	Estimated
R74	$N + CH_3 \to HCN + H_2$	$5.0 \times 10^{-11} e^{-250}$	Estimated
R75	$NH + H \to N + H_2$	$1.7 \times 10^{-12} T^{0.68} e^{-950/T}$	Mayer et al. (1966)
R76	$NH + N \to N_2 + H$	$1.1 \times 10^{-12} T^{0.5}$	Westley (1980)
R77	$2NH + M \to N_2 + H_2 + M$	1.0×10^{-33}	Estimated
R78	$CN + CH_4 \to HCN + CH_3$	$1.0 \times 10^{-11} e^{-857/T}$	Schacke et al. (1977)
R79	$CN + C_2H_6 \to HCN + C_2H_5$	2.0×10^{-11}	Estimated
R80	$CN + C_2H_2 \to HC_3N + H$	5.0×10^{-11}	Schacke et al. (1977)
R81	$HCN + C_2H \to HC_3N + H$	2.2×10^{-12}	Becker and Hong (1983)
R82	$CN + HCN \to C_2N_2 + H$	3.1×10^{-11}	Estimated
R83	$HCN + H + M \to H_2CN + M$	$k_0 = 6.4 \times 10^{-25} T^{-2} e^{-1200/T}$	Estimated
R84	$H_2CN + H \to HCN + H_2$	1.5×10^{-11}	Estimated
R85	$HC_3N + H + M \to H_2C_3N + M$	$k_0 = 6.4 \times 10^{-25} T^{-2} e^{-1200/T}$	Estimated
		$k_\infty = 9.2 \times 10^{-12} e^{-1200/T}$	
R86	$H_2C_3N + H \to C_2H_2 + HCN$	1.5×10^{-11}	Estimated
R87	$C_2N_2 + H + M \to HC_2N_2 + M$	$k_0 = 6.4 \times 10^{-25} T^{-2} e^{-1200/T}$	Estimated
		$k_\infty = 9.2 \times 10^{-12} e^{-1200/T}$	
R88	$HC_2N_2 + H \to 2HCN$	$1.7 \times 10^{-13} e^{-110/T}$	Phillips (1978)
R89a	$O(^1D) \to O + h\nu$	6.7×10^{-3}	Albers (1969)
b	$O(^1D) + N_2 \to O + N_2$	$1.8 \times 10^{-11} e^{107/T}$	Okabe (1978)
R90a	$O(^1D) + CH_4 \to OH + CH_3$	1.4×10^{-10}	DeMore et al. (1982)
b	$\to H_2CO + H_2$	1.4×10^{-11}	DeMore et al. (1982)
R91	$O + CH_2 \to CO + 2H$	8.3×10^{-11}	
R92	$O + CH_3 \to H_2CO + H$	1.3×10^{-10}	Homann et al. (1981)
R93	$OH + CH_4 \to H_2O + CH_3$	$2.4 \times 10^{-12} e^{-1710/T}$	Hampson (1980)
R94	$OH + CH_2 \to CO + H_2 + H$	5.0×10^{-12}	Baulch et al. (1980)
R95	$OH + CH_3 \to CO + 2H_2$	6.7×10^{-12}	Estimated
R96	$OH + CO \to CO_2 + H$	1.4×10^{-13}	Fenimore (1969)
R97	$OH + C_2H_2 + M \to CH_2CO + H + M$	$k_0 = 5.8 \times 10^{-31} e^{1258/T}$	DeMore et al. (1982)
		$k_\infty = 1.4 \times 10^{-12} e^{+388/T}$	Perry and Williamson (1982)
R98	$CH_2 + CO_2 \to CO + H_2CO$	3.9×10^{-14}	Laufer (1981a)
R99	$H + CO + M \to HCO + M$	$2.0 \times 10^{-33} e^{-850/T}$	Pinto et al. (1980)
R100	$H + HCO \to H_2 + CO$	3.0×10^{-10}	Pinto et al. (1980)
R101	$2HCO \to H_2CO + CO$	6.3×10^{-11}	Pinto et al. (1980)
R102	$CH_2 + CO + M \to CH_2CO + M$	$k_0 = 1.0 \times 10^{-28}$	Estimated
		$k_\infty = 1.0 \times 10^{-15}$	Laufer (1981a)

are abundant. Synthesis of alkanes occurs preferentially in the stratosphere, where alkyl radicals are produced by photosensitized dissociation. A comparison between the model computations and observations of the hydrocarbon species is given in table 6.10. The agreement is generally within an order of magnitude.

The important column-integrated production rates are summarized in table 6.11. These rates are comparable to those in the Jovian atmosphere, despite the fact that the solar irradiation is smaller by a factor of 4. The reason is partly due to the extended nature of the atmosphere and the greater rate of photosensitized dissociation of CH_4.

As pointed out above, the removal of H atoms is crucial for ensuring the stability of unsaturated hydrocarbons, which would otherwise be cracked. The most important reactions for producing H atoms in the model are R5b: $CH_4 + h\nu \to CH_2 + 2H$, R7b:

Table 6.8 Boundary conditions for selected species in the photochemical model for Titan

Species[a]	Lower boundary[b]	Upper boundary[c]
H	$\phi = 0$	$v = 2.5 \times 10^4$
H_2	$\phi = 0$	$v = 6.1 \times 10^3$
CH_4	$f = 2.0 \times 10^{-2}$	$\phi = 1.0 \times 10^9$ [d]
N	$\phi = 0$	$\phi = -1.0 \times 10^9$
NH	$\phi = 0$	$\phi = -1.0 \times 10^9$
CH_3	$\phi = 0$	$\phi = -1.0 \times 10^9$
OH	$\phi = 0$	$\phi = 3.3 \times 10^4$
H_2O	$v = 2.0 \times 10^{-4}$	$\phi = -6.1 \times 10^5$
CO	$\phi = 0$	$\phi = 8.8 \times 10^4$

From Yung, Y. L. et al. (1984).
[a] For species not given here, their boundary conditions are zero flux at the upper boundary and maximum deposition velocity at the lower boundary. The symbols f, ϕ, and v refer to mixing ratio, flux (cm^{-2} s^{-1}), and velocity (cm s^{-1}), respectively. The sign convention for ϕ and v is positive for upward flow.
[b] At tropopause $z = 45$ km ($n = 1.1 \times 10^{19}$ cm^{-3}).
[c] At $z = 1160$ km ($n = 1.2 \times 10^9$ cm^{-3}). For all species except H and H_2, for which the upper boundary is at 1425 km ($n = 4.9 \times 10^7$ cm^{-3}).
[d] All fluxes refer to the surface.

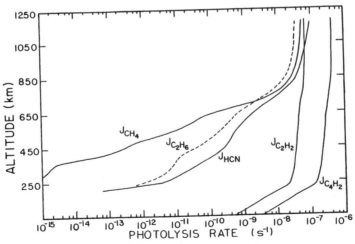

Figure 6.12 Altitude profiles for selected photodissociation coefficients. Total rates (sum of all branches) are given for CH_4, C_2H_2, C_2H_6, C_4H_2, and HCN. After Yung, Y. L. et al., 1984, "Photochemistry of the Atmosphere of Titan: Comparison between Model and Observations." *Astrophys. J.*, **55**, 465.

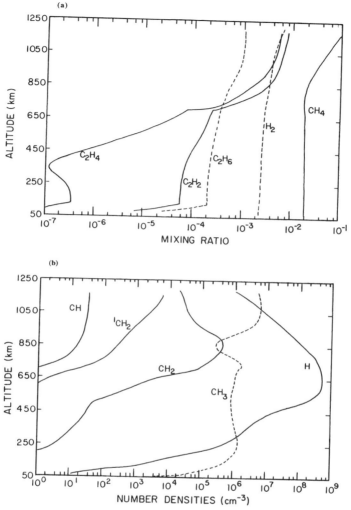

Figure 6.13 (a) Altitude profiles for the mixing ratios of CH_4, C_2H_2, C_2H_4, C_2H_6, and H_2. (b) Altitude profiles for the number densities of H, CH, $^1CH_2$, CH_2, and CH_3. After Yung, Y. L. et al., 1984, "Photochemistry of the Atmosphere of Titan: Comparison between Model and Observations." *Astrophys. J.*, **55**, 465.

$C_2H_4 + h\nu \rightarrow C_2H_2 + 2H$, R34: $CH_2 + CH_3 \rightarrow C_2H_4 + H$, and, in the lower stratosphere, the photolysis of acetylene and polyynes. The column-integrated combined production rate is 1.1×10^{10} cm^{-2} s^{-1}. All H atoms produced below 600 km are effectively scavenged by the polyynes. The escape flux at the upper boundary is 5.5×10^9 cm^{-2} s^{-1}. (Unless otherwise stated, all fluxes refer to the surface of Titan.) The escaping H atoms remain in a torus in the orbit of Titan around Saturn. The column abundance of H atoms above the $\tau = 1$ level for absorption by CH_4 at Lyman α is 3×10^{15} cm^{-2}.

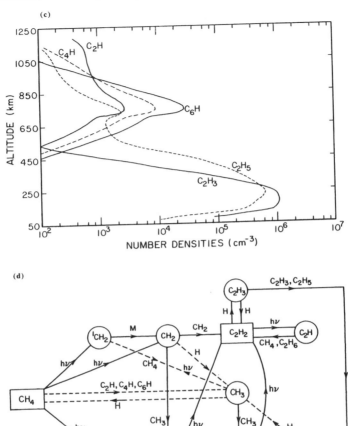

Figure 6.13 (c) Altitude profiles for the number densities of C_2H_3, C_2H_5, C_2H, C_4H, and C_6H. (d) Schematic diagram showing the major pathways for the interconversion of major C_1 and C_2 species. After Yung, Y. L. et al., 1984, "Photochemistry of the Atmosphere of Titan: Comparison between Model and Observations." *Astrophys. J.*, **55**, 465.

Radiative transfer modeling predicts a Lyman α airglow brightness of 600 R, close to the value of 500 R observed by Voyager UVS. This calculation assumes a solar Lyman α flux of 5.6×10^{11} photons cm^{-2} s^{-1} at 1 AU. The most important mechanism for removing H atoms is catalytic recombination to form H_2. Figure 6.14c gives the rates of the important reactions producing and removing H_2 in the model, R5a: $CH_4 + h\nu \rightarrow {}^1CH_2 + H_2$, R7a: $C_2H_4 + h\nu \rightarrow C_2H_2 + H_2$, R6b: $C_2H_2 + h\nu \rightarrow C_2 + H_2$, and R69b: $C_4H_3 + H$. Once formed, H_2 is chemically unreactive in the atmosphere. The fate of 99% of the H_2 formed in Titan's atmosphere is escape from the exosphere.

Table 6.9 Maximum mixing ratios of selected species at the tropopause of Titan

Molecule	Mixing ratio[a]
CH_4	3.0×10^{-2}
C_2H_6	5.1×10^{-7}
C_2H_4	4.1×10^{-6}
C_2H_2	6.9×10^{-10}
CH_3C_2H	3.8×10^{-11}
CH_2CCH_2	1.8×10^{-12}
C_3H_6	6.8×10^{-10}
C_3H_8	1.3×10^{-9}
C_4H_2	7.4×10^{-18}
C_4H_4	2.6×10^{-15}
$1\text{-}C_4H_6$	3.5×10^{-15}
$1,2\text{-}C_4H_6$	3.2×10^{-15}
$1,3\text{-}C_4H_6$	6.0×10^{-13}
C_4H_{10}	3.0×10^{-13}
HCN	2.3×10^{-17}
CO	1.5
CO_2	5.6×10^{-11}
H_2O	0
HCO_2	4.1×10^{-13}

From Yung, Y. L. et al. (1984).
[a] Temperature and number density of the ambient atmosphere at the tropopause assumed in the computations are 70 K and 1.1×10^{19} cm^{-3}, respectively. Vapor pressure data should be regarded as order of magnitude estimates.

The flux is 7.2×10^9 cm^{-2} s^{-1}. The model predicts a mixing ratio of H_2 equal to 2.1×10^{-3}, in good agreement with the abundance of $2 \pm 1 \times 10^{-3}$ deduced from Voyager IRIS observations.

Ethane is produced primarily by the association reaction R35: $2CH_3 \rightarrow C_2H_6$ with rates given in figure 6.14b. Most of the C_2H_6 production occurs in the lower stratosphere as a result of the photosensitized dissociation of CH_4. There is little destruction of C_2H_6 by photolysis or by attack by polyyne radicals. The principal loss mechanism is condensation at the tropopause, followed by deposition on the surface. The downward flux of C_2H_6 through the tropopause is 5.8×10^9 cm^{-2} s^{-1}. The model calculation of C_2H_6 abundance is in good agreement with Voyager IRIS observations.

Ethylene is formed in the upper atmosphere by the reaction R34: $CH_2 + CH_3 \rightarrow C_2H_4 + H$, with minor contributions from R27: $CH + CH_4 \rightarrow C_2H_4 + H$. The column-integrated production rates for these two reactions are 2.1×10^9 and 1.8×10^8, respectively. The altitude profile for the reaction rate of R34 is shown in figure 6.14b. Ethylene is poorly shielded by CH_4 and C_2H_2. Photolysis removes the species rapidly, and its mixing ratio falls off in the stratosphere, as shown in figure 6.13a. The secondary peak in C_2H_4 near 200 km is due to production by disproportionation and exchange reactions R38: $2C_2H_3 \rightarrow C_2H_4 + C_2H_2$ and R41b: $C_2H_3 + C_2H_5 \rightarrow 2C_2H_4$, where the radicals are ultimately derived from photosensitized dissociation of alkanes. The model C_2H_4 is less than IRIS observation by about an order of magnitude, which may result from uncertainties in the photochemical and kinetics data in the model.

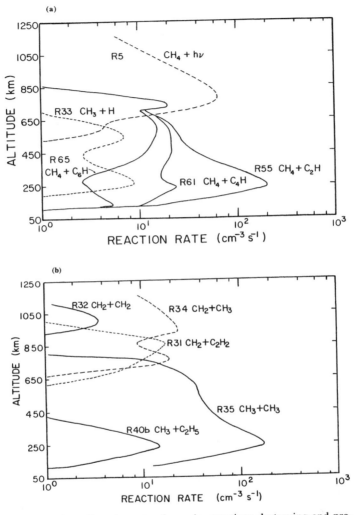

Figure 6.14 (a) Reaction rates for major reactions destroying and producing CH_4: (R5) $CH_4 + h\nu \to$ products, (R55) $CH_4 + C_2H \to CH_3 + C_2H_2$, (R61) $CH_4 + C_4H \to CH_3 + C_4H_2$, (R65) $CH_4 + C_6H \to CH_3 + C_6H_2$, and (R33) $CH_3 + H \to CH_4$. (b) Reaction rates for important synthesis reactions: (R31) $CH_2 + C_2H_2 \to CH_3C_2H$, and CH_2CCH_2 (the sum is shown here), (R32) $2CH_2 \to C_2H_2 + H_2$, (R34) $CH_2 + CH_3 \to C_2H_4 + H$, (R35) $2CH_3 \to C_2H_6$, and (R40b) $CH_3 + C_2H_5 \to C_3H_8$. After Yung, Y. L. et al., 1984, "Photochemistry of the Atmosphere of Titan: Comparison between Model and Observations." *Astrophys. J.*, **55**, 465.

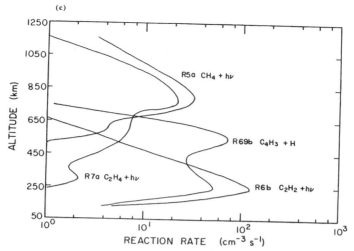

Figure 6.14 (c) Reaction rates for major reactions that produce H_2: (R5a) $CH_4 + h\nu \rightarrow {}^1CH_2 + H_2$, (R6b) $C_2H_2 + h\nu \rightarrow C_2 + H_2$, (R7a) $C_2H_4 + h\nu \rightarrow C_2H_2 + H_2$, and (R69b) $C_4H_3 + H \rightarrow C_4H_2 + H_2$. After Yung, Y. L. et al., 1984, "Photochemistry of the Atmosphere of Titan: Comparison between Model and Observations." *Astrophys. J.*, **55**, 465.

The primary sources of acetylene in the model are the reactions R7a: $C_2H_4 + h\nu \rightarrow C_2H_2 + H_2$ and R32: $2CH_2 \rightarrow C_2H_2 + H_2$. The rate of R7a: is approximately equal to that of formation of C_2H_4 by R34: $CH_2 + CH_3 \rightarrow C_2H_4 + H$. These rates are given in figure 6.14, b and c. Once formed, C_2H_2 is extremely stable against photolytic or chemical destruction. Photodissociation of C_2H_2 does not result in its irreversible destruction. C_2H_2 is used only as a catalyst for the photosensitized dissociation of alkanes. The major loss of C_2H_2 is by condensation. The downward flux is 1.2×10^9 cm^{-2} s^{-1}. The model prediction is within a factor of 2 of the Voyager IRIS observations.

(b) Polyynes

Altitude profiles for the mixing ratios of polyynes C_4H_2, C_6H_2, and C_8H_2 are presented in figure 6.15a. Only diacetylene (C_4H_2) has been observed. The polyynes are produced by insertion reactions, of which the principal ones are R58: $C_2H + C_2H_2 \rightarrow C_4H_2 + H$, R63: $C_4H + C_2H_2 \rightarrow C_6H_2 + H$, R67: $C_6H + C_2H_2 \rightarrow C_8H_2 + H$, R59: $C_2H + C_4H_2 \rightarrow C_6H_2 + H$, R60: $C_2H + C_6H_2 \rightarrow C_8H_2 + H$, and R64: $C_4H + C_4H_2 \rightarrow C_8H_2 + H$, with column-integrated production rates equal to 1.7×10^9, 1.2×10^9, 8.4×10^7, 1.9×10^9, and 1.7×10^8 cm^{-2} s^{-1}, respectively. The polyynes undergo photolysis in the atmosphere, but they are readily recycled, with the net result of photosensitized dissociation of CH_4 and other alkanes. As an example, the major reactions producing and destroying C_4H_2 are shown in figure 6.15b. The model C_4H_2 is an order of magnitude less than that reported by Voyager IRIS. The model predicts comparable concentrations of the higher polyynes, but no observations are available.

Table 6.10 Summary of important model results and comparison with observations

Species	Model abundance	Observed abundance[a]	Remarks
CH_4	2×10^{-2}	$1-3 \times 10^{-2}$	model value imposed as boundary condition at 45 km
H_2	2.1×10^{-3}	$2 \pm 1 \times 10^{-3}$	
C_2H_6	1.7×10^{-5}	2×10^{-5}	
C_2H_4	3.3×10^{-8}	4×10^{-7}	
C_2H_2	4.0×10^{-6}	2×10^{-6}	
CH_3C_2H	1.8×10^{-7} (3.0×10^{-8})	3×10^{-8}	—[b,c]
CH_2CCH_2	5.0×10^{-9}		searched for but not observed by IRIS
C_3H_8	7.9×10^{-7}	$2-4 \times 10^{-6}$	
C_4H_2	7.8×10^{-10}	$10^{-8} - 10^{-7}$	—[b]
HCN	5.6×10^{-7}	$2 \times 10^{-7} - 10^{-6}$	—[b]
HC_3N	5.6×10^{-8}	$10^{-8} - 10^{-7}$	—[b]
C_2N_2	1.9×10^{-8}	$10^{-8} - 10^{-7}$	—[b]
CO	1.8×10^{-4}	6×10^{-5}	—[d]
CO_2	1.5×10^{-9}	1.5×10^{-9}	flux of meteoritic H_2O has been adjusted to yield the correct CO_2 abundance

From Yung, Y. L. et al. (1984).
[a] Unless otherwise stated, concentrations are given in column-averaged mixing ratios above 45 km. To convert mixing ratios into column-integrated abundances above 45 km, multiply the appropriate mixing ratios by 1.9×10^{25} molecules cm^{-2}.
[b] Strong latitudinal asymmetry; more abundant at the north pole, to which the observations refer.
[c] Refers to the case when speculative reactions are included.
[d] The observed abundance for CO quoted here is from Lutz and Owen (1983) and Muhleman et al. (1984).

(c) C_3 and C_4 Compounds

Altitude profiles for the major C_3 species CH_3C_2H, CH_2CCH_2, C_3H_6, and C_3H_8 and their radicals C_3H_2, C_3H_3, C_3H_3 and C_3H_7 are shown in figures 6.16, a and b, respectively. The most abundant C_3 compound is methyl acetylene (CH_3C_2H), which has been detected. The model prediction is somewhat higher than the Voyager IRIS observations. Allene, the other isomer of C_3H_4, is much less abundant. As pointed in chapter 5 (reaction 5.56), allene isomerizes to methyl acetylene, which has a lower energy of formation. The primary source of CH_3C_2H is via the photolysis of C_3H_6, which is formed by the recombination reaction R37b: $C_2H_3 + CH_3 \rightarrow C_3H_6$. The major reactions that destroy CH_3C_2H are photolysis and cracking by H atoms. Very little is transported to the lower atmosphere.

Propane is formed in the model entirely by the recombination reaction R40b: $CH_3 + C_2H_5 \rightarrow C_3H_8$. The efficient rate of formation is due to the high rate coefficient for this class of association reactions (see section 3.5). C_3H_8 is lost by direct photolysis and photosensitized dissociation. The mixing ratio of C_3H_8 in the stratosphere is not constant due to its rapid destruction. Note the difference in the vertical profiles of C_3H_8 between the atmospheres of Titan and Jupiter (see figures 5.18b and 6.16a). There is more C_2H_2 on Titan, resulting in a higher rate of photosensitized dissociation of C_3H_8. The downward flux of C_3H_8 through the tropopause is 1.4×10^8 cm^{-2} s^{-1}.

Table 6.11 Budget for parent molecules CH_4, N_2, and H_2O

Reaction	Flux	Reaction	Flux
Destruction of CH_4 (sum total = 1.5×10^{10})		Destruction of N_2	
R72 $N(^2D) + CH_4$	1.0×10^9	$N_2 + e$	1.0×10^9
R05 $CH_4 + h\nu$	2.9×10^9	N_2+ cosmic rays	3.7×10^7
R53 $CH_4 + C_2$	3.0×10^9	escape[b]	1.6×10^8
R55 $CH_4 + C_2H$	5.1×10^9		
R61 $CH_4 + C_4H$	1.7×10^9	Production of N_2	
R65 $CH_4 + C_6H$	1.0×10^9	R76 N+NH	8.9×10^8
R78 $CH_4 + CN$	4.0×10^7		
		Downward flux of nitrile compounds:	
Production of CH_4 (sum total = 5.9×10^8):		HCN	2.0×10^8
R33 $CH_3 + H$	3.5×10^8	HC_3N	1.7×10^7
R37a $CH_3 + C_2H_3$	2.1×10^8	$2 \times C_2N_2$	1.2×10^7
Escape Fluxes of H and H_2:		Influx of meteoritic H_2O	6.1×10^5
H	5.8×10^9		
H_2	7.2×10^9	Escape of O[b]	3.3×10^4
Downward Fluxes of Hydrocarbons:[b]		$2 \times$ downward flux of CO_2[a]	5.7×10^5
C_2H_6	5.8×10^9		
C_2H_2	1.2×10^9	Net loss of parent molecules from atmosphere	
CH_3C_2H	5.7×10^7	CH_4	1.5×10^{10}
C_3H_8	1.4×10^8	N_2	2.8×10^8
C_4H_{10}	2.3×10^7	H_2O	6.1×10^5

From Yung, Y. L. et al. (1984).
[a] Integrated reaction rates and fluxes (normalized to the surface) are in units of $cm^{-2} s^{-1}$.
[b] At the tropopause; the species transported downward are assumed to be irreversibly lost.
[b] Taken from Strobel and Shemansky (1982).

The C_4 compounds in the model other than C_4H_2 include C_4H_4, 1-C_4H_6, 1,2-C_4H_6, 1,3-C_4H_6, C_4H_8, and C_4H_{10}. Their concentrations are generally low. The principal reactions forming the C_4 compounds are same as those in the Jovian atmosphere. The principal loss mechanism is photolysis and condensation. The most abundant of the species is butane (C_4H_{10}). None of these species has been observed, and their detection remains a challenge to future observations.

6.3.3 Nitrogen Chemistry

A complete set of important reactions related to nitrogen species is summarized in table 6.7. N_2 is almost chemically inert. Photochemistry of nitrogen species is initiated with the breaking of the strong N—N bond. As discussed in section 6.3.1, the thermosphere of Titan is a source of N atoms. Some of the downward-flowing N atoms are removed by recombination to form N_2. The surviving atoms react with hydrocarbon radicals to form HCN and H_2CN:

$$N + CH_2 \rightarrow HCN + H \qquad (6.46)$$

$$N + CH_3 \rightarrow H_2CN + H \qquad (6.47)$$

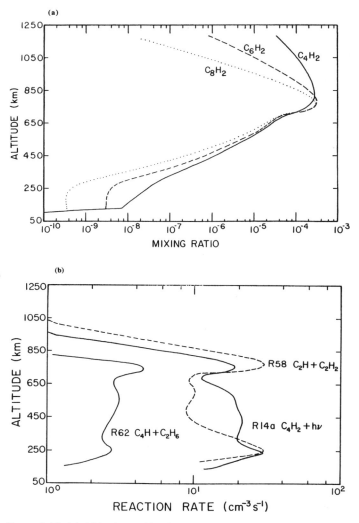

Figure 6.15 (a) Altitude profiles for the mixing ratios of C_4H_2, C_6H_2, and C_8H_2. (b) Reaction rates for selected important reactions producing, destroying, and recycling C_4H_2: (R58) $C_2H_2 + C_2H \rightarrow C_4H_2 + H$, (R14a) $C_4H_2 + h\nu \rightarrow C_4H + H$, and (R62) $C_4H + C_2H_6 \rightarrow C_4H_2 + C_2H_5$. After Yung, Y. L. et al., 1984, "Photochemistry of the Atmosphere of Titan: Comparison between Model and Observations." *Astrophys. J.*, **55**, 465.

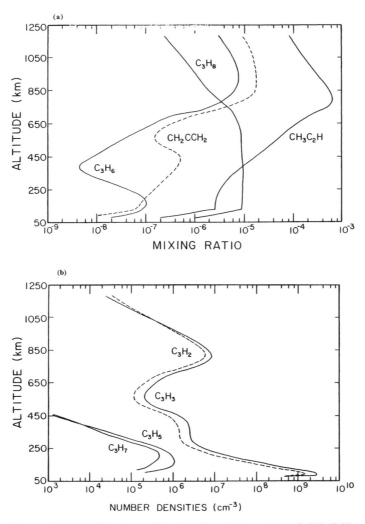

Figure 6.16 (a) Altitude profiles for the mixing ratios of CH_3C_2H, CH_2CCH_2, C_3H_6, and C_3H_8. (b) Altitude profiles for the number densities of C_3H_2, C_3H_3, C_3H_5, and C_3H_7. After Yung, Y. L. et al., 1984, "Photochemistry of the Atmosphere of Titan: Comparison between Model and Observations." *Astrophys. J.*, **55**, 465.

H_2CN is not stable and is readily removed by

$$H_2CN + h\nu \rightarrow HCN + H \tag{6.48}$$

$$H_2CN + H \rightarrow HCN + H_2 \tag{6.49}$$

The end-product is HCN, which is much more stable than H_2CN. Photolysis of HCN produces the reactive nitrile radical CN:

$$HCN + h\nu \rightarrow CN + H \tag{6.50}$$

The most likely fate of CN is to react with CH_4, restoring HCN:

$$CN + CH_4 \rightarrow HCN + CH_3 \tag{6.51a}$$

The net result of reactions (6.50) and (6.51a) is $CH_4 \rightarrow CH_3 + H$. Thus, the photolysis of HCN results in the photosensitized dissociation of CH_4. This accounts for the remarkable stability of HCN in a reducing atmosphere. There may be a small branch for the reaction between CN and CH_4 to proceed by the insertion of CN into CH_4:

$$CN + CH_4 \rightarrow H + CH_3CN \tag{6.51b}$$

The ratio of the rate coefficient for (6.51b) to that of (6.51a) is less than 0.005, but the reaction may be a significant source of methyl cyanide. An alternative source of CH_3CN is the recombination reaction:

$$CN + CH_3 + M \rightarrow CH_3CN + M \tag{6.52}$$

A number of more complex nitrile compounds can be produced from reactions between HCN, CN, and the hydrocarbons:

$$CN + HCN \rightarrow H + C_2N_2 \tag{6.53}$$

$$CN + C_2H_2 \rightarrow H + HC_3N \tag{6.54}$$

$$C_2H + HCN \rightarrow H + HC_3N \tag{6.55}$$

$$CN + HC_3N \rightarrow H + C_4N_2 \tag{6.56}$$

The rate coefficient for (6.53) is slower than expected for this type of reaction. To account for the observed abundance of HC_3N, another scheme has been proposed:

$$N(^2D) + C_2H_2 \rightarrow H + CHCN \tag{6.57}$$

$$CHCN + N \rightarrow H + C_2N_2 \tag{6.58}$$

$$CHCN + CHCN \rightarrow H_2 + C_4N_2 \tag{6.59}$$

where the excited atom $N(^2D)$ is derived from energetic processes in the thermosphere. Most of the above speculative chemistry has not been quantitatively studied in the laboratory.

The boundary conditions for the model are summarized in table 6.8. In addition to the source of odd nitrogen from the thermosphere, absorption of cosmic rays by N_2 in the lower stratosphere is also a source of odd nitrogen. The model includes both types of nitrogen sources. Figure 6.17a summarizes the altitude profiles of the important nitrile compounds HCN, HC_3N, and C_2N_2. The principal source of nitrile compounds is the upper atmosphere. The primary sink is condensation at the tropopause (see table 6.9). Since the abundances of the nitrile compounds are quite high at high altitudes and these compounds all have very low saturation vapor pressure, it is possible that they could be the source of the photochemical aerosol layer observed by the Voyager UVS. However, no quantitative modeling has been carried out to verify this possibility. The concentrations of the odd nitrogen radicals N, NH, and CN are shown in

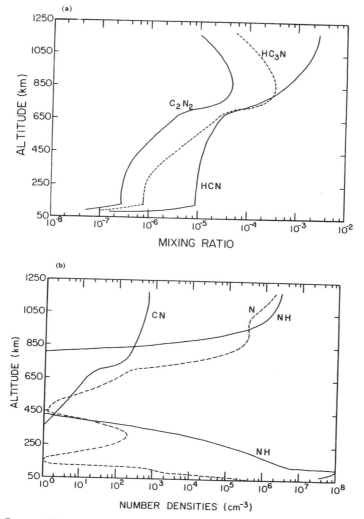

Figure 6.17 (a) Altitude profiles for the mixing ratios of HCN, HC$_3$N, and C$_2$N$_2$. (b) Altitude profiles for the number densities of N, NH, and CN. After Yung, Y. L. et al., 1984, "Photochemistry of the Atmosphere of Titan: Comparison between Model and Observations." *Astrophys. J.*, **55**, 465.

figure 6.17b. The secondary peak in the concentrations of the radicals is due to the cosmic ray source. The major source of odd nitrogen is the downward flux from the thermosphere, 2×10^9 cm^{-2} s^{-1}. The total contribution from cosmic ray dissociation is 7.4×10^7 cm^{-2} s^{-1} in the lower stratosphere. A schematic diagram showing the relation between the nitrogen compounds is given in figure 6.17c. The predicted concentrations are in reasonable agreement with Voyager IRIS observations. The major reactions that destroy odd nitrogen (by returning it to N$_2$) and reactions that produce

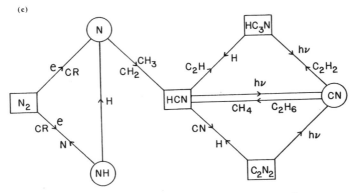

Figure 6.17 (c) Schematic diagram showing the major pathways connecting the nitrogen compounds. After Yung, Y. L. et al., 1984, "Photochemistry of the Atmosphere of Titan: Comparison between Model and Observations." *Astrophys. J.*, **55**, 465.

nitrile compounds are shown in figure 6.18a. A considerable fraction of odd nitrogen produced is destroyed by R76 : $NH + N \rightarrow N_2 + H$. As discussed earlier, the C–N bond is produced mainly by reactions R77: $N + CH_3 \rightarrow H_2CN + H$ and R73: $N + CH_2 \rightarrow HCN + H$. The column-integrated production rates are 2.6×10^8 and 4.6×10^6 cm^{-2} s^{-1}, respectively, for these two reactions. Once formed the C–N bond is extremely stable, and its removal by condensation represents a net sink of atmospheric nitrogen. The flux is 2×10^8 N atoms cm^{-2} s^{-1}. The rates of recycling between nitrile species are shown in figure 6.18b. A sampling of the important recycling reactions includes R17: $HCN + h\nu \rightarrow H + CN$, R79: $CN + C_2H_6 \rightarrow HCN + C_2H_5$, R81: $C_2H + HCN \rightarrow HC_3N + H$, R87 + R88: $C_2N_2 + 2H \rightarrow 2HCN$, and R78: $CN + CH_4 \rightarrow CH_3CN + H$.

6.3.4 Oxygen Chemistry

Two oxygen-bearing molecules, CO and CO_2, have been detected on Titan. Since the principal form of carbon in the outer solar system is CH_4, the existence of a molecule like CO_2 with the maximum state of oxidation is a surprise. The origin of CO may be photochemical or primordial. The origin of CO_2 is certainly photochemical. The mixing ratio of CO_2 in the stratosphere exceeds 10 ppb. Comparison with table 6.9 suggests that the observed CO_2 is far above that allowed by the saturation vapor pressure. The tropopause is a continuous sink of CO_2, and the observed abundance of CO_2 must be maintained by a steady photochemical production of the molecule in the upper atmosphere.

The list of important reactions involving oxygen is given in tables 6.6 and 6.7. There are at least two sources of oxygen on Titan. There is a continuous flux of micrometeoroids into Titan's atmosphere. On burning up in the upper atmosphere, H_2O molecules are released. Dissociation of the H_2O molecules provides a source of oxygen. We show that it is possible to generate CO and CO_2 in Titan's atmosphere using the H_2O derived from micrometeoroids. However, it is possible that the CO

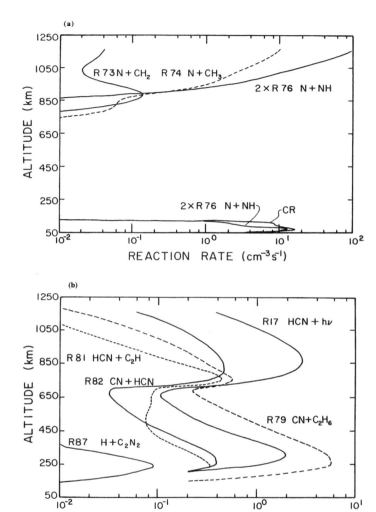

Figure 6.18 (a) Reaction rates for reactions producing and destroying odd nitrogen and those forming the CN bond: CR = production of odd nitrogen (N + NH) by cosmic rays, (R76) N + NH → N_2 + H, (R73) N + CH_2 → HCN + H, and (R74) N + CH_3 → HCN + H_2. The major source of odd nitrogen is downward fluxes of N and NH at the upper boundary, as discussed in section 6.3.3. (b) Reaction rates for selected important reactions recycling nitrile species: (R17) HCN + $h\nu$ → H + CN, (R79) CN + C_2H_6 → HCN + C_2H_5, (R81) C_2H + HCN → HC_3N + H, (R82) CN + HCN → C_2N_2 + H, and (R87) C_2N_2 + 2H → 2HCN. After Yung, Y. L. et al., 1984, "Photochemistry of the Atmosphere of Titan: Comparison between Model and Observations." *Astrophys. J.*, **55**, 465.

on Titan is primordial. The argument derives some support from the recent detection of CO ice on the surface of Triton and Pluto. If CO is primordial and there exists a reservoir of CO in the ice on the surface, then the photolysis of CO provides a source of oxygen. However, we show that it is difficult to use this source of oxygen to oxidize CO further to CO_2.

As mentioned in section 6.3 the sources of reactive oxygen-bearing species are

$$H_2O + h\nu \to OH + H \tag{6.60}$$

$$CO + h\nu \to O + C \tag{6.61}$$

The primary fate of O and OH is to form CO by the following insertion reactions:

$$O + CH_2 \to CO + 2H \tag{6.62}$$

$$O + CH_3 \to H_2CO + H \tag{6.63}$$

$$OH + CH_2 \to CO + H_2 + H \tag{6.64}$$

$$OH + CH_3 \to CO + 2H_2 \tag{6.65}$$

Since H_2CO rapidly dissociates, the final stable product is CO. The carbon atoms produced in (6.61) will most likely react with hydrocarbon radicals:

$$C + CH_2 \to C_2H + H \tag{6.66}$$

$$C + CH_3 \to C_2H_2 + H \tag{6.67}$$

The OH derived from (6.60) can react with CO in the well-known reaction

$$CO + OH \to CO_2 + H \tag{6.68}$$

Therefore, to first order, the photolysis of H_2O results in two chemical schemes:

$$\begin{array}{rl} & H_2O + h\nu \to OH + H \hspace{3em} (6.60) \\ & OH + CH_3 \to CO + 2H_2 \hspace{2em} (6.65) \\ & \underline{(CH_4 + h\nu \to CH_3 + H)} \\ \text{net} & H_2O + CH_4 \to CO + 2H + 2H_2 \hspace{1em} \text{(I)} \end{array}$$

$$\begin{array}{rl} & H_2O + h\nu \to OH + H \hspace{3em} (6.60) \\ & \underline{(CO + OH \to CO_2 + H)} \hspace{2em} (6.68) \\ \text{net} & H_2O + CO \to CO_2 + 2H \hspace{3em} \text{(II)} \end{array}$$

where in scheme (I) the breaking up of CH_4 into CH_3 and H is via photosensitized dissociation. Schemes (I) and (II) offer a simple mechanism for regulating the abundances of CO and CO_2. In steady state the rates of production and loss of CO must balance:

$$\int_0^\infty k_{65}[OH][CH_3]dz = \int_0^\infty k_{68}[OH][CO]dz \tag{6.69}$$

Since H_2O influx and condensation of CO_2 are, respectively, the major source and sink of oxygen, we have

$$\phi(H_2O) = 2\phi(CO_2) \tag{6.70}$$

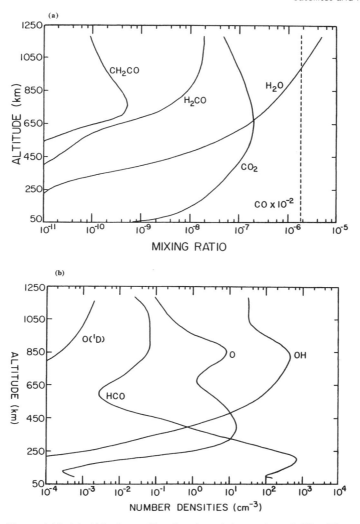

Figure 6.19 (a) Altitude profiles for the mixing ratios of CO, CO_2, H_2O, H_2CO, and CH_2CO. (b) Altitude profiles for the number densities of O, $O(^1D)$, OH, and HCO. After Yung, Y. L. et al., 1984, "Photochemistry of the Atmosphere of Titan: Comparison between Model and Observations." *Astrophys. J.*, **55**, 465.

where $\phi(H_2O)$ is the meteoritic H_2O flux and $\phi(CO_2)$ is the downward flux of CO_2 into the troposphere.

Altitude profiles for the mixing ratios of CO, CO_2, H_2O, H_2CO, and CH_2CO and the number densities of $O(^1D)$, O, OH, and HCO are presented in figure 6.19, a and b, respectively. A schematic diagram showing the relation between the oxygen-bearing compounds is given in figure 6.19c. The downward flux of H_2O at the upper boundary of the model is $6.1 \times 10^5 \, \text{cm}^{-2} \, \text{s}^{-1}$. Other boundary conditions of the model are

(c)

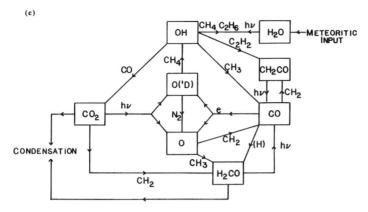

Figure 6.19 (c) Schematic diagram showing the interaction between the oxygen species. After Yung, Y. L. et al., 1984, "Photochemistry of the Atmosphere of Titan: Comparison between Model and Observations." *Astrophys. J.*, **55**, 465.

summarized in table 6.8. The model predicts mixing ratios of 1.8×10^{-4} and 1.5×10^{-9} for CO and CO_2, respectively. The observed value of CO_2 is 6×10^{-5}. The flux of H_2O has been chosen to reproduce the observed CO. The rapid decrease of the mixing ratio profiles for H_2O and CO_2 in the lower stratosphere is due to condensation. CO does not condense at the tropopause, and its abundance is roughly uniform in the upper atmosphere. The concentrations of the radicals shown in figure 6.19b are enhanced in the region of maximum production in the mesosphere. The secondary peak in HCO in the lower stratosphere is due to the reaction between H and CO, with the H atoms being supplied by the photosensitized dissociation of CH_4. The rates of the major chemical reactions that form and destroy CO_2, R96: $CO + OH \rightarrow CO_2 + H$ and R20: $CO_2 + h\nu$, are shown in figure 6.20a. The most important loss of CO_2, however, is condensation at the tropopause. The downward flux is 2.8×10^5 cm^{-2} s^{-1}. The rates of important reactions that form and recycle CO are shown in figure 6.20b. The most important reaction for producing CO is via R95: $OH + CH_3 \rightarrow H_2CO + H$. H_2CO is not stable, and its ultimate fate is to produce CO by photolysis.

In the above analysis we implicitly assume that Titan has no primordial CO and that H_2O is the sole source of oxygen to the satellite. The model predicts the abundances of CO and CO_2 on Titan. Thus, there is much merit in such a simple model. However, if CO is primordial and buffered by a large supply of ice, then its abundance is decoupled from the photochemistry of H_2O and the steady state requirement (6.69) is no longer valid. In this case the model can only predict the abundance of CO_2 in the atmosphere through a modified version of (6.70):

$$\phi(H_2O) = \phi(CO_2) \qquad (6.71)$$

Without carrying out quantitative calculations, we expect the decoupling of CO from H_2O to increase the predicted abundance of CO_2 by a factor of 2 from the difference between (6.70) and (6.71).

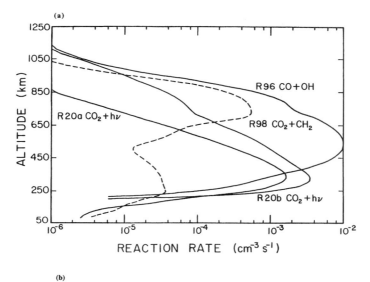

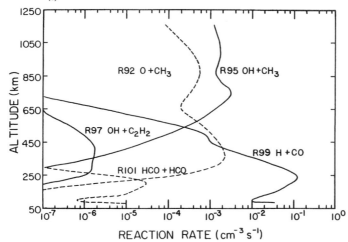

Figure 6.20 (a) Reaction rates for major reactions destroying CO_2: (R20a) $CO_2 + h\nu \rightarrow CO + O$, (R20b) $CO_2 + h\nu \rightarrow CO + O(^1D)$, and (R98) $CO_2 + CH_2 \rightarrow CO + H_2CO$; and producing CO_2: (R96) $CO + OH \rightarrow CO_2 + OH$. (b) Reaction rates for selected important reactions involving O, OH, and HCO: (R92) $O + CH_3 \rightarrow H_2CO + H$, (R97) $OH + C_2H_2 \rightarrow CH_2CO + H$, (R95) $OH + CH_3 \rightarrow CO + 2H_2$, (R99) $H + CO \rightarrow HCO$, and (R101) $2HCO \rightarrow H_2CO + CO$. After Yung, Y. L. et al., 1984, "Photochemistry of the Atmosphere of Titan: Comparison between Model and Observations." *Astrophys. J.*, **55**, 465.

6.3.5 Comparison with Giant Planets

A comparison of the abundances of the principal hydrocarbon species, CH_4, C_2H_6, and C_2H_2, in the giant planets and Titan is summarized in chapter 5 (table 5.14). There is more CH_4, C_2H_6, and C_2H_2 in Titan. The abundance of CH_4 in the atmosphere determines the optical depth of unity at Lyman α. Thus, to first order, the primary photolysis of CH_4 does not depend on the mixing ratio of CH_4. The efficiency of the synthesis of the higher hydrocarbons depends on a large number of factors such as solar flux, temperature, eddy mixing, removal of H atoms, and condensation. A systematic sensitivity study of these factors has been carried out only for Jupiter. We qualitatively discuss the causes of the difference in C_2H_6 and C_2H_2 in the atmospheres of the outer solar system. As discussed in section 6.3.2, the quantum yield for organic synthesis is higher on Titan than in the giant planets because H and H_2 can escape from Titan. The extremely low concentrations of C_2H_6 and C_2H_2 on Uranus are due to the extreme quiescent nature of its upper atmosphere. The higher hydrocarbons synthesized at high altitude are locally destroyed before they can be transported to the lower stratosphere. A more extensive comparison between the composition of Titan and Jupiter is given in figure 6.6. The results support the general picture that the higher hydrocarbons are more abundant on Titan.

6.4 Triton

The Voyager flyby of Triton in 1989 revealed an extremely cold satellite covered by nitrogen ice. The surface temperature was measured at 38 K. The predominantly N_2 atmosphere of 17.5 μbar is in equilibrium with N_2 ice (see figure 6.21). A trace amount of CH_4 in the atmosphere was detected by the Voyager UVS. Post-Voyager observations revealed the presence of CH_4, CO, and CO_2 on the surface of Triton. The variation of vapor pressure of N_2, CH_4, CO and Ar over the temperature range appropriate for the outer solar system is shown in figure 6.21. Thus, the presence of small amounts of CO, CH_4, and Ar at 38 K would be consistent with their respective vapor pressures.

The temperature of the upper atmosphere increases with altitude and reaches 95 K above 400 km. A tropopause may exist at 8 km above the surface. A haze layer near the surface has been detected. Perhaps the most surprising discovery by the Voyager encounter is the ionosphere with peak electron densities in excess of 10^4 cm^{-3}, a value much greater than the upper limit of electron densities in the atmosphere of Titan. The atmosphere of Triton is sufficiently thin that there is considerable influence on the atmospheric composition from both the thermosphere above and the surface below.

6.4.1 Chemical Kinetics

The photochemistry of hydrocarbons, nitrogen, and oxygen species on Triton is similar to that of Titan, and we do not repeat the discussion here. There is more uncertainty in our knowledge of chemical kinetics at the lower temperatures in the atmosphere of Triton. In addition, because of the low temperature, the higher hydrocarbons and the nitrile species readily condense near the surface to form aerosols. As in the case of

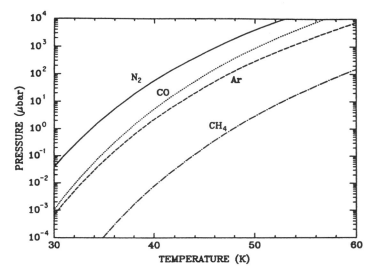

Figure 6.21 Saturation vapor pressures over the pure condensates of selected volatiles as a function of temperature. After Yelle, R. V. et al., 1995, Lower atmospheric structure and surface-atmosphere interactions on Triton, in *Neptune and Triton* (Tucson: The University of Arizona Press), p. 1031.

Titan, the thermosphere of Triton is a source of N atoms, but as we show below, it is the carbon atoms that play a new and remarkable role in Triton's atmosphere.

As mentioned above, Triton has an unusual ionosphere, with an unusually high electron number density. An obvious candidate for the major ion is N^+, formed by photoionization and electron impact via (6.24b) and

$$N_2 + e \rightarrow N^+ + N + 2e \tag{6.72}$$

Additional N^+ may be formed by charge transfer from N_2^+:

$$N_2^+ + N \rightarrow N^+ + N_2 \tag{6.73}$$

where N_2^+ is derived from photoionization (6.24a) or electron impact ionization. However, N^+ is readily removed in the ionosphere by the charge exchange reaction,

$$N^+ + H_2 \rightarrow NH^+ + H \tag{6.74}$$

followed by dissociative recombination,

$$NH^+ + e \rightarrow N + H \tag{6.75}$$

thus leading to the rapid loss of ionization below the ionospheric peak. H_2 in (6.74) is derived from the photolysis of CH_4 in the lower part of the atmosphere. The loss of ions by (6.74) and (6.75) is so drastic that the electron concentrations predicted using solar EUV are an order of magnitude smaller than that detected by Voyager RSS. The early models of Triton's ionosphere all postulated that the principal energy source is precipitation of magnetospheric electrons. The required energy flux is about a factor of 10 higher than that of solar EUV flux. We show that there is another possibility: the chemistry of C atoms and the role of its ion C^+ in the ionosphere.

There are a number of sources of C atoms, the most important of which is

$$CO^+ + e \to C + O \tag{6.76}$$

where the ion CO^+ is derived from

$$CO + h\nu \to CO^+ + h\nu \tag{6.77}$$

$$N_2^+ + CO \to CO^+ + N_2 \tag{6.78}$$

$$N^+ + CO \to CO^+ + N \tag{6.79}$$

Additional contributions to the production of C are from reactions (6.61) and

$$CH_4 + h\nu \to C(^1D) + 2H_2 \tag{6.80}$$

where the quantum yield of (6.80) at Lyman α is 0.4%. The excited C atom formed in (6.80) is readily quenched by

$$C(^1D) + N_2 \to C + N_2 \tag{6.81}$$

C atoms react with N_2 in a three-body reaction:

$$C + N_2 + M \to CNN + M \tag{6.82}$$

It is most likely that CNN would react with N:

$$N + CNN \to CN + N_2 \tag{6.83}$$

followed by

$$N + CN \to C + N_2 \tag{6.84}$$

Note that the net result of the above reactions is the recombination of N atoms using C atoms as a catalyst:

$$
\begin{array}{lr}
C + N_2 + M \to CNN + M & (6.85) \\
N + CNN \to CN + N_2 & (6.86) \\
\underline{N + CN \to C + N_2} & (6.87) \\
net \quad N + N \to N_2 & (\text{III})
\end{array}
$$

This is the principal chemical scheme for the recombination of N atoms. The C atoms are recycled in (III). Since C has a low ionization potential (11.26 eV), most ions will transfer to C in reactions such as

$$N_2^+ + C \to N_2 + C^+ \tag{6.88}$$

$$N^+ + C \to N + C^+ \tag{6.89}$$

Conversely, the transfer reaction for C^+,

$$C^+ + H_2 \to CH^+ + H \tag{6.90}$$

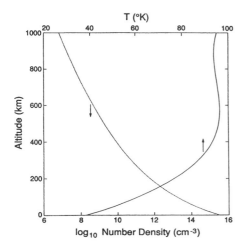

Figure 6.22 Temperature and number density of the atmosphere of Triton as a function of altitude based on Voyager 2 data. After Yung and Lyons (1990).

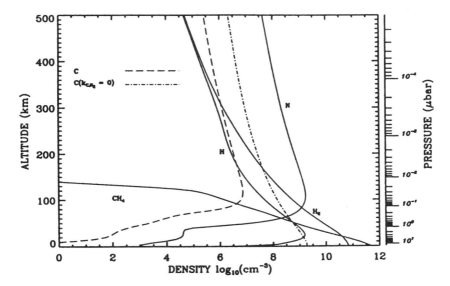

Figure 6.23 Typical globally averaged number density profiles of important minor constituents at ionospheric heights. N atom mixing ratio at 400 km is 0.021. The C density profiles were calculated with (dashed line) and without (dot-dashed line) reactions with N_2. After Strobel, D. F., and Summers, M. E., 1995, Triton's upper atmosphere and ionosphere, in *Neptune and Triton* (Tucson: The University of Arizona Press), p. 1107.

238 Photochemistry of Planetary Atmospheres

is extremely slow. The principal sink for C^+ is charge transfer to CH_4:

$$C^+ + CH_4 \rightarrow C_2H_2^+ + H_2 \quad (6.91a)$$
$$\rightarrow C_2H_3^+ + H \quad (6.91b)$$

The molecular ions produced in (6.91a) and (6.91b) are rapidly removed by dissociative recombination. Since the mixing ratio of CH_4 falls off rapidly with altitude due to its destruction by photolysis, reactions (6.91a) and (6.91b) result in a slow loss of C^+.

6.4.2 Photochemical Model

A model atmosphere of Triton that is consistent with Voyager observations is presented in figure 6.22. The mixing ratios of CH_4 and CO at the surface are 1.7×10^{-4} and 1×10^{-4}, respectively. H and H_2 can escape at almost thermal velocities. Heavier atoms can also escape but at much smaller effusion velocities.

The concentrations of the neutral species computed in the model are presented in figure 6.23. The "parent molecule" of all hydrocarbons, CH_4, falls off rapidly with altitude due to its loss by photodissociation. The rate of photolysis peaks near the surface. The optical depth near Lyman α for the entire column of CH_4 is on the order of unity. The vertical profile of CH_4 computed by the model is consistent with Voyager UVS observations. H, H_2, C_2H_4, and C_2H_2 are the most abundant products

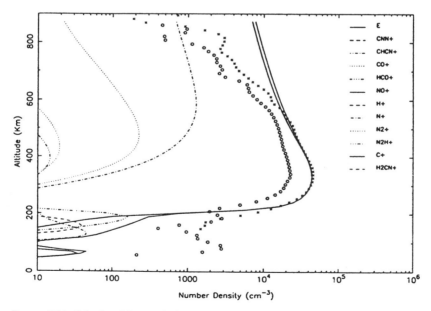

Figure 6.24 Calculated ion and electron densities produced by solar radiation only compared with RSS-measured electron densities for winter, 16° N, dusk (circles), and summer, 45° S, dawn (crosses), occultations. After Lyons, J. R. et al., 1992, "Solar Control of the Upper Atmosphere of Triton." *Science* **256**, 204–206.

of CH_4 photochemistry. However, the concentrations of H and H_2 are limited by their ability to escape from the upper boundary. The higher hydrocarbons are removed by condensation at the surface. Conspicuous by its absence is ethane. There are few methyl radicals produced in the reactions since there is too little C_2H_2 to drive the photosensitized dissociation of CH_4. CO (not shown) has nearly uniform mixing ratio as it has almost the same mass as N_2 and its rate of destruction in the upper atmosphere is trivial compared with the rate of supply from the CO ice on the surface.

The most important atomic species in the model are N, O, and C, produced primarily by photolysis of N_2 and CO. The most interesting consequences of N and O are the production of HCN and CO_2. Removal by condensation may be an important source of CO_2 ice that has been observed on the surface of Triton. The rate of deposition on the surface is 1.3×10^5 cm^{-2} s^{-1}. The high abundance of C atoms results in C^+ being the most abundant ion in the model of the ionosphere, as shown in figure 6.24. The C^+ concentration predicted on the basis of solar photoionization alone is of the correct magnitude to account for the Voyager RSS experiment. The circles and crosses denote the ingress (winter) and egress (summer) observations, respectively. Concentrations of N^+ and N_2^+ are about one and two orders of magnitude smaller than C^+, respectively. In the lower ionosphere there are more complex ions such as HCO^+ and H_2CN^+. The latter ion is believed to be the major ion in the atmosphere of Titan, but its concentration is small in Triton's ionosphere. Since N_2 is the major gas in the atmosphere, the bulk of photoionization results in the production of N_2^+. As pointed out in section 6.4.1, N_2^+ readily transfers charge to N^+ and CO^+, and eventually to the terminal ion C^+. The unusually large concentration of C^+ in the model is due to its inability to transfer charge to a molecular ion. This property appears to be unique to C^+, since the other atomic ions H^+ and N^+ are rapidly removed by charge transfer to molecular ions, followed by dissociative recombination.

6.5 Pluto

Pluto, the ninth planet in the solar system, was discovered in 1930. Our present knowledge of Pluto is at the level of astronomy rather than planetary science. The planet has not been visited by any spacecraft. The orbit of Pluto is highly eccentric ($e = 0.25$) and inclined ($i = 17°$), with an orbital period of 248 yr. The semimajor axis is 39.6 AU, but because of its high eccentricity its perihelion at 29.7 AU is inside the orbit of Neptune. The aphelion is at 49.5 AU, near the edge of the Kuiper Belt. The solar insolation at Pluto varies by about a factor of 3 between aphelion and perihelion. The disk of Pluto cannot be resolved in telescopic observations and, to complicate matters, Pluto has a satellite, Charon, that cannot be resolved from the planet. Thus, for a long time the radius and the density of the planet remained uncertain by large factors. However, a series of observations of the mutual eclipse events of Pluto and Charon (a once per century event that occurred between 1985 and 1990) has yielded highly reliable orbital and physical parameters for the system. Note that the radius of Pluto, 1151 km, is less than Triton's 1360 km. The density of the Pluto-Charon system is 2.03 g cm^{-3}, a value that is typical of an icy body with a rocky core.

Figure 6.25 A comparison of a continuum-adjusted spectrum of Pluto and a model spectrum based on the indicated mixture of ices over the wavelength range of 1.4 to 2.4 μm. The model is an intimate mixture of the components rather than the expected molecular mixture. After Owen, T. C. et al., 1993, "Surface ices and the atmospheric composition of Pluto," *Science* **261**, 745.

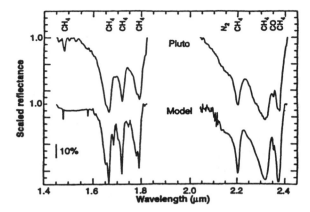

6.5.1 Surface

The high geometric albedo of Pluto suggests that the surface of the planet is covered by a white frost. Figure 6.25 gives the infrared solar reflectance spectrum of the surface of Pluto. Comparison with a model spectrum shown also in figure 6.25 suggests that the bulk composition of the surface frost is N_2, with trace amounts of CH_4 and CO. This composition is similar to that of the surface of Triton with the exception of CO_2, which is present on Triton but not on Pluto. The surface material of Pluto consists of 98% N_2, 1.5% CH_4, 0.5% CO, and less than 0.07% CO_2. For comparison, the surface of Triton consists of 99% N_2, 0.05% CH_4, 0.1% CO, and 0.2% CO_2. The overall chemical composition of the minor constituents is more reducing on Pluto than on Triton. No H_2O ice has been detected on either Pluto or Triton, but its presence has been established on the surface of Pluto's satellite Charon.

6.5.2 Atmosphere

The existence of an atmosphere on Pluto may be inferred from the presence of ices on the surface. A minimum atmosphere would be given by the vapor pressure of N_2 at the temperature of the surface. The best determination of the surface temperature from millimeter wave thermal emissions of the planet is 30–44 K, with a preferred range of 35–37 K. However, the variation of vapor pressure of N_2, CH_4, and CO with temperature over this range is large, as shown in figure 6.21. For example, at 30 K the vapor pressure of N_2 is about 0.1 μbar, but at 44 K the vapor pressure exceeds 500 μbar. Thus, the atmospheric pressure is poorly constrained by the thermal data.

The correct order of magnitude of Pluto's atmosphere has been deduced from the occultation of a star by Pluto in 1988. The atmosphere revealed its presence by the gradual rather than abrupt dimming of the star's light. The simplest interpretation of this beautiful experiment is to assume an isothermal, spherical, and hydrostatic atmosphere. The half-light level of the light curve occurred at 1214 ± 20 km, and at this altitude the atmospheric scale height was 59.7 ± 1.5 km. Assuming that the bulk

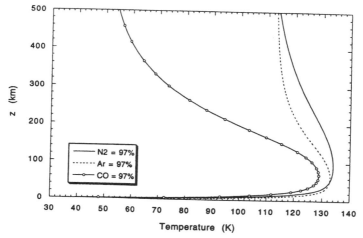

Figure 6.26 Model temperature profiles for the atmosphere of Pluto, with the surface temperature $T_0 = 35$ K, surface pressure $p_0 = 3$ μbar, $\gamma_{CH_4} = 3\%$. "$N_2 = 97\%$," "$Ar = 97\%$," and "$CO = 97\%$" denote atmospheres dominated by N_2, Ar, and CO, respectively. After Strobel D. F. et al., 1996, "On the vertical thermal structure of Pluto's atmosphere," *Icarus* **120**, p. 266.

atmosphere is N_2, the inferred temperature is 117 K. The pressure is 0.78 μbar at the half-light level. There is general agreement on the interpretation of the light-dimming curve from first contact to the half-light level.

A careful examination of the extinction data reveals that the light-dimming curve has two distinct slopes. The isothermal model that provides a satisfactory fit to the data fails as the light ray from the star gets close to the surface of Pluto. There are two models that can account for the change of slope in the data. One model postulates the presence of an aerosol layer with vertical optical depth exceeding 0.19 at 0.64 μm immersed in an isothermal atmosphere. An alternative interpretation is that the discontinuity near the half-light level is caused by a large temperature gradient near the surface of Pluto. At the moment the controversy is not resolved. We briefly comment on the two models. The first model based on an isothermal atmosphere of 117 K is obviously incorrect near Pluto's surface since it cannot exceed 44 K based on the vapor pressure of N_2. Therefore, the existence of a thermal gradient near the surface is not in doubt, especially when compared with the temperature profile of the atmosphere of Triton. But the thermal gradient model requires almost all the heat be deposited in the first scale height of the atmosphere to produce the necessary gradient of 10 K/km. Ironically, the most plausible solution (in our opinion) to the heating problem is to postulate aerosol heating! Thus, the true solution may involve a combination of these two vastly different models. This range of uncertainty is illustrated by the thermal models shown in figure 6.26.

There has been no quantitative modeling of the photochemistry of Pluto's atmosphere. We expect the essential features of the hydrocarbon, nitrogen, and oxygen chemistry to be similar to that of Triton. Since the abundances of CH_4 and CO are an order of magnitude higher on Pluto's surface than on Triton's surface, we expect that the important chemical processes occur at higher altitudes, well above the region of condensation for the photochemical products at the surface. Although Pluto is farther away from the sun than Triton, it has a more extended atmosphere (see section 6.5.3) with an optical cross section that is much larger than the geometric cross section of the planet. Thus, the EUV solar radiation received by the two atmospheres may be comparable.

6.5.3 Extended Atmosphere

Due to the low gravity at the surface of Pluto, the atmosphere is believed to extend well above the surface. Indeed, the optical path of unity for Lyman α may occur at two Pluto radii from the center of the planet. In this case the total flux of EUV radiation received by the planet is about four times that intercepted by the geometric cross section of the planet. Early researchers recognized the possibility of hydrodynamic escape of CH_4 from Pluto, resulting in catastrophic loss of atmosphere and surface material (as in a comet). However, later studies demonstrated that the escape may be limited by the availability of solar EUV energy. The maximum escape flux is of the order of 10^{10} cm^{-2} s^{-1}. There are no observations to date to verify this intriguing hypothesis.

Escape rates of this magnitude are at least two orders of magnitude greater than the corresponding escape rates from the atmospheres of the terrestrial planets. Higher rates of escape from the inner planets are believed to have occurred close to the time of origin of their atmospheres. A similar mechanism, hydrodynamic escape, could have operated on the terrestrial planets. Thus, an understanding of the evolution of the atmosphere of Pluto may provide insights into the early evolution of planetary atmospheres in the inner solar system.

6.6 Unsolved Problems

We list a number of outstanding unresolved problems related to the small bodies in the solar system:

1. Is there a residual atmosphere on the nightside of Io? What is it made of?
2. What is the nature of the interaction between Io's atmosphere and the Jovian magnetosphere?
3. What is the ultimate fate of higher hydrocarbons (derived from CH_4 photochemistry) on the surface of Titan, Triton, and Pluto?
4. Is the origin of CO on Titan primordial or photochemical?
5. What is the chemical scheme that produces CH_3CN in the atmosphere of Titan? In reactions

$$CN + CH_4 \rightarrow HCN + CH_3 \quad (6.51a)$$
$$\rightarrow H + CH_3CN \quad (6.51b)$$

what is the branching ratio k_{51b}/k_{51a}?
6. What is the fate of the condensed hydrocarbons and nitriles when exposed to prolonged ultraviolet radiation in the stratosphere of Titan or the surface of Triton and Pluto?

7. Why is there so little CO relative to CH_4 on Triton and Pluto when observations show that CO is the dominant form of carbon in the molecular clouds?
8. Is there an aerosol layer or an extremely steep thermal gradient near the surface of Pluto?
9. What are the abundances of the photochemically produced species in the atmosphere of Pluto?
10. Is there hydrodynamic escape of gases from the atmosphere of Pluto?
11. Why are the abundant noble gases (e.g., Ar and Ne) absent from the atmospheres of the small bodies?

7

Mars

7.1 Introduction

Mars has been extensively studied by a series of spacecraft since the dawn of the space age: by Mariners 4, 6, 7, and 9 (1965–1972), Mars 2 through 6 (1971–1974), and the two Viking Landers and Orbiters in 1976. The knowledge from spacecraft is supplemented by ground-based observations.

The essential aspects of Mars are summarized in table 7.1. It is a smaller planet than Earth; the radius and mass are, respectively, 53% and 11% of Earth. The surface gravity is 3.71 m s^{-2}, compared with the terrestrial value of 9.82 m s^{-2}. The physical properties and composition of the Martian atmosphere are summarized in tables 7.1 and 7.2; isotopic composition is given in table 7.3. An example of how this knowledge is obtained is illustrated in figure 7.1, showing the mass spectrum obtained by the mass spectrometer experiment on Viking.

The bulk atmosphere is composed of CO_2, with small amounts of N_2 and Ar and a trace amount of water vapor. Located at 1.52 AU from the sun, the mean insolation at Mars is about half that of Earth. As a result, it is a colder planet, with mean surface temperature of 220 K, too cold for water to flow on the surface in the current epoch. The lack of an ocean results in an arid and dusty climate. The obliquity of Mars is 25.2°, close to the terrestrial value of 23.5°; however, Mars has an eccentric orbit, with eccentricity of 0.093. The ratio of incident solar radiation at perihelion to aphelion is 1.45. The large seasonal variation in heating is believed to be responsible for the spectacular global dust storms that can be observed from Earth and have inspired imaginative but erroneous theories about their origin. The polar regions of Mars can

Mars 245

Table 7.1 Astronomical and atmospheric data for mars

Radius (equatorial)	3394.5 km
Mass	6.4185×10^{23} kg
Mean density	3.9335 g cm^{-3}
Gravity (surface, equator)	3.711 m s^{-2}
Semimajor axis	1.52366 AU
Obliquity (relative to orbital plane)	25.19°
Eccentricity of orbit	0.0934
Period of revolution (Earth days)	686.98
Orbital velocity	24.13 km s^{-1}
Period of rotation	$24^h\ 37^m\ 22.663^s$
Atmospheric pressure at surface	5.6 mbar
Mass of atmospheric column	1.50×10^{-2} kg cm^{-2}
Total atmospheric mass	2.17×10^{16} kg
Equilibrium temperature	216 K
Surface temperature	220 K
Annual variation in solar insolation[a]	1.45
Atmospheric scale height ($T = 210$ K)	10.8 km
Adiabatic lapse rate (dry)	4.5 K km^{-1}
Atmospheric lapse rate[b]	2.5 K km^{-1}
Escape velocity	5.027 km s^{-1}

Taken from Kieffer, H. H. et al. (1992).
[a] Based on the ratio of aphelion to perihelion distance squared.
[b] Mean observed value, variable.

be as cold as 125 K, so CO_2 will condense as frost on the surface. In fact, according to the Leighton–Murray model, this is what determines the pressure of the atmosphere. Figure 7.2 shows the seasonal pressure variations at the Viking lander sites for 3.3 Mars years from 1976. Note that the magnitude of the pressure changes is of the order of 20%, compared to the maximum change of 1% on the surface of Earth.

The most fascinating aspect of the photochemistry of Mars is the discovery by McElroy in 1972 that there is self-regulation of the oxidation state of the atmosphere

Table 7.2 Chemical composition of the atmosphere of Mars

Gas	Abundance	Reference and remarks
CO_2	0.9532	(1)
N_2	0.027	(1)
^{40}Ar	0.016	(1)
O_2	1.3×10^{-3}	(1)
CO	7.0×10^{-4}	(1)
H_2O	3.0×10^{-4}	(2); variable
^{36}Ar+^{38}Ar	5.3×10^{-6}	(1)
Ne	2.5×10^{-6}	(1)
Kr	3.0×10^{-7}	(1)
Xe	8.0×10^{-8}	(1)
O_3	3.0×10^{-8}	(3); variable

(1) Owen et al. (1977); (2) Farmer and Doms (1979); (3) Barth (1974).

Table 7.3 Isotopic composition of the atmosphere of Mars

Species	Ratio	Reference
D/H	$9 \pm 4 \times 10^{-4}$	(1)
	$7.8 \pm 0.3 \times 10^{-4}$	(2)
$^{12}C/^{13}C$	90 ± 5	(3)
$^{14}N/^{15}N$	170 ± 15	(3)
$^{16}O/^{17}O$	2655 ± 25	(2)
$^{16}O/^{18}O$	490 ± 25	(3)
	545 ± 20	(2)
$^{36}Ar/^{38}Ar$	5.5 ± 1.5	(4)
$^{40}Ar/^{36}Ar$	3000 ± 500	(5)
$^{129}Xe/^{132}Xe$	$2.5 \pm ^{2}_{1}$	(5)

(1) Owen et al. (1988); (2) Bjoraker et al. (1989); (3) Nier and McElroy (1977); (4) Biemann et al. (1976).

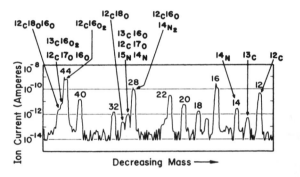

Figure 7.1 Mass spectrum showing various mass peaks (amu) as measured by the neutral mass spectrometer on board Viking 1. From Nier, A. O. et al., 1976, Isotopic composition of the Martian atmosphere *Science*, **194**, 68.

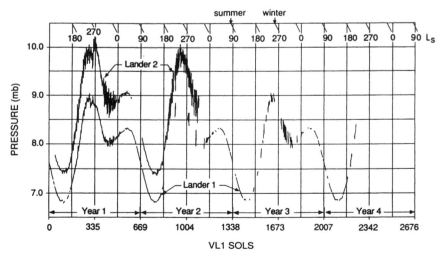

Figure 7.2 Surface pressures recorded by Viking Landers 1 and 2. The years refer to Martian years. From Tillman, J. E., 1988, Mars global atmospheric oscillations—annually synchronized, transient normal-mode oscillations and the triggering of global dust storms *J. Geophys. Res.*, **93**, 9433. SOL = Martian day.

by photochemistry and escape of hydrogen and oxygen from the planet. Mars is the first planet in the solar system whose global chemical environment was shown to be sensitively controlled by trace constituents that can act as catalysts for important chemical cycles. Subsequently, similar catalytic effects were shown to be important in other planetary atmospheres, including that of Earth.

7.2 Photochemistry

7.2.1 Pure CO_2 atmosphere

The major constituent of the Martian atmosphere, CO_2, is readily photolyzed by solar ultraviolet radiation under 2050 Å:

$$CO_2 + h\nu \rightarrow CO + O \tag{7.1}$$

where near the threshold the O atom is in the ground state, $O(^3P)$, but at shorter wavelengths the atom could be in excited states $O(^1D)$ and $O(^1S)$. However, the primary fate of the excited atoms is quenching to the ground state by CO_2 (a small fraction of the excited atoms reacts with H_2O). Once CO_2 is converted into CO and $O(^3P)$, it is difficult to restore it. The reverse of reaction (7.1),

$$CO + O + M \rightarrow CO_2 + M \tag{7.2}$$

is spin-forbidden. The three body rate coefficient at 200 K is 3×10^{-37} cm^3 s^{-1}, a value that is many orders of magnitude smaller than the corresponding rate coefficient for a typical three-body reaction such as

$$O + O + M \rightarrow O_2 + M \tag{7.3}$$

with rate coefficient equal to 2.8×10^{-32} cm^3 s^{-1} at 200 K. Hence, the net results of CO_2 photodissociation are

$$2CO_2 \rightarrow 2CO + O_2 \tag{7.4}$$

Thus, a pure CO_2 atmosphere exposed to solar ultraviolet radiation would have large amounts of CO and O_2 at a ratio of 2 : 1. Of course, O_2 cannot build up indefinitely, and eventually it would dissociate:

$$O_2 + h\nu \rightarrow O + O \tag{7.5}$$

We can construct a self-consistent model with reactions (7.1)–(7.5), assuming that the only path for reversing the result of photodissociation (7.4) is the slow reaction (7.2), with the oxygen atoms being supplied by (7.5). Such a model predicts that the mixing ratios of CO and O_2 are 7.72×10^{-2} and 3.87×10^{-2}, respectively. However, the predictions of CO and O_2 are greater than the observed abundances summarized in table 7.2 by factors of 110 and 30, respectively. In addition, the model CO/O_2 ratio of 2 is significantly larger than the observed ratio of 0.5. It is of interest to point out that in this hypothetical model of Mars, the predicted O_3 mixing ratio at the surface is 10^{-5} and the column-integrated O_3 is 3.4×10^{18} cm^{-2} or 126 DU (1 Dobson unit = 2.69×10^{16} cm^{-2}). This amount of O_3 is sufficient for shielding the surface of Mars from harmful ultraviolet radiation.

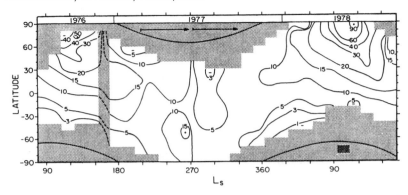

Figure 7.3 Contours of water vapor abundance as a function of latitude and season as measured by the Viking Orbiter. The Earth calendar years are given at the top of the figure. (L_S = areocentric longitude of the sun; $L_S = 0$ is at the vernal equinox for the northerm hemisphere) The units are precipitable microns of water. From Jakosky, B. M., and Farmer, C. B., 1982, "The seasonal and global behavior of water-vapor in the Mars atmosphere—complete global results of the Viking atmospheric water detector experiment." *J. Geophys. Res.* **87**, 2999.

7.2.2 HO_x Catalytic Chemistry

The remarkable stability of CO_2 in the Martian atmosphere appears to be a contradiction of the known chemical kinetics for a pure CO_2 atmosphere. This outstanding problem, the lack of high amounts of CO and O_2 in the Martian atmosphere, is known as the CO_2 stability problem. This problem was first independently solved in two classic papers of McElroy and Donahue (1972) and Parkinson and Hunten (1972). The crucial idea is the recognition that the small amount of water vapor in the atmosphere could play a fundamental role in the photochemistry of the the atmosphere. As listed in table 7.2 the amount of H_2O in the atmosphere of Mars is of the order of a few hundred parts per million and is highly variable both as a function of altitude, latitude, and season. A convenient measure of the column-integrated H_2O amount in the atmosphere is the precipitable micron (pr-μm); 1 pr-μm equals 3.34×10^{18} molecules/cm^2, corresponding to a mixing ratio of 15 ppm if uniformly mixed. The distribution of the H_2O column abundances on Mars over a Martian year is shown in figure 7.3. The complexity of the Martian hydrological cycle is comparable to that of the terrestrial hydrological cycle. The northern polar cap of Mars seems to be a source of water vapor in spring. The dominant flow of volatiles on Mars is from pole to pole as a function of season, whereas on Earth atmospheric water (in the form of vapor or cloud) flows from low latitudes to the high latitudes.

H_2O is photolyzed in the atmosphere by the absorption of solar ultraviolet photons:

$$H_2O + h\nu \rightarrow OH + H \tag{7.6}$$

The OH radical readily attacks CO to form CO_2:

$$CO + OH \rightarrow CO_2 + H \tag{7.7}$$

The recycling of H may be accomplished by the following reactions:

$$H + O_2 + M \rightarrow HO_2 + M \quad (7.8)$$

$$O + HO_2 \rightarrow O_2 + OH \quad (7.9)$$

The sequence of reactions may be summarized as

$$CO + OH \rightarrow CO_2 + H \quad (7.7)$$
$$H + O_2 + M \rightarrow HO_2 + M \quad (7.8)$$
$$O + HO_2 \rightarrow O_2 + OH \quad (7.9)$$
$$\text{net} \quad CO + O \rightarrow CO_2. \quad \text{(Ia)}$$

The net result of chemical scheme (Ia) is the same as (7.2), but it is orders of magnitude faster because each of the reactions that constitute the cycle in (Ia) is an allowed reaction. Note that the HO_x radicals (H, OH, HO_2) are used as a catalyst in the recombination of CO and O. Since the HO_x radicals are not consumed in the chemical scheme, very few molecules are needed. As we show in section 7.3.2, the mole fraction of the sum of the HO_x radicals needed to ensure the stability of the bulk atmosphere of Mars is less than 10^{-9}. All the profoundness and subtlety of photochemistry is contained in the statement that tiny amounts of a reactive species control the bulk composition of the atmosphere. This has the obvious consequence that the composition of the atmosphere is easily perturbed by altering the concentrations of key radical species.

There is a variant of chemical scheme (Ia) that can be summarized as follows:

$$CO + OH \rightarrow CO_2 + H \quad (7.7)$$
$$H + O_3 \rightarrow OH + O_2 \quad (7.10)$$
$$O + O_2 + M \rightarrow O_3 + M \quad (7.11)$$
$$\text{net} \quad CO + O \rightarrow CO_2 \quad \text{(Ib)}$$

In this scheme the OH that is consumed in (7.7) is restored by reaction with O_3 formed in a three-body reaction (7.11). Both chemical schemes (Ia) and (Ib) share a common property that they recombine CO and O to form CO_2. These schemes cannot use O_2 formed in (7.3) for the oxidation of CO. O_2 must first be decomposed by (7.5). However, the following scheme is capable of directly using O_2 for the oxidation of CO:

$$2(CO + OH \rightarrow CO_2 + H) \quad (7.7)$$
$$2(H + O_2 + M \rightarrow HO_2 + M) \quad (7.8)$$
$$HO_2 + HO_2 \rightarrow H_2O_2 + O_2 \quad (7.12)$$
$$H_2O_2 + h\nu \rightarrow 2OH \quad (7.13)$$
$$\text{net} \quad 2CO + O_2 \rightarrow 2CO_2 \quad \text{(II)}$$

Note that in chemical scheme (II) the O–O bond is broken by the photolysis of hydrogen peroxide (7.13).

Chemical schemes (Ia) and (Ib) are more important for a dry atmosphere. Scheme (II) is more important when H_2O is abundant. The rate-limiting step of this scheme, reaction (7.12), is quadratic in the concentration of HO_2 radicals. The relative importance of the two schemes in the current atmosphere is discussed later in section 7.3.2. Early work showed that the three HO_x schemes provide a satisfactory solution to the stability problem. The only significant addition since the 1972 classic papers is the new reaction that breaks the O–O bond by

$$HO_2 + NO \rightarrow NO_2 + OH \qquad (7.14)$$

where the O_2 in HO_2 is derived from (7.8) and NO is derived from the dissociation of N_2 (see section 7.2.7). This introduces two new chemical schemes for the oxidation of CO:

$$
\begin{array}{rlll}
& 2(CO + OH & \rightarrow & CO_2 + H) \qquad (7.7) \\
& 2(H + O_2 + M & \rightarrow & HO_2 + M) \qquad (7.8) \\
& HO_2 + NO & \rightarrow & NO_2 + OH \qquad (7.14) \\
& NO_2 + h\nu & \rightarrow & NO + O \qquad (7.15) \\
& O + HO_2 & \rightarrow & O_2 + OH \qquad (7.9) \\
\hline
net & 2CO + O_2 & \rightarrow & 2CO_2 \qquad (IIIa)
\end{array}
$$

$$
\begin{array}{rlll}
& CO + OH & \rightarrow & CO_2 + H \qquad (7.7) \\
& H + O_2 + M & \rightarrow & HO_2 + M \qquad (7.8) \\
& HO_2 + NO & \rightarrow & NO_2 + OH \qquad (7.14) \\
& O + NO_2 & \rightarrow & O_2 + NO \qquad (7.16) \\
\hline
net & CO + O & \rightarrow & CO_2 \qquad (IIIb)
\end{array}
$$

Note that this scheme uses HO_x and oxides of nitrogen (NO_x) as catalysts. Chemical schemes (IIIa) and (IIIb) are examples of the synergistic coupling between two kinds of catalysts. There are many examples this kind of coupled scheme in the terrestrial atmosphere (see chapter 9). As we show in section 7.3.2, the coupled HO_x–NO_x scheme is less important than the two classical HO_x schemes for the oxidation of CO to CO_2.

7.2.3 O_2 and O_3

We have shown that HO_x and NO_x species can catalytically break the O–O bond in the atmosphere. We now show that the O–O bond can also be formed using HO_x as a catalyst, as in the following scheme:

$$
\begin{array}{rlll}
& O + OH & \rightarrow & O_2 + H \qquad (7.17) \\
& 2(CO_2 + h\nu & \rightarrow & CO + O) \qquad (7.1) \\
& H + O_2 + M & \rightarrow & HO_2 + M \qquad (7.8) \\
& O + HO_2 & \rightarrow & O_2 + OH \qquad (7.9) \\
\hline
net & 2CO_2 & \rightarrow & 2CO + O_2 \qquad (IV)
\end{array}
$$

where the crucial reaction is (7.17) in which the O–O bond is formed. The net result of scheme (IV) is the same as (7.4). Thus, the presence of HO_x radicals promotes the formation of O_2.

As pointed out in section 7.2.1, the column ozone concentration in a hypothetical pure CO_2 atmosphere on Mars could be as high as 126 DU. In a realistic atmosphere with trace quantities of H_2O, the amount of O_3 is smaller by three orders of magnitude. Part of the reason is the loss of oxygen atoms in (7.17) and (7.9), resulting in scheme IV that converts CO_2 to CO and O_2 rather than CO and O. The other reason is that

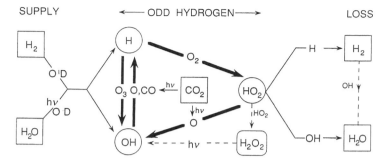

Figure 7.4 Schematic diagram showing the principal pathways for reactions involving hydrogen species. From Nair, H., Allen, M., Anbar, A. D., Yung, Y. L., and Clancy, R. T., 1994, "A Photochemical Model of the Martian Atmosphere." *Icarus* **111**, 124.

(7.10) removes O_3 directly. It is unlikely that the O atom transferred to OH in (7.10) returns to O_3 because the most probable fate of this OH molecule is to react with CO to form CO_2 (7.7). We should note that the photolysis of O_3 only leads to a recycling of O_3 and not a net loss of odd oxygen (odd oxygen $O_x = O + O_3$).

7.2.4 Hydrogen Escape

A schematic showing the principal pathways of the hydrogen species in the Martian atmosphere is presented in figure 7.4. H_2O is the ultimate source of all hydrogen species in the atmosphere. The production of the reactive radical species (HO_x) is initiated by photolysis (7.6) with a minor contribution from

$$H_2O + O(^1D) \rightarrow OH + OH \qquad (7.18)$$

where the excited oxygen atom is derived from the photolysis of O_3 [formed by (7.11)]:

$$O_3 + h\nu \rightarrow O_2 + O(^1D) \qquad (7.19)$$

Once they are produced, the HO_x species participate in the catalytic cycles for the oxidation of CO to CO_2. Ultimately, the HO_x species are destroyed by the reaction that turns the radicals back to H_2O:

$$OH + HO_2 \rightarrow H_2O + O_2 \qquad (7.20)$$

This ultimately limits the amount of HO_x radicals that can exist in the atmosphere.

There is, however, a minor path to remove HO_x by the formation of H_2 via the reaction

$$H + HO_2 \rightarrow H_2 + O_2 \qquad (7.21)$$

The result may be summarized by the following chemical scheme

$$2(H + HO_2 \rightarrow H_2 + O_2) \quad (7.21)$$
$$2(H_2O + h\nu \rightarrow OH + H) \quad (7.6)$$
$$CO + OH \rightarrow CO_2 + H \quad (7.7)$$
$$CO_2 + h\nu \rightarrow CO + O \quad (7.1)$$
$$O + OH \rightarrow O_2 + H \quad (7.17)$$
$$\underline{2(H + O_2 + M \rightarrow HO_2 + M)} \quad (7.8)$$
$$net \quad 2H_2O \rightarrow 2H_2 + O_2 \quad (V)$$

Most of the H_2 produced in scheme (V) is removed by the reactions

$$H_2 + OH \rightarrow H_2O + H \quad (7.22)$$
$$H_2 + O(^1D) \rightarrow OH + H \quad (7.23)$$

and the ultimate fate of the HO_x radicals produced in (7.22) and (7.23) is to return to H_2O. However, a small portion (about 20%) of the H_2 formed in scheme (V) will escape chemical destruction in the lower atmosphere and get transported to the upper atmosphere, where the molecule can be decomposed by ionospheric reactions:

$$CO_2 + h\nu \rightarrow CO_2^+ + e^- \quad (7.24)$$
$$H_2 + CO_2^+ \rightarrow CO_2H^+ + H \quad (7.25)$$
$$\underline{CO_2H^+ + e^- \rightarrow CO_2 + H} \quad (7.26)$$
$$net \quad H_2 \rightarrow 2H \quad (VI)$$

We defer the detailed discussion of ion chemistry to section 7.2.6. Reaction (7.6) produces an H atom in the lower atmosphere of Mars. Its chemical lifetime is short, and it cannot readily diffuse to the top of the atmosphere where it can escape. However, the H_2 molecule produced by scheme (V) is relatively unreactive. It can readily diffuse to the upper atmosphere, where it is decomposed into H atoms by the ion reactions (scheme VI). The hydrogen atoms produced in the ionosphere can diffuse to the exobase and escape thermally from Mars. H_2 can also escape, but its flux is a minor fraction of the total escape flux. The escape of hydrogen from Mars is a well-documented phenomenon. Ultraviolet observations at Lyman α indicate that Mars is surrounded by a corona of escaping H atoms.

In summary, the escape of hydrogen from Mars consists of five steps:

(a) source of hydrogen from the dissociation of H_2O;
(b) production of the long-range carrier of hydrogen, H_2, by scheme (V);
(c) diffusion of H_2 from the lower atmosphere to the upper atmosphere;
(d) decomposition of H_2 into H atoms by scheme (VI), and
(e) thermal escape of H atoms.

These five steps show that the photochemistry of HO_x, transport, ionospheric chemistry, and exospheric dynamics are intimately coupled. The consequence of hydrogen escape for the oxidation state of the atmosphere of Mars is discussed in section 7.2.5.

7.2.5 Oxygen Escape

The escape of hydrogen from Mars has one important implication for the chemistry of the atmosphere. The ultimate source of the escaping hydrogen is H_2O, decomposed according to scheme (V)

$$2H_2O \rightarrow 2H_2 + O_2 \quad (V)$$

Note that the loss of every two molecules of H_2 would leave behind one molecule of O_2. If there were no permanent sink for O_2, the oxygen concentration in the Martian atmosphere would continue to build up. It can be shown that at the current escape rate of hydrogen, it takes only on the order of 10^5 yr to double the amount of O_2 in the Martian atmosphere. Thus, if the Martian atmosphere has reached a steady state for its state of oxidation, there must be a sink for oxygen comparable in magnitude to the escape rate of hydrogen, that is,

$$\phi(O) = 0.5\,\phi(H) \qquad (7.27)$$

where $\phi(O)$ is the escape rate of oxygen atoms (in any form) and $\phi(H)$ is the escape rate of hydrogen atoms (in any form).

Unlike hydrogen, oxygen is too heavy to escape from Mars thermally. However, there is a nonthermal mechanism associated with ion chemistry that can drive the escape of oxygen from Mars. Oxygen atoms produced in the dissociative recombination reaction

$$O_2^+ + e^- \rightarrow O + O \qquad (7.28)$$

may have as much as 2.5 eV per atom, or 7 km s^{-1}. This is higher than the escape velocity of 5 km s^{-1} for Mars. Hence, exothermic oxygen atoms produced above the exobase can escape. From scheme (V), the oxygen escape rate must be half that of hydrogen so that the state of oxidation of the atmosphere remains unchanged.

The question then arises of how these two vastly different mechanisms of escape, one via Jeans escape and the other via dissociative recombination, can affect each other. This problem is known as self-regulation, and we briefly describe it as follows. As discussed in section 7.2.4, the escape rate of hydrogen is controlled by five steps. One of the rate-limiting steps is the production of H_2 in the HO_x chemistry,

$$P(H_2) = \int_0^\infty k_{30}[H][HO_2]\,dz \qquad (7.29)$$

This rate is proportional to the product of two HO_x radicals, H and HO_2. HO_2 is the most abundant of the HO_x radicals in the region of active photochemistry. Its concentration is larger than OH and H, and it is insensitive to O_2. However, reaction (7.8) converts H into HO_2, and this reaction determines the concentration of H atoms in the atmosphere. At high O_2 concentrations [H] and $P(H_2)$ are small, and vice versa.

Consider a hypothetical situation in which the balance in (7.27) is not satisfied. For example, we can imagine a decrease in $\phi(O)$ such that

$$\phi(O) < 0.5\,\phi(H)$$

Initially hydrogen will continue to escape at the same rate. But this would result in a rapid buildup of O_2. By (7.8) and (7.29) the rate of production of H_2, $P(H_2)$, would fall as a result of the increase of O_2. Eventually a new steady state is reached in which $\phi(H)$ would be smaller, so (7.27) is again satisfied. Similar arguments can be applied to the case in which

$$\phi(O) > 0.5\,\phi(H)$$

In this case, the O_2 concentration of the atmosphere will decrease due to its higher rate of loss. This causes an increase in $P(H_2)$ by (7.8) and (7.29), resulting in a greater escape rate of hydrogen, $\phi(H)$. Again, a new steady state would be reached that would satisfy (7.27). The time-dependent response of the atmosphere in which the escape rates of hydrogen and oxygen are initially perturbed from the stoichiometric ratio of (7.27) are presented in section 7.4.1.

7.2.6 Ionosphere

The ionosphere of Mars plays a fundamental role in at least three important aspects of the Martian atmosphere. Ion chemistry is a source of odd nitrogen that is important for the CO_2 stability (schemes IIIa and IIIb). Ion reactions are responsible for breaking H_2 into H atoms and supply a source of escaping atoms. Ionic reactions produce energetic oxygen atoms that can escape from Mars. We briefly describe the essential features of the ionosphere of Mars in this section.

Ion production occurs by ionization of CO_2 at wavelengths below 900 Å (7.24). However, the major ion in the ionosphere is not CO_2^+ but O_2^+ formed by the exchange reaction

$$CO_2^+ + O \rightarrow O_2^+ + CO \quad (7.30)$$

O_2^+ may also be formed in two steps

$$CO_2^+ + O \rightarrow O^+ + CO_2 \quad (7.31)$$
$$CO_2 + O^+ \rightarrow O_2^+ + CO \quad (7.32)$$

The most important pathway for the removal of ions is dissociative recombination by (7.28) and

$$CO_2^+ + e^- \rightarrow CO + O \quad (7.33)$$

A minor loss process is via charge transfer to NO:

$$O_2^+ + NO \rightarrow O_2 + NO^+ \quad (7.34)$$
$$CO_2^+ + NO \rightarrow CO_2 + NO^+ \quad (7.35)$$

followed by dissociative recombination of NO^+:

$$NO^+ + e^- \rightarrow N + O \quad (7.36)$$

The source of NO in the upper atmosphere of Mars is discussed in section 7.2.7. Another minor reaction for removing CO_2^+ is (7.25), which results in chemical cycle (VI), the breaking up of H_2 into H atoms.

Note that (7.28) is capable of producing exothermic oxygen atoms that can escape from Mars. Another reaction that is important for the evolution of the Martian atmosphere is

$$N_2^+ + e^- \rightarrow N + N \quad (7.37)$$

in which the N atoms produced have velocities in excess of the escape velocity from Mars.

7.2.7 Nitrogen Chemistry

Although nitrogen amounts to 2.7% of the atmosphere, it is relatively unreactive unless its strong N—N bond is broken, forming odd nitrogen. It is convenient to group the oxides of nitrogen as NO_x (N + NO + NO_2 + NO_3 + $2 N_2O_5$) and NO_y (NO_x + HNO_2 + HNO_3 + HO_2NO_2). Odd nitrogen is formed primarily by the absorption of extreme ultraviolet solar radiation or photoelectron in the ionosphere or the absorption of comic rays in the lower atmosphere.

N_2 is dissociated by EUV solar radiation:

$$N_2 + h\nu \rightarrow N + N(^2D) \tag{7.38a}$$
$$\rightarrow N^+ + N + e^- \tag{7.38b}$$
$$\rightarrow N_2^+ + e^- \tag{7.38c}$$

where N denotes the nitrogen atom in the ground [$N(^4S)$] state and $N(^2D)$ is the first excited state of the nitrogen atom. For our purpose, we adopt the common assumption that the yields of N and $N(^2D)$ in (7.38) are 0.5 for each branch. The primary fate of $N(^2D)$ and N^+ is to react with CO_2 to form NO:

$$N(^2D) + CO_2 \rightarrow NO + CO \tag{7.39}$$
$$N^+ + CO_2 \rightarrow NO + CO^+ \tag{7.40}$$

The primary fate of N_2^+ is the charge transfer reaction with CO_2 and O:

$$N_2^+ + CO_2 \rightarrow N_2 + CO_2^+ \tag{7.41}$$
$$N_2^+ + O \rightarrow NO^+ + N \tag{7.42}$$

and dissociative recombination to form N atoms (7.37). The N—N bond can also be broken by impact dissociation by photoelectrons [produced in reactions such as (7.24)]

$$N_2 + e^- \rightarrow N^+ + N + 2 e^- \tag{7.43}$$
$$\rightarrow N + N + e^- \tag{7.44}$$

The above energetic processes all ultimately lead to the production of odd nitrogen (N and NO). NO is readily removed by dissociation:

$$NO + h\nu \rightarrow N + O \tag{7.45}$$

The permanent sink for odd nitrogen in the upper atmosphere is the reaction

$$N + NO \rightarrow N_2 + O \tag{7.46}$$

This reaction destroys two odd nitrogen species with the formation of a new N—N bond.

Odd nitrogen that survives the destruction by (7.46) is transported to the lower atmosphere where higher oxides of nitrogen are formed:

$$NO + HO_2 \rightarrow NO_2 + OH \tag{7.47}$$

$$NO_2 + O + M \rightarrow NO_3 + M \tag{7.48}$$

$$NO_2 + NO_3 + M \rightarrow N_2O_5 + M \tag{7.49}$$

The major loss of these higher oxides is by photolysis. However, a substantial fraction of NO_2 is removed (recycled) by (7.16) in scheme (IIIb). The importance of NO_x species in the catalytic chemical schemes for the stability of CO_2 has been discussed in section 7.2.2. A small amount of N_2O may be formed by

$$N + NO_2 \rightarrow N_2O + O \tag{7.50}$$

A number of hydroxyl nitrogen compounds may be formed by

$$NO + OH + M \rightarrow HNO_2 + M \tag{7.51}$$

$$NO_2 + OH + M \rightarrow HNO_3 + M \tag{7.52}$$

$$NO_2 + HO_2 + M \rightarrow HO_2NO_2 + M \tag{7.53}$$

The primary loss process of these species is photolysis. Nitric acid is the most stable of this set of odd nitrogen compounds, and there may be a surface sink due to the formation of nitrate minerals on the surface. There is also a possibility that N_2O_5 may react on the surface of dust or ice particles in the atmosphere to form HNO_3 directly, but no quantitative estimates have been made for these possibilities.

7.3 Model Results

There have not been major conceptual advances in our understanding of the photochemistry of the Martian atmosphere since the classic reports were written in the early 1970s. However, there have been major advances in refining the quantitative spectroscopic and chemical kinetic data. Important examples of these data revisions include the temperature-dependent CO_2 absorption cross sections and updating of the rate coefficients for HO_x and NO_x reactions. In addition, new and better measurements of the constituents of the atmosphere such as O_3 and H_2O are now available. The model results we present are taken from Nair et al. (1994), which uses as input the recent ionospheric model of Fox (1993).

Figure 7.5 presents a model atmosphere of Mars constructed on the basis of the Viking data. The surface pressure is taken to be 6.36 mbar. The temperature at the surface is 214 K, decreases to 139 K in the 70–100 km region, and rises asymptotically to 300 K. The exobase is located at $\sim$200 km. The eddy diffusivity profile, $K(z)$, adopted in the model is given in figure 7.6. Near the surface $K = 10^5$ cm^2 s^{-1}, derived on the basis of aerosol data. The value of K increases with altitude to 10^7 cm^2 s^{-1} at 40 km, remains constant to 70 km, and finally increases to the asymptotic value of 10^8 cm^2 s^{-1}. The choice of the $K(z)$ profile has been made to reproduce the observed concentrations of O and O_2 in the upper atmosphere. The high value of $K(z)$ in the upper atmosphere may be caused by the breaking of upward-propagating atmospheric tides, which are known to be strong on Mars. Of course, above the

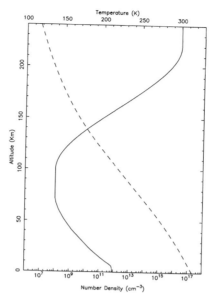

Figure 7.5 Model atmosphere of Mars based on Viking: temperature (solid line) and total number density (dashed line). From Nair, H., Allen, M., Anbar, A. D., Yung, Y. L., and Clancy, R. T., 1994, "A Photochemical Model of the Martian Atmosphere." *Icarus* **111**, 124.

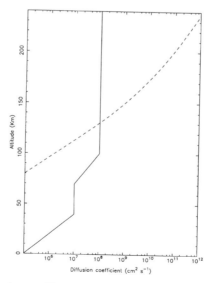

Figure 7.6 Eddy diffusion coefficient (solid line) and molecular diffusion coefficient (dashed line) for atomic oxygen in the background atmosphere. From Nair, H., Allen, M., Anbar, A. D., Yung, Y. L., and Clancy, R. T., 1994, "A Photochemical Model of the Martian Atmosphere." *Icarus* **111**, 124.

Table 7.4 Reactions in the model Mars atmosphere

	Reaction	Rate coefficient	Reference
R1	$O_2 + h\nu \rightarrow 2O$	$1.4 \times 10^{-10}, 1.8 \times 10^{-7}$	1–5
R2	$\rightarrow O + O(^1D)$	$0, 1.0 \times 10^{-6}$	1–5
R3	$O_3 + h\nu \rightarrow O_2 + O$	$2.5 \times 10^{-4}, 3.6 \times 10^{-4}$	2, 6–11
R4	$\rightarrow O_2 + O(^1D)$	$1.5 \times 10^{-3}, 2.2 \times 10^{-3}$	2, 6–11
R5	$\rightarrow 3O$	$7.8 \times 10^{-10}, 2.0 \times 10^{-6}$	2, 6–11
R6	$H_2 + h\nu \rightarrow 2H$	$0, 1.7 \times 10^{-6}$	12, 13
R7	$OH + h\nu \rightarrow O + H$	$4.8 \times 10^{-13}, 5.0 \times 10^{-6}$	14–16
R8	$HO_2 + h\nu \rightarrow OH + O$	$1.1 \times 10^{-4}, 1.7 \times 10^{-4}$	17
R9	$H_2O + h\nu \rightarrow H + OH$	$2.1 \times 10^{-11}, 3.5 \times 10^{-6}$	2, 18
R10	$\rightarrow H_2 + O(^1D)$	$0, 2.4 \times 10^{-7}$	2, 18
R11	$\rightarrow 2H + O$	$0, 2.8 \times 10^{-7}$	2, 18
R12	$H_2O_2 + h\nu \rightarrow 2OH$	$2.0 \times 10^{-5}, 3.2 \times 10^{-5}$	7, 19
R13	$CO_2 + h\nu \rightarrow CO + O$	$7.8 \times 10^{-13}, 6.7 \times 10^{-7}$	1, 2, 20, 21
R14	$\rightarrow CO + O(^1D)$	$0, 2.4 \times 10^{-7}$	1, 2, 20, 21
R15	$2O + M \rightarrow O_2 + M$	$1.1 \times 10^{-27} T^{-2.0}$	17
R16	$O + O_2 + N_2 \rightarrow O_3 + N_2$	$5.0 \times 10^{-35} e^{724/T}$	22
R17	$O + O_2 + CO_2 \rightarrow O_3 + CO_2$	$1.3 \times 10^{-34} e^{724/T}$	See 7.2.1
R18	$O + O_3 \rightarrow 2O_2$	$8.0 \times 10^{-12} e^{-2060/T}$	17
R19	$O + CO + M \rightarrow CO_2 + M$	$1.6 \times 10^{-32} e^{-2184/T}$	23
R20	$O(^1D) + O_2 \rightarrow O + O_2$	$3.2 \times 10^{-11} e^{70/T}$	17
R21	$O(^1D) + O_3 \rightarrow 2O_2$	1.2×10^{-10}	17
R22	$\rightarrow O_2 + 2O$	1.2×10^{-10}	17
R23	$O(^1D) + H_2 \rightarrow H + OH$	1.0×10^{-10}	17
R24	$O(^1D) + CO_2 \rightarrow O + CO_2$	$7.4 \times 10^{-11} e^{120/T}$	17
R25	$O(^1D) + H_2O \rightarrow 2OH$	2.2×10^{-10}	17
R26	$2H + M \rightarrow H_2 + M$	$3.8 \times 10^{-29} T^{-1.3}$	24
R27	$H + O_2 + M \rightarrow HO_2 + M$	$k_0 = 1.3 \times 10^{-27} T^{-1.6}$	17
		$k_\infty = 7.5 \times 10^{-11}$	
R28	$H + O_3 \rightarrow OH + O_2$	$1.4 \times 10^{-10} e^{-470/T}$	17
R29	$H + HO_2 \rightarrow 2OH$	7.3×10^{-11}	17, 25
R30	$\rightarrow H_2 + O_2$	6.5×10^{-12}	17, 25
R31	$\rightarrow H_2O + O$	1.6×10^{-12}	17, 25
R32	$O + H_2 \rightarrow OH + H$	$1.6 \times 10^{-11} e^{-4570/T}$	17
R33	$O + OH \rightarrow O_2 + H$	$2.2 \times 10^{-11} e^{120/T}$	17
R34	$O + HO_2 \rightarrow OH + O_2$	$3.0 \times 10^{-11} e^{200/T}$	17
R35	$O + H_2O_2 \rightarrow OH + HO_2$	$1.4 \times 10^{-12} e^{-2000/T}$	17
R36	$2OH \rightarrow H_2O + O$	$4.2 \times 10^{-12} e^{-240/T}$	17
R37	$2OH + M \rightarrow H_2O_2 + M$	$k_0 = 1.7 \times 10^{-28} T^{-0.8}$	17
		$k_\infty = 1.5 \times 10^{-11}$	
R38	$OH + O_3 \rightarrow HO_2 + O_2$	$1.6 \times 10^{-12} e^{-940/T}$	17
R39	$OH + H_2 \rightarrow H_2O + H$	$5.5 \times 10^{-12} e^{-2000/T}$	17
R40	$OH + HO_2 \rightarrow H_2O + O_2$	$4.8 \times 10^{-11} e^{250/T}$	17
R41	$OH + H_2O_2 \rightarrow H_2O + HO_2$	$2.9 \times 10^{-12} e^{-160/T}$	17
R42	$OH + CO \rightarrow CO_2 + H$	$1.5 \times 10^{-13} (1 + 0.6 P_{atm})$	17
R43	$HO_2 + O_3 \rightarrow OH + 2O_2$	$1.1 \times 10^{-14} e^{-500/T}$	17
R44	$2HO_2 \rightarrow H_2O_2 + O_2$	$2.3 \times 10^{-13} e^{600/T}$	17
R45	$2HO_2 + M \rightarrow H_2O_2 + O_2 + M$	$4.3 \times 10^{-33} e^{1000/T}$	17
R46	$N_2 \rightarrow 2N$		See 7.2.6
R47	$\rightarrow 2N(^2D)$		See 7.2.6
R48	$NO + h\nu \rightarrow N + O$	$3.1 \times 10^{-12}, 1.2 \times 10^{-6}$	13
R49	$NO_2 + h\nu \rightarrow NO + O$	$1.6 \times 10^{-3}, 2.4 \times 10^{-3}$	5
R50	$NO_3 + h\nu \rightarrow NO_2 + O$	$5.3 \times 10^{-2}, 7.2 \times 10^{-2}$	14
R51	$\rightarrow NO + O_2$	$4.4 \times 10^{-3}, 6.0 \times 10^{-3}$	14

Table 7.4 (continued)

	Reaction	Rate coefficient	Reference
R52	$N_2O + h\nu \to N_2 + O(^1D)$	$4.5 \times 10^{-8}, 1.8 \times 10^{-6}$	17
R53	$N_2O_5 + h\nu \to NO_2 + NO_3$	$1.2 \times 10^{-4}, 2.0 \times 10^{-4}$	17
R54	$HNO_2 + h\nu \to OH + NO$	$3.1 \times 10^{-4}, 4.2 \times 10^{-4}$	17
R55	$HNO_3 + h\nu \to NO_2 + OH$	$1.7 \times 10^{-5}, 4.2 \times 10^{-5}$	17
R56	$HO_2NO_2 + h\nu \to HO_2 + NO_2$	$8.7 \times 10^{-5}, 1.4 \times 10^{-4}$	17
R57	$N + O_2 \to NO + O$	$1.5 \times 10^{-11} e^{-3600/T}$	17
R58	$N + O_3 \to NO + O_2$	1.0×10^{-16}	28
R59	$N + OH \to NO + H$	$3.8 \times 10^{-11} e^{85/T}$	28
R60	$N + HO_2 \to NO + OH$	2.2×10^{-11}	29
R61	$N + NO \to N_2 + O$	3.4×10^{-11}	17
R62	$N + NO_2 \to N_2O + O$	3.0×10^{-12}	17
R63	$N(^2D) + O \to N + O$	6.9×10^{-13}	30
R64	$N(^2D) + CO_2 \to NO + CO$	3.5×10^{-13}	31
R65	$N(^2D) + N_2 \to N + N_2$	1.7×10^{-14}	32
R66	$N(^2D) + NO \to N_2 + O$	6.9×10^{-11}	30
R67	$O + NO + M \to NO_2 + M$	$k_0 = 1.2 \times 10^{-27} T^{-1.5}$ $k_\infty = 3.0 \times 10^{-11}$	17
R68	$O + NO_2 \to NO + O_2$	$6.5 \times 10^{-12} e^{120/T}$	17
R69	$O + NO_2 + M \to NO_3 + M$	$k_0 = 2.0 \times 10^{-26} T^{-2.0}$ $k_\infty = 2.2 \times 10^{-11}$	17
R70	$O + NO_3 \to O_2 + NO_2$	1.0×10^{-11}	17
R71	$O + HO_2NO_2 \to OH + NO_2 + O_2$	$7.8 \times 10^{-11} e^{-3400/T}$	17
R72	$O(^1D) + N_2 \to O + N_2$	$1.8 \times 10^{-11} e^{110/T}$	17
R73	$O(^1D) + N_2 + M \to N_2O + M$	$2.8 \times 10^{-35} T^{-0.6}$	17
R74	$O(^1D) + N_2O \to 2 NO$	6.7×10^{-11}	17
R75	$\to N_2 + O_2$	4.9×10^{-11}	17
R76	$NO + O_3 \to NO_2 + O_2$	$2.0 \times 10^{-12} e^{-1400/T}$	17
R77	$NO + HO_2 \to NO_2 + OH$	$3.7 \times 10^{-12} e^{240/T}$	17
R78	$NO + NO_3 \to 2 NO_2$	$1.7 \times 10^{-11} e^{150/T}$	17
R79	$H + NO_2 \to OH + NO$	$2.2 \times 10^{-10} e^{-182/T}$	33
R80	$H + NO_3 \to OH + NO_2$	1.1×10^{-10}	34
R81	$OH + NO + M \to HNO_2 + M$	$k_0 = 4.8 \times 10^{-24} T^{-2.6}$ $k_\infty = 2.6 \times 10^{-10} T^{-0.5}$	17
R82	$OH + NO_2 + M \to HNO_3 + M$	$k_0 = 5.5 \times 10^{-22} T^{-3.2}$ $k_\infty = 4.0 \times 10^{-8} T^{-1.3}$	17
R83	$OH + NO_3 \to HO_2 + NO_2$	2.3×10^{-11}	28
R84	$OH + HNO_2 \to H_2O + NO_2$	$1.8 \times 10^{-11} e^{-390/T}$	28
R85	$OH + HNO_3 \to NO_3 + H_2O$	$7.2 \times 10^{-15} e^{785/T}$	17
R86	$OH + HO_2NO_2 \to H_2O + NO_2 + O_2$	$1.3 \times 10^{-12} e^{380/T}$	17
R87	$HO_2 + NO_2 + M \to HO_2NO_2 + M$	$k_0 = 3.8 \times 10^{-23} T^{-3.2}$ $k_\infty = 1.4 \times 10^{-8} T^{-1.4}$	17
R88	$HO_2 + NO_3 \to O_2 + HNO_3$	9.2×10^{-13}	35, 36
R89	$NO_2 + O_3 \to NO_3 + O_2$	$1.2 \times 10^{-13} e^{-2450/T}$	17
R90	$NO_2 + NO_3 + M \to N_2O_5 + M$	$k_0 = 2.5 \times 10^{-19} T^{-4.3}$ $k_\infty = 2.6 \times 10^{-11} T^{-0.5}$	17
R91	$NO_2 + NO_3 \to NO + NO_2 + O_2$	$8.2 \times 10^{-14} e^{-1480/T}$	17
R92	$O + h\nu \to O^+ + e^-$	$0, 1.1 \times 10^{-7}$	37
R93	$O_2 + h\nu \to O_2^+ + e^-$	$0, 2.5 \times 10^{-7}$	1–5
R94	$CO_2 + h\nu \to CO_2^+ + e^-$	$0, 3.3 \times 10^{-7}$	1, 2, 21, 22
R95	$\to CO + O^+ + e^-$	$0, 2.8 \times 10^{-8}$	1, 2, 21, 22
R96	$O_2^+ + e^- \to 2 O$	$6.6 \times 10^{-5} T_e^{-1.0}$	38
R97	$CO_2^+ + e^- \to CO + O$	3.8×10^{-7}	38
R98	$O^+ + CO_2 \to O_2^+ + CO$	9.6×10^{-10}	39, 40

(continued)

Table 7.4 (continued)

	Reaction	Rate coefficient	Reference
R99	$O + CO_2^+ \rightarrow O_2^+ + CO$	1.6×10^{-10}	39, 40
R100	$\rightarrow O^+ + CO_2$	9.6×10^{-11}	39, 40
R101	$CO_2^+ + H_2 \rightarrow CO_2H^+ + H$	4.7×10^{-10}	39, 40
R102	$CO_2H^+ + e^- \rightarrow CO_2 + H$	3.0×10^{-7}	41
R103	$HO_2 + \text{grain} \rightarrow (HO_2)_{\text{grain}}$		
R104	$(HO_2)_{\text{grain}} + OH \rightarrow H_2O + O_2$		

Units are s^{-1} for photolysis reactions, $cm^3 \, s^{-1}$ for two-body reactions, and $cm^6 \, s^{-1}$ for three-body reactions. Photolysis rate coefficients are given at the ground and at the top of the model atmosphere (240 km). k_0 and k_∞ are the low and high pressure rate coefficients, respectively, for three-body reactions.

References: (1) See references in Yung et al. (1988) and Anbar et al. (1993a), also Nicolet (1984), Lee et al. (1977), Samson et al. (1982); (2) See references in Anbar et al. (1993a), also Taherian and Slanger (1985), Turnipseed et al. (1991), Wine and Ravishankara (1982), Brock and Watson (1980), Sparks et al. (1980), Fairchild et al. (1978); (3) Mentall and Gentieu (1970); R. Gladstone, private communication; (4) Nee and Lee (1984), van Dishoeck and Dalgarno (1984), van Dishoeck et al. (1984); (5) DeMore et al. (1990); (6) See references in Anbar et al. (1993a), also Philips et al. (1977); (7) Schürgers and Welge (1968), DeMore et al. (1990); (8) See references in Yung et al. (1988) and Anbar et al. (1993a), also Kronebusch and Berkowitz (1976), Masuoka and Samson (1980); (9) Lin and Leu (1982); (10) Baulch et al. (1976); (11) Tsang and Hampson (1986); (12) DeMore et al. (1990), Keyser (1986); (13) Allen and Frederick (1982); (14) DeMore et al. (1990), Magnotta and Johnston (1980); (15) Atkinson et al. (1989); (16) Brune et al. (1983); (17) Fell et al. (1990); (18) Piper et al. (1987); (19) Schofield (1979); (20) Ko and Fontijn (1991); (21) Boodaghians et al. (1988); (22) Hall et al. (1988), Mellouki et al. (1988); (23) Samson and Pareek (1985); (24) McElroy et al. (1977); (25) Anicich and Huntress (1986), Anicich (1993); (26) Kong and McElroy (1977a).

homopause (at around 135 km), molecular diffusion becomes more important than eddy diffusion.

The chemical reactions that are essential to the photochemical model are listed in table 7.4. The set of ion reactions is a truncated set taken from a more elaborate model of Fox (1993). We do not compute the source of odd nitrogen from first principles but adopt the output results of the model of Fox.

7.3.1 Ionosphere

A background model of the neutral atmosphere above 120 km is shown in figure 7.7, along with comparisons with measurements from Viking Lander 1. How this model is computed is discussed in section 7.3.2 on the neutral model. The lack of agreement in the lower region is caused by the difference between our temperature profile and that experienced by the lander. The mixing ratio of atomic oxygen at 135 km in the model is 0.8%, in good agreement with the range of 0.5–1.0% deduced from airglow analysis.

The concentrations of the major ions are presented in figure 7.8. These results are from the more elaborate model of Fox (1993), and there is good agreement between theory and the Viking observations. Our own model, computed using the abbreviated reaction set in table 7.4, gives similar predictions as figure 7.8, but the peak ion densities differ by about 50%. The principal ion in the ionosphere is O_2^+. Note that in order to fit the Viking observations (long-dashed line), it is necessary to impose an O_2^+ escape flux of 4.75×10^7 cm^{-2} s^{-1} at the upper boundary. A zero flux boundary condition would predict number densities of O_2^+ (short-dashed line) that

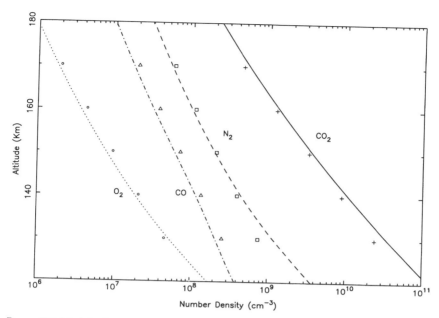

Figure 7.7 Model of upper atmosphere of Mars based on Viking 1 entry measurements of CO_2 (crosses), N_2 (squares), CO (triangles), and O_2 (circles). From Nair, H., Allen, M., Anbar, A. D., Yung, Y. L., and Clancy, R. T., 1994, "A Photochemical Model of the Martian Atmosphere." *Icarus* **111**, 124.

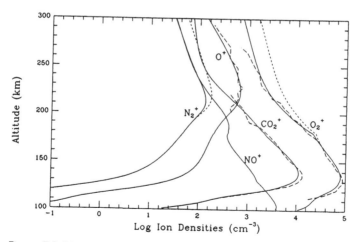

Figure 7.8 Model of ion densities in the upper atmosphere of Mars based on Viking 1 entry measurements of O_2^+, CO_2^+, O^+, NO^+, and N_2^+. From Nair, H., Allen, M., Anbar, A. D., Yung, Y. L., and Clancy, R. T., 1994, "A Photochemical Model of the Martian Atmosphere." *Icarus* **111**, 124.

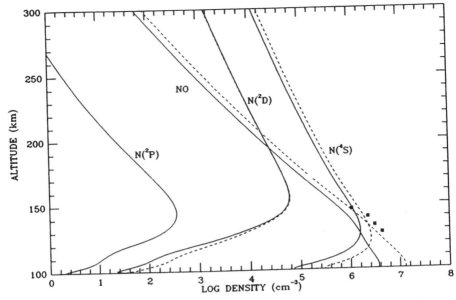

Figure 7.9 Mixing ratios of nitrogen species. From Fox, J. L., 1993, "Production and Escape of Nitrogen from Mars." *J. Geophys. Res.* **98** 3297. The squares are data from Viking.

are higher than the observed values. The minor ions include CO_2^+ (needed for scheme VI), NO^+, O^+, and N_2^+. The concentrations of the major odd nitrogen species are presented in figure 7.9. NO is the major odd nitrogen species at 100 km. Downward transport of this molecule to the lower atmosphere provides the main source of odd nitrogen to the lower atmosphere. The model results for NO are in good agreement with the Viking observations. The solid and dashed lines show the difference between model predictions using the recommended value of 3.4×10^{-11} cm^3 s^{-1} (DeMore et al., 1990) and a slower value of $7.1 \times 10^{-10} e^{-787/T}$ of Davidson and Hanson (1990) for reaction (7.46).

7.3.2 Neutral Atmosphere

The boundary conditions for the model of the neutral atmosphere are stated in table 7.5. The concentrations of the major species CO_2, N_2, and H_2O are fixed at the surface. For all other species, zero flux at the boundary is assumed. At the upper boundary (240 km), an upward flux of 1.2×10^8 cm^{-2} s^{-1} is imposed for oxygen atoms. H and H_2 are allowed to escape at the effusion velocities given by the Jeans formula. The continuity and diffusion equations are solved for all species listed in table 7.5.

The altitude profile for the mixing ratio of H_2O is presented in figure 7.10 (dashed line). Water vapor is a condensible species in the Martian atmosphere. For our adopted temperature profile, the mixing ratio of saturated water vapor is given by the solid line in figure 7.10. The mixing ratio of H_2O in our model is constant between the surface and 25 km. Above this altitude, the mixing ratio of H_2O roughly follows the

Table 7.5 Boundary conditions

	Lower	Upper
O	$\phi = 0$	$\phi = 1.2 \times 10^8$ cm^{-2} s^{-1}
N_2	$f = 0.027$	$\phi = 0$
H	$\phi = 0$	$v = 3.08 \times 10^3$ cm s^{-1}
H_2	$\phi = 0$	$v = 3.39 \times 10^1$ cm s^{-1}
H_2O	$f = 1.35 \times 10^{-4}$	$\phi = 0$
CO_2	$n = 2.05 \times 10^{17}$ cm^{-3}	$\phi = 0$

From Nair et al. (1994).
f denotes mixing ratio, ϕ denotes flux, v denotes velocity, and n denotes number density. Zero fluxes assumed for all other species.

saturation vapor curve with a mean scale height of 4 km. Above 60 km, condensation is not important and H_2O is computed like any other species in the model. The column-integrated water vapor abundance is 8.8 pr-μm.

The altitude profiles for the major species in the model, CO_2, CO, O_2, H_2, and O are presented in figure 7.11a. Above the homopause (135 km) the lighter species (H_2, O) tend to increase due to gravitational separation. Oxygen atoms are more abundant than CO_2 at the exobase. The atomic species O and H are long-lived in the upper atmosphere where they have no chemical sink, but they are rapidly removed in chemical reactions

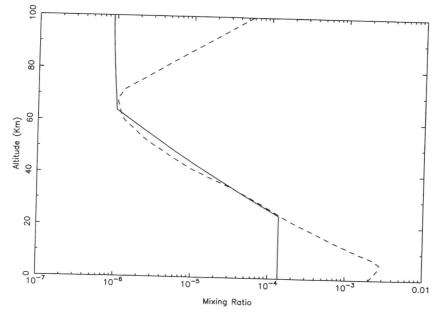

Figure 7.10 Water vapor profile adopted in the model. The column-integrated water vapor abundance is 8.8 pr-μm. The dashed line represents the saturation curve for water vapor. From Nair, H., Allen, M., Anbar, A. D., Yung, Y. L., and Clancy, R. T., 1994, "A Photochemical Model of the Martian Atmosphere." *Icarus* **111**, 124.

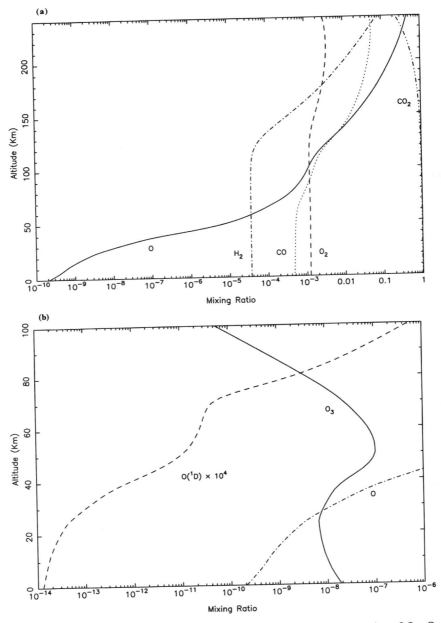

Figure 7.11 (a) Mixing ratios of CO_2, CO, O_2, H_2, and O. (b) Mixing ratios of O_3, O, and $O(^1D)$. From Nair, H., Allen, M., Anbar, A. D., Yung, Y. L., and Clancy, R. T., 1994, "A Photochemical Model of the Martian Atmosphere." *Icarus* **111**, 124.

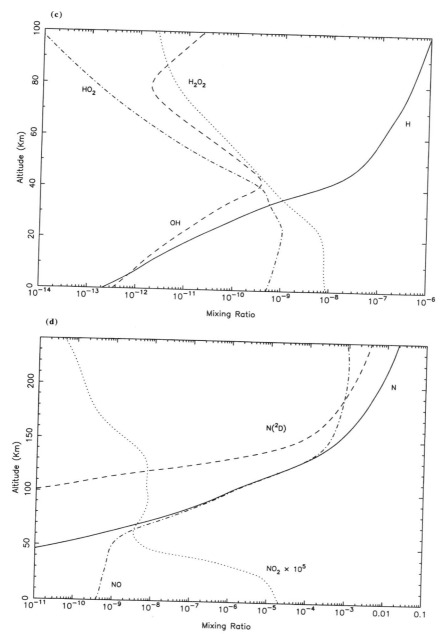

Figure 7.11 (c) Mixing ratios of H, OH, HO$_2$, and H$_2$O$_2$. (d) Mixing ratios of N, NO, NO$_2$, and N(^{2}D). From Nair, H., Allen, M., Anbar, A. D., Yung, Y. L., and Clancy, R. T., 1994, "A Photochemical Model of the Martian Atmosphere." *Icarus* **111**, 124.

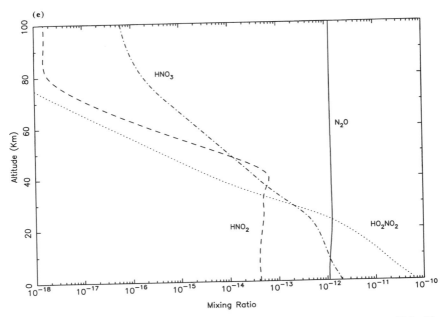

Figure 7.11 (e) Mixing ratios of N_2O, HNO_2, HNO_3, and HO_2NO_2. From Nair, H., Allen, M., Anbar, A. D., Yung, Y. L., and Clancy, R. T., 1994, "A Photochemical Model of the Martian Atmosphere." *Icarus* **111**, 124.

in the lower atmosphere. This accounts for the decrease in their mixing ratios at low altitudes. The mixing ratios of the odd oxygen species, O, $O(^1D)$, and O_3, in the lower atmosphere are shown in figure 7.11b. O atoms are shown here again for comparison with the other odd oxygen species. The model predicts a maximum ozone mixing ratio of 10^{-7}. The column-integrated O_3 abundance is 1.1 μatm (or 0.11 Dobson units). A comparison of the model results and observations is summarized in table 7.6. The agreement is good for O_2 and O_3, but for CO the prediction of the standard model is somewhat low. This implies that the HO_x chemistry is too efficient in removing CO catalytically. The result is an ironic twist of the early models. In these early models, for schemes (Ia) and (Ib) to be effective, an excessively large $K = 10^8$ cm^2 s^{-1} in the lower atmosphere had to be postulated. For scheme (II) to be effective, a large amount

Table 7.6 Comparison between observed and model CO, O_2, and O_3

	f_{CO} ($\times 10^4$)	f_{O_2} ($\times 10^3$)	f_{CO}/f_{O_2}	$\int [O_3]dz$ (μm-amagat)[a]
Observed	6 ± 1.5	1.2 ± 0.2	0.50 ± 0.15	1.5 ± 0.5
Model	4.9	1.3	0.40	1.1

From Nair et al. (1994).
f denotes mixing ratio and $\int [\ ]dz$ denotes column density.
[a] One μm-amagat is equivalent to a column density of 2.69×10^{15} cm^2.

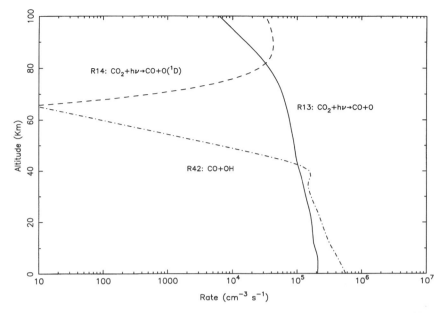

Figure 7.12 Reaction rates of destruction and formation of CO_2. From Nair, H., Allen, M., Anbar, A. D., Yung, Y. L., and Clancy, R. T., 1994, "A Photochemical Model of the Martian Atmosphere." *Icarus* **111**, 124.

of H_2O in the atmosphere must be assumed. We should point out that the low CO is a recent model result caused by the use of the temperature-dependent CO_2 cross sections. These are lower than those used in the early models, and as a result they allow greater penetration of ultraviolet radiation to the lower atmosphere where H_2O is dissociated. A judicious adjustment of the rate coefficients (within 1 σ error bars) of key reactions can resolve this problem, but we shall not go into the details here. Alternatively, the amount of H_2O in the atmosphere may be lower than that adopted in the model.

The mixing ratios of the HO_x species, H, OH, HO_2, and H_2O_2, are presented in figure 7.11c. The bulk of HO_x catalytic chemistry takes place in the lower atmosphere below 50 km. The most reactive radicals are H, OH, and HO_2. It is surprising that the stability of the Martian atmosphere is controlled by trace constituents with abundances less than 1 ppb. H_2O_2 is a temporary reservoir of HO_x. Its abundance is higher because it is less reactive than the other HO_x species.

The altitude profiles of the most abundant NO_x species, N, $N(^2D)$, NO, and NO_2, are shown in figure 7.11d. The principal source of odd nitrogen is the ionosphere, where odd nitrogen is most abundant. NO is the major carrier of NO_x from the upper atmosphere to the lower atmosphere. The concentrations of the other odd nitrogen species, N_2O, HNO_2, HNO_3, and HO_2NO_2, are given in figure 7.11e; We reproduce NO and NO_2 in this figure for comparison with the other odd nitrogen species.

The important reaction rates in the model are briefly discussed. Figure 7.12 presents the rates of destruction (photodissociation) and formation of CO_2. The dissociation of

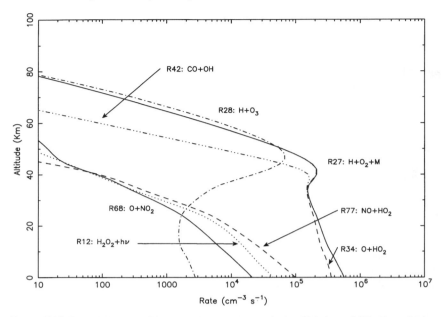

Figure 7.13 Reaction rates of key reactions in the catalytic oxidation of CO. From Nair, H., Allen, M., Anbar, A. D., Yung, Y. L., and Clancy, R. T., 1994, "A Photochemical Model of the Martian Atmosphere." *Icarus* **111**, 124.

CO_2 consists of two parts: a high-altitude branch that produces CO and the excited $O(^1D)$, and another branch that produces CO and O in the ground state at lower altitudes. The column-integrated rate for CO_2 destruction is 1.1×10^{12} cm^{-2} s^{-1}. The destruction of CO_2 in the model is restored by only one reaction, R42: CO + OH, which occurs mainly between the surface and 40 km. The rates of the key reactions in catalytic cycles (I)–(III) are given in figure 7.13. Note that due to the tight balance in the HO_x catalytic cycle (see figure 7.4), the following rates are approximately equal:

$$\int_0^\infty J_{13}[CO_2]\,dz = \int_0^\infty k_{42}[CO][OH]\,dz \quad (7.54)$$

$$= \int_0^\infty k_{34}[O][HO_2]\,dz \quad (7.55)$$

$$= \int_0^\infty k_{27}[H][O_2][M]\,dz \quad (7.56)$$

The contributions of the five catalytic chemical schemes to the oxidation of CO to CO_2 may be deduced from the reaction rates presented in table 7.7. The bulk of the oxidation is carried out by schemes (Ia) (77%) and (Ib) (8%), giving a total of 85%. Scheme (II) (the H_2O_2 path) contributes 8%. The NO_x schemes (IIIa) and (IIIb) contribute 5% and 1.5%, respectively.

The principal source of HO_x is the decomposition of H_2O by photolysis and the reaction with $O(^1D)$. Figure 7.14 presents the rates of these reactions. The rates of

Table 7.7 Column rates of some important reactions in the model

		Reaction		Rate ($cm^{-2}\ s^{-1}$)
CO production	R13	$CO_2 + h\nu$	$\rightarrow CO + O$	9.7×10^{11}
	R14		$\rightarrow CO + O(^1D)$	1.3×10^{11}
HO_x cycling	R42	$CO + OH$	$\rightarrow CO_2 + H$	1.1×10^{12}
	R68	$O + NO_2$	$\rightarrow O_2 + NO$	2.1×10^{10}
	R12	$H_2O_2 + h\nu$	$\rightarrow 2\,OH$	4.8×10^{10}
	R77	$NO + HO_2$	$\rightarrow NO_2 + OH$	8.9×10^{10}
	R27	$H + O_2 + M$	$\rightarrow HO_2 + M$	1.2×10^{12}
	R28	$H + O_3$	$\rightarrow OH + O_2$	9.7×10^{10}
	R34	$O + HO_2$	$\rightarrow OH + O_2$	9.8×10^{11}

From Nair et al. (1994).

recycling of HO_x radicals have been discussed in section 7.2.2 in connection with the catalytic oxidation of CO to CO_2. The principal reaction that destroys HO_x is R40: $OH + HO_2 \rightarrow H_2O + O_2$; its rate is given in figure 7.14. Note that the column-integrated rate of destruction of H_2O is of the order of 10^{10} $cm^{-2}\ s^{-1}$, whereas the corresponding rate of production of CO_2 is of the order of 10^{12} $cm^{-2}\ s^{-1}$. Thus, every HO_x radical derived from H_2O is used about 100 times in the catalytic schemes for the restoration of CO_2 before it is terminated by returning to H_2O via R40: $OH + HO_2 \rightarrow H_2O + O_2$.

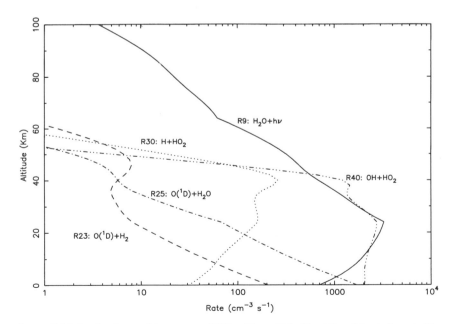

Figure 7.14 Production and loss rates of HO_x. From Nair, H., Allen, M., Anbar, A. D., Yung, Y. L., and Clancy, R. T., 1994, "A Photochemical Model of the Martian Atmosphere." *Icarus* **111**, 124.

270 Photochemistry of Planetary Atmospheres

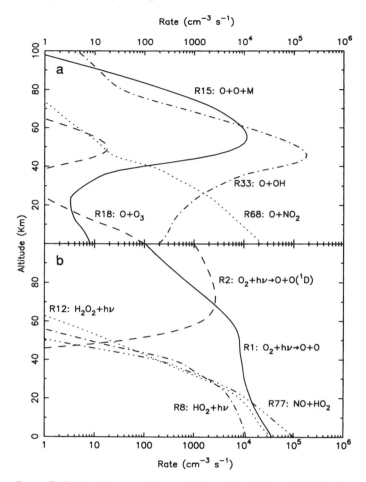

Figure 7.15 Reaction rates for production (a) and loss (b) of the O-O bond. From Nair, H., Allen, M., Anbar, A. D., Yung, Y. L., and Clancy, R. T., 1994, "A Photochemical Model of the Martian Atmosphere." *Icarus* **111**, 124.

The rates of the reactions that lead to the formation and the breaking of the O–O bond are shown in figure 7.15, a and b, respectively. The most important reactions that form O_2 are R33: O + OH and R15: O + O + M. Note that R33 is the rate-limiting step in scheme (IV), forming O_2 from CO_2 (7.17). This reaction is about an order of magnitude more important than the direct recombination of O atoms (R15: 2O + M $\rightarrow$ O_2 + M). Reaction R68: NO_2 + O and the Chapman reaction, R18: O + O_3, are relatively unimportant. The main reactions that destroy the O–O bond are photolysis of O_2, H_2O_2, and HO_2 and reaction R77: NO + HO_2. Of these reactions, the photolysis of O_2 is the most important.

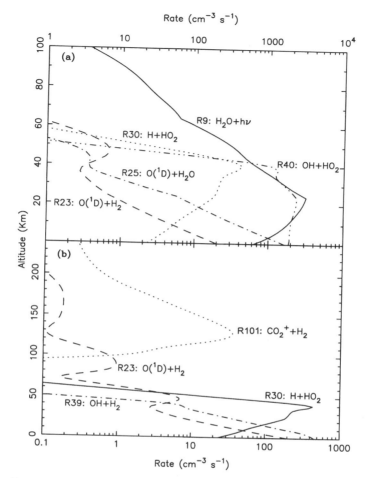

Figure 7.16 Reaction rates for production (a) and loss (b) of H_2. From Nair, H., Allen, M., Anbar, A. D., Yung, Y. L., and Clancy, R. T., 1994, "A Photochemical Model of the Martian Atmosphere." *Icarus* **111**, 124.

One by-product of the HO_x chemistry is H_2 produced by the reaction R30: $H + HO_2$; the rate of this reaction is given in figure 7.16a. The main reactions that destroy H_2 are R39: $H_2 + OH$ and R23: $H_2 + O(^1D)$, whose rates are shown in figure 7.16b. H_2 is a long-lived molecule and can be transported to the upper atmosphere, where it is decomposed by scheme (VI). The rate of the initiation reaction R101: $CO_2^+ + H_2$ is given in figure 7.16b. This reaction provides the main source of hydrogen to drive its escape from Mars. In the ionosphere (about 130 km) H_2 is broken down into H atoms. Some of it flows upward and eventually escapes from Mars. The other portion flows downward and eventually becomes oxidized to H_2O.

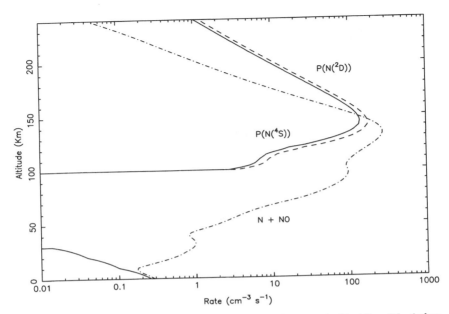

Figure 7.17 Production and loss rates of odd nitrogen. From Nair, H., Allen, M., Anbar, A. D., Yung, Y. L., and Clancy, R. T., 1994, "A Photochemical Model of the Martian Atmosphere." *Icarus* **111**, 124.

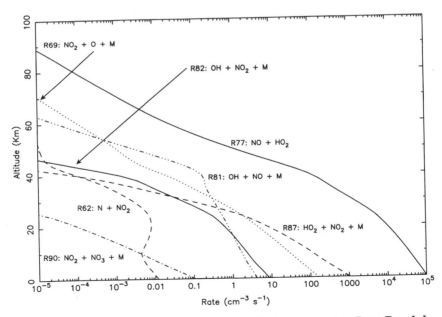

Figure 7.18 Production rates of more complex odd nitrogen species. From Fox, J. L., 1993, "Production and Escape of Nitrogen from Mars." *J. Geophys. Res.* **98**, 3297.

The major sources and sinks of odd nitrogen in the model are given in figure 7.17. The production of N and N(^{2}D) in the upper atmosphere is taken from the ionospheric model of Fox (1993). In the lower atmosphere odd nitrogen is produced by cosmic ray impact. The only reaction that destroys odd nitrogen by conversion to N_2 is reaction R61: $N + NO \rightarrow N_2 + O$. Figure 7.18 presents the rates of the main reactions forming more complex nitrogen compounds from NO: NO_2 formed by R77: $NO + HO_2$; NO_3 formed by R69: $NO_2 + O + M$; N_2O_5 formed by R90: $NO_2 + NO_3 + M$; N_2O formed by R62: $N + NO_2$; HNO_2 formed by R81: $NO + OH + M$; HNO_3 formed by R82: $OH + NO_2 + M$; and HO_2NO_2 formed by R87: $HO_2 + NO_2 + M$. The principal loss mechanism for all these odd nitrogen species is photolysis, and these rates are not shown.

7.4 Evolution

There are at least two rather profound questions concerning the evolution of the Martian atmosphere. First, why is the present atmosphere so thin? The total atmospheric pressure is only 6 mbar. The abundances of the noncondensible gases, N_2 and Ar, are 0.16 and 0.1 mbar, respectively. These abundances are much less than the corresponding terrestrial values of 780 and 9.3 mbar for N_2 and Ar. Second, the surface of Mars is covered by ancient fluvially generated channels, strongly suggesting the existence of a warmer climate that could sustain flow of water. That is very different from the current cold and arid climate. What are the causes of such drastic environmental change? Both of these questions suggest that the Martian atmosphere was denser and warmer in the past. The density of the atmosphere and its climate are of course related. In the current atmosphere the greenhouse effect due to CO_2 is only 6 K. To sustain a warm climate with fluid flow on the surface, a greenhouse effect of 30 K is needed; a CO_2 atmosphere of the order of a bar would be required.

It is now generally accepted that the present Martian atmosphere is the end-product of planetary evolution over the age of the solar system. There are three mechanisms for removing a substantial fraction of the Martian atmosphere. The gravity of Mars is sufficiently small that impacts by small bodies (planetesimals) would have been effective in blowing much of the atmosphere away. This is known as atmospheric cratering, a process that is believed to be more important during the period of heavy bombardment in the early history of the solar system. (This process is not believed to be important for the larger terrestrial planets Earth and Venus.) Another mechanism is the sequestering of volatiles in the polar and subsurface reservoirs. The polar regions of Mars are known to hold large deposits of CO_2 and H_2O ice. The regolith of Mars (up to 1 km deep) may contain large amounts of adsorbed CO_2 or carbonates. The exact amount is difficult to estimate. A third mechanism is the escape of gases from the exosphere of the planet by thermal or nonthermal processes. This last mechanism can be quantitatively modeled, at least in the current epoch. In addition, the exospheric escape leaves behind a cumulative isotopic signature that can be measured.

In this chapter we are primarily concerned with the third mechanism. The rates of escape of gases from the exosphere of the planet and the subsequent isotopic fractionation of the planetary volatile reservoir may be calculated from the model described in section 7.3.

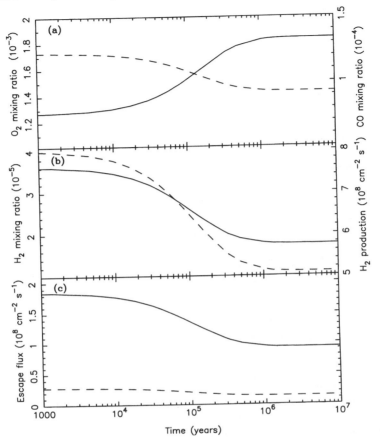

Figure 7.19 Time history of the response of the steady state atmosphere to an abrupt reduction of oxygen flux by a factor of 2 at time = 0. (a) O_2 (solid line) and CO (dashed line) mixing ratios. (b) H_2 mixing ratio (solid line) and H_2 production rate (dashed line). (c) H escape flux (solid line) and H_2 escape flux (dashed line). From Nair, H., Allen, M., Anbar, A. D., Yung, Y. L., and Clancy, R. T., 1994, "A Photochemical Model of the Martian Atmosphere." *Icarus* **111**, 124.

7.4.1 Escape of Gases

Spacecraft observations of Mars at Lyman α have revealed that the planet is surrounded by a corona of H atoms. The escape flux has been deduced to be $1\text{--}2 \times 10^8$ cm^{-2} s^{-1}. In the model described in section 7.3, a mean flux of total hydrogen (H + 2H$_2$), 1.8×10^8 cm^{-2} s^{-1}, has been adopted. This is equivalent to the loss of 2.6×10^{26} H atoms s^{-1} from Mars. According to the hypothesis of McElroy (1972), there is a corresponding escape of O atoms at half the rate of hydrogen escape, so the net result is the loss of H_2O from Mars. As a demonstration of this remarkable relation, we performed a numerical experiment. In the steady state model with the oxygen escape

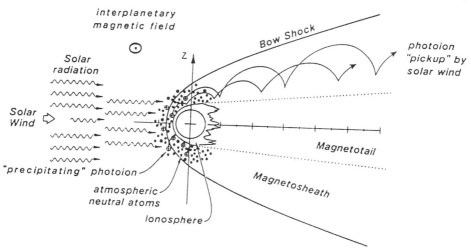

Figure 7.20 Interaction between the solar wind and the upper atmosphere of a nonmagnetic planet. From Luhmann, J. G., and Kozyra, J. U., 1991, "Dayside Pickup of Oxygen Ions Precipitation at Venus and Mars: Spatial Distribution, Energy Deposition and Consequences." *J. Geophys. Res.* **96** 5457.

flux given by table 7.5, we reduced, at time zero, this flux to half. The subsequent evolution of the model atmosphere is shown in figure 7.19. Note that after about 10^5 yr, the model converges to a new steady state characterized by half the rate of hydrogen escape (figure 7.19c). The new steady state atmosphere has more O_2, less CO (figure 7.19a), less H_2, and a lower H_2 production rate. If we assume that this reduced rate of oxygen escape has been uniform over the age of the solar system (4.6 Gyr), the total amount of H_2O lost to space is 1.3×10^{25} molecules/cm^2. This amount is equal to 362 mbar in the atmosphere or 3.9 m of water uniformly spread over the Martian surface. This is probably a lower limit. For comparison, we note that the column-integrated H_2O in the atmosphere today is 8.8 μm, four orders of magnitude less than the amount of water that could have escaped. The atmosphere of Mars is too cold to contain much water vapor. The bulk of water on Mars must be sequestered either at the poles or in the ground. A quantitative estimate based on the analysis of the isotopic composition of H_2O is deferred section 7.4.2.

The escape of nitrogen is driven by the ion recombination reaction (7.37). The rate of loss is 7.5×10^5 N atoms cm^{-2} s^{-1} in the current epoch. At this rate of loss (assumed to be uniform), the total atmospheric inventory of N_2 would be exhausted in 0.51 Gyr, geologically a very short time. This also implies that N_2 must have been more abundant in the past. A quantitative analysis may be carried out using the isotopic data obtained by Viking, as discussed in section 7.4.2.

There is the possibility that greater amounts of material might have been lost from Mars by solar wind–induced sputtering. This mechanism is highly speculative and poorly quantified at present. Due to the lack of an intrinsic magnetic field, the solar wind impinges on the upper atmosphere of Mars (see figure 7.20). Direct sputtering is efficient for light species (such as H and D). However, for the loss of heavier

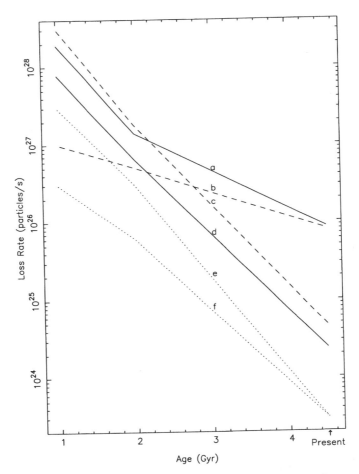

Figure 7.21 Loss rates over the history of the Martian atmosphere. (a) Total loss of H_2O. (b) Loss rate of O by electronic recombination (R96). (c) Loss rate of O by solar wind–induced sputtering. (d) Loss rate of CO_2. (e) and (f) correspond to loss rates (c) and (d) as computed by the model of Luhmann, J. G., Johnson, R. E., and Zhang, M. H. G., 1992, "Evolutionary Impact of Sputtering of the Martian Atmosphere by O^+ Pickup Ions." *Geophys. Res.*, **19**, 2151. From Kass, D. M., and Yung, Y. L., 1995, "Loss of Atmosphere from Mars Due to Solar Wind Induced Sputtering." *Science*, **268**, 697.

Table 7.8 Net escape fluxes from Mars due to solar wind–induced sputtering for three epochs

	6 EUV (1 Gyr)	3 EUV (2 Gyr)	1 EUV (4.5 Gyr)
Sputtered O	3.1×10^{28}	1.8×10^{27}	4.7×10^{24}
Exospheric O	1×10^{27}	5×10^{26}	8×10^{25}
Pickup O^+	3×10^{27}	4×10^{26}	6×10^{24}
Sputtered CO_2	7.8×10^{27}	6.4×10^{26}	2.4×10^{24}
Escaped H_2O	1.9×10^{28}	1.4×10^{27}	8.6×10^{25}

From Kass, D. M., and Yung, Y. L. (1995).

species, the sputtering by solar wind–accelerated ionized exospheric O^+ is more important. The loss rates over the history of Mars are summarized in figure 7.21 and table 7.8. As much as 3 bar of CO_2 and 80 m of water could have been lost from Mars, with most of the loss having occurred in the remote past. The escape rates are higher in the past due to the greater solar wind flux. The uncertainties in these calculations are illustrated by the difference in the loss rates predicted by different models (e.g., between figure 7.21, c and e).

7.4.2 Isotopic Signature

Table 7.3 shows that the D/H ratio of water vapor on Mars is enriched by a factor of 5 relative to terrestrial water. The $^{15}N/^{14}N$ ratio of N_2 on Mars is 1.62 times the terrestrial value. (The isotopic composition of Earth's atmosphere is given in table 9.3.) Note that in both cases the heavier isotope is enriched relative to the lighter isotope. The other isotopic ratios, $^{17}O/^{16}O$, $^{18}O/^{16}O$, and $^{13}C/^{12}C$, are not fractionated relative to their terrestrial counterparts. We adopt the simplifying assumptions that the material outgassed from the interior of Mars has the same isotopic composition as Earth and that the observed isotopic fractionation is caused by difference in the escape efficiencies of the isotopes.

As an illustration of the proposed mechanism of fractionation, let us consider the escape of hydrogen and deuterium from the planet. Let H and D be the total atmospheric abundances of hydrogen and deuterium in the atmosphere. The equations that govern the time evolution of H and D are

$$\frac{dH}{dt} = -AH \tag{7.57a}$$

$$\frac{dD}{dt} = -AfH \tag{7.57b}$$

where we have assumed that the rates of escape of H and D are proportional to their abundances, and f is the fractionation factor describing the efficiency of escape of D compared with H. For $f = 1$ the escape process is not fractionating and D/H ratio would remain constant with time. For $f = 0$ only hydrogen can escape and the D/H ratio would increase at the maximum possible rate. If we assume that the coefficients A and f in equations (7.57a) and (7.57b) are constant, the solution for the evolution of the D/H ratio is

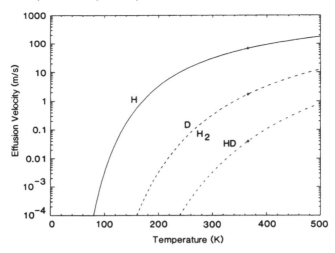

Figure 7.22 Effusion velocities from the exosphere of Mars as a function of exospheric temperature. The crosses indicate the approximate effusion velocities in the present epoch. From Nair, H., Allen, M., Anbar, A. D., Yung, Y. L., and Clancy, R. T., 1994, "A Photochemical Model of the Martian Atmosphere." *Icarus* **111**, 124.

$$\frac{D}{H}(t) = \frac{D}{H}(0)\left[\frac{H(0)}{H(t)}\right]^{1-f} \tag{7.58}$$

where $t = 0$ is the initial time, and $H(0)$ and $H(t)$ denote the total hydrogen reservoir at $t = 0$ and t, respectively. Since hydrogen is lost from the planet we have $H(0) > H(t)$. Equation (7.58) states that there is a corresponding enrichment in the D/H ratio. This result is known as Rayleigh distillation.

The fractionation factor f in (7.57b) is affected by a number of physical and chemical processes in the atmosphere. We briefly discuss the most important of these processes. First, the effusion velocities given by the Jeans escape formula are extremely sensitive to mass difference, as shown in figure 7.22. Second, all the atomic and molecular species are diffusively separated above the homopause according to their own scale heights. This means the lighter species will be enriched at the exobase relative to the bulk atmosphere below. The fractionation factor f for deuterium has been estimated from a detailed model of hydrogen escape. For Mars in the current epoch f is 0.32 i.e.; that is, deuterium is escaping at 32% of the efficiency of hydrogen. Assuming a uniform rate of escape, the observed D/H enrichment implies that the initial reservoir of hydrogen is equal to 3.6 m of water, most of which has escaped, leaving a residue of 0.2 m water on Mars today. This represents a very conservative estimate of water (both for the primordial and for the remaining reservoir). The reservoir sizes would obviously be bigger if escape rates have been faster in the past. Also, this analysis completely ignores loss of volatiles by atmospheric cratering and sequestering in underground reservoirs, because the latter processes are not likely to leave an isotopic signature.

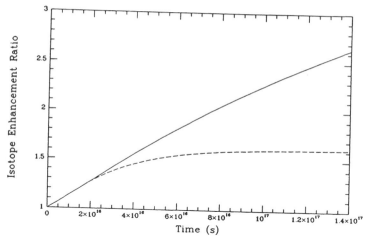

Figure 7.23 Isotopic fractionation for ^{15}N as a function of time. The dashed line is for an initial atmosphere of 2 bar of CO_2 that is lost exponentially with a time constant of 700 Myr. From Fox, J. L., 1993, "Production and Escape of Nitrogen from Mars." *J. Geophys. Res.* **98**, 3297.

A similar analysis may be applied to the oxygen isotopes. All the above results apply if we replace D by the heavy isotope ^{18}O and H by the light isotope ^{16}O. (There is no additional information from analysis of the pair ^{17}O and ^{16}O.) The fractionation factor f equals 0.75 in this case. The lack of an observed isotopic fractionation implies that there is a large oxygen reservoir on Mars, of the order of 1 bar, that is capable of diluting the isotopic enrichment caused by oxygen escape. We cannot distinguish whether this reservoir is CO_2 or H_2O, but from the previous analysis, the oxygen reservoir is probably not H_2O. This leaves CO_2 as the most likely candidate.

A similar analysis may be applied to nitrogen: ^{15}N and ^{14}N. The fractionation factor in this case is 0.82. The results for the evolution of isotopic fractionation of nitrogen are given in figure 7.23. The total enrichment over the age of the solar system $(1.4 \times 10^{17}$ s) is about 2.6, a value significantly higher than the Viking observation of 1.62. The reason is that as we go back in time N_2 became the dominant gas in the atmosphere. This implies large escape rates, hence a greater enrichment factor. One simple way to resolve this discrepancy is to postulate that the Martian atmosphere was denser in the past such that N_2 was "protected" from escaping. The dashed line in figure 7.23 represents the result of a calculation assuming that there was an initial CO_2 atmosphere of 2 bars of CO_2 that is lost exponentially with a time constant of 700 Myr. The predicted isotopic enrichment agrees with the Viking value.

7.5 Terraforming Mars

Our interest in Mars is more than scientific. Of all the planets in the solar system, Mars has a physical and chemical environment that most closely resembles that of the prebiotic Earth and presents the best opportunity for future human habitation.

Although the present average surface temperature of Mars, −60°C, is much colder than that of Earth (15°C), the fluvial features on the surface suggest that Mars was at one time much warmer. Perhaps if we can understand the causes for the ancient warm climate, we can engineer the return of the Martian atmosphere to that state.

The major problem in terraforming Mars is to create an atmosphere that is dense enough to raise the mean surface temperature to around 0°C. To achieve this requires about 1 bar of CO_2 in the atmosphere. From the isotopic composition analysis given in section 7.4, we know that this amount of CO_2 probably exists in the surface reservoirs on Mars today. One efficient way to raise the atmospheric temperature is to inject into the atmosphere molecules that have strong infrared absorption bands, such as the chlorofluorocarbons (CFCs). According to recent estimates, the amount that is needed in the atmosphere is of the order of 10^{-6}. Assuming a CFC lifetime of about 100 years, the rate of industrial production to maintain this amount of CFCs in steady state is 10^8 ton yr^{-1}, which should be compared with the current global production rate of 10^6 ton yr^{-1}. The CFC production must be carried out in situ, since planetary transportation of materials at this rate is impractical. The assumption of a 100 yr lifetime for CFC on Mars assumes that Mars is protected by an ozone layer. Since the CFCs are highly detrimental to O_3, an alternative ultraviolet screen must be provided. The simplest one may be a layer of polysulfur, like that in the atmosphere of Venus. The source material that needs to be injected into the atmosphere is carbonyl sulfide (COS). We estimate that an annual production of 10^7 tons would be sufficient to create an effective ultraviolet screen. The above estimates are somewhat pessimistic because we are thinking in terms of current technology and currently available materials.

Ultimately, the long-term global environment of Mars must be sustained with minimal human input. Once the planet is modified to the point capable of sustaining life (we assume there is no life on Mars now), we could use Mars to test the Gaia hypothesis, which states that the biosphere is a living entity and is capable of modifying the environment for its own welfare. Of course, we do not know what the threshold is for a Gaia runoff of biological modification of a planet. Only an experiment will tell. If successful, Gaia may then be the path for the future colonization of the solar system, and beyond.

But the prospects of this happening soon are not good. Indeed, the Viking landing on Mars may be compared to the historical discovery of North America by the Vikings some five centuries before Columbus. The political, social, and technological developments of Europe at that time were not ready for the colonization of North America. Indeed, the word "colonialism" had not yet been invented. There was a long wait—so long that Columbus had to rediscover what the Vikings did many years earlier. We hope that the contents of this section can serve as an inspiration to a future Columbus of the solar system.

7.6 Unsolved Problems

We list a number of outstanding unresolved problems related to the atmosphere of Mars and its comparison with Earth's atmosphere.

1. Is heterogeneous chemistry important in the "dustiest" atmosphere in the solar system?

2. Are the properties of the high-altitude water ice clouds similar to those of the polar stratospheric clouds on Earth?
3. Do the high-altitude water ice clouds play a role in the NO_x chemistry?
4. Can the HO_x chemistry on Mars and in the terrestrial mesosphere be "fixed" by adjusting of the values of a similar set of reactions such as R42: $CO + OH \rightarrow CO_2 + H$ and R40: $OH + HO_2 \rightarrow H_2O + O_2$? The odd nitrogen model suggests that the principal sink for odd nitrogen, R61: $N + NO \rightarrow N_2 + O$, may be slower at lower temperatures (150 K). What are these implications for laboratory chemical kinetics?
5. How much volatile material has escaped from Mars? By atmospheric cratering? By thermal and nonthermal escape? By solar wind–induced sputtering?
6. How much volatile material is still sequestered on the surface and the polar caps of Mars?
7. Is Mars still degassing reducing gases from the interior?
8. Mars has undergone dramatic climatic variations, as can be seen from seasonal patterns of dust storms and billion-year-old river beds. Were the changes driven by variations of the solar constant or variations of the orbital elements?
9. Is the current arid and dusty atmosphere of Mars a good analog of the terrestrial atmosphere during the last glacial maximum, about 18,000 yr ago?
10. Is there life on Mars today? Was there ever life on Mars?
11. Is it possible (and ethical) to carry out terraforming of Mars and engineer the planet for human habitation?
12. Are the escape rates of oxygen and hydrogen given by the stoichiometric ratio of H_2O today? Is this relation true only when averaged over a Milankovitch climatic cycle of Mars?

8

Venus

8.1 Introduction

Venus has been visited by a number of spacecraft. Those most important in advancing our understanding of the atmosphere include Mariner 10 (1974), Pioneer Venus (1978–1992), and the series of Venera probes by the former USSR (1982–1986). The Pioneer Venus orbiter conducted more than a decade of monitoring of the upper atmosphere of Venus. The spacecraft data are supplemented by telescopic and satellite (in Earth orbit) observations.

The astronomical data for Venus are summarized in table 8.1. Venus is a close but slightly smaller sibling of Earth. The radius is 6051 km, as compared to Earth's 6371 km. The mass is 4.87×10^{24} kg, a little less than Earth's 5.98×10^{24} kg. The gravity of Venus is 8.87 m s^{-2}, as compared to Earth's 9.82 m s^{-2}. The dynamical and orbital parameters of Venus are very different from those of Earth. The rotation of Venus is retrograde, with a period of 225 days. The planet has little obliquity, and its orbit is close to being circular. Thus there is little seasonal variation in insolation over a Cytherian year. Perhaps the greatest surprise about Venus is its dense, dry, and hot atmosphere of 92 bar (see figure 8.1). This is all the more surprising because the planet is completely covered by thick clouds. Figure 8.2 shows the altitude profiles of three modes of cloud particles in the middle atmosphere. The high albedo of Venus implies that the planet receives less energy from the sun than Earth despite its closer proximity to the sun. However, the surface of Venus is hot, with a temperature of 733 K, attributed to the greenhouse effect.

Table 8.1 Astronomical and atmospheric data for Venus

Radius (km)[a]	6051.8 ± 1.0
Mass (kg)[b]	4.869×10^{24}
Mean density (g cm^{-3})[b]	5.24
Gravity (surface, equator) (m s^{-2})[c]	8.88
Escape velocity (km s^{-1})[c]	10.4
Semimajor axis (AU)	0.72333
Eccentricity of orbit[c]	0.0068
Obliquity (deg)[b]	177.36
Sidereal period of revolution (Earth days)[c]	224.701
Mean orbital velocity (km s^{-1})[c]	35.03
Sidereal period of rotation (Earth days, retrograde)[b]	-243.0187
Annual variation in solar insolation[d]	1.028
Atmospheric pressure at 70 km altitude (above cloud tops) (mbar)[e]	30
Temperature at 70 km altitude (K)[e]	238
Atmospheric scale height at 70 km altitude (km)	5.2
Adiabatic lapse rate at 70 km altitude (K km^{-1})	10.2
Atmospheric lapse rate at at 70 km altitude (K km^{-1})[e]	1.5
Atmospheric pressure at surface (bars)[e]	93
Mass of atmospheric column (kg m^{-2})	1.0×10^6
Total atmospheric mass (kg)	4.8×10^{20}
Surface temperature (K)[e]	735
Atmospheric scale height at surface (km)	15.8
Adiabatic lapse rate at surface (K km^{-1})	7.8
Atmospheric lapse rate at surface (K km^{-1})[e]	8

[a] Davies et al. (1992)
[b] The astronomical almanac for the year 1994 (1993)
[c] Beatty and Chaikin (1990)
[d] Based on the ratio of aphelion to perihelion distance squared
[e] Seiff (1983)

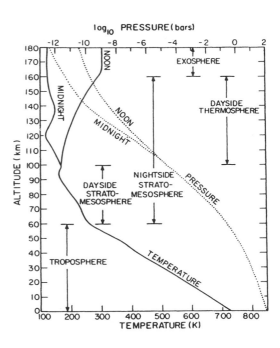

Figure 8.1 Altitude profile of temperature and pressure at 30° latitude for the atmosphere of Venus, based on the standard atmosphere. After Seiff, A., 1983, "Models of Venus' Atmospheric Structure," in Hunten et al. (1983; cited in section 8.1), p. 1045, as quoted by Prinn, R. G., 1985, "The Photochemistry of the Atmosphere of Venus," in Levine (1985; cited in section 8.1). p. 281.

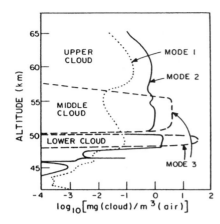

Figure 8.2 Altitude profile of the mass density of three modes of cloud particles in the middle atmosphere of Venus. After Knollenberg, R., and Hunten, D. M., 1980, "The Microphysics of the Clouds of Venus: Results of the Pioneer Venus Particle Size Spectrometer Experiment." *J. Geophys. Res.* **85**, 8039, as quoted by Prinn, R. G., 1985, "The Photochemistry of the Atmosphere of Venus," in Levine (1985; cited in section 8.1). p. 281.

Table 8.2 Composition of the atmosphere of Venus

Species	Mole fraction	References and remarks
CO_2	$96.5 \pm 0.8\%$	
N_2	$3.5 \pm 0.8\%$	
H_2O	2×10^{-5}	surface
	6×10^{-5}	22 km
	1.5×10^{-4}	42 km
Ar	$7 \pm 2.5 \times 10^{-5}$	
CO	$1.7 \pm 1 \times 10^{-6}$	surface
	$3.0 \pm 1.8 \times 10^{-5}$	42 km
	$4.5 \pm 1 \times 10^{-5}$	cloud top
	1×10^{-3}	100 km
		(1)
O_2	$< 3 \times 10^{-7}$	(2); cloud top
He	$12^{+24}_{-8} \times 10^{-6}$	
Ne	$7 \pm 3 \times 10^{-6}$	
HCl	6×10^{-7}	(3); cloud top
COS	3×10^{-7}	(4)
SO_2	$5-100 \times 10^{-9}$	(5)
Kr	2.5×10^{-8}	
SO	$2 \pm 1 \times 10^{-8}$	(6)
HF	$5^{+5}_{-2.5} \times 10^{-9}$	(3)
Xe	1.9×10^{-9}	

Unless otherwise stated, the composition data is taken from von Zahn et al. in Hunten et al., eds. (1983).

(1) See discussion in Yung and DeMore (1982); (2) Trauger, J. T. and Lunine, J. I., 1983, Icarus 55, 272; (3) Connes, P. et al., 1967, Astrophys. J. 147, 1230; (4) Crisp, D. et al., 1990, Nature 345, 508; (5) Zasova, L. V. et al., 1993, Icarus 105, 92; (6) Na, C. Y. et al., 1990, J. Geophys. Res. 95, 7485.

Table 8.3 Isotopic composition of the atmosphere of Venus

Species	Ratio	Remarks[a]
D/H	$1.6 \pm 2 \times 10^{-3}$	PVMS
	$1.9 \pm 6 \times 10^{-3}$	IR
$^{12}C/^{13}C$	86 ± 12	
	88.3 ± 1.6	
$^{14}N/^{15}N$	273 ± 56	
$^{16}O/^{18}O$	500 ± 25	PVMS
	500 ± 80	
$^{20}Ne/^{22}Ne$	11.8 ± 0.7	
$^{35}Cl/^{37}Cl$	2.9 ± 0.3	
$^{36}Ar/^{38}Ar$	5.56 ± 0.62	
	5.08 ± 0.05	
$^{40}Ar/^{36}Ar$	1.03 ± 0.04	
	1.19 ± 0.07	

From von Zahn et al. in Hunten, D. M. et al., eds. (1983).
[a] PVMS=Pioneer Venus Mass Spectrometer, IR=Infrared ground-based observations.

The bulk of our knowledge of the atmosphere of Venus is derived from observations in the middle atmosphere (60–100 km altitude). At cloud-top levels (65 km), ground-based ultraviolet (UV) observations revealed a 4–5 day period, east-to-west circulation that is 60 times faster than the solid surface. The mechanism for generating this superrotation is not well understood. The net result of this rotation is that it gives the upper atmosphere an effective diurnal cycle of 4–5 days.

The composition of the atmosphere of Venus is summarized in table 8.2. Isotopic composition is given in table 8.3. The major constituent of the atmosphere is CO_2 (96.5%). The next most abundant species is N_2 (3.5%). Since Venus is so similar to Earth in size, it is useful to compare the atmospheres of the two planets. Table 8.4 lists the essential features that are compared. The greatest difference between Venus and Earth concerns the amount of water on these two planets. The mixing ratio of water vapor in the lower atmosphere of Venus is variable, with a maximum value of 1.5×10^{-4}. The column-integrated water abundance is equivalent to a layer of 2–10 cm of water, uniformly spread over the surface of the planet. For comparison, Earth contains an average layer of 2.7 km of water, residing mostly in the oceans. Whether Venus was formed dry or whether most of its water has escaped is still the subject of debate. The lack of an ocean on Venus has at least three dramatic consequences for the atmosphere. First, most of the planet's CO_2 remains in the atmosphere, in contrast to Earth, where most of the 50 bars of CO_2 is sequestered as carbonate rock in the sediments. Second, the atmosphere of Venus contains large quantities of SO_2. On Earth, most of the volatile sulfur resides in the ocean as sulfate ions. The presence of this large amount of SO_2 in the atmosphere is largely responsible for the production of the dense H_2SO_4 clouds on Venus. Third, the atmosphere of Venus contains large amounts of HCl. On Earth, the bulk of chlorine is in the form of salt (NaCl) in the oceans. The photochemistry of the atmosphere of Venus is driven by the catalytic chemistry of chlorine in the atmosphere.

Table 8.4 Comparison of the essential aspects of the chemistry of the stratosphere of Earth and Venus

	Earth	Venus
Altitude (km)	20 – 40	60 – 80
Pressure (mbar)	100 – 5	300 – 5
Temperature (K)	200 – 250°	270 – 200
Total chlorine mixing ratio	3×10^{-9}	4×10^{-7}
Total sulfur mixing ratio	$\sim 1 \times 10^{-9}$	$\sim 2 \times 10^{-6}$
Total NO_x mixing ratio	2×10^{-8}	$\leq 3 \times 10^{-8}$
Catalytic chlorine cycle	Important	Important
Inhibitor of chlorine cycle	$Cl + CH_4$	$Cl + H_2$
	$ClO + NO$	$ClO + SO$
Heterogeneous chemistry	Important	Potentially important
Reaction breaking O–O bond	$NO + HO_2$	$SO + HO_2$
		$S + O_2$
		$NO + HO_2$
		$ClCO + O_2$

From DeMore, W. B., and Yung, Y. L. (1982).

There are two fundamental problems in the photochemistry of the atmosphere of Venus. The first is the CO_2 stability problem. The second is the SO_2 oxidation problem. The photochemistry of these two problems is intimately connected via the chemistry of oxygen.

As discussed in chapter 7 regarding Mars, CO_2 is photolyzed by ultraviolet radiation in the atmosphere:

$$CO_2 + h\nu \rightarrow CO + O \tag{8.1}$$

In a pure CO_2 atmosphere the net result, is from (7.37),

$$2CO_2 \rightarrow 2CO + O_2 \tag{8.2}$$

and we expect large quantities of CO and O_2 to be present in the atmosphere at a ratio of 2:1. The observed abundance of CO (from table 8.2) is 4.5×10^{-5} at the cloud tops. O_2 has not been detected despite two decades of searching for it. The upper limit for its mixing ratio is 3×10^{-7}. Thus, the CO/O_2 ratio is greater than 150. For comparison, we note that on Mars the mixing ratio of CO is 7×10^{-4} and the ratio of CO/O_2 is 0.5. Therefore, Venus has a CO_2 stability problem that is more serious than that on Mars because the solar UV flux at Venus is 4.4 times that at Mars. Conversely, the photochemical products, CO and O_2, are much less abundant on Venus than on Mars. This implies that the removal of CO and O_2 (by catalysis) in the atmosphere of Venus must be considerably more efficient.

The most abundant sulfur species in the atmosphere of Venus is SO_2. A most spectacular consequence of the sulfur photochemistry is the formation of H_2SO_4 clouds that completely envelope the planet. The reaction may be schematically written as

$$2SO_2 + O_2 + 2H_2O \rightarrow 2H_2SO_4 \tag{8.3}$$

The vapor pressure of sulfuric acid is low at the temperatures of the upper atmosphere. The primary fate of the H_2SO_4 produced in (8.3) is the formation of aerosols. Note that the consumption of O_2 by SO_2 in (8.3) accounts for the unusually low concentration of O_2 in the atmosphere of Venus. Reaction (8.3) scavenges H_2O in the upper atmosphere. In addition, H_2SO_4 aerosols are highly hygroscopic and serve to desiccate the upper atmosphere even further.

The oxidation of CO to CO_2 (the stability problem) and the oxidation of SO_2 to sulfate on Venus require the assistance of catalytic chemistry. It was recognized in 1971 that chlorine could play a fundamental role in the photochemistry of Venus, prior to the recognition of its importance in the terrestrial stratosphere. The purpose of this chapter is to develop the chemistry of chlorine in some detail.

However, our overall knowledge of the chemistry of Venus is far from complete. The major barriers to further understanding are the lack of reliable observational data on atmospheric composition below the clouds and the lack of an adequate kinetic database for sulfur chemistry.

8.2 Photochemistry

Since CO_2 is the dominant species in the atmospheres of Venus and Mars, the motivations for the stability problem are the same (see section 7.2.1).

8.2.1 HO_x Catalytic Chemistry

Since water vapor had been detected on Venus, the early models attempted to solve the CO_2 stability problem by invoking the HO_x chemistry that had been successfully applied to Mars. The chemistry involves catalytic cycles that are described in section 7.2.2. An obvious source of HO_x radicals is photolysis of H_2O:

$$H_2O + h\nu \rightarrow OH + H \tag{8.4}$$

However, due to the drying of the upper atmosphere by H_2SO_4 aerosols, the H_2O mixing ratio is of the order of 10^{-6}, even though the maximum mixing ratio below the clouds is greater than 1×10^{-4}. A larger source of HO_x is the photolysis of HCl:

$$HCl + h\nu \rightarrow H + Cl \tag{8.5}$$

The Cl atom produced in (8.5) could result in the production of another HO_x radical by

$$Cl + H_2 \rightarrow HCl + H \tag{8.6}$$

The source of H_2 in (8.6) is postulated to be at the surface of the planet, where the equilibrium reaction

$$CO + H_2O \rightarrow CO_2 + H_2 \tag{8.7}$$

turns H_2O into H_2. Hence, the net result is

$$\begin{array}{rl} HCl + h\nu \rightarrow H + Cl & (8.5) \\ Cl + H_2 \rightarrow HCl + H & (8.6) \\ CO + H_2O \rightarrow CO_2 + H_2 & (8.7) \\ \underline{CO_2 + h\nu \rightarrow CO + O} & (8.1) \\ net \quad H_2O \rightarrow 2H + O & (I) \end{array}$$

Note that in chemical scheme (I), chlorine plays the role of a catalyst in the dissociation of H_2O into H and O atoms. The reason that scheme (I) is more effective than the direct photolysis of H_2O (8.4) is that HCl absorbs at longer wavelengths in the UV range, where the solar flux is larger.

The validity of the HO_x chemistry for explaining the stability of CO_2 on Venus was questioned after the reaction

$$Cl + HO_2 \rightarrow HCl + O_2 \tag{8.8}$$

was shown to be extremely fast, with rate coefficient equal to 3.2×10^{-11} cm^3 s^{-1}. This reaction effectively removes the HO_x radicals before they can catalytically oxidize CO to CO_2. However, in view of the importance of heterogeneous chemistry for converting HCl and $ClONO_2$ into labile forms of chlorine (see section 8.2.5 and chapter 10), the HO_x chemistry on Venus should be reexamined.

8.2.2 Chlorine Catalytic Chemistry

Laboratory studies have shown that chlorine is capable of oxidizing CO to CO_2. There are two crucial steps. The first step is the formation of the chloroformyl radical by the three-body reaction

$$Cl + CO + M \rightarrow ClCO + M \tag{8.9}$$

The second step is the formation of the peroxychloroformyl radical by

$$ClCO + O_2 + M \rightarrow ClC(O)O_2 + M \tag{8.10}$$

The structure of this radical is known and may be represented as

$$\begin{array}{c} O \\ \parallel \\ Cl-C \\ \backslash \\ O-O \end{array}$$

The complex is not very stable and may be decomposed by collision:

$$ClC(O)O_2 + M \rightarrow ClCO + O_2 + M \tag{8.11}$$

The empirical rate coefficient for (8.10), taking (8.11) into account, is

$$k = \frac{5.7 \times 10^{-15} \exp(500/T)}{1 \times 10^{17} + 0.05 M} \quad \text{cm}^6 \text{ s}^{-1} \tag{8.12}$$

where M is the number density of the atmosphere. The peroxychloroformyl radical is removed by reactions with O and Cl atoms, resulting in the production of CO_2:

$$ClC(O)O_2 + O \rightarrow CO_2 + O_2 + Cl \tag{8.13}$$

$$ClC(O)O_2 + Cl \rightarrow CO_2 + ClO + Cl \tag{8.14}$$

The net results may be summarized in two catalytic cycles

$$\begin{aligned}
\text{Cl} + \text{CO} + \text{M} &\to \text{ClCO} + \text{M} & (8.9)\\
\text{ClCO} + \text{O}_2 + \text{M} &\to \text{ClC(O)O}_2 + \text{M} & (8.10)\\
\underline{\text{ClC(O)O}_2 + \text{O} \to \text{CO}_2 + \text{O}_2 + \text{Cl}} & & (8.13)\\
net \quad \text{CO} + \text{O} &\to \text{CO}_2 & \text{(IIa)}
\end{aligned}$$

$$\begin{aligned}
\text{Cl} + \text{CO} + \text{M} &\to \text{ClCO} + \text{M} & (8.9)\\
\text{ClCO} + \text{O}_2 + \text{M} &\to \text{ClC(O)O}_2 + \text{M} & (8.10)\\
\text{ClC(O)O}_2 + \text{Cl} &\to \text{CO}_2 + \text{ClO} + \text{Cl} & (8.14)\\
\underline{\text{ClO} + \text{O} \to \text{Cl} + \text{O}_2} & & (8.15)\\
net \quad \text{CO} + \text{O} &\to \text{CO}_2 & \text{(IIb)}
\end{aligned}$$

Note that schemes (IIa) and (IIb) are analogous to HO_x schemes (Ia) and (IIb) described in chapter 7. Reactive chlorine species, derived from (8.5), serve as catalysts for these chemical schemes.

8.2.3 Sulfur Chemistry

Four sulfur species have been firmly identified in the atmosphere of Venus: SO_2, SO, COS, and H_2SO_4 (in aerosols). Their abundances are summarized in table 8.2. The presence of thiozone (S_3) and polysulfur (S_x) in the clouds has been inferred but has not been proved. There are two parts to the chemistry of sulfur species in the atmosphere of Venus. In the deep atmosphere and on the surface, the chemistry is dominated by equilibrium chemistry. Above the cloud tops, the chemistry is driven by photochemistry. Thus, the partitioning of sulfur among the different species represents a chemical tug of war between equilibrium chemistry in the lower atmosphere and photochemistry of the upper atmosphere. We first discuss the photochemistry in the atmosphere above the clouds.

The most reducing species of sulfur that has been observed is COS. In the upper atmosphere it readily undergoes photolysis:

$$\text{COS} + h\nu \to \text{CO} + \text{S}(^1\text{D}) \qquad (8.16)$$

where $S(^1D)$ is the first excited state of the S atom. The most likely fate of $S(^1D)$ is quenching:

$$\text{S}(^1\text{D}) + \text{M} \to \text{S} + \text{M} \qquad (8.17)$$

The S atom gets oxidized to SO by reacting with O:

$$\text{S} + \text{O}_2 \to \text{SO} + \text{O} \qquad (8.18)$$

Further oxidation to SO_2 can proceed via the three-body reaction

$$\text{SO} + \text{O} + \text{M} \to \text{SO}_2 + \text{M} \qquad (8.19)$$

or the self-reaction

$$\text{SO} + \text{SO} \to \text{SO}_2 + \text{S} \qquad (8.20)$$

Note that the net result of (8.18)–(8.20) is the oxidation of S to SO_2 using O_2.

Both SO and SO_2 are readily photolyzed in the upper atmosphere of Venus:

$$SO_2 + h\nu \rightarrow SO + O \qquad (8.21)$$

$$SO + h\nu \rightarrow S + O \qquad (8.22)$$

but because of the fast reactions (8.18) and (8.20), the net result is the breaking of the O–O bond, as shown in the following:

$$\begin{array}{ll} SO_2 + h\nu \rightarrow SO + O & (8.21) \\ SO + SO \rightarrow SO_2 + S & (8.20) \\ \underline{S + O_2 \rightarrow SO + O} & (8.18) \\ \text{net} \quad O_2 \rightarrow 2O & (\text{IIIa}) \end{array}$$

$$\begin{array}{ll} SO + h\nu \rightarrow S + O & (8.22) \\ \underline{S + O_2 \rightarrow SO + O} & (8.18) \\ \text{net} \quad O_2 \rightarrow 2O & (\text{IIIb}) \end{array}$$

Once the strong O–O bond is broken, the oxidation of SO_2 to SO_3 readily takes place by the three-body reaction

$$SO_2 + O + M \rightarrow SO_3 + M \qquad (8.23)$$

followed by the formation of H_2SO_4:

$$SO_3 + H_2O + M \rightarrow H_2SO_4 + M \qquad (8.24)$$

The net result may be summarized by

$$SO_2 + \frac{1}{2}O_2 + H_2O \rightarrow H_2SO_4 \qquad (\text{IV})$$

where the O_2 in scheme (IV) is turned into O by schemes (IIIa) and (IIIb). The efficiency of scheme (IV) for consuming O_2 counterbalances the result of an important catalytic cycle that forms the O–O bond:

$$\begin{array}{ll} ClO + O \rightarrow Cl + O_2 & (8.15) \\ Cl + O_3 \rightarrow ClO + O_2 & (8.25) \\ \underline{O + O_2 + M \rightarrow O_3 + M} & (8.26) \\ \text{net} \quad O + O \rightarrow O_2 & (\text{V}) \end{array}$$

Note that this scheme is primarily responsible for the destruction of odd oxygen in the terrestrial stratosphere by chlorine. It was a puzzle why the atmosphere of Venus contains so little O_2 in view of the catalytic power of scheme (V). The reason is that cycles (IIIa) and (IIIb) are powerful enough to reverse the action of cycle (V), with the net result that O_2 is consumed by scheme (IV).

There is a chemical cycle that results in the simultaneous oxidation of CO to CO_2 and SO_2 to H_2SO_4:

$$Cl + CO + M \rightarrow ClCO + M \tag{8.9}$$
$$ClCO + O_2 + M \rightarrow ClC(O)O_2 + M \tag{8.10}$$
$$ClC(O)O_2 + Cl \rightarrow CO_2 + ClO + Cl \tag{8.14}$$
$$SO_2 + h\nu \rightarrow SO + O \tag{8.21}$$
$$SO_2 + O + M \rightarrow SO_3 + M \tag{8.23}$$
$$SO_3 + H_2O + M \rightarrow H_2SO_4 + M \tag{8.24}$$
$$\underline{ClO + SO \rightarrow Cl + SO_2} \tag{8.27}$$
$$net \quad CO + SO_2 + O_2 + H_2O \rightarrow CO_2 + H_2SO_4 \tag{VI}$$

Note that in the above cycle a crucial reaction is (8.27), the oxidation of SO to SO_2 by reacting with ClO. This chemical scheme consumes O_2, by breaking the strong O-O bond in reaction (8.14).

In addition to oxidation to H_2SO_4, there is another possible fate for sulfur compounds in the atmosphere of Venus: formation of polysulfur. S atoms are generated by the reactions (8.16) and (8.17), (8.20), and (8.22). S_2 may now be formed via

$$S + COS \rightarrow CO + S_2 \tag{8.28}$$
$$S + S + M \rightarrow S_2 + M \tag{8.29}$$

Production of S_n is possible through successive addition reactions such as

$$S + S_2 + M \rightarrow S_3 + M \tag{8.30}$$
$$S_2 + S_2 + M \rightarrow S_4 + M \tag{8.31}$$
$$S_4 + S_4 + M \rightarrow S_8 + M \tag{8.32}$$

S_3 is the sulfur analog of ozone, known as thiozone. As the number of sulfur atoms increases, the polyatomic sulfur compounds tend to have lower saturation vapor pressures. It is convenient to name all sulfur species beyond S_8 "polysulfur" or S_x. In the UV region S_x absorbs strongly, and it is believed to be the principal constituent of the unidentified UV absorber in the upper atmosphere of Venus. Note that reactions (8.28)–(8.32) leading to the formation of S_x compete with reactions (8.18)–(8.20) that lead to the formation of SO_2. The key reactions that determine the rates of formation of oxidized versus reduced sulfur are (8.18) and (8.28). Therefore, production of oxidized sulfur is favored when $k_{18}[O_2] > k_{28}[COS]$, that is,

$$[O_2] > \frac{k_{28}}{k_{18}}[COS] \tag{8.33}$$

The order of magnitude of k_{28}/k_{18} is 10^{-3}, and the mixing ratio of COS in the lower atmosphere is 2.5×10^{-7}. This suggests that the critical mixing ratio of of O_2 given by (8.33) is around 10^{-10}.

When O_2 abundance exceeds this value, production of oxidized sulfur species is favored, and ultimately H_2SO_4 is produced. When O_2 abundance is below this value, production of polysulfur becomes possible. Since the source of O_2 is photolysis of CO_2 in the upper atmosphere and the source of COS is equilibrium chemistry in the lower atmosphere and the surface, we can imagine that the chemistry of the downwelling air parcel and that of the upwelling air parcel could be quite different. This may indeed be the explanation for the patchiness and the transience of the UV markers in the cloud tops of Venus.

8.2.4 Equilibrium Chemistry

The bulk of photochemistry described in the sections 8.2.1–8.2.3 takes place above the cloud tops (about 60 km), in the region of the atmosphere known as the stratomesosphere. As shown in figure 8.1, the troposphere of Venus extends from the surface to about 60 km. The troposphere is shielded from UV radiation by CO_2, clouds, and aerosols. Photochemistry is not expected to be important. However, the high temperature and pressure make the lower atmosphere and the surface a favorable region for equilibrium chemistry. Our knowledge of the chemistry in the deep atmosphere and on the surface of Venus is less secure than that above the cloud tops. The greatest uncertainty is in the kinetic rate coefficients of reactions that drive the composition toward thermodynamic equilibrium.

The consequences of photochemistry in the upper atmosphere of Venus may be summarized by the following stoichiometric reactions:

$$CO_2 + SO_2 + H_2O + h\nu \rightarrow CO + H_2SO_4 \qquad (8.34)$$

$$2CO_2 + COS \rightarrow 3CO + SO_2 \qquad (8.35)$$

$$xCOS + h\nu \rightarrow xCO + S_x \qquad (8.36)$$

where we assume that the "parent species" coming up from the lower atmosphere are CO_2, SO_2, H_2O, and COS, and the final photochemical products are CO, H_2SO_4, and polysulfur (S_x, with $x > 8$). Thus, the net result of photochemistry in the upper atmosphere is the production of species that are far out of equilibrium. Note that the production of CO in reactions (8.34) and (8.35) represents only a small fraction (about 20%) of the photolysis of CO_2. The photolysis of CO_2 (8.3) is a source of oxidant only when CO remains unoxidized. The bulk of CO_2 photolysis is reversed by chemical schemes (IIa) and (IIb), and would not contribute to the oxidation of SO_2 and COS. The rates of the reactions (8.34)–(8.36) are such that all the SO_2 and COS in the bulk atmosphere of Venus would be exhausted in less than 10,000 yr. In this time the buildup of CO would exceed the observed abundance. Therefore, for a steady state composition to be maintained, reactions (8.34)–(8.36) must be reversed in the lower atmosphere or the surface.

The backward reaction of (8.34) is quite simple. In the lower atmosphere, at high temperature, H_2SO_4 is unstable and will decompose into H_2O and sulfur trioxide:

$$H_2SO_4 \rightarrow H_2O + SO_3 \qquad (8.37)$$

SO_3 is a potent oxidant and readily attacks CO:

$$SO_3 + CO \rightarrow SO_2 + CO_2 \qquad (8.38)$$

The net result is the reverse of (8.34):

$$CO + H_2SO_4 \rightarrow CO_2 + SO_2 + H_2O \qquad (8.39)$$

The reversal of (8.35) is more complex, involving pyrite formation on the surface of the planet. First, we have the reaction of SO_2 with carbonate rock to form anhydrite ($CaSO_4$):

$$SO_2 + CaCO_3 \rightarrow CaSO_4 + CO \qquad (8.40)$$

Anhydrite is further converted to pyrite (FeS_2) by

$$2CaSO_4 + FeO + 7CO \rightarrow FeS_2 + 2CaCO_3 + 5CO_2 \qquad (8.41)$$

Equilibrium reaction between pyrite, CO, and CO_2 results in the formation of COS:

$$FeS_2 + CO + CO_2 \rightarrow FeO + 2COS \qquad (8.42)$$

If we sum $2 \times$ (8.40), (8.41), and (8.42), the net result is equivalent to

$$3CO + SO_2 \rightarrow 2CO_2 + COS \qquad (8.43)$$

This is the reverse of (8.35). The reversal of (8.36) in the lower atmosphere of Venus is straightforward. At high temperature polysulfur can reevaporate and dissociate by absorption of visible light, as for instance in

$$S_3 + h\nu \rightarrow S_2 + S \qquad (8.44)$$

The S atoms can readily recombine with CO:

$$S + CO + M \rightarrow COS + M \qquad (8.45)$$

The presence of pyrite on the surface of Venus has been tentatively identified by Magellan. In addition to (8.42), pyrite may also generate H_2S by reacting with CO and H_2O:

$$FeS_2 + CO + 2H_2O \rightarrow FeO + CO_2 + 2H_2S \qquad (8.46)$$

The presence of H_2S was reported by the Pioneer Venus mass spectrometer experiment but has not been confirmed by other studies. Little atmospheric modeling has been carried out for H_2S.

Reactions (8.42) and (8.46) suggest that the surface of Venus is a source of reduced gases for the atmosphere. This chemical activity on the surface is analogous to that on the surface of Earth, where biological processes provide a source of reduced gases such as CH_4 and dimethyl sulfide to the atmosphere.

8.2.5 Nitrogen Chemistry

The only nitrogen compounds that have been detected on Venus are N_2, NO, and NO^+. The photochemistry of nitrogen on Venus is similar to that on Mars and we shall not repeat the discussion of section 7.2.7. However, there is another potential source of odd nitrogen, lightning, that needs to be further explored. The existence of odd nitrogen compounds in quantities greatly in excess of 1 ppb would have interesting consequences for the CO_2 stability and the photochemistry of chlorine. The most important reaction of NO in the atmosphere of Venus is the reaction that breaks the O—O bond:

$$NO + HO_2 \rightarrow NO_2 + OH \qquad (8.47)$$

With this key reaction we can drive two chemical schemes for the oxidation of CO to CO_2:

$$2\text{CO} + \text{O}_2 \to 2\text{CO}_2 \qquad \text{(VIIa)}$$

$$\text{CO} + \text{O} \to \text{CO}_2 \qquad \text{(VIIb)}$$

The details of the coupled NO_x and HO_x chemistry in schemes (VIIa) and (VIIb) are given in section 7.2.2 and are not repeated here. In addition to the oxidation of CO, NO_x can also catalyze the oxidation of SO_2 to sulfate, as in the following sequence of reactions:

$$\text{NO} + \text{HO}_2 \to \text{NO}_2 + \text{OH} \qquad (8.47)$$
$$\text{H} + \text{O}_2 + \text{M} \to \text{HO}_2 + \text{M} \qquad (8.48)$$
$$\text{NO}_2 + h\nu \to \text{NO} + \text{O} \qquad (8.49)$$
$$\text{CO} + \text{OH} \to \text{CO}_2 + \text{H} \qquad (8.50)$$
$$\text{SO}_2 + \text{O} + \text{M} \to \text{SO}_3 + \text{M} \qquad (8.23)$$
$$\underline{\text{SO}_3 + \text{H}_2\text{O} + \text{M} \to \text{H}_2\text{SO}_4 + \text{M}} \qquad (8.24)$$
$$net \quad \text{CO} + \text{SO}_2 + \text{O}_2 + \text{H}_2\text{O} \to \text{CO}_2 + \text{H}_2\text{SO}_4 \qquad \text{(VII)}$$

Note that both CO and SO_2 are oxidized in chemical scheme (VII). There is another process for the oxidation of SO_2 using NO_x as a catalyst, known as the "lead chamber" process in industry. This mechanism has also been proposed for the production of sulfuric acid from SO_2 via the formation of nitrosylsulfuric acid in the lower atmosphere of Venus.

Another interesting possibility is the formation of chlorine nitrate and N_2O_5 by

$$\text{ClO} + \text{NO}_2 + \text{M} \to \text{ClONO}_2 + \text{M} \qquad (8.51)$$

$$\text{NO}_2 + \text{NO}_3 + \text{M} \to \text{N}_2\text{O}_5 + \text{M} \qquad (8.52)$$

On the surface of sulfate aerosols we can have the following heterogeneous reactions:

$$\text{ClONO}_2 + \text{HCl} \to \text{Cl}_2 + \text{HNO}_3 \qquad (8.53)$$

$$\text{HCl} + \text{N}_2\text{O}_5 \to \text{ClNO}_2 + \text{HNO}_3 \qquad (8.54)$$

The net result is the conversion of the unreactive form of chlorine in HCl into more labile forms from which reactive radical species may be easily derived by photolysis:

$$\text{Cl}_2 + h\nu \to 2\text{Cl} \qquad (8.55)$$

$$\text{ClNO}_2 + h\nu \to \text{Cl} + \text{NO}_2 \qquad (8.56)$$

Since the limiting step in the availability of reactive chlorine is the photolysis reaction (8.5), reactions (8.53)–(8.56) greatly enhance the effectiveness of chlorine as a catalyst in the atmosphere of Venus.

We should emphasize that there is no definitive evidence for the occurrence of lightning on Venus. Hence, the importance of the NO_x chemistry outlined above remains purely hypothetical.

8.3 Model Results

The model results presented here emphasize the role of chlorine and sulfur chemistry on Venus. The potential importance of HO_x and NO_x chemistry has been briefly

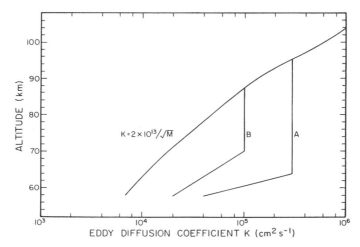

Figure 8.3 The eddy diffusivity profile $K(z)$ for the atmosphere of Venus. The models A and B reflect the range of uncertainty. Based on Model C in Yung, Y. L., and DeMore, W. B., 1982, "Photochemistry of the Stratosphere of Venus: Implications for Atmospheric Evolution." *Icarus* **51**, 199.

discussed in section 8.2, but there will be no detailed modeling results. Our knowledge for the lower atmosphere of Venus remains highly uncertain. The results we present here must be regarded as tentative.

A model of the atmosphere of Venus above the cloud tops is given in figure 8.1. The eddy diffusivity profile used in the one-dimensional model is shown in figure 8.3. At high altitudes the expression for eddy diffusion is

$$K(z) = \frac{2 \times 10^{13}}{\sqrt{M}} \text{ cm}^2 \text{ s}^{-1} \tag{8.57}$$

where M is number density of the ambient atmosphere. This expression is derived empirically using the Pioneer Venus mass spectrometer data. Just above the cloud tops there is an increase in the value of $K(z)$ that is needed in the model for transport of reactive chemical species.

The reactions that are important in the atmosphere of Venus are listed in table 8.5, along with preferred rate coefficients. The ion reactions are the same as in the atmosphere of Mars described in chapter 7 and are not listed in table 8.5.

The results of the ionosphere and of the mesostratosphere are from two different models: one covers the region of the atmosphere from the exobase to about 100 km, and the other spans the range from 100 km to the cloud tops. The (highly speculative) results for the atmosphere below the cloud tops are taken from a third model.

8.3.1 Ionosphere

Number densities of major ions in the upper atmosphere of Venus are presented in figure 8.4. The concentrations predicted by the model are in good agreement with

Table 8.5 List of reactions in the model[a]

	Reaction	Rate coefficient	Reference
R1	$CO_2 + h\nu \to CO + O$	1.5×10^{-12}	1, 2
R2	$HCl + h\nu \to H + Cl$	1.0×10^{-7}	3
R3	$H_2O + h\nu \to H + OH$	2.2×10^{-11}	4, 5, 6
R4a	$O_2 + h\nu \to O + O(^1D)$		4, 7, 8
b	$O_2 + h\nu \to 2O$	7.4×10^{-10}	4, 9, 10
R5a	$O_3 + h\nu \to O_2(^1\Delta) + O(^1D)$	8.6×10^{-3}	7, 10
b	$O_3 + h\nu \to O_2 + O$	4.2×10^{-4}	7, 10
R6	$H_2O_2 + h\nu \to 2OH$	9.9×10^{-5}	10, 11
R7	$ClO + h\nu \to Cl + O$	5.0×10^{-3}	3
R8	$SO + h\nu \to S + O$	2.7×10^{-4}	12
R9	$SO_2 + h\nu \to SO + O$	1.5×10^{-4}	13, 14, 15
R10	$O(^1D) + M \to O + M$	$6.8 \times 10^{-11} e^{117/T}$	16
R11	$O(^1D) + H_2 \to OH + H$	2.0×10^{-10}	17
R12	$O(^1D) + HCl \to OH + Cl$	1.4×10^{-10}	18
R13	$O(^1D) + H_2O \to 2OH$	2.8×10^{-10}	17
R14	$O + CO + M \to CO_2 + M$	$6.5 \times 10^{-33} e^{-2180/T}$	17
R15	$O + H_2 \to OH + H$	$1.6 \times 10^{-11} e^{-4570/T}$	19
R16	$O + HCl \to OH + Cl$	$1.1 \times 10^{-11} e^{-3370/T}$	17
R17	$2O + M \to O_2 + M$	$8.6 \times 10^{-28} T^{-2}$	19, 20
R18a	$O_2(^1\Delta) + M \to O_2 + M$	3.0×10^{-20}	21, 22, 23
b	$O_2(^1\Delta) \to O_2 + h\nu$	2.6×10^{-4}	24
R19	$O_2 + O + M \to O_3 + M$	1.35×10^{-33}	19
R20	$O_3 + O \to 2O_2$	$1.5 \times 10^{-11} e^{-2218/T}$	18
R21	$H + HCl \to H_2 + Cl$	$2.4 \times 10^{-11} e^{-1740/T}$	estimated
R22	$H + O_2 + M \to HO_2 + M$	$3.6 \times 10^{-29} T^{-1}$	17, 20
R23	$H + O_3 \to OH + O_2$	$1.4 \times 10^{-10} e^{-480/T}$	17
R24	$2H + M \to H_2 + M$	$6.6 \times 10^{-31} T^{-0.65}$	20, 25, 26
R25	$OH + CO \to H + CO_2$	$1.4 \times 10^{-13} (1 + P \text{atm})$	18
R26	$OH + H_2 \to H + H_2O$	$1.8 \times 10^{-11} e^{-2330/T}$	17
R27	$OH + HCl \to H_2O + Cl$	$3.0 \times 10^{-12} e^{-425/T}$	17
R28	$OH + O \to H + O_2$	3.8×10^{-11}	17
R29	$OH + O_3 \to HO_2 + O_2$	$1.9 \times 10^{-12} e^{-1000/T}$	17
R30	$2OH \to H_2O + O$	1.8×10^{-12}	17
R31	$HO_2 + O \to OH + O_2$	3.1×10^{-11}	17
R32	$HO_2 + O_3 \to OH + 2O_2$	$1.4 \times 10^{-14} e^{-600/T}$	17
R33a	$HO_2 + H \to 2OH$	3.2×10^{-11}	17
b	$HO_2 + H \to H_2 + O_2$	1.4×10^{-11}	17
c	$HO_2 + H \to H_2O + O$	9.4×10^{-13}	17
R34	$HO_2 + OH \to H_2O + O_2$	8.0×10^{-11}	18
R35	$2HO_2 \to H_2O_2 + O_2$	2.3×10^{-12}	17
R36	$H_2O_2 + O \to OH + HO_2$	$2.8 \times 10^{-12} e^{-2125/T}$	18
R37	$H_2O_2 + OH \to OH + HO_2$	$7.6 \times 10^{-12} e^{-670/T}$	17
R38	$Cl + H_2 \to HCl + H$	$4.7 \times 10^{-11} e^{-2340/T}$	17
R39	$Cl + O_3 \to ClO + O_2$	$2.7 \times 10^{-11} e^{-257/T}$	17
R40	$Cl + OH \to HCl + O$	$1.0 \times 10^{-11} e^{-2970/T}$	16
R41a	$Cl + HO_2 \to HCl + O_2$	$1.8 \times 10^{-11} e^{180/T}$	27
b	$Cl + HO_2 \to ClO + OH$	$2.2 \times 10^{-10} e^{-1000/T}$	27
R42	$Cl + H_2O_2 \to HCl + HO_2$	$1.1 \times 10^{-11} e^{-980/T}$	17
R43	$ClO + CO \to Cl + CO_2$	$1.0 \times 10^{-12} e^{-3700/T}$	19

Table 8.5 (continued)

	Reaction	Rate coefficient	Reference
R44	$ClO + O \rightarrow Cl + O_2$	$7.5 \times 10^{-11} e^{-120/T}$	17
R45	$ClO + OH \rightarrow Cl + HO_2$	9.1×10^{-12}	18
R46	$S + O_2 \rightarrow SO + O$	2.2×10^{-12}	17
R47	$S + CO_2 \rightarrow SO + CO$	1.0×10^{-20}	estimated
R48	$S + O_3 \rightarrow SO + O_2$	1.2×10^{-11}	17
R49	$S + OH \rightarrow SO + H$	3.8×10^{-11}	estimated
R50	$S + HO_2 \rightarrow SO + OH$	3.1×10^{-11}	estimated
R51	$SO + O + M \rightarrow SO_2 + M$	6.0×10^{-31}	19, estimated
R52	$SO + O_2 \rightarrow SO_2 + O$	$6.0 \times 10^{-13} e^{-3300/T}$	17
R53	$SO + O_3 \rightarrow SO_2 + O_2$	$2.5 \times 10^{-12} e^{-1100/T}$	17
R54	$SO + OH \rightarrow SO_2 + H$	1.2×10^{-10}	28
R55	$SO + HO_2 \rightarrow SO_2 + OH$	2.3×10^{-11}	estimated
R56	$SO + ClO \rightarrow SO_2 + Cl$	2.3×10^{-11}	29
R57	$2SO \rightarrow SO_2 + S$	8.3×10^{-15}	30
R58	$SO_2 + O + M \rightarrow SO_3 + M$	$8.0 \times 10^{-32} e^{-1000/T}$	17, 20
R59a	$SO_2 + OH + M \rightarrow HSO_3 + M$	$4.2 \times 10^{-32} e^{1000/T}$	31, 32
b	$SO_2 + OH + M \rightarrow HSO_3 + M$	2.0×10^{-12}	two body limit
R60	$SO_2 + HO_2 \rightarrow SO_3 + OH$	1.0×10^{-18}	17, 33, estimated
R61	$SO_2 + H_2O_2 + \text{aerosol} \rightarrow H_2SO_4$	4.3×10^{-5}	estimated
R62	$SO_2 + Cl + M \rightarrow ClSO_2 + M$	4.6×10^{-33}	34
R63	$SO_2 + ClO \rightarrow SO_3 + Cl$	1.0×10^{-18}	estimated
R64	$SO_3 + SO \rightarrow 2 SO_2$	2.0×10^{-15}	19
R65	$SO_3 + H_2O \rightarrow H_2SO_4$	9.0×10^{-13}	17
R66	$NO + h\nu \rightarrow N + O$		35
R67	$NO_2 + h\nu \rightarrow NO + O$	1.0×10^{-2}	36
R68a	$NO_3 + h\nu \rightarrow NO_2 + O$	2.0×10^{-2}	36
b	$NO_3 + h\nu \rightarrow NO + O_2$	1.0×10^{-2}	36
R69	$HNO + h\nu \rightarrow H + NO$	1.0×10^{-3}	estimated
R70	$HNO_2 + h\nu \rightarrow OH + NO$	2.0×10^{-3}	35
R71	$HNO_3 + h\nu \rightarrow OH + NO_2$	5.4×10^{-5}	36
R72a	$N + O \rightarrow NO + h\nu$	2.5×10^{-17}	37
b	$N + O + M \rightarrow NO + M$	$1.9 \times 10^{-31} T^{-1/2}$	37
R73	$N + O_2 \rightarrow NO + O$	$4.4 \times 10^{-12} e^{-3270/T}$	18
R74	$N + O_3 \rightarrow NO + O_2$	1.0×10^{-15}	18
R75	$N + OH \rightarrow NO + H$	5.3×10^{-11}	18
R76	$N + NO \rightarrow N_2 + O$	2.1×10^{-11}	37
R77	$NO + O + M \rightarrow NO_2 + M$	2.4×10^{-31}	17, 20
R78	$NO + O_3 \rightarrow NO_2 + O_2$	$2.3 \times 10^{-12} e^{-1450/T}$	18
R79	$NO + H + M \rightarrow HNO + M$	$1.5 \times 10^{-32} e^{300/T}$	18
R80	$NO + OH + M \rightarrow HNO_2 + M$	1.3×10^{-30}	17, 20
R81	$NO + HO_2 \rightarrow NO_2 + OH$	$3.5 \times 10^{-12} e^{250/T}$	18
R82	$NO + ClO \rightarrow NO_2 + Cl$	$6.5 \times 10^{-12} e^{280/T}$	18
R83a	$NO_2 + O \rightarrow NO + O_2$	9.3×10^{-12}	17
b	$NO_2 + O + M \rightarrow NO_3 + M$	1.8×10^{-31}	17, 20
R84	$NO_2 + OH + M \rightarrow HNO_3 + M$	5.2×10^{-30}	17, 20
R85	$NO_2 + SO \rightarrow NO + SO_2$	1.4×10^{-11}	17
R86	$HNO + O \rightarrow OH + NO$	1.0×10^{-13}	estimated
R87	$HNO + H \rightarrow H_2 + NO$	1.0×10^{-13}	19
R88	$HNO + Cl \rightarrow HCl + NO$	1.0×10^{-13}	estimated

(continued)

Table 8.5 (continued)

	Reaction	Rate coefficient	Reference
R89	$2HNO \rightarrow N_2O + H_2O$	4.0×10^{-15}	16
R90	$HNO_2 + OH \rightarrow H_2O + NO_2$	6.6×10^{-12}	16
R91	$HNO_3 + OH \rightarrow H_2O + NO_3$	1.5×10^{-14}	18
R92	$Cl_2 + h\nu \rightarrow 2Cl$	2.4×10^{-3}	36
R93	$COCl_2 + h\nu \rightarrow COCl + Cl$	5×10^{-5}	17
R94	$HOCl + h\nu \rightarrow OH + Cl$	4×10^{-4}	36
R95	$NOCl + h\nu \rightarrow NO + Cl$	1.4×10^{-3}	24
R96	$Cl + O_2 + M \rightarrow ClOO + M$	$3.3 \times 10^{-30} T^{-1.3}$	18
R97	$ClOO + M \rightarrow Cl + O_2 + M$	$2.7 \times 10^{-9} e^{-2650/T}$	18
R98	$Cl + H + M \rightarrow HCl + M$	1×10^{-32}	estimated
R99	$Cl + CO + M \rightarrow ClCO + M$	$1.3 \times 10^{-34} e^{1000/T}$	39
R100	$ClCO + M \rightarrow Cl + CO + M$	$6 \times 10^{-11} e^{-2550/T}$	see 8.2.2
R101	$ClCO + O_2 + M \rightarrow ClCO_3 + M$		see 8.2.2
R102a	$ClCO + O \rightarrow Cl + CO_2$	3.0×10^{-11}	estimated
b	$ClCO + O \rightarrow ClO + CO$	3.0×10^{-12}	estimated
R103	$ClCO + H \rightarrow HCl + CO$	1.0×10^{-11}	estimated
R104	$ClCO + Cl \rightarrow Cl_2 + CO$	1.0×10^{-11}	estimated
R105	$ClCO + Cl_2 \rightarrow COCl_2 + Cl$	$6 \times 10^{-13} e^{-1400/T}$	39
R106	$2ClCO \rightarrow COCl_2 + CO$	1.0×10^{-11}	estimated
R107	$2Cl + M \rightarrow Cl_2 + M$	$1.2 \times 10^{-33} e^{900/T}$	19, 20
R108	$Cl_2 + O \rightarrow ClO + Cl$	$4.2 \times 10^{-12} e^{-1370/T}$	16
R109	$Cl_2 + H \rightarrow HCl + Cl$	$1.5 \times 10^{-10} e^{-593/T}$	16
R110	$ClCO_3 + O \rightarrow CO_2 + Cl + O_2$	1.0×10^{-11}	see 8.2.2
R111	$ClCO_3 + Cl \rightarrow CO_2 + Cl + ClO$	1.0×10^{-11}	see 8.2.2
R112	$ClCO_3 + H \rightarrow CO_2 + Cl + OH$	1.0×10^{-11}	see 8.2.2
R113	$Cl + O + M \rightarrow ClO + M$	5.0×10^{-32}	estimated
R114	$ClO + HO_2 \rightarrow HOCl + O_2$	$4.6 \times 10^{-13} e^{710/T}$	18

[a] From Yung, Y. L., and DeMore, W. B., 1982, "Photochemistry of the Stratosphere of Venus: Implications for Atmospheric Evolution." *Icarus* **51**, 199.
See p. 444 for references.

Pioneer Venus data. As in the case of Mars (see section 7.3.1), O_2^+ is the principal ion, followed by CO_2^+, NO^+, O^+, and other minor ions. The ionospheres of Venus and Mars are similar to each other but are different from that of Earth. The terrestrial ionosphere is dominated by a well-developed O^+ upper ionosphere, known as the F1 region. The corresponding F1 region is absent from the upper atmospheres of Venus and Mars, because there are fewer O atoms in the upper atmospheres of Venus and Mars. This is due to the higher eddy diffusion coefficient near the homopause, $K = 10^8$ cm^2 s^{-1}, which is a hundred times larger than the corresponding terrestrial value.

8.3.2 Mesostratosphere

Table 8.6 summarizes the boundary conditions of the one-dimensional model used in the study of Venus. The lower boundary of the model is at 58 km near the cloud tops, where the mixing ratios of some of the long-lived species are fixed from the observations. All other species, including the short-lived radical species, we assume are lost from the lower boundary at the maximum deposition velocity allowed by eddy diffusion. The upper boundary is at 110 km, where the fluxes are zero for all

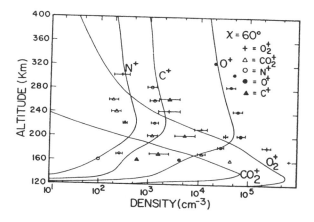

Figure 8.4 Model and measured densities of ions in the upper atmosphere of Venus. The model (lines) is for solar zenith angle of 60°. The data are from Pioneer Venus ion mass spectrometer. After Nagy, A. F. et al., 1980, "Model Calculations of the Dayside Ionosphere of Venus: Ionic Composition." *J. Geophys. Res.* **85**, 7795.

species except for O and CO, for which downward fluxes, equal to the integrated CO_2 dissociation rate above 110 km, are imposed. The continuity equations are solved for all important carbon, oxygen, HO_x, chlorine, and sulfur species.

Altitude profiles for the mixing ratios of the most important minor species in the model, CO, O_2, and SO_2, are presented in figure 8.5a, along with comparisons with observations. Note that in the upper atmosphere (above 80 km) CO and O_2 are produced at a ratio of 2:1 according to (8.2). The mixing ratios of both CO and O_2 drop rapidly with decreasing altitude. We explain later in this section why O_2 decreases more rapidly than CO at the level of the cloud tops. SO_2 in the model decreases rapidly with increasing altitude. The vertical distribution of these species suggests that the upper atmosphere is a net source for CO and O_2 and that the lower atmosphere is a net source of SO_2. The rate of CO_2 photolysis, the reaction producing CO and O (and ultimately O_2), is given in figure 8.5b. The principal reactions responsible for the oxidation of CO and SO_2 are also shown. It is clear that the bulk of the oxygen produced is consumed by CO oxidation. Oxidation is catalyzed primarily by chlorine,

Table 8.6 Boundary conditions for chemical species solved in the one-dimensional model

Species	Lower boundary	Upper boundary
CO	$f = 4.5 \times 10^{-5}$	$\phi = -1.0 \times 10^{12}$
O	$v = -2.0 \times 10^{-2}$	$\phi = -1.0 \times 10^{12}$
O_2	$v = -2.0 \times 10^{-2}$	$\phi = 0$
SO_2	$f = 4.0 \times 10^{-6}$	$\phi = 0$
Cl_2	$v = -2.0 \times 10^{-2}$	$\phi = 0$
H_2	$v = -2.0 \times 10^{-2}$	$\phi = 0$
HCl	$f = 8 \times 10^{-7}$	$\phi = 0$

From Yung and DeMore (1982). For short-lived species not explicitly listed here photochemical equilibrium is assumed (flux = 0) at both boundaries. The symbols f, ϕ, and v denote, respectively, mixing ratio, flux (cm^2 sec^{-1}), and velocity (cm sec^{-1}). The sign convention for ϕ and v is positive for upward flow. The maximum deposition velocity at the lower boundary is given by $v = -K/H$, where K = eddy diffusion coefficient and H = scale height.

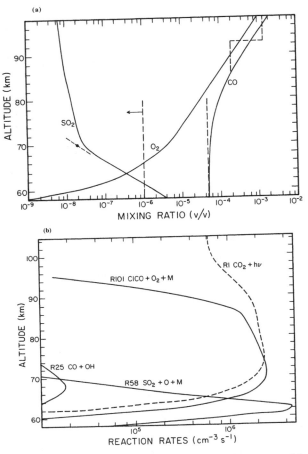

Figure 8.5 (a) Abundances of SO_2, CO, and O_2 computed in a model of the atmosphere of Venus and comparison with observations (dashed lines). (b) Reaction rates of major reactions destroying CO_2, producing CO_2, and oxidizing SO_2. Based on Model C in Yung, Y. L., and DeMore, W. B., 1982, "Photochemistry of the Stratosphere of Venus: Implications for Atmospheric Evolution." *Icarus* **51**, 199.

the rate-limiting step being the reaction $ClCO + O_2$. Catalysis by HO_x (via the reaction $CO + OH$) is insignificant. Within the upper consumption of oxygen by SO_2 (to form sulfate) is important. This additional mechanism for the removal of oxygen is largely responsible for the rapid decay of O_2 above the cloud tops in figure 8.5a. The ratio of the column-integrated concentrations of CO to O_2 is 16 cloud in the model, as compared to the observed lower limit of 150. This suggests that the removal of O_2 in the atmosphere of Venus is even more efficient than that computed in the model. Note that this ratio is 0.5 in the atmosphere of Mars, where the primary sink of O_2 is oxidation of CO.

Upper Atmosphere

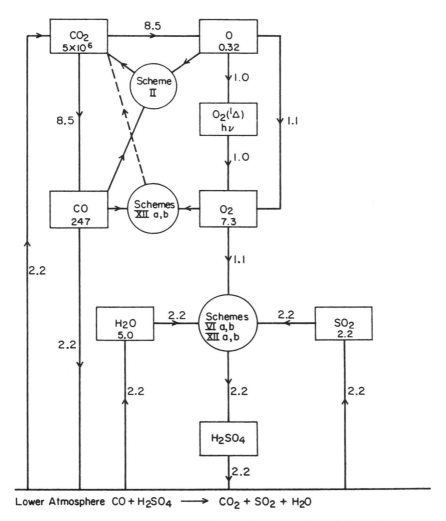

Figure 8.6 Schematic diagram summarizing the budget and flow of the major oxygen-bearing species. The column abundances above the cloud tops are in units of 10^{18} cm^{-2}. The fluxes are in units of 10^{12} cm^{-2} s^{-1}. Based on Model C in Yung, Y. L., and DeMore, W. B., 1982, "Photochemistry of the Stratosphere of Venus: Implications for Atmospheric Evolution." *Icarus* **51**, 199.

A schematic diagram summarizing the budget and flow of the major oxygen-bearing species is given in figure 8.6. As pointed out in section 8.2.4, the net result of photochemistry above the cloud tops is given by reaction (8.34). The photochemical products, CO and H_2SO_4, are transported to the lower atmosphere, where reaction (8.34) is reversed. As a check on the model, we can compare the model flux of H_2SO_4 to

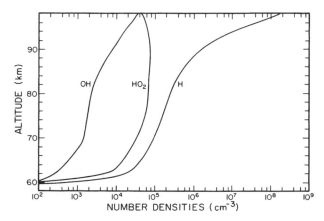

Figure 8.7 Concentrations of HO_x (H, OH, HO_2) in the model. Based on Model C in Yung, Y. L., and DeMore, W. B., 1982, "Photochemistry of the Stratosphere of Venus: Implications for Atmospheric Evolution." *Icarus* **51**, 199.

the lower atmosphere with that deduced from Pioneer Venus observations. From the aerosol data we estimate a column density of H_2SO_4 equal to 7×10^{18} cm^{-2}. The mean radius of the particles is 1 μm, with Stokes falling velocity of 2.7×10^{-2} cm s^{-1}. The downward flux of H_2SO_4 is 1×10^{12} cm^{-2} s^{-1}, a value that is within a factor of 2 of our model prediction shown in figure 8.6.

Altitude profiles for the major HO_x species in the model are presented in figure 8.7. The principal HO_x radical is H, followed by HO_2 and OH. The HO_x concentrations in the atmosphere of Venus are orders of magnitude less than those in the atmosphere of Mars. The mixing ratios of the reactive chlorine species are given in figure 8.8. The

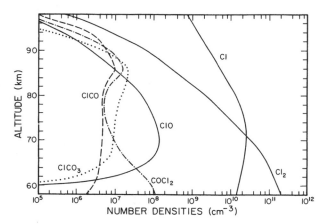

Figure 8.8 Concentrations of ClO_x (Cl, ClO, Cl_2, ClCO, $COCl_2$, $ClCO_3$) in the model.

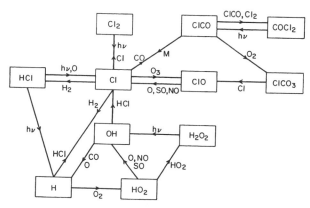

Figure 8.9 Schematic diagram showing the major sources, sinks, and recycling pathways for HO_x, NO_x, and ClO_x. Based on Model C in Yung, Y. L., and DeMore, W. B., 1982, "Photochemistry of the Stratosphere of Venus: Implications for Atmospheric Evolution." *Icarus* **51**, 199.

most abundant chlorine compounds are Cl_2 and Cl, followed by $COCl_2$ and ClO. Small concentrations of the radicals $ClCO$ and $ClC(O)O_2$ are responsible for the catalytic oxidation of CO to CO_2. Figure 8.9 shows a schematic diagram showing the major sources, sinks, and recycling pathways for HO_x and ClO_x. As shown in figure 8.10, the ultimate source of all HO_x and ClO_x radicals is the breakup of HCl, primarily by photolysis. The reaction O + HCl makes a minor contribution. The major sink of the reactive radicals is the reaction $Cl + HO_2$. The reaction $H + Cl_2$ accounts for a minor fraction of the loss of radicals. The mixing ratios of HCl and H_2 are shown in figure 8.11. H_2 is a photochemical product in the upper atmosphere, produced by the reaction H + HCl. It is rapidly removed by the reaction $Cl + H_2$ to form HCl

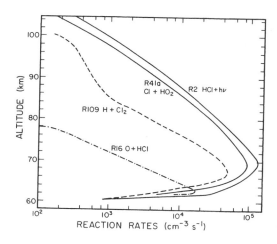

Figure 8.10 Reaction rates of major reactions producing and destroying HO_x and ClO_x radicals. Based on Model C in Yung, Y. L., and DeMore, W. B., 1982, "Photochemistry of the Stratosphere of Venus: Implications for Atmospheric Evolution." *Icarus* **51**, 199.

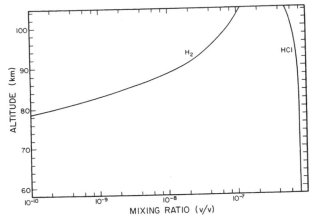

Figure 8.11 Abundances of H_2 and HCl computed in the model. Based on Model C in Yung, Y. L., and DeMore, W. B., 1982, "Photochemistry of the Stratosphere of Venus: Implications for Atmospheric Evolution." *Icarus* **51**, 199.

again. The low concentration of H_2 in the lower atmosphere is the result of the high concentration of chlorine. In the mesosphere, H_2 becomes more abundant, and the conversion of HCl to H_2 causes the mixing ratio of HCl to decrease. H_2 flows to the thermosphere, where it is decomposed into H atoms, some of which eventually escape from the planet.

Figure 8.12a shows the number densities of the major sulfur species. SO_2 is the dominant sulfur species in the model. The photochemical products include SO, SO_3, and S. The abundance of SO_2 falls off rapidly with altitude due to its conversion to sulfate. Its scale height in this region is 1–3 km, as compared with the mean atmospheric scale height of 6 km. As shown in figure 8.5a, the model reproduces both the abundance and slope of SO_2 in the lower mesostratosphere. The model predicts a mixing ratio of SO of the order of 10^{-9}, in agreement with recent UV measurements taken from rocket experiments. The major reactions that destroy and produce SO_2 are shown in figure 8.12b. The main loss reaction is photolysis. The reactions SO + O and SO + ClO are mainly responsible for restoring SO to SO_2.

The major oxygen species in the model, O, O_2 and O_3, are given in figure 8.13a. Of these three species of oxygen, only O atoms and the excited state $O_2(^1\Delta)$ have been positively identified in the upper atmosphere from airglow emissions. There are only upper limits for the abundances of O_2 (in the ground state) and O_3. The column density of O_3 in the model is 8×10^{15} cm^{-2}, or about 0.3 DU. The amount of O_3 on Venus is many orders of magnitude less than that on Earth (200–300 DU) but is comparable to that on Mars (0.1–1 DU). The reason is the lesser amount of O_2 on Venus and the catalytic effect of chlorine radicals for destroying ozone. We discuss in section 8.2.2 that a crucial problem in the photochemistry of oxygen on Venus is the formation and the breaking of the O—O bond. Figure 8.13b shows the major reactions that produce

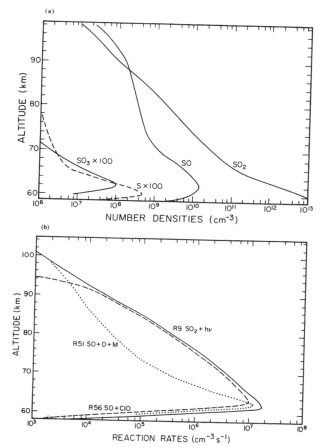

Figure 8.12 (a) Concentrations of major sulfur species (S, SO, SO_2, SO_3) in the model. (b) Reaction rates of major reactions producing and destroying SO_2. Based on Model C in Yung, Y. L., and DeMore, W. B., 1982, "Photochemistry of the Stratosphere of Venus: Implications for Atmospheric Evolution." *Icarus* **51**, 199.

O_2. In the upper atmosphere recombination is mostly via the three-body reaction $O + O + M \rightarrow O_2 + M$, with a minor contribution from the reaction $O + OH \rightarrow O_2 + H$. The bulk of the production of O_2 is via catalytic scheme (V). In the terrestrial stratosphere this scheme is known as the Molina-Rowland cycle for the removal of odd oxygen $(O + O_3)$. The rate-limiting step of this cycle is the key reaction $ClO + O$, shown in figure 8.13b. The rates of the principal reactions that break the O—O bond are given in figure 8.13c. As discussed in section 8.2.2 the bulk of the bond breaking is via the the two reactions between Cl and O and the peroxychloroformyl radical, resulting in the oxidation of CO and SO_2 [chemical schemes (IIa), (IIb) and (VI)]. In the region just above the cloud tops S atoms are effective in breaking the O—O bond.

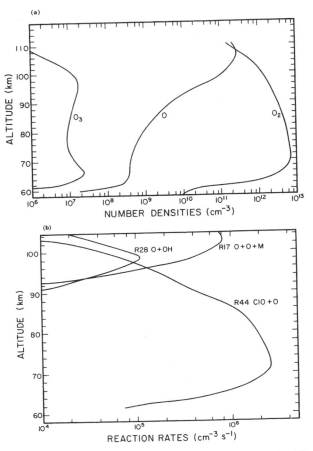

Figure 8.13 (a) Concentrations of major oxygen species (O, O_2, O_3) in the model. (b) Reaction rates of major reactions forming the O–O bond. Based on Model C in Yung, Y. L., and DeMore, W. B., 1982, "Photochemistry of the Stratosphere of Venus: Implications for Atmospheric Evolution." *Icarus* **51**, 199.

The reaction $S + O_2 \rightarrow SO + O$ is the rate-limiting step for chemical schemes (IIIa) and (IIIb).

Although the search for O_2 on Venus has been unsuccessful, the atmosphere is known to be an intense source of infrared radiation arising from the radiative decay of the excited state, $O_2(^1\Delta)$, at 1.27 μm. The observed intensity of 1 MR (1 mega-Rayleigh = 10^{12} photons cm^{-2} s^{-1}) is surprising because it is of the same order of magnitude as the total photolysis rate of CO_2. The excited oxygen is produced in the following reactions:

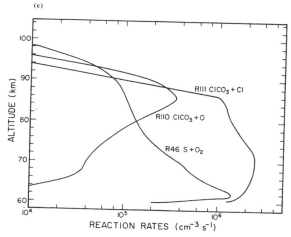

Figure 8.13 (c) Reaction rates of major reactions breaking the O—O bond. Based on Model C in Yung, Y. L., and DeMore, W. B., 1982, "Photochemistry of the Stratosphere of Venus: Implications for Atmospheric Evolution." *Icarus* **51**, 199.

$$O_3 + h\nu \rightarrow O + O_2(^1\Delta) \tag{8.58}$$

$$O + O + M \rightarrow O_2(^1\Delta) + M \tag{8.59}$$

$$HO_2 + O \rightarrow OH + O_2(^1\Delta) \tag{8.60}$$

$$Cl + O_3 \rightarrow ClO + O_2(^1\Delta) \tag{8.61}$$

$$ClO + O \rightarrow Cl + O_2(^1\Delta) \tag{8.62}$$

The excited state of oxygen is lost by radiative decay or quenching:

$$O_2(^1\Delta) \rightarrow O_2 + h\nu \tag{8.63}$$

$$O_2(^1\Delta) + M \rightarrow O_2 + M \tag{8.64}$$

Quenching of $O_2(^1\Delta)$ by CO_2 is inefficient. As a result, a large fraction of the energy in $O_2(^1\Delta)$ is radiated away by (8.63). The emission rates deduced from the model are given in table 8.7. Due to the low quantum yields of reactions (8.59)–(8.62) for producing $O_2(^1\Delta)$, the model cannot account for the observed emissions. We should point out that the only process known to have the highest quantum yield for the formation of $O_2(^1\Delta)$ in laboratory studies is recombination of oxygen atoms on the surface of pyrex glass. In the mesosphere of Venus this may correspond to the recombination of O atoms on the surface of meteoritic dust. However, there is no independent evidence to support this speculation.

The major predictions of the model, including the abundances of the major species and production rates, are summarized in table 8.8.

Table 8.7 Reactions capable of producing $O_2(^1\Delta)$ emissions in the atmosphere of Venus and estimated column emission ratio

Reaction	Emission rate (MR)
$O_3 + h\nu$	0.18
$O + O$	0.72
$O + HO_2$	0.06
$Cl + O_3$	0.37
$ClO + O$	0.72
$Cl + ClOO$	0.2

1 mega-Rayleigh (MR) = 10^{12} photons cm^{-2} s^{-1}. All estimates are taken from Yung, Y. L., and DeMore, D. M., 1982, *Icarus* **51**, 199, except for the last reaction proposed by Leu, M. T., and Yung, Y. L., 1987, *Geophys. Res. Lett.* **14**, 949.

Table 8.8 Concentrations of major species and production rates predicted by the model

Physical quantity	Model
f_{CO} (62 km)	4.8×10^{-5}
f_{CO} (100 km)	2.5×10^{-3}
f_{O_2} (62 km)	1.5×10^{-7}
CO column abundance	2.8×10^{20}
O_2 column abundance	1.8×10^{19}
Ratio of CO to O_2 column abundances	16
Scale height of SO_2 (70 km)	2.5
f_{SO_2} (70 km)	3.2×10^{-8}
R1:$CO_2 + h\nu$	7.6×10^{12}
R25:$CO + OH$	4.9×10^{12}
R101:$ClCO + O_2$	5.0×10^{12}
R102:$ClCO + O$	1.2×10^{12}
$\phi_{SO_2} = -\phi_{CO}$	1.4×10^{12}
R27:$OH + HCl$	3.0×10^9
ϕ_{H_2}	0
ϕ_{Cl_2}	-3.0×10^9
$O_2(^1\Delta)$ dayglow	0.9×10^{12}
$O_2(^1\Delta)$ nightglow	0.76×10^{12}

From Yung, Y. L., and DeMore, W. B. (1982).
The column abundances and column production rates are integrals from 58 to 110 km in units of cm^{-2} and cm^{-2} sec^{-1}, respectively. f_{CO} has been fixed at 58 km to equal 4.5×10^{-5}. The column abundance of CO_2 is 5×10^{24} cm^{-2}

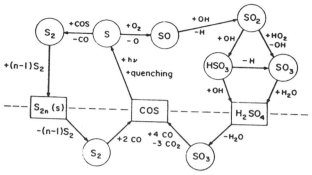

Figure 8.14 Schematic diagram illustrating the coupled cycles of reduced and oxidized species of sulfur in the atmosphere of Venus. Transient species are enclosed in circles, and stable species in rectangles. The dashed line represents the cloud base. After Prinn, R. G., 1985, "The Photochemistry of the Atmosphere of Venus," in Levine (1985; cited in section 8.1), p. 281.

(a)

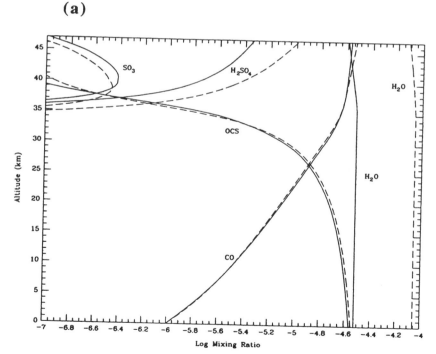

Figure 8.15 (a) Abundances of H_2O, CO, COS, SO_3, and H_2SO_4 in the lower atmosphere of Venus for a dry model (solid lines) and a wet model (dashed lines). After Krasnopolsky, V. A., and Pollack, J. B., 1994, "H_2O-H_2SO_4 System in Venus' Clouds and OCS, CO and H_2SO_4 Profiles in Venus' Troposphere." *Icarus* **109**, 58.

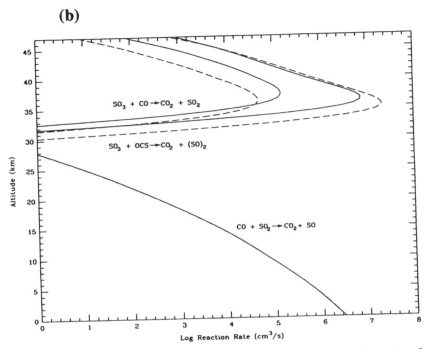

Figure 8.15 (b) Reaction rates of major reactions involved in the transformation of species shown in (a). The solid and dashed lines refer to the dry and wet models in (a). After Krasnopolsky, V. A., and Pollack, J. B., 1994, "H_2O-H_2SO_4 System in Venus' Clouds and OCS, CO and H_2SO_4 Profiles in Venus' Troposphere." *Icarus* **109**, 58.

8.3.3 Lower Atmosphere

Our knowledge of reduced sulfur in the model is much less secure. Figure 8.14 illustrates the coupled cycles of reduced and oxidized species of sulfur in the atmosphere of Venus. The dashed line represents the cloud base. Above this line photochemistry dominates; below this line thermodynamic equilibrium chemistry becomes important, especially at the surface.

A model of the lower atmosphere of Venus is shown in figure 8.15a. Due to the uncertainty in our knowledge of H_2O in the atmosphere, two models have been used: a wet model (dashed lines) and a dry model (solid lines). The mixing ratio of COS falls with altitude since the source is at the surface and the sink is near the upper boundary of the model. The species H_2SO_4, SO_3, and CO are produced above the cloud tops and are destroyed by equilibrium chemistry in the lower atmosphere. This accounts for their decrease as we approach the surface. The rates of the major reactions destroying CO and COS in the model are given in figure 8.15b. We must emphasize the tentative nature of this model because most of the rate coefficients for the key reactions are either uncertain or unknown.

8.4 Evolution

The definitive evidence that the atmosphere of Venus has evolved extensively came from the observed value of the D/H ratio, which is about 100 times the terrestrial value (see table 8.3). This implies that Venus must have lost at least the equivalent of 100 times more water than its present reservoir. The loss of hydrogen from Venus is determined by the efficiency of the escape processes, the rate at which hydrogen is transferred from the lower atmosphere to the upper atmosphere where the escape processes operate. In this section we discuss the mechanics of hydrogen escape in some detail.

An important process that is unique to Venus is chemical drying of the middle atmosphere by the H_2SO_4 aerosols, which are extremely hygroscopic. This is analogous to the cold trap that operates at the tropical tropopause of the terrestrial atmosphere to freeze-dry the air that enters the stratosphere. We should note that the chemical drying mechanism is effective only when the condition

$$[H_2SO_4] > [H_2O] \tag{8.65}$$

is satisfied. It may not have been as effective in the past when water was more abundant than H_2SO_4, violating (8.65). The evolution of water on Venus is thus closely tied to the chemistry of sulfur in the atmosphere. Relatively little work has been done on this subject.

8.4.1 Escape of Gases

The gravity of Venus is sufficiently strong that even H atoms cannot escape thermally from Venus. The principal mechanisms for the escape of hydrogen and other light gases are nonthermal, including charge exchange, collisional ejection, and solar wind impact.

The most important process for the escape of hydrogen is via the resonant charge exchange reactions that take place in the exosphere:

$$H(cold) + O^+ \rightarrow H^+(hot) + O \tag{8.66}$$

$$H^+(hot) + O \rightarrow H(hot) + O^+ \tag{8.67}$$

In this process the thermal (cold) H atom charge exchanges with O^+ to form H^+. H^+ is accelerated to the high temperature in the plasma by electromagnetic forces. The hot H^+ can transfer charge with O to produce a hot H atom. This atom may now escape if its velocity vector is pointed upward.

H atoms in the exosphere may also be accelerated by collision with hot atoms that are produced in the dissociative reaction

$$O_2^+ + e \rightarrow O^* + O^* \tag{8.68}$$

This reaction is exothermic by 6.98 eV, producing nonthermal O atoms with $v_o = 6.49$ km s^{-1}. If one of the atoms is in the excited ^1D state, v_o becomes somewhat less, 5.5 km s^{-1}. Unlike the case for Mars, O* produced in reaction (8.68) is not energetic enough to escape from Venus. However, the elastic collision

$$O^* + H \rightarrow O^* + H^* \tag{8.69}$$

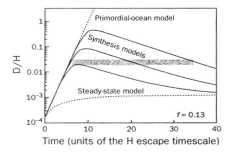

Figure 8.16 Evolution of D/H ratio in the atmosphere of Venus. The observational constraints are shown by the gray bar. After Gurwell, M. A., 1995, "Evolution of Deuterium on Venus." *Nature* **378**, 22.

is efficient at transferring momentum to the H atom, and H* can attain velocities in excess of the escape velocity of 10.2 km s^{-1}. The collision is also capable of ejecting heavier atoms:

$$O^* + X \rightarrow O^* + X^* \tag{8.70}$$

where X = D, ^{3}He or ^{4}He if we take into account the thermal energy of the ion in (8.68). The efficiency is of course less than that for (8.69).

Venus does not have a magnetic field. As a result, the solar wind interacts directly with the exosphere of the planet. The solar wind is composed primarily of protons. Direct sputtering of the atmosphere by the solar wind is important only for the light gases. Momentum transfer from protons to heavy atoms such as O is inefficient. However, there is an indirect solar wind–induced process that may play a fundamental role in the evolution of Venus. The solar wind carries its own magnetic field. In flowing past Venus the magnetic field generates an electric field that can accelerate the O$^+$ ions to keV energies. Reimpacting the atmosphere, the energetic O$^+$ ions can now efficiently sputter material (including heavy atoms) off the exosphere of Venus.

The rates of escape of hydrogen due to charge exchange, collisional ejection, and solar wind impact are 1×10^7, 3×10^6, and 1×10^5 atoms cm^{-2} s^{-1}, respectively, for the present epoch. This rate of escape implies that the e-folding time for the loss of H$_2$O on Venus is only 500 Myr, a geologically insignificant length of time. This leads to the suggestion that water on Venus may be resupplied by cometary impacts. The rate of loss of oxygen might have been higher in the past due to a more active sun that emitted more UV and corpuscular radiation. The only constraint we have on the total amount of hydrogen that has escaped is the D/H ratio, which is discussed in section 8.4.2.

8.4.2 Isotopic Signature

An analysis of D/H fractionation due to the difference in the escape efficiencies of H and D is given in section 7.4.2. The crucial fractionation factor f is close to 0 for charge exchange but is as high as 0.47 for collisional ejection. By weighting the various escape processes, the mean value for f has been estimated to be 0.13. The evolution of the D/H ratio in the atmosphere of Venus is summarized in figure 8.16. The time axis is given in units of the escape lifetime of H (85–800 Myr). There are essentially two opposing models. One is the dissipation of a massive primordial ocean, leading to very large fractionation of D. The other is an equilibrium model

Table 8.9 Estimated total H_2O buried inside the growing solid Earth, Venus, and Mars

	Earth	Venus	Mars
Radius at which impact deyhdration starts, R_1 (km)	1150	1210	1370
Mass at which deyhdration starts (10^{22} kg)	3.5	3.9	4.2
Amount of H_2O buried inside R_1 (10^{20}kg)	1.2	1.3	1.4
Radius at which complete deyhdration starts, R_2(km)	3190	3330	>3390
Mass between R_1 and R_2 (10^{23} kg)	7.1	7.7	5.9
Amount of H_2O buried between R_1 and R_2 (10^{21} kg)	1.2	1.3	1.3
Total buried H_2 (10^{21} kg)	1.3	1.4	1.4

Model by Liu (1988) based on the impact dehydration model proposed by Lange and Ahrens (1984), assuming that the H_2O content of the infalling materials is 0.33 wt%.

with the water on Venus being supplied by comets. This model leads to predicted D/H ratio that is too low. A time-dependent synthesis model is shown in figure 8.16. The observational constraints are marked by the gray bar. The nominal model predicted an initial global ocean of 85 m on Venus. The minimal model (within limits of the gray bar) had a 5.4 m primordial ocean.

8.5 Mystery of the Missing Water

Earth is unique in the solar system in having massive oceans of water. The total amount of water in the oceans is 1.5×10^{21} kg. The entire near-surface geochemical reservoirs (oceans + crust) contain about 2×10^{21} kg of water. Water is deficient on Venus by a factor of 10^4 to 10^5 relative to Earth. Escape can account for a factor of at least 100. The rest of the difference must be accounted for by a speculative theory of hydrodynamic escape or a theory of planetary evolution. We briefly describe an explanation based on the latter idea.

It is now generally accepted that the origin of water in the planets is dehydration of hydrous minerals during formation. The bulk of infalling material has composition similar to that of CI chondrites that contain up to 3 wt% H_2O. Once the accreting planetary body exceeds a critical size, impact velocities become sufficiently high that devolatization can start. The critical radii, R_1, for Earth, Venus, and Mars are about 20%, 20%, and 40% of their present radii, respectively (see summary of these values in table 8.9). Complete devolatization (total loss of H_2O to space) can occur when the planets grow to another critical radius, R_2. The values for R_2 given in table 8.9 are roughly 50% of the present radii for the Earth and Venus; Mars never reached R_2. Estimates of water buried inside the growing terrestrial planets are summarized in table 8.9 assuming that the infalling body contains 0.33 wt% water. Thus, the amount of sequestered water is of the order of that in the terrestrial oceans. This is a conservative estimate because we have assumed that the impacts are between solid bodies.

When the growing planetary body reaches a certain fraction of its final size, it may be covered by a magma ocean. Once this magma ocean is formed, subsequent

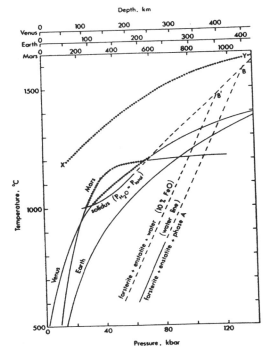

Figure 8.17 Thermal models for Earth, Venus, and Mars compared with a solidus of peridotite and the water line determined experimentally. The curve X–Y represents a hypothetical temperature profile before solidification. After Liu, L. G., 1988, "Water in the Terrestrial Planets and the Moon." *Icarus* **74**, 98.

accretion proceeds via solid-liquid rather than solid-solid impacts. The infalling objects may penetrate to great depths in the magma ocean, and all the released volatiles would be dissolved in the melts. The total amount of water trapped in the terrestrial planets is at least a few times the mass of Earth's oceans.

The above discussion shows that the terrestrial planets all should have acquired massive amounts of water of the order of terrestrial oceans. Why, then, are these oceans of water absent from Venus and Mars? The release of water in the interior of planets is determined by the equilibrium between minerals:

$$5Mg_2SiO_4 + 3H_2O \leftrightarrow 3MgSiO_3 + Mg_7Si_2O_{14}H_6 \qquad (8.71)$$
$$\text{forsterite} \quad \text{water} \quad \text{enstatite} \quad \text{phase A}$$

Figure 8.17 shows the "water line" and the solidus, derived empirically from experiments. A hypothetical temperature profile of the planets before solidification is given by the curve X–Y. Models of thermal profile for the terrestrial planets at present are shown. Venus has a high surface temperature, and this results in a higher temperature in the interior. Mars has lower gravity than Earth; the same pressure is attained at much greater depth, where the temperature is higher. Note that Mars and Venus, but not Earth, intercept the solidus between 20 and 60 kbar. Earth's temperature curve intercepts the solidus at about 100 kbar. Thus, the amount of water that can be expelled from the interior of Venus and Mars is significantly less than that of Earth.

8.6 Unsolved Problems

We list a number of outstanding unresolved problems related to the atmosphere of Venus and its comparison with those of Mars and Earth.

1. Why is Venus deficient in water? If the bulk of water is inside the planet, is there any observational evidence?
2. Is heterogeneous chemistry of chlorine compounds on the surface of sulfuric acid aerosols important?
3. Is lightning a source of NO_x in the lower atmosphere of Venus?
4. Does NO_x chemistry play a role in the photochemistry in the mesostratosphere?
5. What is the nature of the unidentified UV absorber?
6. What is the detailed mechanism for the production of reduced sulfur species on Venus?
7. Is SO_2 geochemically stable on Venus? Must it be resupplied by volcanism?
8. What is the evolutionary history of water on Venus?
9. Is Venus volcanically active in the current epoch?
10. How does the atmosphere interact with the interior of the planet? Are there tectonic processes on Venus?
11. Why are the noble gas contents greatly enhanced on Venus relative to the other terrestrial planets?

9

Earth

Imprint of Life

9.1 Introduction

Earth is the largest of the four terrestrial planets, three of which have substantial atmospheres. The astronomical and orbital parameters are summarized in table 9.1. Our planet has an obliquity of 23.5°, giving rise to well-known seasonal variations in solar insolation. The orbital elements are slightly perturbed by other planets in the solar system (primarily Jupiter), with time scales from 20 to 100 kyr, and these changes are believed to cause the advance and retreat of ice sheets. The last glacial maximum (LGM) occurred 18 kyr ago, at which time the planet was colder by several degrees centigrade on average. At present Earth is in an interglacial warm period.

The origin of Earth may not be very different from that of the other terrestrial bodies. However, three properties may be unique to this planet. One is the formation of the Moon, probably via collision between Earth and a Mars-sized body. Second is the release of a huge amount of water from the interior (see discussion in section 8.5). Third, Earth is endowed with a large magnetic field that protects it from direct impact by the solar wind.

Seventy percent of Earth's surface is covered by oceans, which have a mean depth of 3 km. There is so much water that Arthur C. Clarke proposed that "Ocean" might be a better name for our planet than "Earth." The enormous body of water became the cradle of life as early as 3.85 Gyr ago. The present terrestrial environment is the end-product of billions of years of evolution driven by the hydrological cycle

Table 9.1 Greenhouse effects in planetary atmospheres

	Surface pressure (relative to Earth)	Main greenhouse gases	Surface temperature in absence of greenhouse effect (K)	Observed surface temperature (K)	Warming due to greenhouse effect (K)
Venus	90	> 90% CO_2	227	750	523
Earth	1	~ 0.04% CO_2 ~ 1% H_2O	255	288	33
Mars	0.007	> 90% CO_2	216	226	10

From Intergovernmental Panel on Climate Change (1990).

and global biogeochemical cycles, in addition to the slower forces of geodynamics and geochemistry. The massive hydrological cycle and the biogeochemical cycles that operate on Earth are absent from other planets in the solar system. Mars in the remote past might have had a milder climate with liquid water on the surface, but the planet dried up a few eons ago. There is to date no observational evidence for the hypothetical oceans (composed of liquid hydrocarbons) on Titan. Life on a planetary scale equivalent to the terrestrial biosphere does not exist elsewhere in the solar system.

Perhaps the most distinctive character of Earth is the imprint of life and the fact that it has sustained life for at least 3.8 Gyr. Today the global planetary environment is genial and conducive to life. The surface of the planet is protected from harmful ultraviolet radiation by an ozone layer at an altitude of about 20–30 km. The existence of the ozone layer, which makes advanced life on Earth possible, is caused by the abundance of atmospheric oxygen, which is in turn a product of the biosphere. As pointed out in chapters 7 and 8, the abundances of O_3 on Mars and Venus are about two orders of magnitude less than that on Earth. Minor constituents of the atmosphere (e.g., H_2O and CO_2) provide a greenhouse effect of about 30 K, an amount that is just about right to maintain the mean temperature of Earth at a comfortable 285 K. By comparison we may note that the surface temperatures of Mars and Venus are 250 and 750 K, respectively. Table 9.1 shows that Mars has too little greenhouse effect; Venus has too much. It is surprising that the Earth has apparently never been completely frozen, nor has it been so hot as to destroy life, despite large changes in the solar constant and the composition of the atmosphere. Did the biosphere play a role in mitigating the excesses of climatic changes? Did the biosphere actively promote a physical and chemical environment that is most felicitous for life? These questions bring us into the heart of a controversial theory: the Gaia hypothesis.

The theme of this chapter is the imprint of life on the global terrestrial environment, as compared to other planets. The purpose is to prepare the proper background and perspective for an assessment of the effects of anthropogenic activities on the global environment (chapter 10).

Table 9.2 Chemical composition of the terrestrial troposphere

Gas	Abundance[a]	Source(s)	Sink(s)	Comments
N_2	78.084%	Denitrifying bacteria	Nitrogen-fixing bacteria	
O_2	20.946%	Photosynthesis	Respiration and decay	
H_2O	<4%	Evaporation	Condensation	variable
Ar	9340 ppm	Outgassing (^{40}K)	—	
CO_2	350 ppm	Combustion, biology	Biology	
Ne	18.18 ppm	Outgassing	—	
^4He	5.24 ppm	Outgassing (U, Th)	Escape	
CH_4	1.7 ppm	Biology and agriculture	Reaction with OH	+1%/yr
Kr	1.14 ppm	Outgassing	—	
H_2	0.55 ppm	H_2O photolysis	H atom escape	
N_2O	~320 ppb	Biology	Photolysis (stratosphere)	
CO	125 ppb	Photochemistry	Photochemistry	
Xe	87 ppb	Outgassing	—	
O_3	~10–100 ppb	Photochemistry	Photochemistry	
HCl	~1 ppb	Derived from sea salt	Rainout	
Isoprene, etc.	~1–3 ppb	Foliar emissions	Photooxidation	
C_2H_6, etc.	~3–80 ppb	Combustion, biomass burning, grasslands	Photooxidation	
H_2O_2	~0.3–3 ppb	Photochemistry	Photochemistry	
C_2H_2, etc.	~0.2–3 ppb	Combustion, biomass burning, oceans	Photooxidation	
C_2H_4, etc.	~0.1–6 ppb	Combustion, biomass burning, oceans	Photooxidation	
C_6H_6, etc.	~0.1–1 ppb	Anthropogenic	Photooxidation	
NH_3	0.1–3 ppb	Biology	Wet and dry deposition	
HNO_3	~0.04–4 ppb	Photochemistry (NO_x)	Rainout	
CH_3Cl	612 ppt	Ocean, biomass burning	Reaction with OH	
COS	500 ppt	Biology	Photodissociation	
NO_x	~30–300 ppt	Combustion, biology	Photooxidation	
CF_2Cl_2 (F12)	300 ppt	Anthropogenic	Photolysis (stratosphere)	+5.1%/yr
$CFCl_3$ (F11)	178 ppt	Anthropogenic	Photolysis (stratosphere)	+5.1%/yr
CH_3CCl_3	157 ppt	Anthropogenic	Reaction with OH	+4.4%/yr
CCl_4	121 ppt	Anthropogenic	Photolysis (stratosphere)	+1.3%/yr
CF_4 (F14)	69 ppt	Anthropogenic	Photolysis (upper atmosphere)	+2.0%/yr
$CHClF_2$ (F22)	59 ppt	Anthropogenic	Reaction with OH	+10.9%/yr
H_2S	30–100 ppt	Biology	Photooxidation	
$C_2Cl_3F_3$ (F113)	30–40 ppt	Anthropogenic	Photolysis (stratosphere)	+11.5%/yr
CH_2Cl_2	30 ppt	Anthropogenic	Reaction with OH	
CH_2ClCH_2Cl	26 ppt	Anthropogenic	Reaction with OH	
CH_3Br	12 ppt	Ocean, marine biota	Reaction with OH	
SO_2	20–90 ppt	Combustion	Photooxidation	marine air
$CHCl_3$	16 ppt	Anthropogenic	Reaction with OH	
CS_2	~15 ppt	Anthropogenic	Photooxidation	
$C_2Cl_2F_4$ (F114)	14 ppt	Anthropogenic	Photolysis (stratosphere)	
C_2H_5Cl	12 ppt	Anthropogenic	Reaction with OH	
$CHClCCl_2$	7.5 ppt	Anthropogenic	Reaction with OH	
$(CH_3)_2S$	5–60 ppt	Biology	Photooxidation	marine air
C_2ClF_5 (F115)	4 ppt	Anthropogenic	Photolysis (stratosphere)	
C_2F_6 (F116)	4 ppt	Anthropogenic	Photolysis (upper atmosphere)	

Table 9.2 (*continued*)

Gas	Abundance	Source(s)	Sink(s)	Comments
$CClF_3$ (F13)	3.3 ppt	Anthropogenic	Photolysis (stratosphere)	
CH_3I	~2 ppt	Ocean, marine biota	Photolysis (troposphere)	
$CHCl_2F$ (F21)	1.6 ppt	Anthropogenic	Reaction with OH	
$CClF_2Br$	1.2 ppt	Anthropogenic	Photolysis (stratosphere)	+20%/yr

From Fegley, B. Jr. (1995).
[a] Abundances by volume in dry air. ppm = parts per million, ppb = parts per billion, ppt = parts per trillion.

Table 9.3 Isotopic composition of the noble gases in the terrestrial atmosphere and of terrestrial standards for isotopic analyses

Isotopic ratio	Observed value	Comments[a]
D/H	$(1.5576 \pm 0.0005) \times 10^{-4}$	SMOW
$^3He/^4He$	$(1.399 \pm 0.013) \times 10^{-6}$	
$^{12}C/^{13}C$	89.01 ± 0.38	
$^{14}N/^{15}N$	272.0 ± 0.3	air
$^{16}O/^{17}O$	2681.80 ± 1.72	SMOW
$^{16}O/^{18}O$	498.71 ± 0.25	SMOW
$^{20}Ne/^{22}Ne$	9.800 ± 0.080	
$^{21}Ne/^{22}Ne$	$(2.899 \pm 0.025) \times 10^{-2}$	
$^{35}Cl/^{37}Cl$	3.1273 ± 0.1990	
$^{36}Ar/^{38}Ar$	5.320 ± 0.002	
$^{40}Ar/^{36}Ar$	296.0 ± 0.5	
$^{78}Kr/^{84}Kr$	$(6.087 \pm 0.002) \times 10^{-3}$	
$^{80}Kr/^{84}Kr$	$3.960 \pm 0.002\%$	
$^{82}Kr/^{84}Kr$	$20.217 \pm 0.021\%$	
$^{83}Kr/^{84}Kr$	$20.136 \pm 0.021\%$	
$^{86}Kr/^{84}Kr$	$30.524 \pm 0.025\%$	
$^{124}Xe/^{132}Xe$	$(3.537 \pm 0.0011) \times 10^{-3}$	
$^{126}Xe/^{132}Xe$	$(3.300 \pm 0.017) \times 10^{-3}$	
$^{128}Xe/^{132}Xe$	$7.136 \pm 0.009\%$	
$^{129}Xe/^{132}Xe$	$98.32 \pm 0.12\%$	
$^{130}Xe/^{132}Xe$	$15.136 \pm 0.012\%$	
$^{131}Xe/^{132}Xe$	$78.90 \pm 0.11\%$	
$^{134}XE/^{132}Xe$	$38.79 \pm 0.06\%$	
$^{136}Xe/^{132}Xe$	$32.94 \pm 0.04\%$	

From Fegley, B. Jr. (1995).
[a] SMOW = standard mean ocean water.

9.2 Gaia Hypothesis

9.2.1 Gaia: Imprint of Life

The chemical composition of the terrestrial atmosphere and its isotopic composition are summarized in tables 9.2 and 9.3, respectively. The chemical composition of the oceans is given in table 9.4. The composition of the terrestrial atmosphere departs

Table 9.4 Chemical composition of seawater

Atomic No.	Element	Dissolved form	Mean concentration	Comments
1	H	H_2		biogenic or hydrothermal origin
2	He	Dissolved He	1.9 nmole/kg	non-nutrient dissolved gas
3	Li		178 µg/kg	conservative
4	Be		0.2 ng/kg	increases with depth
5	B	Inorganic boron	4.4 mg/kg	conservative
6	C	Total CO_2	2200 µmole/kg	nutrient
7	N	N_2	590 µmole/kg	non-nutrient dissolved gas
		NO_3	30 µmole/kg	nutrient
8	O	Dissolved O_2	150 µmole/kg	biologically controlled profile
9	F	Fluoride	1.3 mg/kg	conservative
10	Ne	Dissolved Ne	8 nmole/kg	non-nutrient dissolved gas
11	Na		10.781 g/kg	conservative
12	Mg		1.28 g/kg	conservative
13	Al		1 microg/kg	nutrientlike profile
14	Si	Silicate	110 µmole/kg	nutrient
15	P	Reactive phosphate	2 µmole/kg	nutrient
16	S	Sulfate	2.712 g/kg	conservative
17	Cl	Chloride	19.353 g/kg	conservative
18	Ar	Dissolved Ar	15.6 µmole/kg	non-nutrient dissolved gas
19	K		399 mg/kg	conservative
20	Ca		415 mg/kg	conservative (1st. approx.)
21	Sc		<1 ng/kg	
22	Ti		<1 ng/kg	
23	V		<1 µg/kg	conservative
24	Cr		330 ng/kg	nutrientlike profile
25	Mn		10 ng/kg	nutrientlike profile
26	Fe		40 ng/kg	nutrientlike profile
27	Co		2 ng/kg	nutrientlike profile
28	Ni		480 ng/kg	nutrientlike profile
29	Cu		120 ng/kg	nutrientlike profile
30	Zn		390 ng/kg	nutrientlike profile
31	Ga		7–60 ng/kg	
32	Ge		5 ng/kg	correlated with silicate
33	As	As(V)	2 µg/kg	nutrientlike profile
34	Se	Total Se	170 ng/kg	correlated with phosphate
35	Br	Bromide	67 mg/kg	conservative
36	Kr	Dissolved Kr	3.7 nmole/kg	non-nutrient dissolved gas
37	Rb		124 microg/kg	conservative
38	Sr		7.8 mg/kg	correlated with phosphate
39	Y		13 ng/kg	conservative (1st. approx.)
40	Zr		<1 µg/kg	
41	Nb		1 ng/kg	
42	Mo		11 µg/kg	conservative
44	Ru		~1 ng/kg	
45	Rh			
46	Pd		0.2–0.7 pmole/kg	
47	Ag		3 ng/kg	
48	Cd		70 ng/kg	correlated with phosphate
49	In		0.2 ng/kg	
50	Sn		0.5 ng/kg	anthropogenic

Table 9.4 (continued)

Atomic No.	Element	Dissolved form	Mean concentration	Comments
51	Sb		0.2 µg/kg	conservative
52	Te	Total Te	0.6–1.3 pmole/kg	
53	I		59 µg/kg	correlated with phosphate
54	Xe		0.5 nmole/kg	non-nutrient dissolved gas
55	Cs		0.3 ng/kg	conservative
56	Ba		11.7 µg/kg	nutrientlike profile
57	La		4 ng/kg	nutrientlike profile
58	Ce		4 ng/kg	nutrientlike profile
59	Pr		0.6 ng/kg	nutrientlike profile
60	Nd		4 ng/kg	nutrientlike profile
62	Sm		0.6 ng/kg	nutrientlike profile
63	Eu		0.1 ng/kg	nutrientlike profile
64	Gd		0.8 ng/kg	nutrientlike profile
65	Tb		0.1 ng/kg	nutrientlike profile
66	Dy		1 ng/kg	nutrientlike profile
67	Ho		0.2 ng/kg	nutrientlike profile
68	Er		0.9 ng/kg	nutrientlike profile
69	Tm		0.2 ng/kg	nutrientlike profile
70	Yb		0.9 ng/kg	nutrientlike profile
71	Lu		0.2 ng/kg	nutrientlike profile
72	Hf		<8 ng/kg	
73	Ta		<2.5 ng/kg	
74	W		<1 ng/kg	
75	Re		7.2–7.4 ng/kg	conservative
76	Os			
77	Ir			
78	Pt			
79	Au		50–150 fmole/liter	
80	Hg		6 ng/kg	
81	Tl		12 ng/kg	correlated with silicate
82	Pb		1 ng/kg	conservative
83	Bi		10 ng/kg	anthropogenic
90	Th		<0.7 ng/kg	
92	U		3.2 µg/kg	conservative

From Fegley, B. Jr. (1995).

radically from that of a lifeless planet. This can best be appreciated from figure 9.1a by comparing the composition of the present atmosphere with a hypothetical atmosphere on an abiotic Earth. Note that in the absence of life, CO_2 would be the dominant gas in the atmosphere, with total pressure of about 200 mbar. The higher abundance (by more than three orders of magnitude compared to the present) is the result of the slower removal rate of atmospheric CO_2 by geochemical processes in the absence of life. Therefore, the composition of the terrestrial atmosphere would, to first order, resemble the sister planets Mars and Venus. N_2 and O_2 would be minor constituents, with abundances orders of magnitudes below those at present. The disappearance of most of the O_2, an extremely reactive gas, is not surprising. However, the loss of most of the N_2, a very unreactive gas, is somewhat surprising. The reason is that the

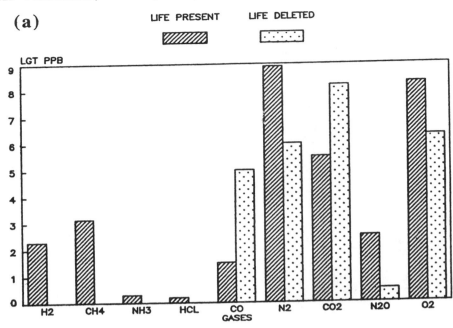

Figure 9.1 (a) Partial pressure of reactive gases of Earth's atmosphere at present (with life) and those calculated to result from reactions were life extinguished. The concentrations are expressed as log base 10 of quantities in parts per billion by volume. After Margulis, L., and Lovelock, J. E., 1989, "Gaia and Geognosy," in Rambler et al. (1989; cited in section 9.1), and Lovelock, J. E., and Margulis, L., 1974, "Atmospheric Homeostasis, by and for the Biosphere: The Gaia Hypothesis." *Tellus* **26**, 1.

stable form of nitrogen on an abiotic Earth is nitrate ion in the oceans. The trace gases such as CH_4, N_2O, NH_3, HCl, and H_2 would either vanish from the atmosphere or have their abundances greatly reduced. Only the concentrations of CO would greatly increase, due to the greater source from CO_2 photolysis. The time constant for the complete obliteration of the imprint of life from Earth's atmosphere is of the order of millions of years, a geologically insignificant length of time.

The intense disequilibrium of the present atmosphere with respect to an abiotic Earth, as illustrated in figure 9.1a, suggests that the biosphere must be continuously producing prodigious amounts of gases that are out of thermodynamic equilibrium with the chemistry and geochemistry of the rest of the planet. Figure 9.1b presents estimates of the biologically produced fluxes of these gases. Note the simultaneous production of an oxidizing gas, O_2, as well as a highly reducing gas, CH_4. In the absence of life, all production rates will drop by many orders of magnitude to the level that can be sustained by photochemical and geochemical reactions. For gases such as CH_4 and NH_3 the production rates on an abiotic Earth are close to zero because it is very difficult to synthesize such complex molecules by atmospheric chemistry in a CO_2-dominated atmosphere. We may take the absence of these gases

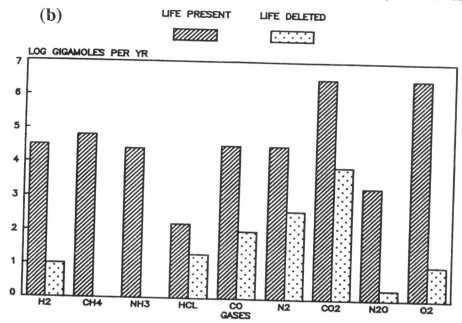

Figure 9.1 (b) Fluxes of reactive gases of Earth's atmosphere at present (with life) and those calculated to result from reactions were life extinguished. The fluxes are expressed as log base 10 of fluxes in 10^9 moles per year. After Margulis, L., and Lovelock, J. E., 1989, "Gaia and Geognosy," in Rambler et al. (1989; cited in section 9.1), and Lovelock, J. E., and Margulis, L., 1974, "Atmospheric Homeostasis, by and for the Biosphere: The Gaia Hypothesis." *Tellus* **26**, 1.

in the atmospheres of Mars and Venus as proof that CH_4 and NH_3 would be absent on an abiotic Earth.

The most important reaction in the biosphere is photosynthesis, represented schematically as

$$CO_2 + H_2O + h\nu \rightarrow CH_2O + O_2 \qquad (9.1)$$

where $(CH_2O)_n$ represents carbohydrate. The global photosynthesis rate is 3×10^{15} moles of carbon per year, or 142 Gt-C yr^{-1} (1 Gt = 10^{15} g). The mean rate of destruction of CO_2 is equivalent to 4.4×10^{13} molecules cm^{-2} s^{-1}. For comparison, we may note that the photolysis rates of CO_2 on Mars and Venus are 1×10^{12} and 7×10^{12} cm^{-2} s^{-1}, respectively. The biosphere is much more efficient at removing CO_2 than ultraviolet photolysis, because (9.1) is a double photon process that can use visible light. At this rate all the CO_2 in the terrestrial atmosphere can be removed by photosynthesis in 5 yr. Of course, most of the CO_2 consumed in (9.1) is restored by respiration and decay, a process that nearly balances photosynthesis. Photosynthesis is the heartbeat of the entire biosphere. It is the ultimate source of nearly all organic matter as well as oxygen. Reaction (9.1) stores up 124.2 kcal of chemical energy for each mole of carbon synthesized into organic matter. The total biospheric power is 1.7×10^{14} W, or about 0.1% of the solar energy incident on Earth. For comparison,

Table 9.5 Biogeochemical cycles in prokaryotes

Element	Process	Examples of organisms and summary of reactions
Carbon	CO_2 fixation	$CO_2 + H_2 \rightarrow (CH_2O)_n + A_2$ (A = O, S) photoautotrophs: cyanobacteria, purple and green sulfur bacteria chemoautotrophs: sulfur- and iron-oxidizing bacteria
	Methanogenesis	$COO^- + H_2 \rightarrow CH_4$ methanogenic bacteria
	Methylotrophy	$CH_4 + O_2 \rightarrow CO_2$ methylotrophic bacteria
	Fermentation	$(CH_2O)_n \rightarrow CO_2$ aerobic heterotrophic bacteria
	Respiration	$(CH_2O)_n + O_2 \rightarrow CO_2$ aerobic heterotrophic bacteria
Sulfur	Sulfur reduction	$SO_4^{2-} + H_2 \rightarrow H_2S$ sulfur reducing bacteria
	Sulfur oxidation	$H_2S \rightarrow S^o$ purple and green sulfur phototrophs $S^o + O_2 \rightarrow SO_4$ sulfur-oxidizing bacteria
Nitrogen	N_2 fixation	$N_2 + H_2 \rightarrow NH_4$ phototrophic bacteria, nitrogen-fixing heterotrophic bacteria
	Nitrification	$NH_4^- + O_2 \rightarrow NO_2, NO_3$ nitrifying bacteria
	Denitrification	$NO_2, NO_3 \rightarrow NO_2, N_2$ denitrifying heterotrophic bacteria

From Rambler, M. B. et al. (1987).

the power that drives all human activities in 1980 is 1×10^{12} W. We discuss in sections 9.4 and 9.5 the impact of photosynthesis on the chemical composition and climate of Earth.

The biota may be classified into three categories by their functions: producers, consumers, and decomposers of organic compounds. The primary producers are plants, algae, and photosynthetic bacteria. They synthesize organic compounds from CO_2 and H_2O using solar energy (9.1) and provide the ultimate source of energy to fuel the biosphere. The consumers, such as animals, protozoans, and some bacteria, do not synthesize organic compounds from inorganic compounds, but they can transform organic compounds into other organic or inorganic compounds. Decomposers feed on dead organisms by transforming the organic compounds back to inorganic compounds, usually CO_2 and H_2O. Examples of decomposers include fungi and many bacteria.

The activities of the biosphere have a major impact on the cycling of the biogeochemical elements: H, O, C, N, S, and P. Table 9.5 lists some of the important net reactions in the biosphere for transforming carbon, nitrogen, and sulfur compounds. Hydrogen and oxygen are involved in these reactions. Phosphorus appears to have no important linkage to the atmosphere. The details of the hydrogen, oxygen, carbon, nitrogen, and sulfur cycles will be discussed in separate sections in this chapter.

9.2.2 Gaia: Planetary Homeorrhesis

We first explain the difference between homeostasis and homeorrhesis. A homeostatic system maintains specified variables at relatively constant levels despite perturbing influences. If, however, the specified variables are not fixed but change progressively with time, the system is said to be homeorrhetic. The planetary environment of Earth has progressively changed with time. There are at least three driving forces. First, the solar constant is believed to have increased by about 40% since the origin of the solar system. Second, substantial escape of hydrogen from the planet has occurred and is still occurring. Third, there has been a profound biological evolution from simple to complex organisms, and from marine to terrestrial ecosystems. A key question about the global environment is why it has remained so well regulated for the benefit of the biosphere. Has the biosphere played the role of homeorrhesis for the planetary environment?

The interaction between the environment and the biosphere is a problem of such enormous complexity that a fundamental understanding based on the rigorous principles of physics and chemistry is probably very difficult, if not impossible. We use an idealized model by Lovelock to illustrate the role of the biosphere in homeorrhesis. Consider the problem of the "faint sun." According to accepted theories of the evolution of stars, in the main sequence of which the sun is a typical member, the solar luminosity has been steadily increasing by about 40% since the sun's origin. The mean surface temperature (T) of a planet like the Earth is determined by the energy balance equation

$$\epsilon \sigma T^4 = \frac{1}{4}(1-a)F \tag{9.2}$$

where ϵ is the emissivity, σ is the Stefan-Boltzmann constant, a is the planetary albedo, and F is the solar constant. Consider a particularly simple model of Earth with no greenhouse effect ($\epsilon = 1$). The surface temperature is, from (9.2),

$$T = \left[\frac{(1-a)F}{4\sigma}\right]^{0.25} \tag{9.3}$$

This planet has no biology. The only feedback on the change of climate is physical, via the change of phase of water and its impact on the albedo (a). The results are presented in figure 9.2a. The dotted line represents the monotonic increase in temperature as the solar luminosity increases for a planet with constant albedo. The line ABCDE is from a model that allows for feedback due to phase changes of water. From A to B, the planet is completely covered by ice. Between B and C the ice is removed as the planet becomes warmer. The lower albedo results in a rapid warming of the planet. However, as the planet gets even warmer, there is more water vapor in the atmosphere and clouds can form. The higher albedo of clouds slows the rate of warming between C and D. Beyond D, there is no further albedo change. This simple model illustrates the relation between the surface temperature of the planet and the solar constant when there is no albedo feedback (AB and DE), positive feedback (BC), and negative feedback (CD).

Now let us consider a planet on which life is possible. For simplicity, let "life" consist of two kinds of daisies: a light species (with high albedo) and a dark species (with low albedo). The feedback between the environment and this biosphere is now governed by the laws of biology. The growth rate of living organisms is optimum in a

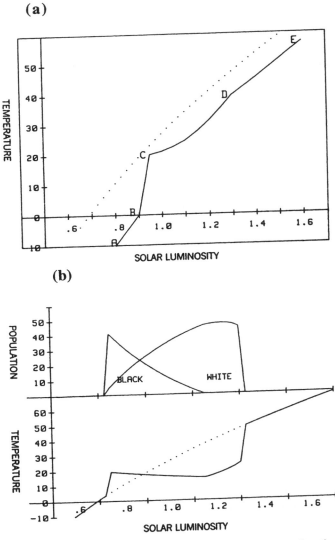

Figure 9.2 (a) The temperature (°C) on a lifeless planet as a function of changing solar luminosity. The dotted (solid) line gives the results for a dry (wet) planet. See section 9.2:2 for explanation of A–E. (b) The population of black and white daisies and temperature (°C) on a daisy planet as a function of changing solar luminosity. The dotted line gives the temperature of a dry, lifeless planet. After Margulis, L., and Lovelock, J. E., 1989, "Gaia and Geognosy," in Rambler et al. (1989; cited in section 9.1), and Lovelock, J. E., and Margulis, L., 1974, "Atmospheric Homeostasis, by and for the Biosphere: The Gaia Hypothesis." *Tellus* **26**, 1.

narrow range of environmental parameters such as temperature and water. We would expect that as the solar constant gets larger, the light daisy will fare better because it reflects more sunlight and consequently can remain cool. The dark daisy will fare poorly in this environment. The net result is a higher population of light daisies and a cooler planet. Figure 9.2b shows the evolution of the two populations of daisies as the solar constant increases, as well as the corresponding temperature of the planet. Note that the temperature has been maintained at a roughly constant value by adjusting the relative populations of the light and dark daisies. The biological homeorrhesis keeps the temperature range of the planet to within 10°C while the solar constant increases from 0.7 to 1.3 times its present value. For comparison, the temperature change is more than 40°C for an abiotic planet. Eventually Gaia's control of the planet fails: When the solar luminosity rises above 130% of the present value, even the light daisies perish and the planet becomes lifeless. This shows that Gaia can regulate the planetary environment only within certain limits.

The above is an artificial model of a hypothetical interaction between the environment and the biosphere. The actual interaction is much more complex and is not known. However, there is evidence that the existence and evolution of life on this planet may have had a profound impact on the climate by regulating the amount of CO_2 in the atmosphere (see section 9.4).

9.3 Hydrological Cycle

The bulk of water (1.5×10^{24} g, equivalent to 440 atmospheres) on our planet resides in the oceans. A smaller amount of water is stored on land, mainly in glaciers. The atmosphere contains 1.3×10^{19} g of water vapor. The mean volume mixing ratio is 2×10^{-3}, making it the fourth most abundant constituent after N_2, O_2, and Ar. The latitude-altitude profile of the mixing ratio of water vapor in the atmosphere is presented in figure 9.3a. The rapid decrease of water vapor with latitude and altitude is due to its strong dependence on temperature. Between -30° and 30°C, the saturation vapor pressure of water is approximated by

$$P(T) = 1.94 \times 10^9 e^{-5344/T} \text{mbar} \tag{9.4}$$

where T is given in degrees K. Air entering the stratosphere is freeze-dried at the cold tropopause with the result that the middle atmosphere is highly depleted of water vapor. The slight increase in the mixing ratio of water above 30 km is due to a chemical source derived from the oxidation of CH_4. The latitudinal variation of precipitation is shown in figure 9.3b. The maximum is in the tropics. The decrease in water vapor in mid and high latitudes is caused by the lower temperature. Water plays a major role in the energy budget of the atmosphere. About 20% of the solar energy incident on the planet is used to evaporate water from the surface. This energy is released to the atmosphere when water vapor condenses and returns to the surface as rain. Thus, the hydrological cycle is the primary heat engine for driving atmospheric motion. The rate of evaporation of water must be balanced by the rate of precipitation. The planetary mean rainfall rate is 100 cm yr^{-1}. The decrease at higher latitudes in this rate is due to the lower rate of evaporation at lower temperatures. The evaporation rate as a function of temperature is approximated by

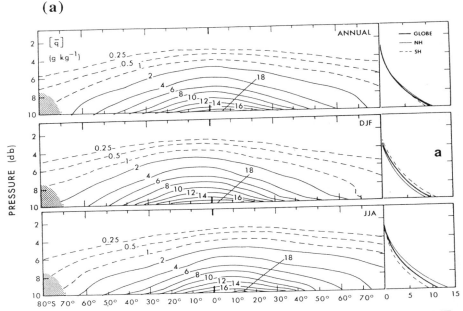

Figure 9.3 (a) Zonally averaged specific humidity (g kg^{-1}) for mean annual, winter (DJF), and summer (JJA) conditions. The horizontal axis is latitude. After Peixoto, J. P., and Oort, A. H., 1992, *Physics of Climate* (New York: American Institute of Physics).

$$E(T) = 1.2 \times 10^9 e^{-4620/T} \text{ cm yr}^{-1} \tag{9.5}$$

The mean residence time of water in the atmosphere is 11 days. Rainout provides an efficient mechanism for the removal of aerosols, dust, and soluble species from the atmosphere. The annual amount of rain falling on land is 9.9×10^{19} g, and this is essential for sustaining the biosphere. A large fraction of this water (62%) is reevaporated into the atmosphere. The remaining 38% returns to the oceans in river runoffs. The hydrological cycle is the primary agent of mechanical and chemical erosion of the continents. The combined erosion rate is 0.81 cm kyr^{-1} of solids and is a major source of cations in the oceans. A side effect of chemical weathering is the removal of atmospheric CO_2 by the dissolution of calcium and magnesium silicates:

$$CO_2 + CaSiO_3 \rightarrow CaCO_3 + SiO_2 \tag{9.6}$$

$$CO_2 + MgSiO_3 \rightarrow MgCO_3 + SiO_2 \tag{9.7}$$

The rate of removal is estimated to be $1.9 \pm 0.2 \times 10^{14}$ g-C yr^{-1}. These reactions are very important for the control of atmospheric CO_2 over geologic time.

It is convenient to divide the ocean into two layers. The uppermost several hundred meters of ocean is known as the thermocline. It exchanges dissolved gases and heat with the atmosphere in less than a year. The bulk of the ocean beneath the thermocline does not undergo rapid exchange with the atmosphere. The rate of exchange of water between the upper and the deep oceans is about 50 Sv (1 Sverdrup = 10^6 m^3 s^{-1}, or

(b)

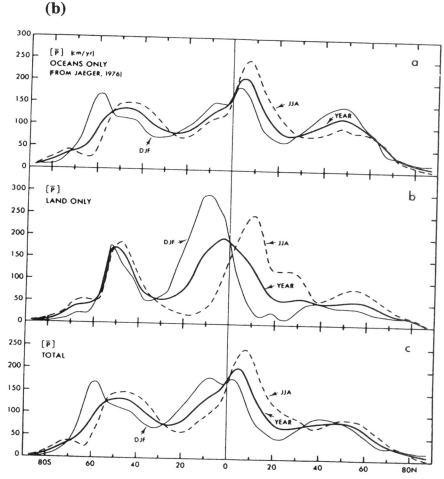

Figure 9.3 (b) Zonally averaged precipitation rate (cm yr^{-1}) for (a) ocean area, (b) land area, and (c) total area for mean annual, winter (DJF), and summer (JJA) conditions. After Peixoto, J. P., and Oort, A. H., 1992, *Physics of Climate* (New York: American Institute of Physics).

1.5×10^{21} g yr^{-1}). The time constants for ventilating the deep oceans are very long, on the order of 100 yr for the Atlantic and 1000 yr for the Pacific.

The water in the ocean exchanges with the interior of Earth via subduction and hydrothermal vents. The flux of water is estimated to be 1×10^{18} g yr^{-1}. At this rate it takes a million years to recycle all the water in the ocean once. This process is also important for the transport of heat from Earth's interior and accounts for 68% of the total heat flow of 3.55×10^{13} W from the interior.

The troposphere exchanges water with the middle atmosphere via the flow of air across the tropopause. The annual mass flux is 10^{15} g. The residence lifetime of water in the stratosphere is 2 yr. One of the most important consequences of water in the

330 Photochemistry of Planetary Atmospheres

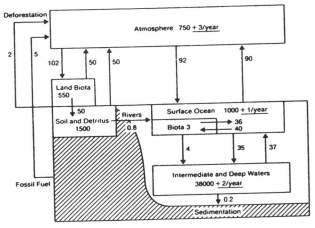

Figure 9.4 Global carbon reservoirs and fluxes. Units are gigatons of carbon (Gt-C) and Gt-C yr^{-1} for reservoirs and fluxes, respectively. After Intergovernmental Panel on Climate Change, 1990, *Climate Change: The IPCC Scientific Assessment*, and 1992, *Climate Change 1992: Supplement to the IPCC Scientific Assessment* (Cambridge: Cambridge University Press).

stratosphere is the formation of polar stratospheric clouds (PSCs) in the polar regions during winter. They serve as sites for heterogeneous chemistry leading to enhanced destruction of ozone.

9.4 Carbon Cycle

Like the other terrestrial planets, Earth is endowed with a large inventory of carbon, most of which resides in carbonate rocks. If all this carbon were released as CO_2, it would generate an atmosphere exceeding 50 bar. Carbon plays two fundamental roles on this planet: as a principal regulator of the planet's climate via its greenhouse effect, and as the primary source of chemical energy that drives the entire biosphere, via the conversion of CO_2 to organic matter in photosynthesis.

The major reservoirs of carbon and the rates of interreservoir transfer are shown in figure 9.4. The units for the reservoirs are gigatons of carbon; the fluxes are in Gt yr^{-1}. The mole fraction of CO_2 in the atmosphere in 1990 was 353 ppmv (parts per million by volume), equivalent to 750 Gt of carbon (1 ppmv = 2.12 Gt-C or 7.8 Gt-CO_2). The land and ocean biota contain 550 and 3 Gt of carbon, respectively. Averaged over the surface of the planet, the total living biosphere is equivalent to 0.11 g C cm^{-2}. The total rate of photosynthesis is 142 Gt-C yr^{-1}, with contributions of 102 and 40 Gt yr^{-1} from land and oceans, respectively. Thus, the mean lifetime of CO_2 in the atmosphere is 5.3 yr before it is used in photosynthesis. The turnover time of carbon in the land biota is 5.4 yr. The ocean has an extremely small living biomass of 3 Gt-C, which is more than two orders of magnitude less than the land biomass. This large difference is probably due to the need for land plants to have strong structures. The rate of photosynthesis in the ocean is about 40% of that on land. Thus, marine biota have a turnover time of 27 days.

The land has a large reservoir of detritus and dead organic material in the soil, estimated to be 1500 Gt-C. The time for exchange between this reservoir and the atmosphere is much longer, on the order of 30 yr. The bulk of the mobile carbon reservoir is in the oceans, discussed in section 9.4.1.

9.4.1 CO_2 in the Ocean

The fate of CO_2 in the oceans involves complex interactions between solution chemistry, biochemistry, and geochemistry. These interactions are further complicated by the exchange of carbon between the upper oceans, where photosynthesis plays a major role in transforming inorganic carbon into organic carbon, and the deep oceans, where decomposition of marine organic carbon takes place.

We first consider a particularly simple chemical model of the ocean involving carbonate species only. On dissolving in water, CO_2 forms carbonic acid:

$$CO_2 + H_2O \leftrightarrow H_2CO_3 \qquad Kh \qquad (9.8)$$

where the equilibrium constant for the reaction is Kh, the constant of Henry's Law. Carbonic acid dissociates into bicarbonate and carbonate ions:

$$H_2CO_3 \leftrightarrow H^+ + HCO_3^- \qquad K_1 \qquad (9.9)$$

$$HCO_3^- \leftrightarrow H^+ + CO_3^{2-} \qquad K_2 \qquad (9.10)$$

These ions interact with ions derived from the dissociation of water:

$$H_2O \leftrightarrow H^+ + OH^- \qquad Kw \qquad (9.11)$$

and other positive ions, collectively defined as "alkalinity" (A):

$$[A] = \sum \text{negative ions} - \sum \text{positive ions} \qquad (9.12)$$

with HCO_3^-, CO_3^{2-}, OH^-, and H^+ excluded from the definition of A. Note that all the constants Kh, K_1, K_2, and Kw are physical constants that can be determined in the laboratory. However, the alkalinity depends on the concentrations of metal ions such as Ca^{2+} and Mg^{2+}, which are determined by the geochemistry of ocean water and transport of cations from the continents via river runoffs. The above system has five unknowns, $[H_2CO_3]$, $[HCO_3^-]$, $[CO_3^{2-}]$, $[H^+]$, and $[OH^-]$, and five equations (9.8)–(9.12). The system can be solved once the atmospheric CO_2 and the ocean alkalinity are prescribed. Note that the partitioning of carbonate species in solution is such that $[HCO_3^-] : [CO_3^{2-}] : [H_2CO_3] = 130:10:1$. The existence of these three carbonate species accounts for the extraordinary solubility of CO_2 in water. The bulk of oceanic carbon resides in bicarbonate ions with concentrations of about 2×10^{-2} mole/liter. For comparison, the concentrations of N_2 and O_2 in seawater are much less, approximately 6×10^{-4} and 2×10^{-4} mole/liter, respectively, even though their abundances in the atmosphere greatly exceed that of CO_2. One simple consequence of the aforementioned partitioning of carbon species is the difficulty of detecting changes in oceanic carbon in response to changes of CO_2 in the atmosphere. We can derive the approximate relation

$$\frac{\Delta[HCO_3^-]}{[HCO_3^-]} = R\frac{\Delta(P_{CO_2})}{P_{CO_2}} \quad (9.13)$$

where P_{CO_2} is the partial pressure of atmospheric CO_2 and R is of the order of 0.2. Therefore, a doubling of atmospheric CO_2 will result in only a 20% increase in ocean bicarbonate concentration.

A more realistic model of carbonate chemistry in the ocean must include the effect of $CaCO_3$ and the influence of biology (via the precipitation of $CaCO_3$). In solution $CaCO_3$ dissociates into ions:

$$CaCO_3 \leftrightarrow Ca^{2+} + CO_3^{2-} \qquad K_3 \quad (9.14)$$

Combining (9.9), (9.10), and (9.14), we have

$$CaCO_3 + H_2CO_3 \leftrightarrow Ca^{2+} + 2HCO_3^- \quad (9.15)$$

The forward reaction is a weathering reaction. Proceeding from left to right, reaction (9.15) consumes one molecule of CO_2. However, if $CaCO_3$ is saturated and precipitates (as is the case in the ocean), reaction (9.15) proceeds from right to left and CO_2 is released. Because the weathering products eventually enter the ocean, weathering of carbonate rock is not a permanent sink of atmospheric CO_2. Only the weathering of silicate rocks, as given by (9.6) and (9.7), provides a permanent sink for CO_2. The impact of (9.14) on carbonate speciation may be incorporated by including Ca^{2+} in the alkalinity:

$$[A]_T = [A] + 2[Ca^{2+}] \quad (9.16)$$

where $[A]_T$ is the total alkalinity and $[A]$ is the alkalinity as defined by (9.12). The bulk of the ocean is supersaturated in calcium carbonate. Calcite formation can occur by the backward reaction of (9.15) and is greatly facilitated by marine organisms. However, in the deep ocean calcite is undersaturated and readily dissolves by the forward reaction of (9.15).

Figure 9.5a presents the depth profiles of ΣCO_2 (sum of dissolved CO_2 and H_2CO_3) and alkalinity in the ocean. Note the depletion near the surface of the ocean. This is due to the formation of calcite by living organisms in the euphotic zone (first 100 m). The predominance of biological activity in the upper part of the oceans is also evident from the concentrations of PO_4^{3-} and $\delta^{13}C$ shown in figure 9.5, a and b. PO_4^{3-} is a limiting nutrient in the ocean, and its depletion near the surface is the result of its being used by marine biota. Biota have a slight preference for using the lighter isotope of carbon in photosynthesis, leading to the enrichment of the heavier isotope ^{13}C in the carbon reservoir in the surface water. Note that the light carbon is stored in the organic matter produced by photosynthetic phytoplanktons.

The ultimate fate of calcite that forms the shells of marine organisms is shown in figure 9.5c. For illustration, we chose planktonic foraminifera. These microorganisms (size of the order of 100 μm) live in the euphotic zone (first 100 m). On death they sink to the bottom of the ocean, where the calcite in their shells can dissolve. The "lysocline" is defined as the region where the calcite dissolution increases rapidly with depth. The lysocline is usually a few hundred meters above the calcium carbonate

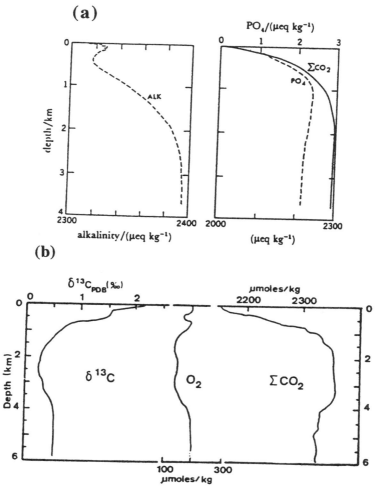

Figure 9.5 (a) Global average profiles of alkalinity, ΣCO_2, and dissolved phosphate in the ocean. PDB = Standard based on belemnite from the Peedee Formation in South Corolina. After Sarmiento, J. L., Toggweiler, J. R., and Najjar, R., 1988, "Ocean Carbon-Cycle Dynamics and Atmospheric pCO$_2$." *Phil. Trans. Roy. Soc. Lond. A* **325**, 3. (b) Profile of ^{13}C content of dissolved inorganic carbon ($\delta^{13}C$), dissolved oxygen (O$_2$), and total dissolved inorganic carbon (ΣCO_2) from a midlatitude Pacific Ocean station (17°S, 172°W). After Hsu, K. J., and McKenzie, J. A., 1985, "A Strangelove Ocean in the Earliest Tertiary," in *The Carbon Cycle and Atmospheric CO$_2$: Natural Variations Archean to Present*, E. T. Sundquist and W. S. Broecker, editors (Geophys. Monogr. 32; Washington, D.C.: American Geophysical Union), p. 487.

(c)

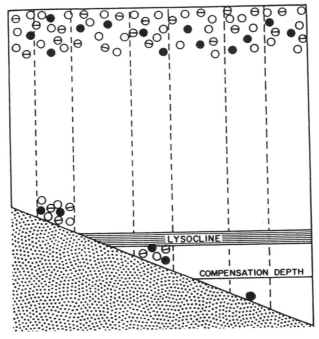

Figure 9.5 (c) Selective dissolution of planktonic foraminifera species at depth. Open circles (*Globigerinoides ruber*) are easily dissolved; circles marked with a bar (*Globigerina bulloides*) are more resistant; dark circles (*Globorotalia tumida*) are most resistant to dissolution. After Be, A. W. H., 1977, "An Ecological Zoogeographic Taxonomic Review of Recent Planktonic Foraminifera," in *Ocean Micropalaeontology* A. T. S. Ramsay, editor (New York: Academic Press).

compensation depth (CCD). The latter is generally defined to be the depth below which the amount of calcium carbonate is less than 10%. Twenty-two species are ranked by their ability to dissolve in the deep ocean. Most calcite shells can be preserved in the sediments in the shallow oceans. Below the lysocline few species can be preserved; below the CCD all but the last two species listed in figure 9.5c can survive the dissolution. Note that the burial and preservation of foraminifera provide a record of the paleoenvironment of the planet.

The rapid conversion of CO_2 to calcite and organics in surface waters followed by their sinking to the deep ocean exerts a major influence on the partitioning of CO_2 between the atmosphere and the oceans. The process is known as the biological pump. Without this pump, the concentration of atmospheric CO_2 in equilibrium with the deep oceans could have been as high as 450 ppmv in the preindustrial atmosphere. The fate of the biologically produced calcite in the deep oceans has important consequences for the carbon budget in near-surface reservoirs. If the biogenic calcite is dissolved, it remains in the oceans and eventually exchanges with the atmosphere. If it is buried in the sediments, it is permanently sequestered until it is recycled by tectonic processes and released by volcanic outgassing. This is important for regulating the amount of CO_2 in surface reservoirs over geological time (see section 9.4.2).

9.4.2 CO_2 and Climate: Precambrian

The two most important regulators of climate on Earth are H_2O and CO_2. The partitioning of water among different reservoirs has a most profound impact on climate, but this partitioning is itself largely determined by the temperature. Water, by itself, is not the driver of climatic changes. Its role in climate lies in amplifying climatic changes through strong nonlinear feedbacks. Carbon dioxide is a major greenhouse molecule in the terrestrial atmosphere. The effectiveness of CO_2 in controlling the climate of a planet is most dramatically demonstrated by the high temperature (750 K) on the surface of Venus (see table 9.1). Since the abundance of CO_2 in the terrestrial atmosphere is determined by geochemistry and biochemistry, changes in the planet's geochemistry and biosphere could have a major impact on climate via the regulation of atmospheric CO_2 abundance. It is believed that CO_2 has played such a role in the past climate of Earth.

The major energy source for the solar system is the sun. The sun's luminosity (L) has gradually increased according to the relation

$$L(t) = \left[1 + 0.4\left(1 - \frac{t}{t_0}\right)\right]^{-1} L_0 \qquad (9.17)$$

where t_0 is the present age of the sun (4.6 Gyr), and L_0 is the present luminosity. Thus, the initial solar constant was about 40% lower than the present value. The reduced solar constant would imply that Earth was completely frozen, a result in conflict with known geological evidence (e.g., sedimentary rocks). A resolution of this paradox is to postulate that the CO_2 content of the atmosphere has been changing with time to compensate for the changing solar constant. Figure 9.6 presents estimates of P_{CO_2} over geologic time. There could have been as much as 1 bar of CO_2 in the primitive Earth's atmosphere. The values of P_{CO_2} declined gradually over the eons to the present value of less than 10^{-3} bar. There is no strong observational evidence to support the hypothesis of greatly enhanced CO_2 abundances in the past. However, there is at least one indirect indicator of high P_{CO_2} from the analysis of paleosols at 2.5 Gyr, although the paleosol-derived value is considerably less than the estimates of a climate model.

The question arises as to whether this change in CO_2 over Earth's history is due to geochemical or biological causes. If the control is purely geochemical, the question is how the rates were adjusted so as to compensate exactly for the change in solar luminosity. If the control is biological, this brings us back to the fundamental premise of the Gaia hypothesis. The biosphere has taken on the function of homeorrhesis. In section 9.4.1 we discuss the effect of the biological pump in transferring carbon from the surface oceans (which exchange readily with the atmosphere) to the deep oceans and the sediments. On land, weathering reactions (9.6) and (9.7) are greatly enhanced by biological activities. In the soil, the CO_2 concentrations are 10–40 times greater

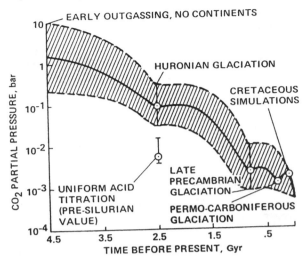

Figure 9.6 Estimated P_{CO_2} over geologic time. The solid line is a best-guess model. The shaded envelope represents the climatologically reasonable range of P_{CO_2}. The point labeled "uniform acid titration" is the pre-Silurian CO_2 pressure estimated by Holland and Zbinden (1986) from paleosol data. After Kasting, J. F., 1987, "Theoretical Constraints on Oxygen and Carbon Dioxide Concentrations in the Precambrian Atmosphere." *Precambrian Res.* **34**, 205.

than that in the atmosphere due to the presence of roots of plants and decaying organic material.

There is no doubt that the biosphere today plays a major role in reducing the amount of CO_2 in the atmosphere. According to the Gaia hypothesis, the biosphere may indeed have evolved since its origin to counteract the problem of the increasing luminosity of the sun. However, there is a finite limit to the power of Gaia. The abundance of CO_2 is now very low, while the luminosity of the sun continues to increase. Further decrease of CO_2 by biological activity may be difficult because photosynthesis itself stops when P_{CO_2} falls below a threshold level of about 150 ppmv. Thus, there is a point beyond which the Gaian control of the global environment would fail and the planet would become abiotic again. Fortunately, this ultimate doomsday is 30–300 Myr away.

The Gaia hypothesis remains a controversial concept. The critic could argue that the biosphere has no "purpose" and that it is impossible for the lower forms of life that dominate the biosphere (microbes and plants) to "know" what is good for the planet. Incidentally, for the species with advanced intelligence (*Homo sapiens*), the possession of the knowledge of the planetary environment does not always lead to action that is beneficial to the planet. We shall refrain from getting involved in this controversy, but we shall point out a simple geochemical reason why Earth cannot remain for long in a frozen state. The argument is as follows.

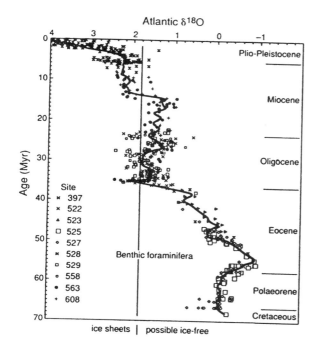

Figure 9.7 Benthic $\delta^{18}O$ measurements, from deep sea drilling sites, spanning the past 70 Myr. The standard for $\delta^{18}O$ is PDB. After Raymo, M. E., and Ruddiman, W. F., 1992, "Tectonic Forcing of Late Cenozoic Climate." *Nature* **359**, 117.

Suppose that for some reason (e.g., a faint sun), Earth is completely frozen. The hydrological cycle would cease to operate. Chemical weathering, as represented by (9.6) and (9.7), and the producing and sinking of calcite shells in the oceans, would stop. Thus, there would be no sink for atmospheric CO_2. Meanwhile, the volcanic source of CO_2 is independent of its removal on the surface and could supply 0.1×10^{15} Gt-C yr^{-1}. At this rate, it is possible to generate 1 bar of CO_2 in only 20 Myr on a frozen Earth. Earth would probably defrost before this massive CO_2 buildup in the atmosphere could be completed. Thus, geochemistry alone is capable of resolving the faint sun paradox without invoking the intervention of Gaia.

9.4.3 CO_2 and Climate: Phanerozoic

The temperature of Earth during the Cenozoic can be deduced from the $\delta^{18}O$ values preserved in foraminifera, as shown in figure 9.7. The most prominent change in Earth's climate its steady cooling by about 12°C since the Paleocene. It is generally agreed that this cooling is caused by a substantial decrease in atmospheric CO_2, although the precise reason for the CO_2 decline is still under debate.

We present a simple model for atmospheric CO_2 for the Phanerozoic. For this purpose the carbon inventories are represented by a three-box model (see figure 9.8a): atmospheric and oceanic carbon, carbonate in the sediments, and organic carbon. The atmosphere and the oceans contain a minute fraction of the total exchangeable carbon on this planet. We consider three principal reactions that govern the rate of transfer of carbon between the major reservoirs:

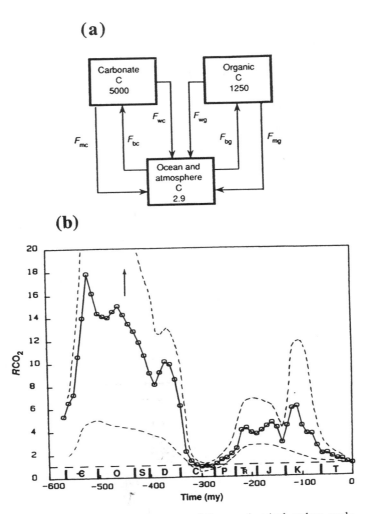

Figure 9.8 (a) Schematic diagrams of the geochemical carbon cycle. Masses of carbon are in units of 10^{18} moles. F_{wc}, F_{wg} = rate of release of carbon to the ocean-atmosphere system by weathering of carbonates (c) and organic matter (g); F_{mc}, F_{mg} = rate of release of carbon to the ocean-atmosphere system by the metamorphic and volcanic breakdown of carbonates (c) and organic matter (g); F_{bc}, F_{bg} = rate of burial of carbon as carbonates (c) and organic matter (g) in sediments. (b) Plot versus time of R_{CO_2}, the ratio of mass of CO_2 in the atmosphere at time t to that in the present atmosphere, based on a mass balance model of carbon. The curve represents the best-estimate model. Dashed lines show the envelope of approximate error. After Berner, R. A., 1990, "Atmospheric Carbon Dioxide Levels over Phanerozoic Time." *Science* **249**, 1382.

Table 9.6 Inputs and outputs of carbon and their isotopic composition

	Rate (10^{15} g yr^{-1})	$\delta^{13}C$(‰)
Inputs to ocean-atmosphere system		
Oxidation of organic carbon	0.09 ± 0.02	-25 ± 5
Weathering of carbonates	0.16 ± 0.04	0 ± 2
Degassing due to igneous and metamorphic processes	0.09 ± 0.03	-7 ± 3
Total inputs	0.34 ± 0.09	-8 ± 3
Outputs from the ocean-atmosphere system		
Deposition of organic carbon	0.12 ± 0.03	-23 ± 3
Deposition of carbonate carbon	0.22 ± 0.04	0 ± 1
Total outputs	0.34 ± 0.07	-8 ± 2

From Holland, H. D. (1978).

$$CO_2 + CaSiO_3 \leftrightarrow CaCO_3 + SiO_2 \tag{9.18}$$

$$CO_2 + MgSiO_3 \leftrightarrow MgCO_3 + SiO_2 \tag{9.19}$$

$$CO_2 + H_2O \leftrightarrow CH_2O + O_2 \tag{9.20}$$

Proceeding from left to right, these are weathering and photosynthesis reactions (9.6), (9.7), and (9.1), respectively. Sequestering the products in the sediments provides a net sink for CO_2 in all three reactions. The backward reactions of (9.18) and (9.19) occur in metamorphism and magmatism, and CO_2 is returned to the atmosphere by volcanic outgassing. The reverse reaction of (9.20) occurs when the organic carbon in sedimentary rocks is oxidized by atmospheric oxygen. The latter process is essentially the same as respiration and decay of organic matter, albeit at a much slower rate. The current rates of transfer of carbon between the major reservoirs are estimated in table 9.6. These rates are very small (0.34 Gt yr^{-1}), and the time constant for cycling the carbon through the major reservoirs once is on the order of 10 Myr. It is therefore surprising that the total amount of carbon in the atmosphere and oceans is determined by such small net transfer rates and not by the recycling rates that are many orders of magnitudes higher (see figure 9.4). This circumstance makes the measurement of the net transfer rates and their time history difficult. Thus, much of our information about CO_2 in the past must be deduced from indirect evidence, such as isotopic composition of carbon, history of tectonic activity and rates of weathering. The CO_2 concentrations in the atmosphere, reconstructed from a model that incorporates the relevant geochemical data, are given in figure 9.8b. The units are relative to the present CO_2 concentration. The result shows that there has been a general decline of atmospheric CO_2 by about a factor of 30 since 100 Myr ago. Since CO_2 is a major greenhouse molecule, this decrease in CO_2 can account for the cooling of Earth.

The main reasons for this decline in CO_2 are a reduction in the planet's magmatism and an increase in the rate of weathering. The former is driven by the energy source derived from the interior of the planet; the latter is controlled by the hydrological cycle that is driven by solar energy. Thus, the planet's global environment is ultimately controlled by its internal energy and energy from the sun. We note that the two fundamental processes controlling the long-term cycling of carbon, magmatism and weathering, are independent. Is there any feedback mechanism that can prevent the system from extreme fluctuations in P_{CO_2}? The possibility of a feedback mechanism seems to lie in the weathering portion of the cycle. Weathering is affected by temperature, rain, plants, and topography. For instance, a higher P_{CO_2} results in a warmer climate with more vegetation, which increases the rate of weathering and removes CO_2 from the atmosphere, resulting in a negative feedback. Enhanced magmatism (source of CO_2) could lead to mountain building; the resulting steeper topography increases the rate of physical weathering and, consequently, the rate of chemical weathering (sink of CO_2). There are major uncertainties in our present understanding of the controls and feedbacks in the regulation of atmospheric CO_2. Some biological feedbacks may be attributed to Gaian homeorrhesis, but proof of this appears to be difficult.

9.4.4 CO_2 and Climate: Pleistocene and Pliocene

Perhaps the most dramatic development in the climate of the Plio-Pleistocene was the occurrence of ice ages during which Earth was colder by several degrees, ice sheets covered much of North America and northern Europe, and the mean sea level was lower by 120 m. Figure 9.9 presents the global mean temperature of Earth since the Pleistocene on three time scales: (a) the last million years, (b) the last 10,000 yr, and (c) the last 1000 yr. The temperature variations have been estimated using isotopic data from deep sea cores (the CLIMAP project) and isotopic measurements of water ice in ice cores taken from polar regions. Variations in the time series in figure 9.9a are believed to be caused by astronomical forcing. The theory, developed by Milankovitch early in the twentieth century, states that the ice ages were caused by small changes in insolation at high latitudes due to secular variations of the position of the Earth in its orbit around the sun at vernal equinox, the tilt of Earth's rotational axis, and the ellipticity of Earth's orbit. These variations have dominant periods of about 20, 40, and 100 kyr and are recorded in the observed data. The amplitudes of the forcing function are, however, different from those of the temperature variations. This may be explained by the nonlinear dynamics of ice growth and collapse.

The changes in insolation due to astronomical forcing are very small. It is now known that the Milankovitch effects are greatly enhanced by greenhouse gases such as CO_2 and CH_4. Figure 9.10 shows the correlation between temperature, CO_2, and CH_4 during the deglaciation of the penultimate glacial maximum (PGM) about 140 kyr ago. The results suggest that during the PGM the atmospheric CO_2 and CH_4 were only two-thirds and one-half of the interglacial values, respectively. It is not clear from the generally good correlation in figure 9.10 that the greenhouse gases drove climatic changes. The only piece of circumstantial evidence in favor of this possibility is in the $\delta^{18}O$ values of atmospheric O_2 shown in figure 9.10. Without going into the details, we interpret the $\delta^{18}O$ (of O_2) data to reflect change in ice volume. The PGM results suggest that the ice sheets melt *after* the rise of CO_2, CH_4, and temperature. This tentative conclusion has not been confirmed by independent observations or for other glaciation periods. We should point out that even if it is proven that CO_2 and other greenhouse gases drive climatic changes, questions will remain concerning what drives the changes in the gases and how the gases respond to changes in the orbital parameters of the planet.

Earth: Imprint of Life 341

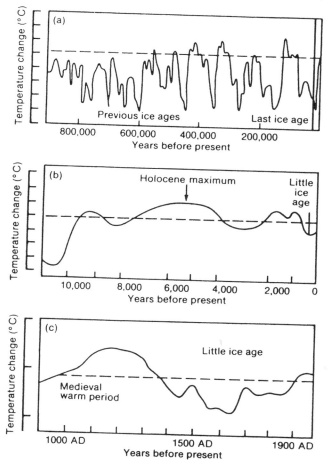

Figure 9.9 Schematic diagrams of global temperature variations for the last million years. Note the different time scales in each panel. The dashed lines represent conditions near the beginning of the twentieth century. After Houghton et al. (1991).

The most prominent climatic anomaly in the last 10,000 yr is the Holocene Optimum that occurred about 5000–6000 yr ago. The global mean temperature during this period was about 1°C warmer than present. The difference cannot be attributed to CO_2 or solar insolation, but must be attributed to some internal variables of the climate system such as vegetation or cloud cover, or perhaps ocean circulation. There is no accepted explanation for the high temperature of the Holocene Optimum.

In the last 1000 yr the most interesting feature in the climate record is the Little Ice Age, a prolonged period of cooling that extended from 1500 to 1900 A.D. The coldest part of this period, 1700 and 1800s, was characterized by a marked lack of sunspots. However, quantitative changes in the solar constant cannot be deduced from the sunspot numbers, and there is no explanation for the observed climatic change.

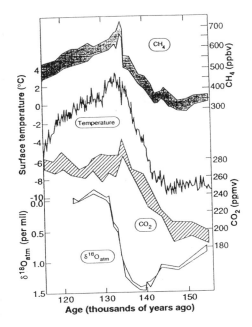

Figure 9.10 Changes in atmospheric CH_4, temperature, CO_2, and $\delta^{18}O$ of atmospheric oxygen from the Vostok ice core during the penultimate deglaciation. After Raynaud, D., et al., 1993, "The Ice Record of Greenland Gases." *Science* **259**, 926.

The most intriguing part of climatic change is not the gradual change in response to an applied external forcing but the sudden changes arising from internal dynamics. Figure 9.11 shows an example of such changes (as revealed by $\delta^{18}O$, a proxy for temperature, and dust, a proxy for aridity) around the time of the LGM. The most dramatic change occurred at about 10–11 kyr ago, known as the Younger Dryas. During this time Earth was just emerging from the ice age, but for about a thousand years the climate reverted back to that of an ice age. Then it recovered and the warming continued. The cause of this sudden change is attributed to a shift in the North Atlantic conveyor, a current that is largely responsible for transporting heat from the tropical ocean to the polar ocean in the Atlantic. Figure 9.11 also shows a large number of

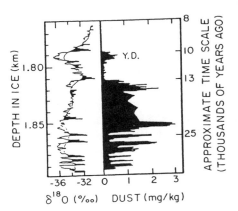

Figure 9.11 Oxygen isotope and dust record for the period from about 8000 to 40,000 yr B.C.. in the Dye 3 Greenland ice core. The Younger Dryas cold event occurred at about 11,000 B.C. After Hammer, C. U. et al., 1985, "Continuous Impurity Analysis along Dye 3 Deep Core," in *Greenland Ice Core: Geophysics, Geochemistry, and the Environment*, C. C. Langway et al., editors (American Geophysical Union Monogr. 33; Washington, D.C.: American Geophysical Union), p. 90.

rapid climate oscillations during the ice age. These events are known as Dansgaard-Oeschger events. The cause of these rapid "quantum jumps" in the climate system is not completely understood. We may note the occurrence of greatly enhanced amounts of dust during the ice age. Dust is characteristic of cold, windy, and arid environments. It is not known whether dust is a diagnostic of the climatic conditions or if it had an active role in inducing climatic change.

In addition to Milankovitch forcing, variations in greenhouse gases, and possible secular changes in the solar constant, we must include volcanic eruptions as a possible trigger for climate changes. A major volcanic eruption has immediate impact on the radiation budget of the planet via the dust cloud raised in the atmosphere. This effect, however, is local and short-lived and will dissipate in a week. The greater and sustained impact comes from the small amount of SO_2 injected into the stratosphere, where the SO_2 is converted to sulfuric acid aerosols. These aerosols can stay in the upper atmosphere for up to 2 yr and can profoundly alter Earth's albedo. There are well-documented perturbations of Earth's climate after the eruptions of Krakatoa, Agung, St. Helens, El Chichon, and Pinatubo. Figure 9.12 shows the major volcanic eruptions and estimated optical depths during the past 2100 yr as recorded in an ice core in Greenland.

The quasi-periodic advance and retreat of ice sheets and glaciers over much of Earth during the Plio-Pleistocene may have exerted a profound influence on the evolution of one insignificant (in terms of total biomass) species in the biosphere, Homo sapiens (Hs). In the past few million years these global-scale environmental catastrophes may have stimulated a fourfold growth in mass and complexity of the central nervous system of Hs, to the point where abstract reasoning and logical construction and development are possible. We can only speculate on the origin of human consciousness. It may be the result of efforts to improve the odds of survival in the face of inexorable climatic changes. Lower animals have the ability to recognize cause and effect and learn from experience. The human brain, being larger, has the ability to store complex patterns of environmental change and thus evolve the ability to manipulate these patterns in the mind and eventually develop the capability to predict the future, or at least the trend of future events. Note that once this capability is developed, it acts as its own positive feedback for its further development. Each new advance of the ice eliminates the portion of Hs deficient in this capability, and each retreat of the ice allows the repopulation of the deglaciated areas by the more intelligent descendants of the survivors. Thus, the glacial and interglacial alternations serve to distill Hs according to intelligence. The net product, after several million years, is a human mind capable of grasping the secret of not only the terrestrial environment but also the solar system and the cosmos at large. We shall not enter into the debate whether this evolution of the Hs has a "purpose" or not.

According to the Gaia hypothesis, we are part of the biosphere and Hs must ultimately contribute to the welfare of the biosphere. At the moment this conclusion is not obvious. In fact, possessing advanced intelligence, our species has succeeded in building an a civilization with possibly adverse impact on the global environment. Some of these problems will be addressed in chapter 10. Is this a repudiation of the Gaia hypothesis? Will advanced human intelligence, originally developed for improving the chances of human survival in a harsh environment, be a force of destruction of the global environment?

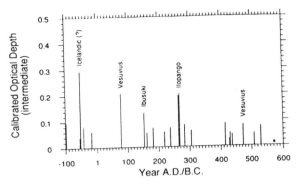
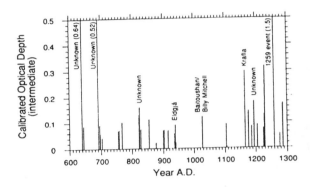
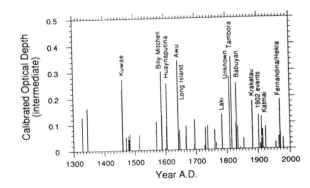

Figure 9.12 Optical depth of dust estimated from measurements of sulfate derived from explosive volcanism over the last 2100 yr in the Greenland Ice Sheet Project 2 ice core. After Zielinski, G. A., 1995, "Stratospheric Loading and Optical Depth Estimates of Explosive Volcanism over the Last 2100 Years Derived from Greenland Ice Sheet Project 2 Ice Core." *J. Geophys. Res.* **100**, 20937.

9.5 Rise of Oxygen

Perhaps the single most spectacular feature of the terrestrial atmosphere is the enormous amount of O_2 and the ozone layer associated with it. Oxygen makes advanced life possible. Ozone shields the entire biosphere from harmful ultraviolet radiation. The levels of oxygen and ozone on prebiotic Earth have been deduced from geochemical, geological, and paleontological records and atmospheric chemistry. It is surprising that over half of the history of the planet there was less than 1% of the present atmospheric level (PAL) of oxygen and less than 0.1 PAL of ozone. The earliest forms of life, the photosynthetic stromatolites, appeared as early as 3.8 Gyr B.P., and yet it took some 2 Gyr before oxygen became a major component of the atmosphere.

There are two sources of oxygen: dissociation of H_2O, followed by the escape of hydrogen, and photosynthesis, followed by the burial of organic carbon. The first source is permanent because the escape of hydrogen to space is an irreversible part of planetary evolution over the history of the solar system. This process is capable of generating small amounts of O_2 on Mars and Venus. On Mars the photochemical source of oxygen is probably responsible for the production of the highly oxidized magnetite (main chromophore for the red color of the planet's surface), and the low observed $[CO]/[O_2]$ ratio of 0.5 (the expected value based on stoichiometry is 2). On Venus the escape rate of hydrogen is much smaller and the bulk atmosphere is much more massive. There is no evidence that the photochemical source of oxygen has any impact on the surface. On Earth we show in section 9.5.1 that the escape rate of hydrogen today has relatively little impact on the atmospheric oxygen content. However, the rates might have been greater in the past and the integrated result might be substantial.

The second source, photosynthesis, is much larger, but the organic carbon is only temporarily sequestered in the sediments [this is the forward reaction of (9.20)]. Magmatism and volcanism will bring the reduced carbon in contact with atmospheric oxygen, destroying both by the reverse of (9.20).

9.5.1 Escape of Hydrogen

The best demonstration of the production of O_2 arising from the escape of hydrogen is the large oxygen excess on Mars relative to CO_2 stoichiometry. The expected CO/O_2 ratio in a pure CO_2 atmosphere is 2. On Mars the observed CO/O_2 ratio is 0.5, suggesting an additional source of O_2. This source has been identified as chemical scheme (V) from chapter 7:

$$2H_2O \rightarrow 2H_2 + O_2 \qquad (I)$$

The conversion of H_2O into H_2 and O_2 is initiated by the photolysis of H_2O and assisted by the photochemistry of CO_2. Early workers (Berkner and Marshall) considered the photolysis of H_2O to be the rate-limiting step for the generation of O_2 by (I) and computed enormous rates of production of O_2. This is, however, incorrect. As shown in section 7.2.4, most of the H_2 produced in (I) is oxidized back to H_2O by reactions with OH and $O(^1D)$:

$$H_2 + OH \rightarrow H_2O + H \qquad (9.21)$$

$$H_2 + O(^1D) \rightarrow OH + H \qquad (9.22)$$

The ultimate source of oxygen used in the oxidation of H_2 is O_2 produced in (I). Thus, the reactions initiated by (9.21) and (9.22) result in the reversal of (I). Only a small fraction of the H_2 produced in (I) survives the back reaction to H_2O and gets transported to the upper atmosphere, where it undergoes decomposition via reactions similar to scheme (VI) of chapter 7:

346 Photochemistry of Planetary Atmospheres

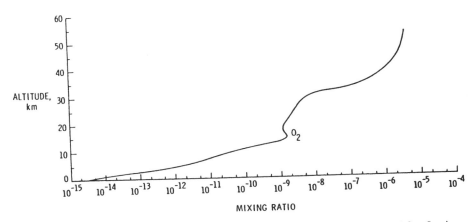

Figure 9.13 The vertical distribution of oxygen in the prebiotic atmosphere. After Levine, J. S., 1985, "The Photochemistry of the Early Atmosphere," in *The Photochemistry of Atmospheres*, J. S. Levine, editor (New York: Academic Press).

$$N_2 + h\nu \rightarrow N_2^+ + e \qquad (9.23)$$
$$H_2 + N_2^+ \rightarrow N_2H^+ + H \qquad (9.24)$$
$$\underline{N_2H^+ + e \rightarrow N_2 + H} \qquad (9.25)$$
$$net \quad H_2 \rightarrow 2H \qquad (II)$$

Note that this chemical scheme is important today in the upper atmosphere of Triton, where the H_2 is derived from the dissociation of CH_4. Once decomposed into atoms, the H atoms can escape from the exosphere of Earth. Three processes of escape are known: thermal (Jeans) escape, ion exchange, and polar wind. The thermal escape is discussed in section 1.4; ion exchange has been discussed in section 8.4.1. We briefly discuss here escape by the polar wind mechanism. In a planet such as Earth, protected by a magnetic field, the field lines are normally closed, making it difficult for ions to escape. However, in the polar region some field lines become open due to the interaction with the solar wind. Ions that diffuse into the polar region where the field lines are open may escape along the field lines into space.

The detailed mechanisms of escape for hydrogen from the primitive atmosphere of Earth have not been worked out. If we use the present Earth as an analog, the total escape rate today is 3×10^8 atoms cm^{-2} s^{-1}, with contributions from thermal escape, ion exchange, and polar wind being 10%, 40%, and 50%, respectively. At this rate it is possible to generate 1 bar of O_2 over the age of the solar system. However, we have reasons to believe that the rate of escape of hydrogen must have been much higher in the past. According to the isotopic data of water obtained from the deep sea samples, as much as 35% of the terrestrial ocean might have escaped in the past. This speculation remains to be substantiated.

Although it is difficult to compute the amount of O_2 that can be accumulated from the escape of hydrogen, it is not hard to estimate the minimum amount of O_2 in the abiotic atmosphere. The amount of steady state atmospheric O_2 that can be generated photochemically from H_2O and CO_2 is shown in figure 9.13. Note that though the O_2 mixing ratio exceeds 1 ppmv in the middle atmosphere above 30 km, it declines rapidly in the lower atmosphere. The reason is that O_2 is not stable against reactions that restore it to CO_2 and H_2O.

9.5.2 Oxygen cycle

The amount of dead organic matter in the ocean is large (3000 Gt-C) compared with the living marine biomass (3 Gt-C). This large reservoir is formed by the rain of organic detritus from the biologically productive surface oceans at the rate of 4 Gt-C yr^{-1}. This is equivalent to 1.3×10^{12} O_2 molecules cm^{-2} s^{-1}, or about 10^4 times the rate that O_2 is generated by the escape of hydrogen from the present atmosphere. If the deep oceans were anoxic, all this organic carbon would be buried in the sediments. At this rate all the O_2 in the present atmosphere could be generated in 0.3 Myr. The actual rate of generation of O_2 is much less since the oceans are not anoxic (at least in the present epoch) and is equal to 0.2 Gt-C yr^{-1}. Even in anoxic environments, not all the organics can be preserved. The bulk of organic material (95%) is returned to the atmosphere as CH_4, which is then readily oxidized to CO_2 (see chapter 10). The net rate of oxygen generation given the processes described above must be balanced by a similar rate of oxygen loss, yielding a lifetime for atmospheric oxygen of 6 Myr.

We argue that the production of oxygen by the burial of photosynthetically derived organic carbon is an inevitable consequence of the physical properties of carbon and oxygen: the enormous solubility of CO_2 in water and the extreme insolubility of O_2 in water. According to figure 9.4, the partitioning of CO_2 between the atmosphere and the ocean is 1:50; that is, 50 times as much CO_2 resides in the ocean (mainly in the form of HCO_3^-) as in the atmosphere. For O_2 the partitioning between the atmosphere and the ocean is 140:1; that is, the atmosphere contains 140 times as much O_2 as the ocean. The difference is therefore a factor of 7000 in the partitioning of CO_2 and O_2 between the atmospheric and oceanic reservoirs. Since it is so difficult to deliver O_2 to the aquatic medium, there is a tendency for it to become anoxic, thus creating an ideal environment for the burial and preservation of organic carbon. Today, the bottoms of the oceans are not anoxic, but in the past there have been periods of global anoxia. Earth's surface always has a substantial fraction of area under marshes and wetland that are largely anoxic and conducive to the preservation of organic carbon. Thus, we view the rise of oxygen on this planet as the inevitable consequence of the production of organic carbon and the ease of getting it buried. We may regard the enormous atmospheric concentration of O_2 as a way the system tries to make up for the insolubility of O_2 in water. In other words, if O_2 were 7000 times more soluble in water than CO_2, then its mixing ratio would be 7000 times less, making its abundance the same order of magnitude as CO_2. But then the bulk of the oxygen would be in the ocean rather than in the atmosphere. All advanced life would be marine rather than continental.

The burial of organic carbon in the sediments produces and regulates atmospheric O_2, but its time constant is of the order of millions of years. For shorter time scales other regulation mechanisms must be important. The photosynthetic time constant for atmospheric O_2 is only 1000 yr. It is believed that forest fires act as regulator of O_2. Laboratory experiments have shown that spontaneous combustion of even living plants can start when the atmospheric O_2 concentrations exceed about 25%. Conversely,

348 Photochemistry of Planetary Atmospheres

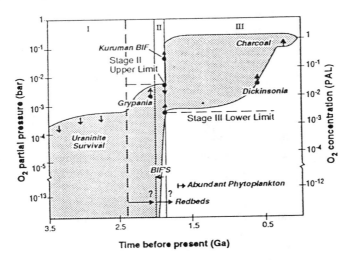

Figure 9.14 Estimated change of oxygen levels over geologic time. The shaded area represents the range permitted by data. The dashed line indicates a possible earlier date for the beginning of stage II. (Ga = 10^9 yr) After Kasting, J. F., 1993, "Earth's Early Atmosphere." *Science* **259**, 920.

vegetation will not burn when the O_2 level falls below 15%. Thus, it is unlikely that atmospheric O_2 ever exceeded 25% in the history of the atmosphere.

The history of the rise of O_2 in the atmosphere can be surmised only from indirect evidence and is presented in figure 9.14. Note that for the first 2 Gyr the abundance of O_2 was less than 1%, and the ozone shield generated by the photochemistry of oxygen was not sufficient to protect life on land. The long time required for the rise

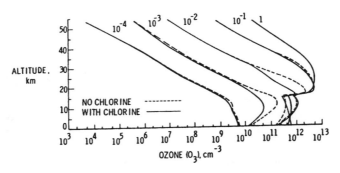

Figure 9.15 The vertical distribution of ozone as a function of atmospheric oxygen level, represented in terms of present atmospheric level (PAL). After Levine, J. S., 1985, "The Photochemistry of the Early Atmosphere," in *The Photochemistry of Atmospheres*, J. S. Levine, editor p. 3 (New York: Academic Press).

of atmospheric O_2 is attributed to the large quantities of reduced material present on the surface of the early Earth, such as ferrous iron. The initial production of O_2 was consumed by the oxidation of ferrous to ferric iron, depositing the well-known banded iron formations. It was not until the reduced iron inventories were exhausted that there was a net accumulation of O_2 in the atmosphere.

The abundance and distribution of ozone at different atmospheric oxygen levels are shown in figure 9.15. It is a matter of conjecture how early life on Earth coped with the lack of an ozone layer. One obvious solution is in the depth of the oceans. Ocean water attenuates radiation, such that by 100 m depth only 1% of the incident solar radiation gets transmitted. In addition, the filtering is more efficient for ultraviolet and infrared radiation. Another possibility has been demonstrated in the laboratory using algae: the buildup of a layer of algal mats such that the dead algae above provides shielding for the living algae beneath.

9.6 Nitrogen Cycle

The nitrogen cycle of the biosphere is closely related to the carbon cycle. The C/N ratios are 92 and 6.7 for terrestrial and marine plants, respectively. Hence, the nitrogen cycle, as shown in figure 9.16, is to first order a scaled down carbon cycle. To sustain the rate of photosynthesis as shown in figure 9.4, the demands for nitrogen by the land and ocean biosphere are 600 and 4700 Mt-N yr^{-1}, respectively. The bulk of nitrogen in the surface reservoir resides in the atmosphere as N_2, a rather inert molecule chemically. To be useful biologically, the strong N—N bond must first be broken, a process known as nitrogen fixation. In the absence of biology, lightning provides a background source of fixed nitrogen. In the lightning fireball the air parcel temperature is raised to 10,000 K, resulting in the production of NO via the following equilibrium chemistry:

$$N_2 + O_2 \leftrightarrow 2NO \qquad (9.26)$$

$$N_2 + 2CO_2 \leftrightarrow 2NO + 2CO \qquad (9.27)$$

The first reaction is more important in the present atmosphere with O_2 as the oxidant. The second reaction is more important in the primitive terrestrial atmosphere before the rise of O_2. This reaction is potentially important also in the present atmospheres of Mars and Venus. On cooling the equilibrium composition given by reactions (9.26) and (9.27) is "frozen" at about 2000 K and NO is released into the atmosphere. NO is subsequently oxidized by atmospheric reactions to higher oxides of nitrogen and eventually gets rained out as nitrate. This global source is estimated to be 5–10 Mt-N yr^{-1}. The principal source of fixed nitrogen is biological, via microbes that live in symbiosis with land plants and marine organisms. The total input of fixed nitrogen to the land biosphere is about 200 Mt-N yr^{-1}, of which 100 Mt is fixed biologically, 65 is via rain, and 40 is from manmade fertilizers (the subject of anthropogenic perturbations is addressed in chapter 10). The total input into the ocean biosphere is 85 Mt-N yr^{-1}, of which 30 Mt is fixed in situ, 45 is from rain, and 10 is from river runoffs.

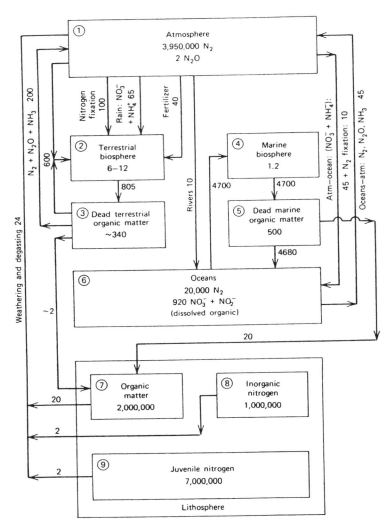

Figure 9.16 Major aspects of the global nitrogen cycle. Nitrogen is shown in units of 10^{15} g; transfer rates, in units of 10^{12} g yr^{-1}. After Holland, H. D., 1978, *The Chemistry of the Atmosphere and the Oceans* (New York: Wiley).

The return of fixed nitrogen to N_2 is known as denitrification, and this must balance the rate of fixation. We note that the net rates of fixation and denitrification are considerably less than those required by photosynthesis. Thus, fixed nitrogen is efficiently recycled, especially in the ocean. The denitrification process has an alternative minor product in addition to N_2. Production of nitrous oxide, N_2O, accounts for about 10% of total denitrification. N_2O is inert in the troposphere and is ultimately destroyed in the stratosphere, where its decomposition gives rise to the principal source of oxides

Table 9.7 Estimates of global emission to the atmosphere of gaseous sulfur compounds

Source	Annual flux (Tg)
Anthropogenic (mainly SO_2 from fossil fuel combustion)	80
Biomass burning (SO_2)	7
Oceans (DMS)	40
Soils and plants (H_2S, DMS)	10
Volcanoes (H_2S, SO_2)	10
Total	147 ± 1.4

From Intergovernmental Panel on Climate Change (1990).
The uncertainty ranges are estimated to be about 30% for the anthropogenic flux and a factor of 2 for the natural fluxes.

of nitrogen. This topic is pursued in greater detail in chapter 10. There is a net loss of nitrogen from the atmosphere-ocean reservoirs when the organic material is buried in the sediments, at a rate of 22 Mt yr^{-1}, which must be balanced by the recovery of nitrogen from the weathering of sedimentary rocks. The lifetime for cycling atmospheric N_2 by this process is 2×10^8 yr, a value that is about two orders of magnitude larger than that for O_2.

9.7 Minor Elements

In addition to the major elements carbon, oxygen, and nitrogen, organic matter also contains smaller amounts of other elements such as sulfur and phosphorus. As part of the biospheric cycle of organic matter, the trace elements are cycled through the atmosphere. The marine organisms have unusual abilities to concentrate large amounts of halogens in their bodies, and on death some of the halogens are released to the atmosphere as methyl halides.

9.7.1 Sulfur cycle

The bulk of sulfur in the surface reservoirs of the planet resides in the ocean as sulfate, with mean concentration of 2.71 g kg^{-1}. The total sulfur content of the oceans amounts to 1.3×10^{21} g, or about a quarter of the total mass of the entire atmosphere. The total amount of sulfur in the atmosphere, about 3.6×10^{12} g, is utterly trivial compared with the oceanic reservoir. Estimates of the global emissions of gaseous compounds to the atmosphere are summarized in table 9.7. The sum total of the emission rates is 147 Mt-S yr^{-1}, with the largest contribution derived from combustion of fossil fuel. We discuss the anthropogenic impact of sulfur emission on the atmosphere in chapter 10. The largest natural source of atmospheric sulfur appears to be dimethyl sulfide (DMS) emitted by marine organisms. Volcanoes are sporadic sources; their influence is small on the average but can be large in the years following an eruption.

The ultimate fate of all reduced sulfur species (such as H_2S and DMS) is oxidation to SO_2, followed by the oxidation of SO_2 to SO_3. The terminal product is H_2SO_4 aerosol. A schematic of the chemical pathways is shown in figure 9.17. The presence

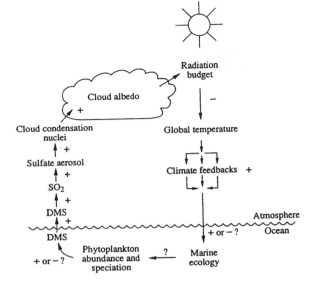

Figure 9.17 Climate feedback loop linking oceanic phytoplankton, emission of dimethyl sulfoxide (DMS), atmospheric sulfur, cloud albedo, and climate. The plus and minus signs indicate the sign of the feedback. After Andreae, M. O., 1995, "Climatic Effects of Changing Atmospheric Aerosols," in *World Survey of Climatology. Vol. 16: Future Climates of the World*, A. Henderson-Sellers, editor (Amsterdam: Elsevier).

of H_2SO_4 aerosols in the atmosphere can have two possible profound impacts on the climate of the planet. First, the aerosols are good scatterers of visible sunlight, contributing to the planetary albedo. Second, the aerosols serve as condensation nuclei for cloud droplets; the potential influence of this mechanism on cloud formation and its subsequent impact on climate are enormous. Indeed, it has been proposed by Lovelock that DMS may be an instrument of Gaia for regulating the planet's climate. The argument is as follows. Suppose there is an increase in the surface temperature of Earth. This would result in an enhancement and release of DMS by marine organisms. (Marine productivity increases with temperature when nutrients are not the limiting factor.) Oxidation of DMS in the atmosphere leads to an increase in marine aerosols and cloudiness. The enhanced albedo would reduce the fraction of sunlight absorbed by the planet, resulting in a lower temperature. Thus, homeorrhesis of surface temperature is achieved. The theory is currently under scrutiny by the scientific community, and no final verdict on the outcome is available.

Most of the sulfur compounds have short lifetimes in the atmosphere and will not have an impact on the stratosphere. However, the biogenic species COS and CS_2 are remarkably stable and will diffuse to the stratosphere, where they are decomposed by ultraviolet radiation and provide a source of sulfur for the Junge layer, a thin aerosol layer composed of H_2SO_4 particles of about 1 μm in radius. The Junge layer is optically thin and does not have a direct impact on the albedo of the planet. However, the sulfate aerosols serve as condensation nuclei for PSCs; hence there is an indirect impact on the radiation balance of the stratosphere. Both the aerosol particles and the PSCs play an important role in the heterogeneous chemistry of the lower stratosphere.

A large volcanic eruption, such as El Chichon in 1982 and Pinatubo in 1991, is capable of injecting more than 10 Mt-S into the stratosphere, mostly as SO_2. SO_2 is rapidly converted into H_2SO_4 aerosols, with concentrations exceeding the background aerosol contents by an order of magnitude. This sulfate layer is global and has a mean

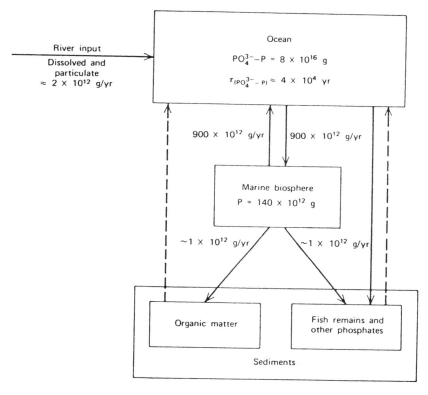

Figure 9.18 The marine geochemistry of phosphorus. After Holland, H. D., 1978, *The Chemistry of the Atmosphere and the Oceans* (New York: Wiley).

lifetime of 2 yr. The main effects on the planetary environment are cooling of the surface of Earth by about 1 °C and a decrease in ozone by a few percent. In the polar stratosphere, Pinatubo aerosols are believed to be responsible for record low ozone values over Antarctica in 1992 and 1993.

9.7.2 Phosphorus Cycle

Land plants obtain phosphorus from the weathering of rocks, and the deficiency of phosphorus is not a serious limitation. However, in the oceans the only net source of phosphorus is river input, estimated to be 2 Mt-P yr^{-1}. The inventories and cycling rates of phosphorus in the oceans are shown in figure 9.18. The bulk of phosphorus, 80 Gt-P, is in the form of PO_4^{3-}. A minor portion, 140 Mt-P, resides in the marine biosphere. The C/P ratio for phytoplankton is close to 106. Thus, from the magnitude of the carbon cycle (figure 9.4), we may estimate that the demand for phosphorus in marine photosynthesis is 900 Mt-P yr^{-1}. Most of this phosphorus is efficiently recycled by the marine biosphere. However, a small residue, 2 Mt-P yr^{-1}, is buried as part of the organic remains in the sediments. This minuscule but inexorable removal rate would eventually deplete the oceans of all phosphorus in 40 kyr were it not for the input from rivers. The phosphorus cycle releases no known gaseous compounds, and the atmosphere plays no role in the phosphorus cycle.

The short time constant associated with the depletion of phosphorus has been the basis of an interesting theory of the ocean's role in causing the ice age. The time constant for phosphorus is intriguingly close to the Milankovitch time scales for orbital changes. According to Broecker's idea, the oceans periodically receive a large input of phosphorus at the end of an ice age when warm climate returns and the continental shelves are re-submerged under water. This enhances the marine productivity, depleting the atmosphere of CO_2 via the biological pump and sending the planet into another ice age. Eventually, the surplus phosphorus is used up and the marine productivity slows down. The biological pump now becomes less effective, and CO_2 is released back to the atmosphere, marking the end of an ice age. The cycle now starts again with a new injection of phosphorus from the coastal shelves. This imaginative theory of the biological modulation of Earth's climate has not been verified by available data.

On a planetary scale phosphorus is the limiting nutrient that intimately couples the biospheric productivity to the geochemical cycle of sedimentation, subduction, mountain building, and weathering. The former is driven by the energy from the sun, the latter by solar energy and energy from the interior of Earth. Is the total productivity of the biosphere ultimately limited by the geochemical cycle of phosphorus? We can imagine a planet like Mars that initially had a biosphere, with P as the limiting nutrient. As the planet cools, the rate of turnover of the sediments is greatly reduced, trapping most of the available phosphorus in the sediments. Eventually, the biosphere dies from the lack of tectonic activities for recycling phosphorus.

9.7.3 Organic Halogen

The oceans contain abundant supplies of halogens. The concentrations of Cl^-, Br^-, and I^- are 19.4 g kg^{-1}, 67 mg kg^{-1} and 59 μg kg^{-1}, respectively. The direct transfer of halogens from the ocean to the atmosphere is small and includes for small amounts present in sea spray. Most of the halogen in the marine aerosols remains immobilized and eventually returns to the ocean by rainout. However, a small fraction may be released to the atmosphere by

$$NaCl + HNO_3 \rightarrow NaNO_3 + HCl \qquad (9.28)$$

$$NaCl + N_2O_5 \rightarrow NaNO_3 + ClNO_2 \qquad (9.29)$$

Similar reactions also apply to salts of Br and I. The release of chlorine into the atmosphere as HCl would not initiate any active chlorine chemistry because HCl is eventually removed by rainout to the oceans. However, if reactive Cl atoms were released via (9.29) followed by the photolysis of $ClNO_2$, there would be an impact on tropospheric chemistry: Cl atoms can remove hydrocarbons by hydrogen abstraction. The subject is still at a somewhat speculative stage at present.

The oceans are a source of three important organic halides, CH_3Cl, CH_3Br, and CH_3I, with atmospheric concentrations equal to 600, 20, and 1 parts per trillion by volume, respectively. CH_3I is readily removed by photolysis in the troposphere. Its lifetime is only a few days and the molecule has no known impact on the chemistry of

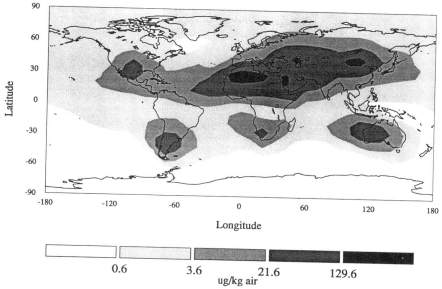

Figure 9.19 Dust concentrations in spring in the lowest layer of a general circulation model. After Tegen, I., and Fung, I., 1994, "Modeling of Mineral Dust in the Atmosphere—Sources, Transport, and Optical-Thickness." *J. Geophys. Res.* **99**, 22897.

the stratosphere. Iodine chemistry may be important for destroying O_3 in the marine boundary layer. The principal loss mechanism for CH_3Cl and CH_3Br is by reaction with OH:

$$CH_3Cl + OH \rightarrow CH_2Cl + H_2O \qquad (9.30)$$

$$CH_3Br + OH \rightarrow CH_2Br + H_2O \qquad (9.31)$$

The lifetimes of CH_3Cl and CH_3Br in the troposphere are about 1–2 yr, long enough for the molecules to survive to the stratosphere, where they are the principal sources of natural chlorine and bromine. The impact of halogens on the ozone layer is addressed in chapter 10.

9.7.4 Dust

Iron has been recently identified as a limiting nutrient in the ocean. Since iron in its trivalent oxidized form is insoluble in water, it is difficult to transport it to the open oceans. Atmospheric dust is the major source of iron to large expanses of the world's oceans. Figure 9.19 presents the dust concentrations in the atmosphere. The largest sources of dust are in the northern hemisphere, mainly in the Sahara Desert. The southern hemisphere has less continental area, so the dust contents of the atmosphere are much lower than those in the northern hemisphere.

The role of dust in climate change on Earth is poorly understood, but dust (in greater concentrations) is known to play a crucial role in the Martian atmosphere. It is of interest to note that the atmospheric dust loading increased by more than an order of magnitude during the LGM, as shown in figure 9.11. What caused the changes and what the consequences were of having enhanced amounts of dust in the atmosphere remain unanswered at present.

9.8 Unsolved Problems

We list a number of outstanding unresolved problems related to the terrestrial atmosphere:

1. Is Earth's physical and chemical environment regulated by Gaia? Are the physical, chemical, biochemical, and geochemical processes part of Gaia or a contradiction of Gaia?
2. What determines the amount of O_2 in the atmosphere? What has been the amount of O_2 throughout the history of the planet?
3. What sets the ultimate limit of the primary productivity of the biosphere? What are the historical values of primary productivity? Is phosphorus the ultimate limiting nutrient for the biosphere?
4. What was the escape rate of hydrogen from the planet in the past?
5. What determined the composition of the atmosphere during the LGM? How was more than 200 Gt-C transferred between the atmosphere and the oceans from glacial to interglacial periods?
6. Does the biosphere regulate the climate via production of DMS?
7. Does the biosphere regulate the ozone layer via production of methyl halides?
8. The nitrogen cycle has a minor by-product, N_2O. To what extent does the biosphere use N_2O to regulate the ozone layer?
9. The carbon cycle has a minor by-product, CH_4. To what extent is the chemistry of the troposphere and the stratosphere regulated by the biosphere via CH_4?
10. Is halogen chemistry an important part of the chemistry of the troposphere?
11. Was the ozone layer modulated by organic halogens before the rise of human influence?
12. What is the role of dust in supplying minerals to the oceans and in climate change?

10

Earth

Human Impact

10.1 Introduction

For eons the global environment of the planet has shaped the biosphere, and has been in turn shaped by the biosphere. According to the Gaia hypothesis, the overall impact of the biosphere on the global environment has been beneficial for the development and sustenance of life. However, this harmonious relationship between the biosphere and the environment has been disturbed with the emergence of one species, *Homo sapiens*, in the biosphere in the last million years. Our species has the potential to cause major disruptions in the global environment, thus threatening the integrity of the biosphere and its own survival. Some of these adverse effects are already known. The crucial challenge facing the future of humanity is to achieve a fundamental understanding of the global environment and to arrive at a new harmony between ourselves and nature.

The adverse impact of humans on the local environment has been known for some time in human history. Urban pollution is an ancient problem. Land degradation and destruction of natural habitats were the probable causes of the earliest recorded migration of the Chinese people during the Shang Dynasty around 1500 B.C. (about the time of Moses) in the valleys of the Yellow River. However, until recently there has been relatively little anthropogenic impact on the global environment. There are at least two major global environmental problems that have been identified to date: the CO_2 greenhouse effect and the global ozone depletion. These problems lie at the heart of what makes Earth a habitable planet. As discussed in chapter 9, Earth's atmosphere is responsible for a greenhouse effect of about 30°C, without which the surface of the planet would be too cold to allow water to flow. A doubling of atmospheric

CO_2 would increase the mean surface temperature of the planet by 2–3 °C. There would also be major shifts in the patterns of precipitation. Although the anticipated climatic changes caused by CO_2 are small compared to the variations in climate in the geological history of the planet, the rate of change is unprecedented and could result in major social and economical disruptions. As discussed in chapter 9, life on land became possible only after an ozone shield had developed. As far as we know, no advanced living organism can survive the harsh radiation environment on Earth's surface in the absence of this ultraviolet screen. In 1985 the Antarctic ozone hole was discovered. During the month of October, the column abundance of ozone in the polar vortex dropped by more than 50% as compared with climatological values in the 1950s. New record lows were set almost every year since 1985. The cause of the ozone hole is now known: the catalytic destruction of ozone by chlorine derived from chlorofluorocarbons (CFCs). There is evidence that similar but weaker ozone depletion also occurs in the spring of the northern polar region. Satellite data also show that there is ozone depletion on the decadal time scale. The total amount of chlorine in the atmosphere is only of the order of parts per billion by volume (ppbv), and yet it exerts a decisive control over the ozone layer. This is part of the beauty and subtlety of atmospheric chemistry that we wish to elucidate in this chapter, which emphasizes the above two global environmental problems. Other related problems are discussed only when they are important to these two fundamental issues.

10.1.1 Population and Standard of Living

We should note that primitive humans had limited impact on the global environment. The global environmental problems arose only when we developed "advanced" civilizations, first agriculture and then industry. These advances resulted in a rapid growth in the world population as well as a higher standard of living for the average person. Figure 10.1 shows the growth of population in historical times. The world population was small until the development of agriculture, after which it increased substantially. The next dramatic increase occurred after the industrial revolution. Today the total world population is about 5 billion and is expected to exceed 10 billion by the end of the next century. The total living human biomass, about 0.1 Gt-C (1 gigaton = 10^{15} g), is insignificant compared with the land biomass of 450 Gt-C (figure 9.4). However, the demand for natural resources by modern humans is very large. Figure 10.2 shows the per capita energy consumption throughout human history. Primitive humans only needed to eat, with a per capita requirement of 2500 kcal/day. This threshold of human subsistence is equivalent to 125 W (the power of a small light bulb). With the advance of human civilization, the food requirement per capita did not increase greatly, but the energy requirement increased dramatically. As shown in figure 10.2, the bulk of energy in modern society is spent on housing, industries, and transportation. The per capita rate of energy use is close to 125,000 kcal/day (7 kW), or 50 times that of primitive humans. Thus, the problem facing the future of humankind is not just the increase of population but also the increase in the standard of living.

The human energy requirement may best be appreciated by comparing it with the most abundant renewable energy source—solar energy. The planetary average solar constant is 340 W m^{-2}, of which 240 W m^{-2} is absorbed. Since the average human being has a cross section of 0.1 m^2, the total available solar power is 24 W, which

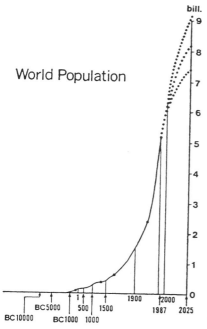

Figure 10.1 Growth of world population since prehistoric times. The vertical axis is in billions. After Kondo, J., 1991, "Plenary Lecture: Research and Policy on the Environment in Japan," in *The Global Environment*, p. 1. K. Takeuchi and M. Yoshino, editors (Berlin: Springer-Verlag).

is substantially below the subsistence threshold. Therefore, even at the subsistence level, humans must obtain food from the biosphere. In the 1980s, the global annual grain yield was 1.5 Gt. In comparison, the total catch of fish was only 0.1 Gt in 1988. The energy requirements of primitive humans were rather modest and could be met by using renewable resources (such as wood) from the biosphere. However, the enormous demands of modern humans cannot be met by the biosphere alone. Today, the bulk of the energy demand of modern society is satisfied by burning fossil fuels, remnants of the planet's biosphere from the ancient past (hundreds of million of years ago). The heat of combustion of fossil fuel (mostly carbon) is 10^4 cal g^{-1}. One single human requires the burning of 4.5 tons of fossil fuel per year to satisfy energy needs (if all the energy is derived from fossil fuels). This has resulted in a per capita emission of 16 tons of CO_2 per year into the atmosphere. Figure 10.3 illustrates the statistics of CO_2 emission by representative countries in 1985. Note that the world's average is about 4 tons of CO_2 per capita per year. The total emission in 1985 was close to 5 Gt-C, a value that must be compared with the carbon cycle shown in figure 9.4. The cumulative anthropogenic emission of CO_2 is 100 Gt-C in 20 years. Note that this is a nontrivial fraction of the total of 750 Gt-C of atmospheric CO_2 shown in figure 9.4. This demonstrates our ability to change the concentration of a major atmospheric constituent. We note that the global average energy consumption per capita is four times less that of the world's most advanced society, the United States of America. Thus, there is considerable pressure for the world average value to grow as the developing countries acquire the knowledge and the capital to modernize. The problem will obviously be exacerbated by population growth.

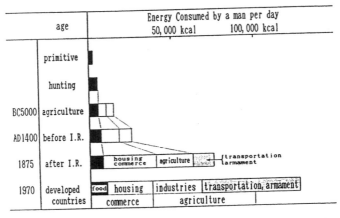

Figure 10.2 Energy requirements for a human being from primitive to modern times. I. R. = Industrial Revolution. After Kondo, J., 1991, "Plenary Lecture: Research and Policy on the Environment in Japan," in *The Global Environment*, p. 1. K. Takeuchi and M. Yoshino, editors (Berlin: Springer-Verlag).

One of the amenities of a high standard of living is the ability to regulate the temperature of the local microenvironment for comfort and for food storage using air conditioning and refrigeration. By the middle of this century, this technology had been perfected. A fluid is needed for the exchange of heat between the microenvironment and the outside atmosphere. The most widely used fluids are CFC-11 ($CFCl_3$) and CFC-12 (CCl_2F_2), manufactured from methane and halogens. There appear to be no natural sources of these compounds. They have the ideal property of being nontoxic and nonflammable. Figure 10.4 shows their steady production in recent history. The recent release rates are 350 and 450 kt yr^{-1} for CFC-11 and CFC-12, respectively. This amounts to the release of less than 0.2 kg of CFCs per capita per year. But the extreme chemical inertness of the CFCs in the lower atmosphere proved to be disastrous. The only place where the CFCs are destroyed is in the stratosphere, where the compounds undergo photolysis, thereby releasing their chlorine atoms to the ozone layer. Subsequent catalytic chemistry leads to efficient destruction of ozone and the depletion of the ozone layer, the most dramatic manifestation of which is the annually occurring Antarctic ozone hole phenomenon.

The two most visible global environmental problems of the last two decades were both caused by the pursuit of a high standard of living. The cost to the atmosphere is the emission of the order of 10 tons of CO_2 and 1 kg of CFCs per capita per year. It is a somber fact that the majority of humans have not attained this high standard of living but are actively pursuing this worthwhile goal. There appear to be no simple answers to questions such as: What should be the upper limit of standard of living? Who has the right to pursue a higher standard of living? Our moral obligation to preserve the global planetary environment may be the only restraint we can impose on our otherwise unbounded desires.

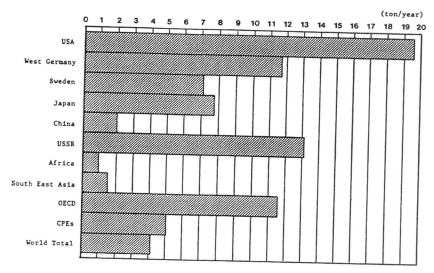

Figure 10.3 Per capita CO_2 emission in 1985. OECD = average of developed countries; CPEs = average of the former USSR and eastern Europe. After Kondo, J., 1991, "Plenary Lecture: Research and Policy on the Environment in Japan," in *The Global Environment*, p. 1. K. Takeuchi and M. Yoshino, editors (Berlin: Springer-Verlag).

In the case of ozone depletion the evidence that the ozone layer has been damaged is incontrovertible. The principal culprits (CFCs) have also been identified beyond reasonable doubt. This has resulted in a general consensus by the international community on the regulation of the emission of CFCs and related halogen compounds to the atmosphere. We discuss some of the details in section 10.4.4. In the case of global warming, the signal is much weaker, and it is harder to separate the anthropogenic perturbations from the natural fluctuations. Nevertheless, there is general consensus in the scientific community that warming between 0.5°C and 1 °C has occurred in the global mean temperature over the last century. However, it is much harder to argue that this warming is necessarily bad for the planet. Nor is it clear how regulation may be carried out for a source that is so diffuse and spread out over the globe.

10.1.2 Sustainable World

There are three different attitudes one can take about the future of humankind's happiness on this planet. There is Malthus, the pessimist; Lovelock, the optimist; the biblical Joseph, the rationalist. Malthus in his classic treatise published in 1798 predicted a pessimistic future for mankind facing population explosion and food shortage. But Malthus did not anticipate the development of science and technology for advancing agriculture. Today the crop yield per unit area of farmland is three times that of the time of Malthus through the use of chemical fertilizers and pesticides. There is potential for further improvement in the yields of agricultural products from genetic engineering. Thus, two centuries after Malthus, his dire predictions of food shortage

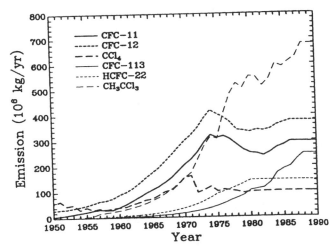

Figure 10.4 Rate of chlorofluorocarbon emissions since the 1950s. After Weisenstein, D. K., Ko, M. K. W., and Sze, N. D., 1992 "The Chlorine Budget of the Present-Day Atmosphere." *J. Geophys. Res.* **97**, 2547.

have not been realized. Lovelock, in contrast, takes a more optimistic view of the flexibility and adaptability of the biosphere as a whole, which he believes functions as a living organism. Some of his ideas are discussed in chapter 9 and are not repeated here. We believe that Lovelock may be correct, but only on a very long time scale, from thousands to millions of years. On the short time scale, of the order of decades and centuries, it is doubtful that the biosphere can respond to counteract adverse environmental perturbations. The last position, that of Joseph the rationalist, corresponds to that taken in this book. According to Genesis, Joseph predicted correctly the coming climatic changes and, by taking proper action in advance, steered the kingdom of Egypt safely through the period of hardship. The key elements of Joseph's success were correct prediction, personal integrity and credibility (based on his other actions in Egypt), authority, and decisive action.

It is unrealistic to require that the world be restored to its pristine state. In fact, it would be difficult to define this pristine state for the global environment. Over the history of the planet the global environment has undergone profound changes. In chapter 9 we pointed out the rise of O_2 in the atmosphere about 2 Gyr ago, when the atmosphere switched from the anaerobic to the aerobic state. Land plants developed about 450 Myr ago, providing the continents with a vegetative cover. It is hard to imagine that the atmosphere was anaerobic for half of the history of the planet and that the land biosphere has developed only in the last one-tenth of that history. Drastic environmental changes had occurred and yet the biosphere succeeded in adapting to the new global environment. A new sustainable world replaced the old one. In the terminology of Lovelock, this is known as homeorrhesis (note the fine distinction between this and homeostasis; see section 9.2.2). What will constitute a new sustainable world for the future of advanced civilization on this planet is not currently resolved. The problem will remain unresolved until our basic knowledge of the planet's land, atmosphere, and oceans becomes better defined.

It is perhaps fortunate that population growth and achieving a high standard of living do not reinforce each other. A country burdened with a large population cannot easily achieve a high standard of living, and a society that has achieved a high standard of living (and high level of literacy) usually does not increase its population. Hence, there is hope that a new level of equilibrium may be reached, as in the age of agriculture, with a stable world population on a sustainable planet.

10.2 Global Warming

We have only a short record of the quantitative measurements of the surface temperature of Earth based on instruments. Figure 10.5 shows the instrument-based mean surface temperature of the northern hemisphere, southern hemisphere, and the globe from 1861 to 1989 relative to the 1951–1980 climatological average. The data all consistently show an increasing trend. A linear fit between 1890 and 1989 gives values of 0.47°C (northern hemisphere), 0.53°C (southern hemisphere), and 0.50°C (globe) in 100 yr. There is a period of rapid increase in temperature during the 1920s. A cooling trend appears in the northern hemisphere data in the 1960s. Both hemispheres exhibit rapid warming trends since 1970. The deduced warming trend is consistent with the expected global warming due to the increase of greenhouse gases. The short-term fluctuations, such as that in the 1970s, are not understood but are believed to be associated with large-scale ocean dynamics.

In addition to an increase of the surface temperature, other changes are expected to be associated with global warming. These include changes in precipitation, cloudiness, soil moisture, and frequency of storms. We do not have a sufficiently accurate century-old global record of these climate variables to check for the "fingerprints" of global warming and to calibrate our climate models. The best evidence to date is circumstantial evidence based on a limited dataset of the temperature of the upper troposphere and the stratosphere. Figure 10.6 shows the temperature anomalies in the troposphere and the lower stratosphere from 1958 through 1989. Note that while there is no well-defined trend in the lower atmosphere, the upper troposphere and the lower stratosphere show trends of -0.5 and -2°C, respectively. This is consistent with the predictions of the greenhouse effect. The same molecules that raise the surface temperature of the planet by blocking the infrared emission are also efficient radiators in the upper atmosphere, where their enhanced radiation contributes to a cooling of the atmosphere. However, no quantitative comparison between the observations and model calculations has been made.

10.2.1 Greenhouse Gases

The principal greenhouse molecules in the terrestrial atmosphere are summarized in table 10.1, along with their current and preindustrial concentrations, rates of accumulation, and lifetimes. The five gases listed in table 10.1, CO_2, CH_4, CFC-11, CFC-12, and N_2O, are believed to be primarily responsible for the 100 yr temperature trend deduced from figure 10.5. Water vapor is actually the most important greenhouse

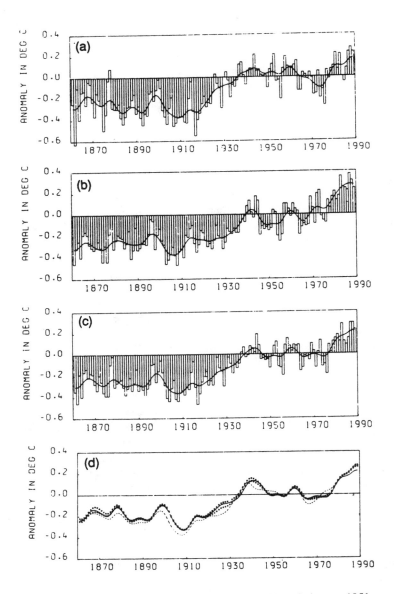

Figure 10.5 Mean surface temperature 1861–1989, relative to 1951–1980: (a) Northern hemisphere. (b) Southern hemisphere. (c) Globe. (d) Frozen grid analyses for the globe. After J. T. Houghton et al., 1991, editor, Intergovernmental Panel on Climate Change, 1990, *Climate Change: the IPCC Scientific Assessment*, and 1992, *Climate Change 1992: Supplement to the IPCC Scientific Assessment* (Cambridge: Cambridge University Press).

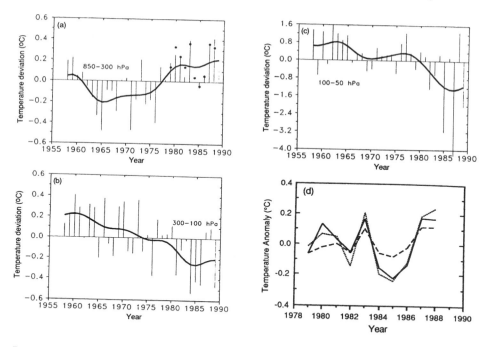

Figure 10.6 Temperature anomalies in the troposphere and lower stratosphere 1958–1989: annual global values for 850–300 mbar (a) and 300–100 mbar (b), and annual values for Antarctic (60° S to 90° S) for 100–50 mbar (c). d. annual global values from tropospheric satellite (solid), radiosonde (dot) and surface (dash) data. After J. T. Houghton et al., 1991, editor, Intergovernmental Panel on Climate Change, 1990, *Climate Change: the IPCC Scientific Assessment*, and 1992, *Climate Change 1992: Supplement to the IPCC Scientific Assessment* (Cambridge: Cambridge University Press).

molecule in the atmosphere. However, its concentration in the atmosphere is determined largely by the temperature itself. A change of 10°C would result in an order of magnitude change in the water vapor concentration in the atmosphere. Therefore, when we study the greenhouse effect of the atmosphere we do not consider water vapor as an independent input parameter. Its concentration is determined by the strong temperature feedback. A small fraction (of the order of 1%) of atmospheric water appears as clouds, which account for 20% of the reflectivity of the planet. Cloud feedback is a major uncertainty in climate modeling, as we change the temperature and the amount of water vapor in the atmosphere. Another major uncertainty is the response of the ocean currents to a change in the hydrological cycle.

The rise of greenhouse gases in the atmosphere is now briefly discussed. The increase of atmospheric CO_2 is due to industrial activity. The CFCs have no natural sources. Due to their long lifetimes (65–130 yr), almost all the released CFCs have remained in the atmosphere. Their presence will last well into the next century even if all production were to cease now. The increase of atmospheric CH_4 is mainly the

Table 10.1 Summary of key greenhouse gases influenced by human activities

Parameter	CO_2	CH_4	CFC-11	CFC-12	N_2O
Preindustrial atmospheric concentration (1750–1800)	280 ppmv	0.8 ppmv	0	0	288 ppbv
Current atmospheric concentration (1990)[a]	353 ppmv	1.72 ppmv	280 pptv	484 pptv	310 ppbv
Current rate of annual atmospheric accumulation	1.8 ppmv (0.5%)	0.015 ppmv (0.9%)	9.5 pptv (4%)	17 pptv (4%)	0.8 ppbv (0.25%)
Atmospheric lifetime[b] (yr)	(50–200)	10	65	130	150

From Intergovernmental Panel on Climate Change (1990).
Ozone is not included in the table because of lack of precise data. ppmv = parts per million by volume; ppbv = parts per billion by volume; pptv = parts per trillion by volume.
[a] The current (1990) concentrations have been estimated based upon an extrapolation of measurements reported for earlier years, assuming that the recent trends remained approximately constant.
[b] For each gas listed, except CO_2, the "lifetime" is defined here as the ratio of the atmospheric content to the total rate of removal. This time scale also characterizes the rate of adjustment of the atmospheric concentrations if the emission rates are changed abruptly. CO_2 is a special case since it has no real sinks, but is merely circulated between various reservoirs (atmosphere, ocean, biota). The "lifetime" of CO_2 given here is a rough indiction of the time it would take for the CO_2 concentration to adjust to changes in the emissions.

result of agricultural activities. The cause for the increase in N_2O remains obscure but is probably associated with the perturbation of the nitrogen cycle.

Figure 10.7, a and b, presents the data for atmospheric CO_2 in the last 250 yr. The pristine CO_2 concentrations are deduced from measurements of air bubbles trapped in ice in Antarctica. The measurements since 1958 are from air samples taken in Mauna Loa, Hawaii. The high temporal resolution of the latter dataset reveals the seasonal cycle of CO_2 caused by photosynthesis, respiration, and decay. These figures suggest that the preindustrial CO_2 concentration was 280 ppmv and that there has been a steady rise of CO_2 in the atmosphere. The pace accelerated during this century such that by 1990, the mean CO_2 concentration reached 353 ppmv, or 26% over the preindustrial value. The cause of this increase of CO_2 is known: The principal source is combustion of fossil fuels. Figure 10.7c shows the global annual emissions of CO_2 since 1860. The average rate of increase in emissions between 1860 and 1910 and between 1950 and 1970 was about 4% per year. In the 1980s the mean emission rate was between 5.3 and 5.7 Gt-C yr^{-1}. The cumulative release (fossil fuel plus cement manufacturing) from 1850 to 1987 is estimated to have been 200 ± 20 Gt-C, a value that must be compared with the total of 750 Gt-C in atmospheric CO_2 today. In addition, land use and deforestation also contribute to the release of CO_2 to the atmosphere. As summarized in table 10.2, the mean rate of emission in the decade 1980–1989 was 7 Gt-C yr^{-1}. The accumulation in the atmosphere and uptake by the ocean account for 3.4 and 2.0 Gt-C yr^{-1}, respectively, amounting to a total of 5.4 Gt-C yr^{-1}. There is thus a missing 1.6 ± 1.4 Gt-C yr^{-1} that is not accounted for. The problem of the missing sink is not resolved at present. One possibility is the fertilization effect caused by the increase of CO_2 and the increasing use of chemical fertilizers (see chapter 9).

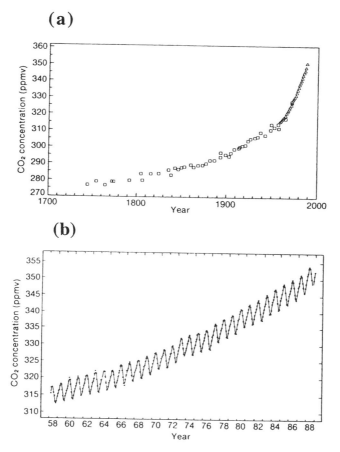

Figure 10.7 CO_2 concentrations for the past 250 years. (a) The early data are from air trapped in ice in Antarctica. The recent data data are from Mauna Loa, Hawaii, taken by Keeling et al. (1989). (b) Recent data (monthly mean) from Mauna Loa. After J. T. Houghton et al., 1991, editor, Intergovernmental Panel on Climate Change, 1990, *Climate Change: the IPCC Scientific Assessment*, and 1992, *Climate Change 1992: Supplement to the IPCC Scientific Assessment* (Cambridge: Cambridge University Press).

We have a hint of this possibility from the seasonal amplitude of the CO_2 variations shown in figure 10.7d. There is clearly an increase in the seasonal cycle of CO_2, suggesting enhanced biospheric productivity.

As summarized in table 10.1, the principal CFCs are CFC-11 and CFC-12. Table 10.3 lists an extended set of halocarbons that are important in the atmosphere. Although not important for the greenhouse effect, some of these compounds (e.g., those containing bromine) are known to affect ozone. They have been included for

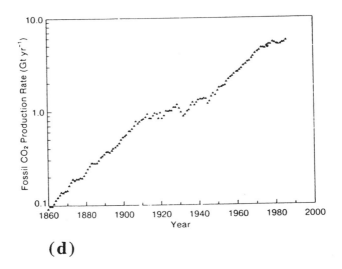

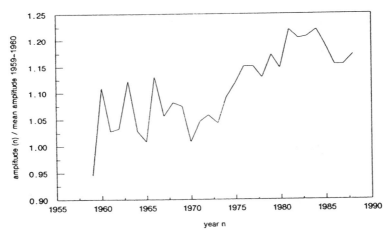

Figure 10.7 CO_2 concentrations for the past 250 years. (c) Global annual emission of CO_2 from fossil fuel combustion. After J. T. Houghton et al., 1991, editor, Intergovernmental Panel on Climate Change, 1990, *Climate Change: the IPCC Scientific Assessment*, and 1992, *Climate Change 1992: Supplement to the IPCC Scientific Assessment* (Cambridge: Cambridge University Press). (d) Relative amplitude of the seasonal cycle of CO_2 at Mauna Loa. After Bacastow, R. B., Keeling, C. D., and Whorf, T. P., 1985, "Seasonal Amplitude Increase in Atmospheric CO_2 Concentrations at Mauna Loa, Hawaii, 1959–1982." *J. Geophys. Res.* **90**, 10529.

Table 10.2 CO_2 emissions and accumulation in the atmosphere (Gt-C yr^{-1})

Emissions from fossil fuels into the atmosphere	5.4 ± 0.5
Emissions from deforestation and land use	1.6 ± 1.0
Accumulation in the atmosphere	3.4 ± 0.2
Uptake by the ocean	2.0 ± 0.8
Net imbalance	1.6 ± 1.4

From Intergovernmental Panel on Climate Change (1990).

future reference. Note that methyl chloride (CH_3Cl) is the major nonanthropogenic organic halogen in the atmosphere. The present atmosphere contains a total of 4.5 ppbv of chlorine. In the preindustrial era the total chlorine was 0.6 ppbv, all derived from CH_3Cl. Both CH_3Cl and CH_3Br are primarily produced by marine organisms and biomass burning. CH_3Br is also released from fumigants used in agriculture. The lifetimes of the halocarbons differ greatly. The fully halogenated species have longer lifetimes, in excess of tens of years. The ones containing a hydrogen atom (known as HCFCs) have much shorter lifetimes. The chemistry of halocarbons in the atmosphere is discussed in section 10.3. All halocarbons, with the exception of CH_3Cl and CH_3Br, are rapidly increasing in the atmosphere. Figure 10.8 shows the concentrations of the most important halocarbons measured at Cape Grim in Tasmania (a remote site) during the period 1978–1989. As stated above, the cause of the increase is known and can be quantitatively related to industrial emission rates shown in figure 10.4.

The concentration of CH_4 in the atmosphere since 1600 is shown in figure 10.9a. The early data are deduced from measurements of air bubbles trapped in ice in Greenland and Antarctica. Good atmospheric data are available only for the last two decades. There was a steady rise in the concentration of atmospheric CH_4 from the preindustrial value of 0.8 ppmv to 1.72 ppmv in 1990, with most of the increase occurring

Table 10.3 Halocarbon concentrations and trends (1990)

Halocarbon	Mixing ratio (pptv)	Annual rate of increase		Lifetime (yr)
		pptv	%	
CCl_3F (CFC-11)	280	9.5	4	65
CCl_2F_2 (CFC-12)	484	16.5	4	130
$CClF_3$ (CFC-13)	5			400
$C_2Cl_3F_3$ (CFC-113)	60	4–5	10	90
$C_2Cl_2F_4$ (CFC-114)	15			200
C_2ClF_5 (CFC-115)	5			200
CCl_4	146	2	1.5	50
$CHClF_2$ (HCFC-22)	122	7	7	15
CH_3Cl	600			1.5
CH_3CCl_3	158	6	4	7
$CBrClF_2$ (halon 1211)	1.7	0.2	12	25
$CBrF_3$ (halon 1301)	2	0.3	15	110
CH_3Br	10–15			1.5

From Intergovernmental Panel on Climate Change (1990).

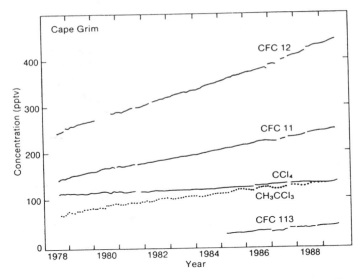

Figure 10.8 Halocarbon concentrations measured at Cape Grim, Tasmania, from 1978 to 1989. After J. T. Houghton et al., 1991, editor, Intergovernmental Panel on Climate Change, 1990, *Climate Change: the IPCC Scientific Assessment*, and 1992, *Climate Change 1992: Supplement to the IPCC Scientific Assessment* (Cambridge: Cambridge University Press).

in this century. Figure 10.9b shows details of the seasonal and latitudinal variations of CH_4 during the 1980s. These data reveal a latitudinal gradient from a high northern hemispheric value to a lower southern hemispheric value. The seasonal variations in the two hemispheres are 180° out of phase. The principal sources and sinks of CH_4 are summarized in table 10.4, along with estimates of uncertainties. The total source strength of CH_4 is 525 Tg yr^{-1} (1 Tg = 10^{12} g or 1 Mt), of which the natural sources (wetlands, termites, oceans, etc.) account for 175 Tg CH_4 yr^{-1}, or one-third of the total emission. All the known important sources are on land, and the oceans appear to be unimportant. The large rate of increase of atmospheric methane (0.9% per year) is clearly the result of accelerated human agricultural and industrial activities. This is consistent with the latitudinal gradient in CH_4 concentrations that reflects the dominance of the northern hemisphere as the source region.

As mentioned in chapter 9, CH_4 is an important part of the carbon cycle. Table 10.4 shows that this part of the carbon cycle has now been seriously perturbed by humans. In addition to being an important greenhouse molecule, CH_4 plays fundamental roles in the chemistry of the troposphere and the stratosphere, a subject that is discussed in section 10.3.2. One consequence of CH_4 chemistry in the stratosphere is the production of water vapor in the stratosphere. Thus, there is an additional radiative effect derived from the increase of H_2O in the upper atmosphere.

The concentrations of atmospheric N_2O for the past 2000 yr deduced from ice core samples are shown in figure 10.10a and recent data from direct sampling of

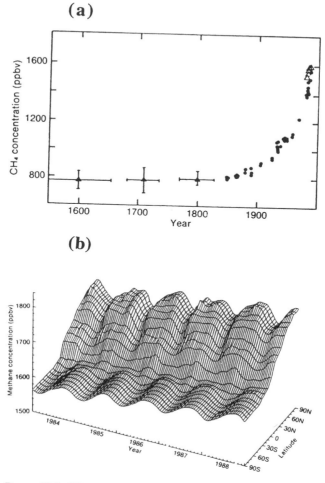

Figure 10.9 CH_4 concentrations in the atmosphere. (a) Variations in the past centuries from air trapped in ice. (b) Global distribution of atmospheric CH_4 and recent trend. After J. T. Houghton et al., 1991, editor, Intergovernmental Panel on Climate Change, 1990, *Climate Change: the IPCC Scientific Assessment*, and 1992, *Climate Change 1992: Supplement to the IPCC Scientific Assessment* (Cambridge: Cambridge University Press).

the atmosphere in figure 10.10b. The data reveal that the preindustrial concentration of N_2O was 288 ppbv and that there has been an increase in the last two centuries to the level of 310 ppbv in 1990. The high-precision modern data taken since 1977 show an increasing trend of 0.2–0.3% per year. The total atmospheric inventory of N_2O in 1990 was 1500 Tg-N. The only known major sink for N_2O is destruction by photolysis in the stratosphere, yielding a lifetime of 150 yr. The subsequent chemistry

Table 10.4 Estimated sources and sinks of methane (Tg CH_4)

	Annual Release	Range
Source		
Natural wetlands (bogs, swamps, tundra, etc.)	115	100–200
Rice paddies	110	25–170
Enteric fermentation (animals)	80	65–100
Gas drilling, venting, transmission	45	25–50
Biomass burning	40	20–80
Termites	40	10–100
Landfills	40	20–70
Coal mining	35	19–50
Oceans	10	5–20
Freshwaters	5	1–25
CH_4 hydrate destabilization	5	0–100
Sink		
Removal by soils	30	15–45
Reaction with OH in the atmosphere	500	400–600
Atmospheric increase	44	40–48

From Intergovernmental Panel on Climate Change (1990).

of N_2O in the stratosphere is important for the ozone layer, a subject that is discussed in section 10.4.3. In steady state a source strength of about 10 Tg-N yr^{-1} must be supplied by the Earth's surface. The sources and sinks of N_2O are summarized in table 10.5. There are still major uncertainties in our current understanding of the budget of N_2O, in particular in the role of the ocean. It is not clear whether N_2O is produced in the ocean from nitrification in the surface waters, or denitrification in the deep, oxygen-deficient waters. The oceanic reservoir of N_2O is estimated to be between 900 and 1100 Tg-N, comparable to that of the atmosphere. Therefore, any exchange between the atmosphere and the ocean would have a serious impact on the N_2O concentrations in the atmosphere.

As stated in chapter 9, N_2O is part of the nitrogen cycle that has been perturbed by human activities. Although we cannot quantify the causes for the observed increase in atmospheric N_2O, there is no doubt that this is the result of anthropogenic activities.

10.2.2 Greenhouse Effect

To first order the thermal properties of the atmosphere may be divided into two regimes. The stratosphere is in radiative equilibrium; that is, its temperature is determined by the balance between solar heating and infrared radiation. The temperature structure of the troposphere is determined by convection. The rate of change of temperature with altitude is known as the lapse rate, between -9 and -6 °C km^{-1}, depending on whether the air is dry or moist. Thus, a quantitative evaluation of the impact of a greenhouse molecule on the surface temperature of the planet is rather difficult. This kind of modeling is at best a highly parameterized procedure, making comparisons between models difficult. However, there is a simple quantity, known as radiative forcing, that is commonly adopted by the modeling community as a funda-

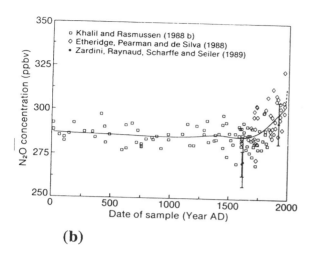

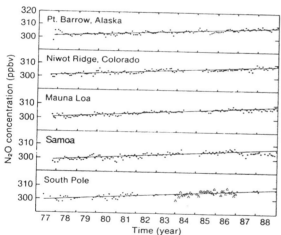

Figure 10.10 N$_2$O concentrations in the atmosphere. (a) Last two millennia from air trapped in ice. (b) Recent atmospheric measurements. After Intergovernmental Panel on Climate Change, 1990, *Climate Change: the IPCC Scientific Assessment*, and 1992, *Climate Change 1992: Supplement to the IPCC Scientific Assessment* (Cambridge: Cambridge University Press).

mental quantity for evaluating the impact of a greenhouse molecule in the atmosphere. This is defined as the change in net radiative flux ΔF (in W m^{-2}) at the tropopause corresponding to a volumetric change of trace species from C_0 to C:

$$\Delta F = f(C_0, C) \tag{10.1}$$

Note that the tropopause has been chosen because that is the region where the atmosphere makes a transition from radiative to convective equilibrium. The addition (or the increase) of a greenhouse gas to the atmosphere alters the downward radiative flux at the tropopause and would force the lower atmosphere (and the surface) to come to a new equilibrium.

Figure 10.11 presents the radiative forcing due to increases of greenhouse gases between 1765 and 1990, as summarized in table 10.6. Note that the radiative forcing

Table 10.5 Estimated sources and sinks of nitrous oxide

	Range (TgN per year)
Source	
Oceans	1.4–2.6
Soils	
tropical forests	2.2–3.7
temperate forests	0.7–1.5
Combustion	0.1–0.3
Biomass burning	0.02–0.2
Fertilizer (including ground-water)	0.01–2.2
Total	4.4–10.5
Sink	
Removal by soils	?
Photolysis in the stratosphere	7–13
Atmospheric increase	3–4.5

From Intergovernmental Panel on Climate Change (1990).

has increased by 2.5 W m^{-2} since the preindustrial era (table 10.7). The bulk of the contribution to radiative forcing is from CO_2, followed by smaller contributions from CH_4 and other trace gases. For comparison, we note that the radiative forcing due to a doubling of atmospheric CO_2 is 5 W m^{-2} and that the mean solar flux absorbed by the planet is 240 W m^{-2}. As we show in section 10.2.4, a radiative forcing of 2.5 W m^{-2} corresponds to an increase in the mean surface temperature of about 0.7°C, a value that is of the right magnitude to explain the secular increase in surface temperature over the last century (figure 10.5).

The computation of the radiative forcing involves detailed knowledge of molecular spectroscopy and radiative transfer. It is surprising from the above discussion that trace

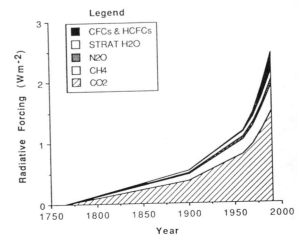

Figure 10.11 Changes in radiative forcing (W m^{-2}) due to increases in greenhouse gases since the preindustrial era. After J. T. Houghton et al., 1991, editor, Intergovernmental Panel on Climate Change, 1990, *Climate Change: the IPCC Scientific Assessment*, and 1992, *Climate Change 1992: Supplement to the IPCC Scientific Assessment* (Cambridge: Cambridge University Press).

Table 10.6 Trace gas concentrations from 1765 to 1990

Year	CO_2 (ppmv)	CH_4 (ppbv)	N_2O (ppbv)	CFC-11 (ppbv)	CFC-12 (ppbv)
1765	279.00	790.00	285.00	0	0
1900	295.72	974.10	292.02	0	0
1960	316.24	1272.0	292.62	0.0175	0.0303
1970	324.76	1420.9	298.82	0.0700	0.1211
1980	337.32	1569.0	302.62	0.1575	0.2725
1990	353.93	1717.0	309.68	0.2800	0.4844

From Intergovernmental Panel on Climate Change (1990).

gases such as CH_4, N_2O, and the CFCs can exert so much radiative forcing relative to CO_2. The reason is that the principal absorption bands of CO_2 at 15 μm are highly saturated, so additional CO_2 molecules are not efficient in causing further change in radiative forcing. Conversely, the other trace molecules have strong absorption bands in the 7–15 μm "window" region for terrestrial thermal radiation. On a per molecule basis relative to CO_2, the effectiveness of the trace species CH_4, N_2O, CFC-11, and CFC-12 are 21, 206, 12,400, and 15,800, respectively.

Our knowledge about the future of our global environment is much less certain than that of the past. To extrapolate into the future, we have to create a scenario. For example, we may define a "business as usual" scenario (scenario BaU), in which all the nations of the world decide on no regulation for the emission of greenhouse gases. This is obviously one extreme possibility. Another possibility is that the civilized nations of the world may all agree to some austere regulations to limit the emission of greenhouse gases after a certain target date (scenario D). The radiative forcings corresponding to these two scenarios between 1985 and 2100 are shown in figure 10.12, a and b. The year 1765 is taken as the reference year for zero radiative forcing. The estimated global mean temperature is shown in figure 10.12c. The predicted increases in radiative forcing (surface temperature) are 9.9 W m^{-2} (4.2°C) for scenario BaU and 4.3 W m^{-2} (2°C) for scenario D.

We should point out that there is a wide dispersion in the predicted effects due to an increase of greenhouse gases. The best-documented case is the extensive general circulation model (GCM) studies of the effect of a doubling of atmospheric CO_2. Figure 10.13 shows the predictions of 17 GCM models for mean surface warming and

Table 10.7 Forcing in Wm^{-2} due to changes in trace gas concentrations (W m^{-2})

Year	Sum	CO_2	CH_4 direct	Strat H_2O	N_2O	CFC-11	CFC-12	Other CFCs
1765–1900	0.53	0.37	0.10	0.034	0.027	0.0	0.0	0.0
1765–1960	1.17	0.79	0.24	0.082	0.045	0.004	0.008	0.005
1765–1970	1.48	0.96	0.30	0.10	0.054	0.014	0.034	0.021
1765–1980	1.91	1.20	0.36	0.12	0.068	0.035	0.076	0.048
1765–1990	2.45	1.50	0.42	0.14	0.10	0.062	0.14	0.085

From Intergovernmental Panel on Climate Change (1990).

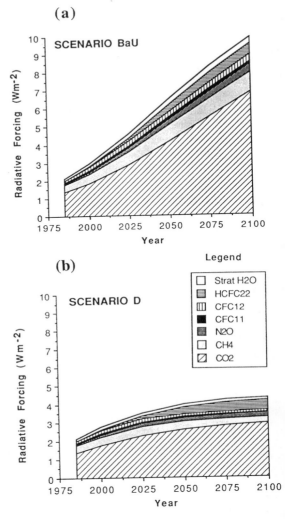

Figure 10.12 Projections of radiative forcing (W m^{-2}) in the next century. (a) Business as usual scenario. (b) Strict regulation scenario (□). After Intergovernmental Panel on Climate Change, 1990, *Climate Change: the IPCC Scientific Assessment*, and 1992, *Climate Change 1992: Supplement to the IPCC Scientific Assessment* (Cambridge: Cambridge University Press).

percentage change in precipitation. The range of temperature change for a doubling of CO_2 (radiative forcing of 5 W m^{-2}) is from 1.9°C to 5.2°C, with most of the model values lying between 3.5°C and 4.5°C. There is a corresponding range of (well-correlated) uncertainty in the change in global mean precipitation. The mean increase in precipitation is from zero to 15%, with most of the model results around 10%. The main uncertainties in the GCM studies arise from the lack of fundamental understanding of the hydrological cycle, in particular the role of clouds in climate change. It is unlikely that these uncertainties can be removed in the near future.

All GCMs predict greater changes at high latitudes in the winter. The time-latitude diagram of the zonally averaged temperature change in a particular model (model 11 in figure 10.13) is shown in figure 10.14. According to this model, the increase of the mean surface temperature is 4.2°C. Most of the increase occurs in the winter season,

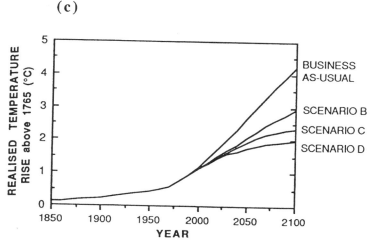

Figure 10.12 Projections of radiative forcing (W m^{-2}) in the next century. (c) Simulations of increase in global mean temperature from 1850 to 2100 for scenarios described in (a) and (b). (The other two cases correspond to intermediate scenarios.) After Intergovernmental Panel on Climate Change, 1990, *Climate Change: the IPCC Scientific Assessment*, and 1992, *Climate Change 1992: Supplement to the IPCC Scientific Assessment* (Cambridge: Cambridge University Press).

especially in the polar winter, where the changes can exceed 10°C. The changes in the summer season are relatively small. The models also predict a cooling of the stratosphere, as shown in figure 10.15 based on the same model as figure 10.14.

10.2.3 Aerosols

As discussed in section 9.7.1, the sulfur cycle is dominated by anthropogenic input of SO_2 from fossil fuel combustion and biomass burning. This may best be illustrated by the ratio of total sulfate simulated in a model to that in the natural background (figure 10.16). Note that the enhancement factor is as high as 10 over a wide area in the northern hemisphere and exceeds 2 for most of the northern hemisphere. The impact of this manmade aerosol layer on climate is difficult to assess. A recent model estimated that the net effect was to reduce the radiative forcing due to the increase of greenhouse molecules in the last century from 2.2 W m^{-2} to 1.7 W m^{-2}. This may account for the slight difference in the rate of global warming between the northern and the southern hemispheres. The climatic impact associated with volcanic aerosols is well documented. For example, the recent eruptions of Agung (1962), El Chichon (1982), and Pinatubo (1991) injected large amounts of SO_2 into the stratosphere. The subsequent formation of a sulfate layer in the stratosphere caused a cooling of Earth's surface. Figure 10.17 shows the observed and modeled temperature changes. The general agreement between model and observations provides good circumstantial evidence for the validity of the climate models.

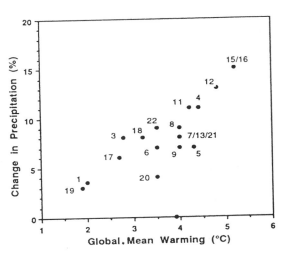

Figure 10.13 Changes in precipitation and surface temperature predicted by 17 models. The numbers refer to different models. After Intergovernmental Panel on Climate Change, 1990, *Climate Change: the IPCC Scientific Assessment*, and 1992, *Climate Change 1992: Supplement to the IPCC Scientific Assessment* (Cambridge: Cambridge University Press).

10.2.4 Climate of Recent Past

Given the complexity of the climate system, the study of the past provides a useful guide to the future. In chapter 9 we discuss the climate history of the planet over geological time in terms of compositional changes and the greenhouse effect. However, our knowledge of Earth in the remote past is very uncertain. Thus, most of the modeling of paleoclimate does not constitute a validation of our current climate models. The exception may be the last glacial maximum (LGM) about 18 kyr ago and the penultimate glacial maximum (PGM) about 140 kyr ago. These periods are sufficiently

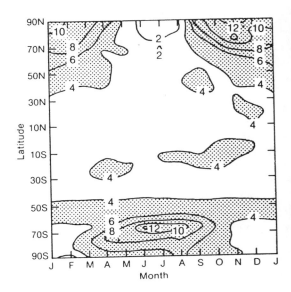

Figure 10.14 Surface temperature changes at different latitudes and time of the year predicted by the GISS model. The shaded areas have changes in excess of 4°C. After Hansen, J. E. et al., 1984, "Climate Sensitivity: Analysis of Feedback Mechanisms," in *Climate Processes and Climate Sensitivity*, J. E. Hansen and T. Takeuchi, editors (Geophysical Monogr. 29; Washington, D.C.: American Geophysical Union), p. 130.

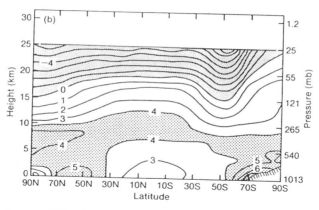

Figure 10.15 Altitude-latitude diagram of changes in zonally averaged temperatures predicted by the GISS model. After Hansen, J. E. et al., 1984, "Climate Sensitivity: Analysis of Feedback Mechanisms," in *Climate Processes and Climate Sensitivity*, J. E. Hansen and T. Takeuchi, editors (Geophysical Monogr. 29; Washington, D.C.: American Geophysical Union), p. 130.

close to modern times that a record of the paleoenvironments may be reconstructed from geological and geochemical data.

Figure 10.18 shows the local temperature and CH_4 and CO_2 concentrations deduced from Antarctic ice cores in the last 160 kyr. Note the generally good agreement between the temperature and the concentrations of the principal greenhouse gases, CO_2 and CH_4. For example, during the coldest periods (the LGM and the PGM), the temperatures were about 10°C lower than they were in the interglacial period. At the same time CH_4 concentrations fell from 600 to 380 ppbv; CO_2 concentrations decreased from 280 to 190 ppmv. Thus, there is a well-established correlation between temperature and the greenhouse gases. However, we must emphasize that this does not imply that changes in the greenhouse gases drive climatic changes. It is difficult to determine which is the cause and which is the effect. The only circumstantial evidence we have is that the CO_2 increase preceded the rise of temperature during the PGM (see section 9.4.4). Without inquiring into the ultimate cause of the ice ages, we can ask a more restricted question: Can we account for the temperature changes in the LGM by the observed changes of the greenhouse gases? Quantitative modeling of the climate of the LGM has been carried out by GCM studies. The results of one model (the GISS model) incorporating CO_2 and other known changes as well as water vapor feedbacks are summarized in figure 10.19. The change in global mean temperature is 4.7°C. Note the large changes due to feedback of the hydrological cycle (water vapor, ice, and clouds). Change in CO_2 only accounts for 0.6°C. CH_4 has not been included in this model but is expected to contribute less than 0.2°C to global mean temperature. The change of the global mean surface temperature during the LGM is still controversial but is believed to have been between 4°C and 7°C. The GISS

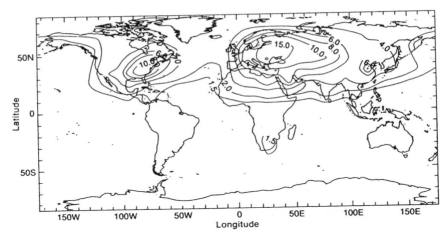

Figure 10.16 Ratio of simulated sulfate concentrations at 900 mbar based on total emission, to that due to natural emission. After Intergovernmental Panel on Climate Change, 1990, *Climate Change: the IPCC Scientific Assessment*, and 1992, *Climate Change 1992: Supplement to the IPCC Scientific Assessment* (Cambridge: Cambridge University Press).

model could account for the bulk of the temperature change between the LGM and the present.

As pointed out in section 9.4.4, the climate system in the past has exhibited rapid "quantum jumps." The nature of these sudden drastic changes in climate with time constants less than 100 yr is not well understood but is believed to be the result of a major reorganization of the ocean circulation system. None of the observed quantum behavior of paleoclimate can be explained by the greenhouse effect alone. Internal dynamical instabilities and strongly nonlinear coupling between the atmosphere and the ocean may be important parts of the intricate mechanism of climate. All these are poorly understood at the moment.

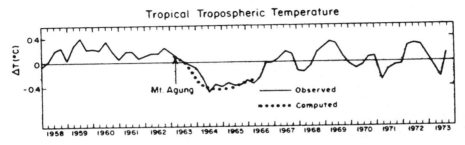

Figure 10.17 Perturbation of surface temperature due to stratospheric aerosols after the eruption of Mount Agung in 1963. After Hansen, J. E., Wang, W.-C., and Lacis, A. A., 1978, "Mount Agung Eruption Provides Test of a Global Climate Perturbation." *Science* **199**, 1065.

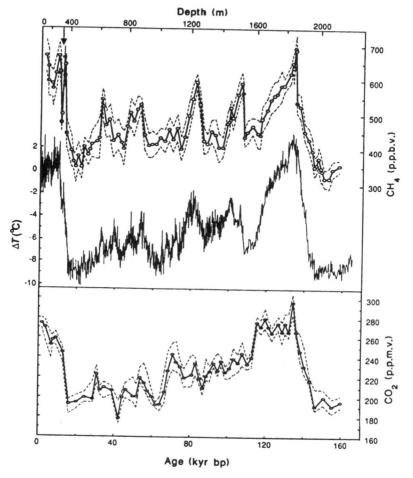

Figure 10.18 Variations in local temperature, CO_2, and CH_4 during the last 160,000 yr. The dashed lines indicate the range of uncertainty in the measurements. After Intergovernmental Panel on Climate Change, 1990, *Climate Change: the IPCC Scientific Assessment*, and 1992, *Climate Change 1992: Supplement to the IPCC Scientific Assessment* (Cambridge: Cambridge University Press).

10.3 Tropospheric Chemistry

The troposphere contains 90% of the mass of the atmosphere. Most of the trace species in the atmosphere have their origin in the continental or the oceanic biosphere. A large number of trace gases (e.g., CO, CH_4, C_2H_6, H_2, CH_3Cl, CH_3Br, H_2S, CH_3CCl_3) first undergo reactions in the troposphere. Only species that are not totally destroyed in the troposphere ultimately get transported to the stratosphere.

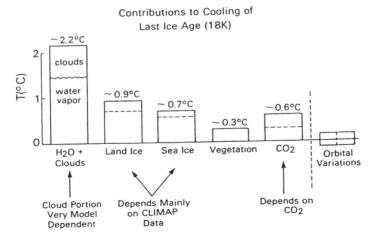

Figure 10.19 Contributions to the cooling of the global mean temperature during the last glacial maximum, based on the GISS model. After Hansen, J. E. et al., 1984, "Climate Sensitivity: Analysis of Feedback Mechanisms," in *Climate Processes and Climate Sensitivity*, J. E. Hansen and T. Takeuchi, editors (Geophysical Monogr. 29; Washington, D.C.: American Geophysical Union), p. 130.

10.3.1 Oxidants

The principal oxidant in the troposphere is the hydroxyl radical (OH). The mechanism of its formation is summarized in figure 10.20. The precursor molecule is O_3. Photolysis in the tail of the Huggins bands yields an excited oxygen atom $O(^1D)$:

$$O_3 + h\nu \to O_2(^1\Delta) + O(^1D) \tag{10.2}$$

Most of $O(^1D)$ atoms are quenched to the ground state by ambient air molecules. However, a small fraction will react with H_2O to form OH:

$$O(^1D) + H_2O \to OH + OH \tag{10.3}$$

The principal removal reactions for OH are

$$OH + CO \to H + CO_2 \tag{10.4}$$

$$OH + CH_4 \to H_2O + CH_3 \tag{10.5}$$

But these reactions are not permanent sinks of HO_x (H, OH, HO_2, and H_2O_2). The primary fate of the H atom produced in (10.4) is to form HO_2:

$$H + O_2 + M \to HO_2 + M \tag{10.6}$$

followed by

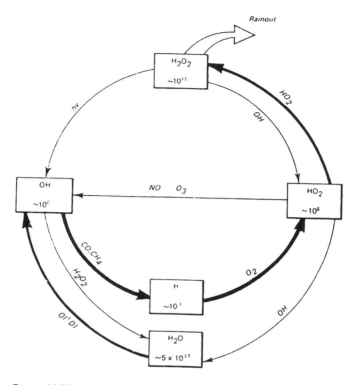

Figure 10.20 Major reaction pathways for HO_x radicals in the troposphere. The concentration given in each box is in molecules/cm³. Bold arrows denote major chemical pathways. Adapted from Sze, N. D., 1977, "Anthropogenic CO Emissions: Implications for the Atmospheric CO-OH-CH₄ Cycle." *Science* **195**, 673.

$$HO_2 + O_3 \rightarrow OH + 2O_2 \qquad (10.7)$$

$$HO_2 + NO \rightarrow OH + NO_2 \qquad (10.8)$$

$$HO_2 + HO_2 \rightarrow H_2O_2 + O_2 \qquad (10.9)$$

$$H_2O_2 + h\nu \rightarrow OH + OH \qquad (10.10)$$

Thus, the OH consumed in (10.4) is recycled by (10.6)–(10.10). There is, however, a permanent sink for HO_x if the H_2O_2 formed in (10.9) is removed by

$$H_2O_2 + OH \rightarrow H_2O + HO_2 \qquad (10.11)$$

$$H_2O_2 + \text{rain} \rightarrow \text{products} \qquad (10.12)$$

HO_x may also be removed via the formation of HNO_3:

$$OH + NO_2 + M \rightarrow HNO_3 + M \qquad (10.13)$$

followed by

$$HNO_3 + \text{rain} \rightarrow \text{products} \qquad (10.14)$$

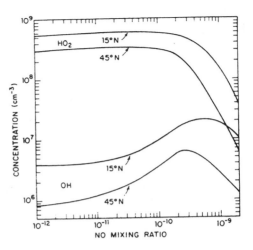

Figure 10.21 Concentrations of OH and HO_2 at noon at the surface as a function of NO abundance. The calculations were performed for equinox conditions. After Logan, J. A. et al., 1981, "Tropospheric Chemistry: A Global Perspective." *J. Geophys. Res.* **86**, 7210.

We have not followed the fate of the OH consumed in the oxidation of CH_4 in (10.5). This is the subject of section 10.3.2.

Since ozone is the precursor of OH, a natural question arises as to what is the origin of O_3 in the troposphere. Transport from the ozone layer in the stratosphere is an obvious source and probably accounts for a portion of ozone present in the remote background atmosphere in the southern hemisphere. There is, however, a simple chemical scheme for producing ozone in the troposphere, driven by the oxidation of CO to CO_2:

$$OH + CO \to H + CO_2 \qquad (10.4)$$

$$H + O_2 + M \to HO_2 + M \qquad (10.6)$$

$$HO_2 + NO \to OH + NO_2 \qquad (10.8)$$

$$NO_2 + h\nu \to NO + O \qquad (10.15)$$

$$\underline{O + O_2 + M \to O_3 + M \qquad (10.16)}$$

$$\text{net} \quad CO + 2O_2 \to CO_2 + O_3 \qquad (I)$$

Note that the HO_x radicals and NO_x (NO and NO_2) are used as catalysts in the production of O_3. The crucial step that breaks the O—O bond is reaction (10.8). In the clean background atmosphere the concentrations of NO_x are low and in situ production of O_3 is not important. However, in a polluted atmosphere CO and NO_x levels are high from industrial emissions, and scheme (I) is an important source of O_3.

Since NO_x is a catalyst for the production of O_3, we expect that a higher NO_x concentration would result in a higher concentration of HO_x. Figure 10.21 shows the concentrations of OH and HO_2 in a model as a function of NO mixing ratio. Note that OH increases with NO until the latter reaches the level of about 300 pptv. For higher concentrations of NO, the OH concentrations start to decrease because of the destruction of HO_x via the formation of HNO_3 (10.13), followed by rainout of HNO_3 (10.14).

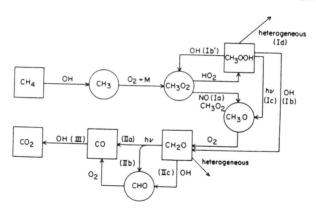

Figure 10.22 Major reaction pathways for the oxidation of atmospheric methane. After Logan, J. A. et al., 1981, "Tropospheric Chemistry: A Global Perspective." *J. Geophys. Res.* **86**, 7210.

Chemical scheme (I) implies that an O_3 molecule is produced for each CO molecule that is oxidized to CO_2. Thus, the production of O_3 (and hence the production of OH) is intimately connected to the oxidation of hydrocarbons, an important source of CO. This subject is pursued in section 10.3.2.

We should point out that in addition to the principal oxidant OH, the troposphere also contains other oxidants such as $O(^1D)$, NO_3, and H_2O_2. The concentrations of $O(^1D)$ in the atmosphere are too small to be important. NO_3 is formed primarily at night and may be important as a nighttime oxidant for higher hydrocarbons. H_2O_2 is unreactive in the gas phase but could be important in the chemistry of cloud drops. The possibility of a significant role for tropospheric halogen radicals such as Br and Cl has been suggested, but to date no definitive evidence supports this interesting hypothesis.

10.3.2 CH_4 and CO

CH_4 is the most abundant hydrocarbon in the terrestrial atmosphere. We discussed its sources and radiative properties in section 10.2.1. In this section we address its role in tropospheric chemistry, in particular its crucial relation to HO_x chemistry. Figure 10.22 is schematic diagram summarizing the principal chemical pathways leading to the complete oxidation of CH_4 to CO_2. The destruction of CH_4 is initiated by the reaction with OH (10.5). The methyl radical reacts with O_2 to form the peroxymethyl radical:

$$CH_3 + O_2 + M \rightarrow CH_3O_2 + M \tag{10.17}$$

There are three possible fates for the CH_3O_2 radical in the atmosphere:

$$CH_3O_2 + NO \rightarrow CH_3O + NO_2 \tag{10.18}$$

$$CH_3O_2 + HO_2 \rightarrow CH_3OOH + O_2 \tag{10.19}$$

$$CH_3O_2 + CH_3O_2 \rightarrow CH_3OH + H_2CO \tag{10.20}$$

where in (10.20) we have listed only the principal branch (60%) of CH_3O_2 disproportionation. The methyl hydroperoxy radical is removed by photolysis,

$$CH_3OOH + h\nu \rightarrow CH_3O + OH \qquad (10.21)$$

or by rainout,

$$CH_3OOH + rain \rightarrow products \qquad (10.22)$$

CH_3OH will most likely be rained out from the atmosphere. The primary fate of the methoxy radical is to react with O_2, yielding formaldehyde:

$$CH_3O + O_2 \rightarrow H_2CO + HO_2 \qquad (10.23)$$

H_2CO is removed by photolysis:

$$H_2CO + h\nu \rightarrow HCO + H \qquad (10.24a)$$
$$\rightarrow H_2 + CO \qquad (10.24b)$$

Additional removal mechanisms include reaction with OH and rainout:

$$H_2CO + OH \rightarrow HCO + H_2O \qquad (10.25)$$
$$H_2CO + rain \rightarrow products \qquad (10.26)$$

The formyl radical is removed by reaction with O_2:

$$HCO + O_2 \rightarrow CO + HO_2 \qquad (10.27)$$

The terminal product of the oxidation chain is CO_2, formed by reaction (10.4). There is a crucial question in the oxidation chain of CH_4 on whether the destruction of CH_4 is a source or sink of HO_x. This may be more easily seen if we carefully examine one possible chemical path of oxidation:

$$
\begin{array}{lr}
OH + CH_4 \rightarrow H_2O + CH_3 & (10.5) \\
CH_3 + O_2 + M \rightarrow CH_3O_2 + M & (10.17) \\
CH_3O_2 + NO \rightarrow CH_3O + NO_2 & (10.18) \\
CH_3O + O_2 \rightarrow H_2CO + HO_2 & (10.23) \\
H_2CO + h\nu \rightarrow HCO + H & (10.24a) \\
HCO + O_2 \rightarrow CO + HO_2 & (10.27) \\
OH + CO \rightarrow H + CO_2 & (10.4) \\
NO_2 + h\nu \rightarrow NO + O & (10.15) \\
O + O_2 + M \rightarrow O_3 + M & (10.16) \\
2(H + O_2 + M \rightarrow HO_2 + M) & (10.6) \\
\hline
\text{net} \quad CH_4 + 2OH + 6O_2 \rightarrow CO_2 + H_2O + O_3 + 4HO_2 & (IIa)
\end{array}
$$

This chemical scheme implies that each molecule of CH_4 consumes $2HO_x$ and produces O_3 and $4HO_x$. The net gain of odd hydrogen is $2HO_x$. In scheme (IIa), we have chosen branch (10.24a) for the photolysis of H_2CO. If we choose the other branch (10.24b) for the photolysis of H_2CO, the net result would be

$$CH_4 + 2OH + 4O_2 \rightarrow CO_2 + H_2O + H_2 + O_3 + 2HO_2 \qquad (IIb)$$

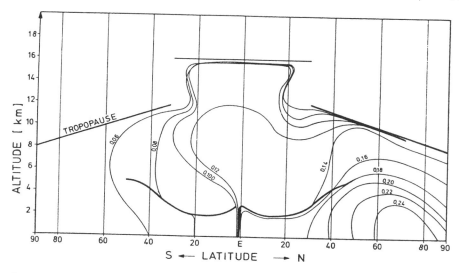

Figure 10.23 CO concentrations (ppmv) in the troposphere over the Atlantic Ocean. Thick lines represent the tropopause and the trade wind inversion. After Seiler, W., 1974, "The Cycle of Atmospheric CO." *Tellus* **26**, 116.

In this case one CH_4 molecule consumes $2HO_x$ and produces O_3 and $2HO_x$. There is no net gain or loss of HO_x. Chemical schemes (IIa) and (IIb) are but two of the many possibilities involving the interaction between CH_4, O_3, and HO_x. We do not explore all the consequences of the different pathways.

In section 10.3.1 we showed that the OH concentrations in the troposphere may be seriously perturbed by NO_x emission. The concentrations of OH may also be perturbed by industrial sources of CO via reaction (10.4). Figure 10.23 shows the concentrations of CO in the troposphere. The background atmosphere in the southern hemisphere contains 50 ppbv of CO. However, the mixing ratio in the midlatitudes in the northern hemisphere exceeds 200 ppbv. The major natural source of CO is oxidation of hydrocarbons, of which CH_4 is the most important single contributor. The major industrial source is the incomplete combustion of fossil fuels. Table 10.8 summarizes our current knowledge of the sources of atmospheric CO.

10.3.3 Tropospheric Lifetime

The lifetime of a molecule in the troposphere is crucial in determining its impact on atmospheric chemistry. Species with short lifetimes affect only local photochemistry. There is little impact away from the immediate source region. Species with lifetimes that are long compared with atmospheric transport can have an impact on the global environment.

We may roughly classify the molecules of the atmosphere into four types according to their lifetimes in the troposphere. The classification is not intended to be exclusive. The first type, which includes SO_2 and HNO_3, is highly soluble in water. The major loss mechanism is by rainout, so the mean lifetime in the atmosphere is on the order of

Table 10.8 Sources of carbon monoxide (Tg yr^{-1})

Source	Anthropogenic	Natural	Global	Range
Directly from combustion				
Fossil fuels	500	—	500	400–1000
Forest clearing	400	—	400	200–800
Savanna burning	200	—	200	100–400
Wood burning	50	—	50	25–150
Forest fires	—	30	30	10–50
Oxidation of Hydrocarbon				
Methane	300	300	600	400–1000
Nonmethane hydrocarbons	90	600	690	300–1400
Other Sources				
Plants	—	100	100	50–200
Oceans	—	40	40	20–80
Total	1500	1100	2600	2000–3000

From Intergovernmental Panel on Climate Change (1990).

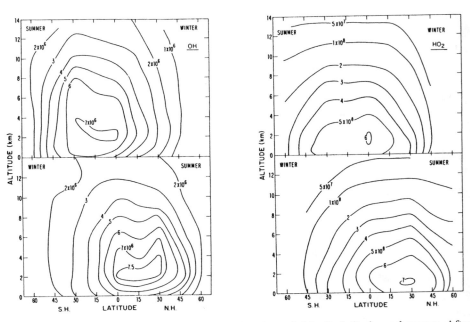

Figure 10.24 OH concentrations at noon as a function of altitude, latitude, and season. After Logan, J. A. et al., 1981, "Tropospheric Chemistry: A Global Perspective." *J. Geophys. Res.* **86**, 7210.

10 days. The second type, including CH_3I and H_2CO, dissociates at wavelengths above the O_3 cutoff at 310 nm. Since the solar flux at longer wavelengths is intense, molecules that have absorption bands at such wavelengths tend to have very short lifetimes. The third type, which includes CH_4 and CO, reacts with OH. Typical OH concentrations in the atmosphere computed using models that incorporate the chemistry described in sections 10.3.1 and 10.3.2 are shown in figure 10.24. Note that OH concentrations are of the order of 10^6 cm^{-3} and vary with season. The lifetime of a molecule that is destroyed by OH is usually in the range of months to years. Finally, there are molecules that are not removed by rainout, photolysis at long wavelength, or reaction with OH. These molecules have very long lifetimes. They are ultimately destroyed in the upper atmosphere above the ozone layer. The impact of these molecules on the chemistry of the stratosphere is discussed in section 10.4.

10.4 Stratospheric Chemistry

The existence of the ozone layer in the stratosphere is a unique feature of Earth. The bulk of ozone resides between the tropopause (about 15 km) and 35 km. The total mass of ozone in the atmosphere equals 4 Gt. The mass density of the average ozone column is 10^{-3} g cm^{-2}, equivalent to the mass density for a thin plastic sheet. The ozone layer is primarily responsible for filtering out harmful solar UV radiation below 300 nm. Figure 10.25a shows the transmissivity of Earth's atmosphere in the UV. Below 290 nm the UV radiation is known as UV-C, which the ozone layer completely filters out. UV-B (between 290 and 320 nm) is partially filtered by ozone. Beyond 320 nm the UV radiation is known as UV-A, most of which can reach Earth's surface. The relative biological damage due to exposure to UV radiation is shown in figure 10.25b. The response of the surface UV flux to a change in the overhead ozone column is nonlinear. Figure 10.25c shows the radiation amplification factor for DNA-effective

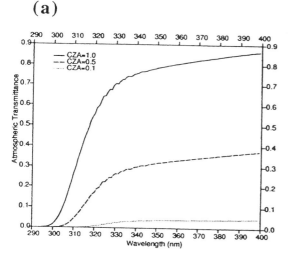

Figure 10.25 (a) Transmittance of the atmosphere as a function of wavelength. The results are for a typical ozone profile in January at 57° N for three cosines of the solar zenith angle (CZA). The column abundance of ozone is 350 Dobson units (DU). 1 DU corresponds to 10^{-3} cm of gas at standard temperature and pressure (S.T.P.). After Forster, P. M., and Shine, K. P., 1995, "A Comparison of Two Radiation Schemes for Calculating Ultraviolet Radiation" *Q. J. Roy. Meteorol. Soc.* **121**, 1113.

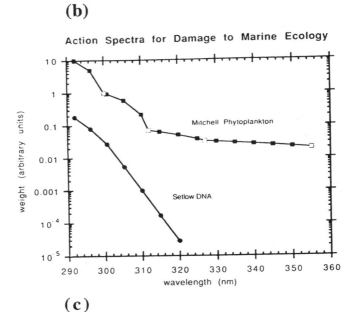
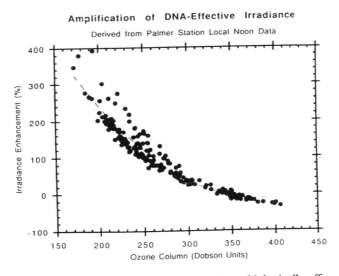

Figure 10.25 (b) Action spectra used to estimate biologically effective UV doses. The upper and lower curves refer to inhibition of photosynthesis in Antarctic phytoplankton and DNA damage, respectively. (c) The radiation amplification factor for DNA-effective surface irradiance computed for noon at Palmer Station during the Austral springs 1988 and 1990. (b) and (c) after Lubin, D. et al., 1992, "A Contribution Toward Understanding the Biospherical Significance of Antarctic Ozone Depletion." *J. Geophys. Res.* **97**, 7817.

surface irradiance computed for noon at Palmer Station during the Austral springs 1988 and 1990. Note that a 50% reduction in the ozone column results in a 300% enhancement in DNA-effective irradiance at the surface.

The absorption of UV energy by ozone heats the atmosphere in the ozone layer, creating a large temperature gradient. The thermal stratification of the stratosphere is in fact caused by ozone itself. One consequence of this stratification is the inhibition of vertical motion, thereby creating a stable and quiescent environment for the storage of ozone.

A fundamental understanding of the distribution of stratospheric ozone requires an understanding of photochemistry, radiation, and dynamics and their interactions. This is a problem of considerable intricacy and complexity, but due to recent intensive research efforts, considerable progress has been made. Despite the great advance of current knowledge, we believe that the future may continue to hold surprises. It is perhaps a sobering fact that the Antarctic ozone hole was never predicted before its discovery in 1985.

The crux of stratospheric chemistry is concerned with the sources and sinks of O_3. The core of the photochemistry of O_3 is the Chapman chemistry involving pure oxygen species, O, O_2, and O_3. The importance of the catalytic chemistry of HO_x, NO_x, and halogens for destroying O_3 is now recognized. The problem acquires a deeper sense of urgency when some of these catalytic agents are known to be derived from our industrial activities. Heterogeneous chemistry plays an important role in the chemistry of the polar stratosphere. The recognition of our adverse impact on the ozone layer prompted an intensive effort in the photochemistry and chemical kinetics of simple molecules of importance to the stratosphere. This extensive database is summarized in four tables in the appendix of this chapter. Photolysis reactions in Earth's atmosphere are presented in table 10.A1. Binary and ternary reactions are given in tables 10.A2 and 10.A3, respectively. Equilibrium constants for selected reactions are listed in table 10.A4.

10.4.1 Chapman Chemistry

First proposed by Chapman in 1930, this theory gives the correct first-order explanation of the ozone layer, which was discovered at the time. The basic concepts of the Chapman theory are as beautiful as they are simple. Consider an atmosphere containing O_2. Absorption of solar UV radiation below 240 nm leads to photolysis:

$$O_2 + h\nu \rightarrow O + O \qquad J_1 \qquad (10.28)$$

The O atom combines with O_2 in a three body reaction forming O_3:

$$O + O_2 + M \rightarrow O_3 + M \qquad k_1 \qquad (10.16)$$

Ozone is removed by photolysis:

$$O_3 + h\nu \rightarrow O_2 + O \qquad J_2 \qquad (10.29)$$

This is the reaction that is responsible for absorbing the bulk of UV solar radiation in the stratosphere. Note that reaction (10.29) is not a permanent sink for ozone, because the most likely fate of the O atom it produces is to react with O_2 via reaction (10.16) and restore O_3. It is convenient to define odd oxygen, O_x, as the sum of O and O_3. Reactions such as (10.16) and (10.29) turn one odd oxygen into another odd oxygen. There is no net production or destruction of O_x. We may consider the (chemically) net nothing cycle as a "catalytic cycle" for converting solar UV energy into thermal energy:

$$O + O_2 + M \rightarrow O_3 + M \quad (10.16)$$
$$O_3 + h\nu \rightarrow O_2 + O \quad (10.29)$$
$$\text{net} \quad \text{UV radiation} \rightarrow \text{thermal energy}$$

The beauty of the ozone layer is that the absorber (O_3) is not efficiently destroyed by the UV radiation, and this accounts for the extraordinary effectiveness of ozone for blocking UV radiation. Ultimately ozone (odd oxygen) is removed by the reactions

$$O_3 + O \rightarrow O_2 + O_2 \qquad k_2 \qquad (10.30)$$
$$O + O + M \rightarrow O_2 + M \qquad k_3 \qquad (10.31)$$

The set of five reactions listed above, (10.16) and (10.28)–(10.31), is known as Chapman chemistry. It can be shown that reaction (10.31) is not important in the part of the stratosphere where ozone concentrations are maximum. In this case there are approximate analytic solutions for the concentrations of O and O_3 as defined by the Chapman chemistry:

$$[O] = \sqrt{\frac{J_1 J_2}{k_1 k_2 M}} \qquad (10.32)$$

$$[O_3] = \sqrt{\frac{k_1 J_1 M}{J_2 k_2}} [O_2] \qquad (10.33)$$

Note that the expression for O_3 concentration in (10.33) implies the existence of an ozone layer. At high altitudes, the values of M and $[O_2]$ are small, implying that $[O_3]$ will be small. In the deep atmosphere J_1 is small (all the photons are absorbed by the overlying O_2 column), and $[O_3]$ will also be small. Maximum values of $[O_3]$ are attained in the stratosphere.

Chapman chemistry provides a generally accurate model of ozone concentrations (to within a factor of 2) above the ozone peak. However, there are at least two weaknesses of the theory. First, in the lower stratosphere below the ozone peak the lifetime of O_3 becomes long compared with transport. Dynamics play a crucial role in determining the seasonal and latitudinal distributions of ozone. Figure 10.26 presents the zonally averaged observed column abundance of atmospheric ozone as a function of latitude and time of the year. Note the low column densities in the tropics (where O_3 is produced by Chapman chemistry) and the higher column densities in the polar regions. This distribution is the opposite of what a photochemical theory of ozone would predict. The spatial and temporal variations of ozone, as displayed in figure 10.26, are the result of the dynamical motion of the lower atmosphere.

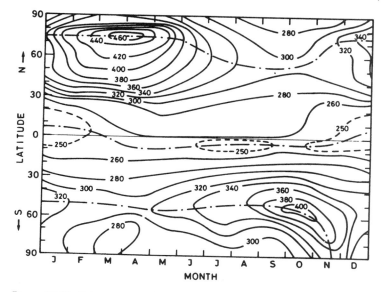

Figure 10.26 Column ozone abundances, in Dobson units (DU), as a function of latitude and season. After Brasseur, G., and Solomon, S., 1984, *Aeronomy of the Middle Atmosphere* (Boston: Dordrecht).

A curious feature of the Chapman theory is the extremely slow rate coefficient for reaction (10.30) that is responsible for removing odd oxygen:

$$k_2 = 8.0 \times 10^{-12} e^{-2060/T} \tag{10.34}$$

At the typical stratospheric temperature of 250 K, we have $k_2 = 2.1 \times 10^{-15}$ cm^3 s^{-1}, a value that is some four orders of magnitude below that of the gas kinetic rate. According to (10.33), we owe the existence of the ozone layer to the small value of k_2. If k_2 were at the gas kinetic limit value, the entire ozone layer would only be about 1% of the present ozone layer. The consequences would obviously be catastrophic for higher forms of life on this planet. Due to the small value of k_2, the loss of O_x in the Chapman scheme is inefficient. This is responsible for the existence of the ozone layer but it opens the possibility for alternative pathways for the destruction of O_x. The most important of these additional sinks of O_x is the catalytic chemistry to be described in sections 10.4.2–10.4.7.

10.4.2 HO_x Chemistry

Although water vapor is an abundant molecule in the troposphere, with mixing ratios in the range of 1%, the stratosphere is extremely dry. The air parcels that enter the stratosphere have to go through the tropical tropopause where the air is freeze-dried. Satellite data suggest that the air entering the stratosphere contains about 3 ppmv of H_2O. Another major source of water in the stratosphere is the oxidation of CH_4 derived from the biosphere. The mean mixing ratio of CH_4 in the troposphere is 1.7 ppmv.

Since CH_4 is long-lived and does not condense at the tropopause, it readily enters the stratosphere, where on oxidation [the chemistry is similar to schemes (IIa) and (IIb)] it is converted into 3.4 ppmv of H_2O. Thus, the concentration of H_2O increases with altitude in the stratosphere. HO_x radicals are generated by reaction (10.3) between $O(^1D)$ and H_2O, where the $O(^1D)$ is derived from the photolysis of O_3 (10.2). OH reacts with O and O_3:

$$OH + O \to O_2 + H \qquad (10.35)$$

$$OH + O_3 \to HO_2 + O_2 \qquad (10.36)$$

The H atom produced in (10.35) readily combines with O_2 via (10.6) to form HO_2, or it may react with O_3:

$$H + O_3 \to OH + O_2 \qquad (10.37)$$

HO_2 may also react with NO (10.8) or with O or O_3:

$$HO_2 + O \to OH + O_2 \qquad (10.38)$$

$$HO_2 + O_3 \to OH + 2O_2 \qquad (10.39)$$

The reaction

$$OH + HO_2 \to H_2O + O_2 \qquad (10.40)$$

provides a major sink for HO_x in the stratosphere. In the lower stratosphere, HO_x is also removed (catalytically) by reactions with oxides of nitrogen (see section 10.4.3 on the chemistry of odd nitrogen compounds), including (10.13) and

$$HO_2 + NO_2 + M \to HO_2NO_2 + M \qquad (10.41)$$

$$OH + HNO_3 \to H_2O + NO_3 \qquad (10.42)$$

$$OH + HO_2NO_2 \to H_2O + NO_2 + O_2 \qquad (10.43)$$

The catalytic destruction of ozone (O_x) by HO_x radicals may be summarized by three cycles:

$$\begin{array}{lrr} & OH + O \to O_2 + H & (10.35) \\ & H + O_3 \to OH + O_2 & (10.37) \\ \hline net & O + O_3 \to O_2 + O_2 & (\text{IIIa}) \end{array}$$

$$\begin{array}{lrr} & OH + O_3 \to HO_2 + O_2 & (10.36) \\ & HO_2 + O \to OH + O_2 & (10.38) \\ \hline net & O + O_3 \to O_2 + O_2 & (\text{IIIb}) \end{array}$$

$$\begin{array}{lrr} & OH + O_3 \to HO_2 + O_2 & (10.36) \\ & HO_2 + O_3 \to OH + 2O_2 & (10.39) \\ \hline net & O_3 + O_3 \to O_2 + O_2 + O_2 & (\text{IIIc}) \end{array}$$

HO_x catalysis, (IIIa) and (IIIb), is most important in the upper stratosphere. Scheme (IIIc) is more important in the lower stratosphere.

A schematic diagram summarizing the chemical pathways connecting the major HO_x species is given in figure 10.27a. The concentrations of the major HO_x species (OH, HO_2) are shown in figure 10.27b.

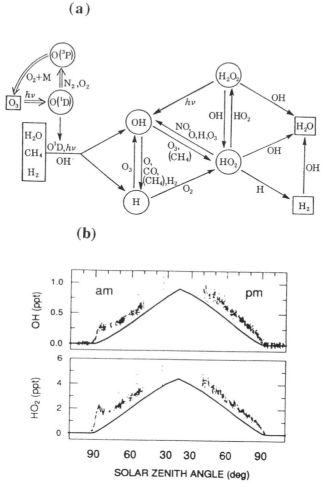

Figure 10.27 (a) Schematic diagram summarizing the HO_x species and major chemical pathways. After Logan, J. A. et al., 1978, "Atmospheric Chemistry: Response to Human Influence." *Phil. Trans. Roy. Soc. Lond.* **290**, 187. (b) Diurnal variations of OH and HO_2 measured at 19 km, 37°N, in May 1993, during the Stratospheric Photochemistry, Aerosols, and Dynamics Expedition (SPADE). The lines are calculations using a photochemical model. After Salawitch, R. J. et al., 1994 "The Diurnal Variation of Hydrogen, Nitrogen, and Chlorine Radicals: Implications for Heterogeneous Production of HNO_2." *Geophys. Res. Lett.* **21**, 2551.

10.4.3 Odd Nitrogen Chemistry

The NO_x that is derived from anthropogenic pollutants has a short lifetime in the troposphere and does not get transported to the stratosphere. The principal source of odd nitrogen in the stratosphere is derived from the long-ranged carrier N_2O, which is biologically produced (see section 10.2.1). In the stratosphere N_2O reacts with $O(^1D)$ to form nitric oxide:

$$N_2O + O(^1D) \rightarrow NO + NO \qquad (10.44)$$

There is a minor source of NO_x in the lower stratosphere arising from the absorption of galactic cosmic rays. This source is only about 10% of the N_2O source, but it is strongly modulated by the solar cycle. NO readily reacts with O_3:

$$NO + O_3 \rightarrow NO_2 + O_2 \qquad (10.45)$$

Whether this reaction results in a net destruction of O_3 depends on the fate of nitrogen dioxide. If NO_2 photolyzes via (10.15), then the net result of (10.45) and (10.15) is the conversion of one O_x (O_3) into another O_x (O). There is no net destruction of odd oxygen. However, if NO_2 reacts with O, there will be a net loss of O_x:

$$\begin{aligned} NO + O_3 &\rightarrow NO_2 + O_2 & (10.45)\\ NO_2 + O &\rightarrow NO + O_2 & (10.46)\\ \hline \text{net} \quad O + O_3 &\rightarrow O_2 + O_2 & (IV) \end{aligned}$$

There are a number of more complex odd nitrogen compounds that can be formed from NO_2: nitrogen trioxide (NO_3), and dinitrogen pentoxide (N_2O_5)

$$NO_2 + O_3 \rightarrow NO_3 + O_2 \qquad (10.47)$$

$$NO_2 + NO_3 + M \rightarrow N_2O_5 + M \qquad (10.48)$$

and nitric acid (HNO_3) (10.13) and pernitric acid (HO_2NO_2) (10.41). These compounds are chemically less reactive than NO and NO_2 and serve as reservoir species for odd nitrogen. They are removed by photolysis and thermal decomposition,

$$NO_3 + h\nu \rightarrow NO_2 + O \qquad (10.49)$$

$$HNO_3 + h\nu \rightarrow NO_2 + OH \qquad (10.50)$$

$$HO_2NO_2 + h\nu \rightarrow NO_2 + HO_2 \qquad (10.51)$$

$$N_2O_5 + M \rightarrow NO_2 + NO_3 + M \qquad (10.52)$$

and by reactions with OH, (10.42) and (10.43). The heterogeneous conversion of N_2O_5 to HNO_3 is discussed in section 10.4.7 on the Antarctic ozone hole.

The major chemical sink of odd nitrogen in the stratosphere is the reaction

$$NO + N \rightarrow N_2 + O \qquad (10.53)$$

where the N atom is derived from

$$NO + h\nu \rightarrow N + O \qquad (10.54)$$

This sink is important in the upper stratosphere where NO photolysis can occur. The bulk of odd nitrogen in the lower stratosphere has no sink via gas phase chemical reactions. The major removal mechanism for odd nitrogen is via transport to the troposphere. Heterogeneous removal on the surface of aerosols and the polar stratospheric clouds (PSCs) may also contribute to its loss.

The major chemical pathways determining the interconversion between odd nitrogen compounds are summarized in the schematic diagram in figure 10.28a; the concentrations of odd nitrogen species are shown in figure 10.28b.

10.4.4 Chlorine Chemistry

The major carriers of chlorine to the stratosphere are CH_3Cl and the CFCs, discussed in sections 10.2.1 and 10.2.2. The chemistry of the halocarbons following their breakdown (by photolysis or reaction with reactive radicals such as OH) in the stratosphere is similar to that for CH_4 in the troposphere described in section 10.3.2. Here we examine the breakdown of CF_2Cl_2 as an example. In the stratosphere the molecule undergoes photolysis, releasing a Cl atom:

$$CF_2Cl_2 + h\nu \rightarrow CF_2Cl + Cl \qquad (10.55)$$

The radical CF_2Cl can now be readily destroyed in the following sequence of reactions:

$$CF_2Cl + O_2 + M \rightarrow CF_2ClO_2 + M \qquad (10.56)$$

$$CF_2ClO_2 + NO \rightarrow CF_2ClO + NO_2 \qquad (10.57)$$

$$CF_2ClO + O_2 \rightarrow COF_2 + ClO_2 \qquad (10.58)$$

Note that the destruction of CF_2Cl_2 results in the production of Cl and ClO. The ultimate fate of COF_2 is photolysis, followed by the formation of HF, a terminal product in the stratosphere. The release of the reactive form of chlorine as Cl and ClO leads to the catalytic destruction of ozone by

$$\begin{array}{ll} Cl + O_3 \rightarrow ClO + O_2 & (10.59) \\ \underline{ClO + O \rightarrow Cl + O_2} & (10.60) \\ net \quad O + O_3 \rightarrow O_2 + O_2 & (V) \end{array}$$

The efficiency of chlorine for destroying O_3 is mitigated somewhat by its reaction with NO:

$$ClO + NO \rightarrow Cl + NO_2 \qquad (10.61)$$

Note that the cycle

$$\begin{array}{ll} Cl + O_3 \rightarrow ClO + O_2 & (10.59) \\ ClO + NO \rightarrow Cl + NO_2 & (10.61) \\ \underline{NO_2 + h\nu \rightarrow NO + O} & (10.15) \\ net \quad O_3 \rightarrow O + O_2 & \end{array}$$

converts one O_x into another O_x and is a null cycle for the destruction of odd oxygen.

The active forms of chlorine may be converted into chemically less active reservoir species by

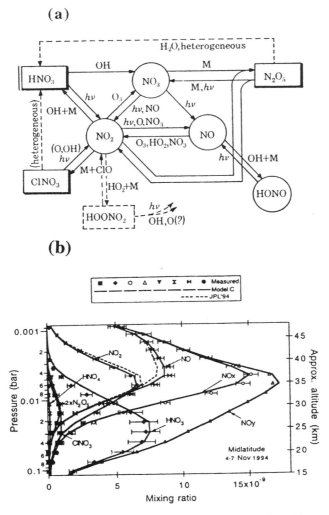

Figure 10.28 (a) Schematic diagram summarizing the odd nitrogen species and major chemical pathways. The dashed line indicates conversion of N_2O_5 by heterogeneous chemstry. After Logan, J. A. et al., 1978, "Atmospheric Chemistry: Response to Human Influence." *Phil. Trans. Roy. Soc. Lond.* **290**, 187. (b) Partitioning of odd nitrogen species in the stratosphere. Symbols represent measurements made by the Atmospheric Trace Molecule Spectroscopy (ATMOS) instrument during the ATLAS-3 Space Shuttle mission at midlatitudes on 4–7 November 1994. After Michelsen, H. A. et al., 1996, "Stratospheric Chlorine Partitioning: Constraints From Shuttle-Borne Measurements of HCl, $ClNO_3$, and ClO." *Geophys. Res. Lett.* **23**, 2361.

$$Cl + CH_4 \rightarrow HCl + CH_3 \tag{10.62}$$
$$Cl + HO_2 \rightarrow HCl + O_2 \tag{10.63}$$
$$ClO + HO_2 \rightarrow HOCl + O_2 \tag{10.64}$$
$$ClO + NO_2 + M \rightarrow ClONO_2 + M \tag{10.65}$$

The reservoir species do not react with O_3. They can be converted back to the reactive radical species by the reactions

$$OH + HCl \rightarrow H_2O + Cl \tag{10.66}$$
$$HOCl + h\nu \rightarrow OH + Cl \tag{10.67}$$
$$ClONO_2 + h\nu \rightarrow Cl + NO_3 \tag{10.68a}$$
$$\rightarrow ClO + NO_2 \tag{10.68b}$$

A schematic diagram showing the interconversion between the inorganic chlorine species is shown in figure 10.29a. The concentrations of the principal inorganic chlorine species are shown in figure 10.29b.

10.4.5 Bromine Chemistry

The major carrier of bromine to the stratosphere is CH_3Br. The sources of CH_3Br are production in the oceans and anthropogenic emissions. It is mainly destroyed by OH in the troposphere, but enough CH_3Br survives to provide a significant source of bromine to the stratosphere. The breakdown of CH_3Br is believed to be similar to that of CH_4 (see section 10.3.2), and the details are not shown here. With the release of the reactive radicals, Br and BrO, ozone may be destroyed by the following synergistic coupling between bromine and chlorine:

$$Br + O_3 \rightarrow BrO + O_2 \tag{10.69}$$
$$Cl + O_3 \rightarrow ClO + O_2 \tag{10.59}$$
$$\underline{BrO + ClO \rightarrow Br + Cl + O_2} \tag{10.70a}$$
$$net \quad O_3 + O_3 \rightarrow O_2 + O_2 + O_2 \tag{VI}$$

Note that this catalytic scheme involves reactions with O_3 and not with O. For reasons discussed in section 10.4.7, this makes the scheme more important in the lower atmosphere.

The photochemistry of bromine is similar to that of chlorine. Much of it is poorly known and is inferred by analogy with the chemistry of chlorine. The reservoir species, HBr, HOBr, and $BrONO_2$, may be formed by

$$Br + HO_2 \rightarrow HBr + O_2 \tag{10.71}$$
$$BrO + HO_2 \rightarrow HOBr + O_2 \tag{10.72}$$
$$BrO + NO_2 + M \rightarrow BrONO_2 + M \tag{10.73}$$

The reservoir species are much less stable than their chlorine counterparts and are readily destroyed by

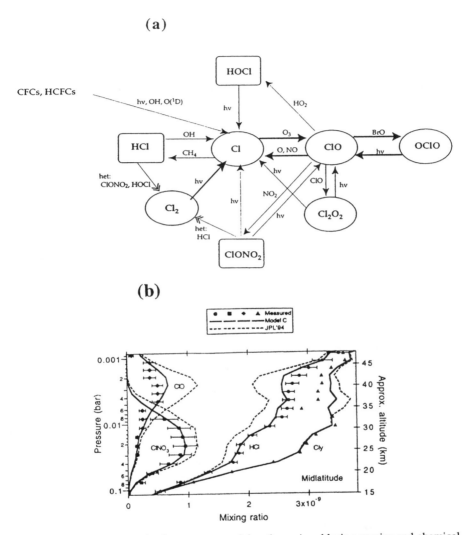

Figure 10.29 (a) Schematic diagram summarizing the major chlorine species and chemical pathways. The conversion of $ClONO_2$ and HCl to Cl_2 proceeds via a hetergeneous (het) reaction. After Jaegle, L., 1996, *Stratospheric Chlorine and Nitrogen Chemistry*, Ph.D. Thesis, California Institute of Technology. (b) Partitioning of inorganic chlorine species in the stratosphere. The triangles represent the sum of HCl(sunset) + $ClONO_2$(sunset) + ClO(midmorning). Lines labeled "Cly" represent the abundance of total inorganic chlorine necessary to match the measured sum of these gases. Model C (solid lines) uses kinetics rate coefficients that are different from the recommendations of JPL'94. The measurements are taken by the Atmospheric Trace Molecule Spectroscopy (ATMOS) instrument. After Michelsen, H. A. et al., 1996, "Stratospheric Chlorine Partitioning: Constraints From Shuttle-Borne Measurements of HCl, $ClNO_3$, and ClO." *Geophys. Res. Lett.* **23**, 2361.

Earth: Human Impact 401

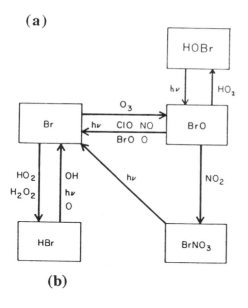

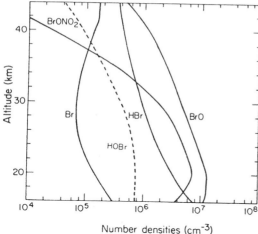

Figure 10.30 (a) Schematic diagram summarizing the major bromine species and chemical pathways. (b) Altitude profiles for major inorganic bromine species in the stratosphere computed with total bromine equal to 20 parts per trillion by volume. After Yung, Y. L. et al., 1980, "Atmospheric Bromine and Ozone Perturbations in the Lower Stratosphere." *J. Atmos. Sci.* **37**, 339.

$$\text{OH} + \text{HBr} \rightarrow \text{H}_2\text{O} + \text{Br} \qquad (10.74)$$

$$\text{HOBr} + h\nu \rightarrow \text{OH} + \text{Br} \qquad (10.75)$$

$$\text{BrONO}_2 + h\nu \rightarrow \text{Br} + \text{NO}_3 \qquad (10.76a)$$

$$\rightarrow \text{BrO} + \text{NO}_2 \qquad (10.76b)$$

Figure 10.30a presents a schematic diagram illustrating the interconnections between the major bromine species. The concentrations of the major bromine species are shown in figure 10.30b.

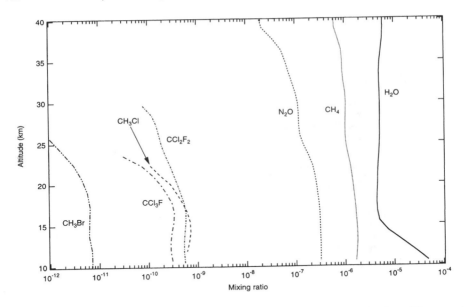

Figure 10.31 Altitude profiles for major source species in the upper atmosphere. The concentrations of H_2O, CH_4, N_2O, CH_3Cl, $CFCl_3$, and CF_2Cl_2 are taken from measurements by the Atmospheric Trace Molecule Spectroscopy (ATMOS) instrument. The CH_3Br concentrations were taken over India (18° N, 77° W) on 9 April 1990. After Lal, S. et al., 1994, "Vertical Distribution of Methyl Bromide over Hyderabad, India." *Tellus* **46B**, 373.

10.4.6 Overview of Catalytic Chemistry

The altitude profiles of the most important source molecules in the stratosphere, H_2O, CH_4, N_2O, CH_3Cl, $CFCl_3$, CF_2Cl_2, and CH_3Br, are shown in figure 10.31. The molecules are the "parent molecules" of HO_x, NO_x, ClO_x, and BrO_x radicals in the stratosphere. Note that all these molecules have higher mixing ratios in the troposphere near the source region at the surface of the planet. Their mixing ratios fall with altitude in the stratosphere. The destruction of the source molecules releases the active radicals, which can then react with ozone. With the exception of H_2O, all source molecules are irreversibly destroyed in the stratosphere, and they must be replenished by transport from the lower atmosphere.

Figures 10.32a and 10.32b show the concentrations of O_3 and O in the stratosphere. The rates of the rate-limiting reactions for the destruction of odd oxygen by the Chapman reaction, HO_x, odd nitrogen, chlorine, and bromine are shown in figure 10.32c. Destruction by NO_x is most important in the midst of the ozone layer. The Chapman reaction and destruction by chlorine become more important above this level. Destruction by HO_x is more important in the upper stratosphere and the lower stratosphere. Loss via bromine chemistry is relatively more important in the lower stratosphere.

Having reviewed the catalytic destruction of ozone by HO_x, NO_x, and halogen compounds, we recognize a simple pattern in the nature of this catalytic chemistry. There are four integral components: the long-lived carrier, the reactive radical species

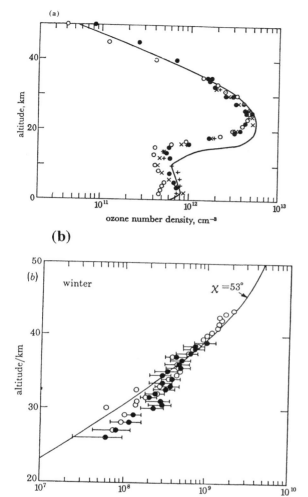

Figure 10.32 (a) Vertical distribution of ozone at midlatitude. The solid line is the result of a model, and the data points are measurements. After Logan, J. A. et al., 1978, "Atmospheric Chemistry: Response to Human Influence." *Phil. Trans. Roy. Soc. Lond.* **290**, 187. (b) Altitude profile of O for 30° N in winter with solar zenith angle of 53°. The data are from Anderson, J. G., 1975, "The Measurement of Atomic Oxygen and Hydroxyl in the Stratosphere," in *Proceedings of the Fourth Conference on the Climatic Impact Assessment Program* (4–7 February; U.S. Dept. of Transportation Report no. DOT-TSC-OST-75-38), p. 458.

that destroy O_3, the reservoir species that sequester the reactive radicals, and the removal mechanism. Note that there is a wide range in the lifetimes of the reservoir species. For example, $ClONO_2$ is photolyzed in a day, but HCl has a lifetime in excess of a month in the lower stratosphere.

We may now also understand why there is relatively little direct impact on O_3 by the chemistry of fluorine and sulfur. In the case of fluorine, the major reservoir species is HF, an extremely stable molecule that does not dissociate or react with OH in the stratosphere. Once formed, HF is the terminal product of stratospheric fluorine and there is no further chemical reactivity between HF and other stratospheric species. In the case of sulfur, the bulk of SO_2 and DMS produced in the biosphere is destroyed in the troposphere. The only long-lived carriers of sulfur in the absence of volcanic eruptions are COS and CS_2. But the most stable reservoir species of sulfur

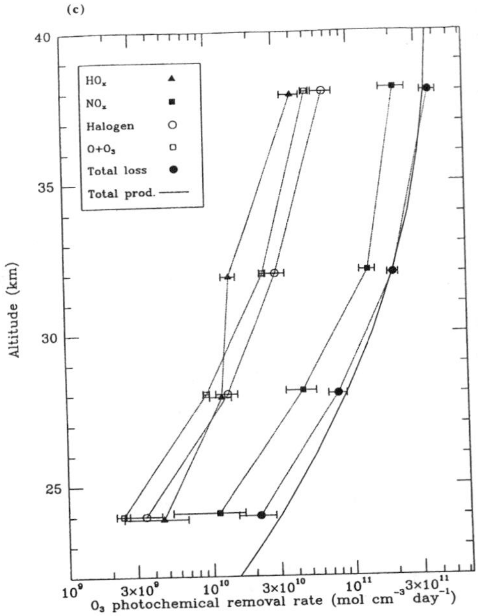

Figure 10.32 (c) Production and loss rates of odd oxygen in the stratosphere. Error bars represent 1-σ total accuracy based on uncertainties associated with measurements of radicals. After Jucks, K. W. et al., 1996, "Ozone Production and Loss Rates Measurements in the Middle Stratosphere." *J. Geophys. Res.* **101**, 28785.

in the stratosphere is H_2SO_4 aerosol, which will not undergo further reaction with stratospheric species. The aerosol surfaces do provide sites for condensation of water in the formation of PSCs and for heterogeneous reactions. Thus, there may be an indirect impact on ozone (see section 10.4.8).

10.4.7 Ozone Hole

The possible destruction of ozone associated with human activities was first suggested in the early 1970s. The early theories focused on catalytic cycles of the form

$$\begin{array}{r} X + O_3 \rightarrow XO + O_2 \\ XO + O \rightarrow X + O_2 \\ \hline net \quad O + O_3 \rightarrow O_2 + O_2 \end{array} \quad \text{(VII)}$$

where X = H, OH, NO, and Cl [chemical schemes (IIIa), (IIIb), (IV) and (V)]. From figure 10.32c we expect that most of the ozone destruction would occur in the upper stratosphere, where O_3 is chemically controlled. We therefore expect that most of the adverse anthropogenic impact would be confined to the photochemically active region in the upper stratosphere. The bulk of ozone that resides in the lower stratosphere (including the polar stratosphere) is dynamically controlled and would not be directly affected by chemical destruction. The inadequacy of this "classical" view was clearly revealed by its failure to predict, and subsequently to explain, the ozone hole phenomenon discovered in 1985. Figure 10.33a shows the decadal decline of ozone column abundance in the Austral spring in Antarctica. Later balloon data, shown in figure 10.33b, revealed that the bulk of the ozone loss occurred in the lower stratosphere. In some places 90% of the initial O_3 was removed.

(a)

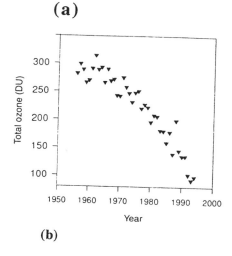

(b)

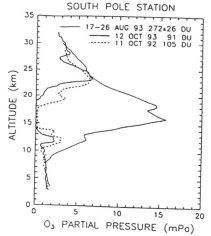

Figure 10.33 (a) Lowest daily value of total ozone observed over Halley Bay (76° S, 27° W) in October for the years between 1956 and 1994. After Jones, A. E., and Shanklin, J. D., 1995, "Continued Decline of Total Ozone over Halley, Antarctica, since 1985." *Nature* **376**, 409. (b) Comparison of the South Pole vertical ozone profiles: predepletion (solid line) and postdepletion (dashed line). Adapted from Hofmann, D. J. et al., 1994, "Record Low Ozone at the South Pole in the Spring of 1993." *Geophys. Res. Lett.* **21** 421.

There followed a surge of activity concerning the crucial question of what key factors control stratospheric ozone. New theoretical ideas as well as laboratory results were advanced. A series of polar campaigns, AAOE (Airborne Antarctic Ozone Expedition; 1987), AASE 1 and 2 (1989, 1991–1992), and SPADE (Stratospheric Photochemistry, Aerosols, and Dynamics Expedition; 1992–1993), were organized to probe the chemistry and dynamics of the polar stratosphere and the lower stratosphere. The new results can briefly be summarized as follows:

1. The major loss of ozone does not occur in the photochemically active region, but rather in the lower stratosphere where O_3 is supposed to be dynamically controlled (in the gas phase model);
2. heterogeneous chemistry plays a fundamental role in repartitioning the odd nitrogen (NO_x) and active chlorine species (Cl_y) such that the destructive power of halogen is greatly enhanced; and

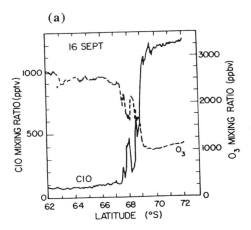

Figure 10.34 (a) Measurements of O_3 and ClO on 16 September 1987 during the Airborne Antarctic Ozone Experiment (AAOE). After Anderson, J. G., W. H. Brune, and M. H. Proffitt, 1989, "Ozone Destruction by Chlorine Radicals within the Antarctic Vortex: The Spatial and Temporal Evolution of ClO-O_3 Anticorrelation Based on *in situ* ER-2 data." *J. Geophys. Res.* **94**, 11465–11479.

3. the influence of PSCs may extend far beyond the polar regions of the stratosphere and affect the midlatitude atmosphere via the export of "processed" air.

Note that the new picture of stratospheric ozone that emerges is one that places greater emphasis on heterogeneous chemistry and the interaction between dynamics and chemistry in the lower stratosphere. This makes the assessment of the anthropogenic impact on the ozone layer much more difficult than previously thought.

The heterogeneous reactions found to be important on the surface of particles are

$$ClONO_2 + HCl \rightarrow Cl_2 + HNO_3 \tag{10.77}$$

$$ClONO_2 + H_2O \rightarrow HOCl + HNO_3 \tag{10.78}$$

$$HCl + N_2O_5 \rightarrow ClNO_2 + HNO_3 \tag{10.79}$$

$$HCl + HOCl \rightarrow Cl_2 + H_2O \tag{10.80}$$

$$N_2O_5 + H_2O \rightarrow 2HNO_3 \tag{10.81}$$

Note that the net result of heterogeneous chemistry is to release chlorine from the less active forms (HCl and $ClONO_2$) and convert them into the labile forms (Cl_2 and HOCl). Reactive odd nitrogen is also converted into the inactive form (HNO_3). The enhanced concentrations of the active radical chlorine species lead to the destruction of O_3 by the following catalytic scheme:

$$2(Cl + O_3 \rightarrow ClO + O_2) \tag{10.59}$$
$$ClO + ClO + M \rightarrow Cl_2O_2 + M \tag{10.82}$$
$$Cl_2O_2 + h\nu \rightarrow Cl + ClOO \tag{10.83}$$
$$ClOO + M \rightarrow Cl + O_2 \tag{10.84}$$
$$\text{net} \quad O_3 + O_3 \rightarrow O_2 + O_2 + O_2 \tag{VIII}$$

Note that this scheme is nonlinear in ClO. In addition, it is not of the same type as (VII) in which O atoms are involved in the catalytic cycle. The enhanced level of chlorine can now also drive the synergistic BrO-ClO catalytic cycle (VI). Dramatic loss of ozone can now occur in the spring in the Antarctic polar stratosphere. The

(b)

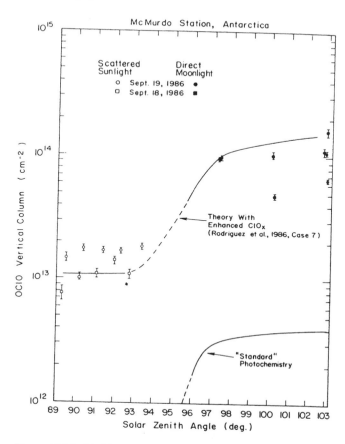

Figure 10.34 (b) Observed twilight (solar zenith angle less than 93°) and moonlight OClO vertical column abundances on 18 and 19, September 1986, along with model calculations for both "standard" (gas phase only) and heterogeneous chemistry. After Solomon, S. et al., 1987, "Visible Spectroscopy at McMurdo Station, Antarctica 2. Observations of OClO." *J. Geophys. Res* **92**, 8329.

isolation of the polar air mass in the vortex prevents the influx of O_3-rich air from lower latitudes. The result is the Antarctic ozone hole phenomenon. Figure 10.34a shows the remarkable anticorrelation between O_3 and ClO as measured on an aircraft flying across the boundary of the polar vortex. The enhancement of BrO is revealed in the unique product OClO produced in the minor branch (figure 10.34b)

$$BrO + ClO \rightarrow Br + OClO \qquad (10.70b)$$

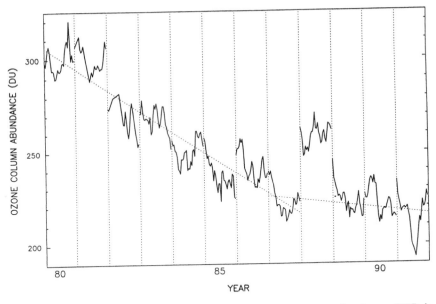

Figure 10.35 The vortex-averaged area-weighted column ozone abundance (DU) in the vortex from day 260 to day 290 (mid-September to mid-October), the season of minimum column ozone, each year. The steep dotted line is the linear least square fit to the data during this month from 1980 to 1987. The slope is 3.8% (11.3 DU) per year. The less steep dotted line is similar fit to data from 1987 to 1991 but excluding 1988. The slope is 0.9% (2.0 DU) per year. After Jiang, Y., Yung, Y. L., and Zurek, R. W., 1996, "Decadal Evolution of the Antarctic Ozone Hole." *J. Geophys. Res.* **101**, 8985.

A combination of circumstances make the Antarctic ozone hole unique. The first is the formation of a compact vortex in the polar stratosphere in the winter induced by radiative cooling. The cooling promotes the formation of PSCs. Type I PSCs, consisting of nitric acid trihydrate, are formed when the temperature falls below 195 K; type II PSCs (water ice) can form at temperatures below 187 K. In the Antarctic polar stratosphere both types of PSCs are readily formed. By comparison, the northern polar stratosphere is much more turbulent, disturbed by planetary waves propagating from the troposphere (the topography and land-sea contrast induce more wave activities in the northern hemisphere). The cold temperatures in the northern polar vortex do not last long enough or get cold enough for condensation of type II PSCs. In the Austral spring the Antarctic polar vortex is very stable and lasts much longer than the Arctic polar vortex, allowing more destruction of O_3. The evolution of the Antarctic ozone hole from 1979 to 1991 is shown in figure 10.35. Note the trend of increasingly lower amounts of ozone with time, with the exception of 1988, the year of perturbed dynamics. The export of O_3-poor air to the lower latitudes can cause an apparent decrease of ozone at lower latitudes. In addition, air parcels that travel through the polar regions may undergo "processing," that is, conversion of reservoir species to reactive species. Figure 10.36 shows the loss of O_3 over the

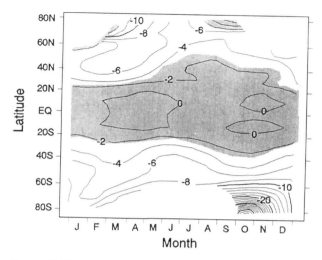

Figure 10.36 Trend (%/decade) in total column ozone obtained from Total Ozone Mapping Spectrometer (TOMS) as a function of latitude and season. Data extend from November 1978 to March 1991. There are no data in the polar night. After Stolarski, R. S., Bojkov, R., Bishop, L., Zerefos, C., Staehelin, J., and Zawodny, J., 1992, "Measured Trends in Stratospheric Ozone." *Science* **256**, 342–349.

last decade. The average is of the order of 4% per decade, with enhanced losses at polar latitudes.

10.4.8 Volcanic Perturbation

Volcanic eruptions episodically inject large amounts of SO_2 into the stratosphere. The estimated stratospheric inputs by the two recent volcanoes, El Chichon and Pinatubo, are 10 and 30 Mt of SO_2, respectively. By comparison, the total sulfur content of

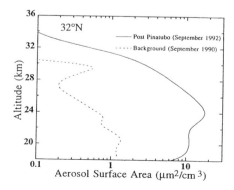

Figure 10.37 Total surface aerosol area density profile at 32° N in September 1990 (background) and in September 1992 (after the eruption of Mt. Pinatubo), as measured by the Stratospheric Aerosols and Gas Experiment (SAGE). After Jaegle, L., 1996, *Stratospheric Chlorine and Nitrogen Chemistry*, Ph.D. Thesis, California Institute of Technology, and is based on data taken by McCormick, M. P., and Veiga, R. E., 1992, "SAGE II Measurements of Early Pinatubo Aerosols." *Geophys. Res. Lett.* **19**, 155.

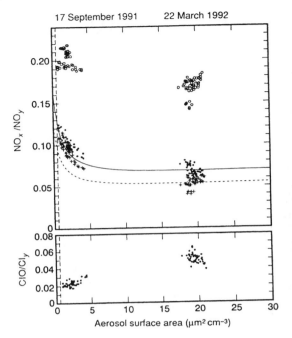

Figure 10.38 Scatter plot of observed NO_x/NO_y and ClO/Cl_y with observed areosol surface area density (solid circles) in high and low aerosol conditions. Gas phase only (open circles) and heterogeneous cases (crosses) model calculations are included. The vertical dashed line represents background areosol surface area density. The curves represent the dependence on surface area in two heterogeneous models. After Fahey, D. W. et al., 1993, "In Measurements Constraining the Role of Sulfate Aerosols in Midlatitude Ozone Depletion." *Nature* **363**, 509.

the stratosphere during quiescent periods is of the order of 1 Mt equivalent of SO_2, the bulk of it residing in H_2SO_4 aerosols. In the stratosphere SO_2 gets oxidized by reacting with OH:

$$SO_2 + OH + M \rightarrow HSO_3 + M \qquad (10.85)$$

$$HSO_3 + O_2 \rightarrow SO_3 + HO_2 \qquad (10.86)$$

$$SO_3 + H_2O + M \rightarrow H_2SO_4 + M \qquad (10.87)$$

Note that the HO_x radicals act as a catalyst in the oxidation of SO_2 to H_2SO_4. Once formed, the vapor pressure of H_2SO_4 at stratospheric temperatures is so low that it rapidly condenses into aerosol particles. This is the terminal product. The only removal mechanism is transport to the troposphere, followed by rainout.

Although there is no direct effect of H_2SO_4 aerosols on stratospheric chemistry, the surfaces of H_2SO_4 aerosols act as sites for heterogeneous reactions (10.77)–(10.81). The rate-limiting step in a heterogeneous reaction is often the adsorption of a molecule onto an aerosol surface:

$$X + \text{aerosol} \rightarrow X(\text{ads}) \qquad J \qquad (10.88)$$

with a unimolecular rate coefficient given by

$$J = \frac{1}{4}\gamma v A \qquad (10.89)$$

where γ is the sticking coefficient, v is the molecular speed, and A is the total surface area of aerosols per unit volume. Figure 10.37 shows the surface area per unit volume

(A) before and after the Pinatubo eruption. Note that there was an order of magnitude increase in the values of A due to the effect of volcanic aerosols. The impact of this on one aspect of stratospheric chemistry, the partitioning between NO_2 and HNO_3, is shown in figure 10.38. The effect of the aerosols is to decrease the NO_x/NO_y ratio, while increasing the ClO/Cl_y ratio. The net result is an enhancement of the effectiveness of chlorine for destroying ozone.

10.5 Unsolved Problems

We list a number of outstanding unresolved problems related to the problems of the global environment:

1. Is there a missing sink of CO_2, and if so, what is it ?
2. Is the biosphere responding to the increase of CO_2 and surface temperature?
3. What are the causes for the increase in atmospheric CH_4?
4. What are the causes for the increase in atmospheric N_2O?
5. What is an "acceptable" level of ozone depletion?
6. What fraction of the long term secular changes in stratospheric ozone is caused by natural fluctuations and what fraction may be attributed to anthropogenic influence?
7. Is it possible (or wise) to alter the global planetary environment for the welfare of man by environmental engineering?

There is an ultimate unsolved problem challenging *Homo sapiens*, posed by Professor Bruce Murray of the California Institute of Technology. For the first time in human history, humanity will have to manage the planetary environment and our interaction with it. The frontier days of mining the forests and seas, and of dumping waste products into politically unrepresented environments, are rapidly approaching an end. The inexorable growth of human civilization since the onset of agriculture has been financed by consumption and degradation of much of the biospheric potential of soils and forests. Well before the end of the next century, the world will have consumed nearly all the easily extracted and transported oil that has fueled the rapid global economic growth since the middle of this century. Well before the end of the next century, much of the natural capital Man the Toolmaker inherited will be gone forever. There will be no chance to start over. All that is uncertain is whether we will have used our collective intelligence to transform that inherited natural capital into a self-sustaining social and economic system no longer dependent on exploitation of natural resources.

Appendix

Appendix 10.1 Photodissociation reactions in Earth's atmosphere at the equator in January.

	Photodissociation coefficient[a] (s^{-1})		
Reaction	1 km	25 km	80 km
$O_2 + h\nu \rightarrow 2O$	0.0	1.2×10^{-12}	3.0×10^{-9}
$O_2 + h\nu \rightarrow O + O(^1D)$	0.0	0.0	4.7×10^{-10}

(continued)

Appendix 10.1 (continued)

Reaction	Photodissociation coefficient[a] (s^{-1})		
	1 km	25 km	80 km
$O_3 + h\nu \rightarrow O_2 + O$	2.1×10^{-4}	2.1×10^{-4}	6.2×10^{-4}
$O_3 + h\nu \rightarrow O_2 + O(^1D)$	1.2×10^{-5}	1.5×10^{-5}	3.7×10^{-3}
$HO_2 + h\nu \rightarrow OH + O$	0.0	5.5×10^{-7}	2.8×10^{-4}
$H_2O + h\nu \rightarrow H + OH$	0.0	2.4×10^{-12}	1.9×10^{-6}
$H_2O + h\nu \rightarrow H_2 + O(^1D)$	0.0	0.0	2.1×10^{-7}
$H_2O + h\nu \rightarrow 2H + O$	0.0	0.0	2.5×10^{-7}
$H_2O_2 + h\nu \rightarrow 2OH$	3.4×10^{-6}	3.3×10^{-6}	4.8×10^{-5}
$NO + h\nu \rightarrow N + O$	0.0	7.4×10^{-11}	2.4×10^{-6}
$NO_2 + h\nu \rightarrow NO + O$	4.6×10^{-3}	4.7×10^{-3}	4.7×10^{-3}
$NO_3 + h\nu \rightarrow NO_2 + O$	7.7×10^{-2}	7.8×10^{-2}	7.8×10^{-2}
$NO_3 + h\nu \rightarrow NO + O_2$	6.2×10^{-3}	6.3×10^{-3}	6.3×10^{-3}
$N_2O + h\nu \rightarrow N_2 + O(^1D)$	0.0	1.8×10^{-9}	5.0×10^{-7}
$N_2O_5 + h\nu \rightarrow NO_2 + NO_3$	2.1×10^{-5}	1.0×10^{-6}	9.4×10^{-5}
$N_2O_5 + h\nu \rightarrow NO + NO_3 + O$	1.2×10^{-6}	1.4×10^{-6}	2.2×10^{-4}
$HNO_2 + h\nu \rightarrow OH + NO$	1.1×10^{-3}	1.1×10^{-3}	1.0×10^{-3}
$HNO_3 + h\nu \rightarrow NO_2 + OH$	2.9×10^{-7}	5.9×10^{-7}	6.2×10^{-5}
$HO_2NO_2 + h\nu \rightarrow HO_2 + NO_2$	1.2×10^{-6}	2.0×10^{-6}	1.5×10^{-4}
$HO_2NO_2 + h\nu \rightarrow OH + NO_3$	6.5×10^{-7}	1.1×10^{-6}	8.0×10^{-5}
$CH_3O_2NO_2 + h\nu \rightarrow CH_3O_2 + NO_2$	5.3×10^{-6}	8.5×10^{-6}	2.3×10^{-4}
$NH_3 + h\nu \rightarrow NH_2 + H$	0.0	4.1×10^{-7}	5.4×10^{-5}
$Cl_2 + h\nu \rightarrow 2Cl$	1.2×10^{-3}	1.3×10^{-3}	1.4×10^{-3}
$ClO + h\nu \rightarrow Cl + O$	6.2×10^{-5}	8.9×10^{-5}	3.0×10^{-3}
$ClOO + h\nu \rightarrow ClO + O$	0.0	4.3×10^{-10}	4.6×10^{-3}
$OClO + h\nu \rightarrow ClO + O$	4.2×10^{-2}	4.5×10^{-2}	4.3×10^{-2}
$ClO_3 + h\nu \rightarrow ClO + O_2$	2.8×10^{-4}	2.8×10^{-4}	2.8×10^{-4}
$Cl_2O + h\nu \rightarrow Cl + ClO$	5.1×10^{-4}	5.6×10^{-4}	1.9×10^{-3}
$Cl_2O_2 + h\nu \rightarrow Cl + ClOO$	7.1×10^{-4}	7.8×10^{-4}	3.1×10^{-3}
$Cl_2O_3 + h\nu \rightarrow ClO + ClOO$	2.7×10^{-4}	4.0×10^{-4}	1.0×10^{-2}
$ClNO + h\nu \rightarrow Cl + NO$	1.6×10^{-3}	1.6×10^{-3}	2.6×10^{-3}
$ClNO_2 + h\nu \rightarrow Cl + NO_2$	1.8×10^{-4}	2.0×10^{-4}	7.8×10^{-4}
$ClONO + h\nu \rightarrow Cl + NO_2$	2.2×10^{-3}	2.4×10^{-3}	4.0×10^{-3}
$ClNO_3 + h\nu \rightarrow Cl + NO_3$	4.0×10^{-6}	4.0×10^{-6}	2.1×10^{-4}
$ClNO_3 + h\nu \rightarrow ClO + NO_2$	2.2×10^{-5}	2.2×10^{-5}	2.1×10^{-4}
$ClNO_3 + h\nu \rightarrow O + ClONO$	2.5×10^{-6}	2.1×10^{-6}	3.6×10^{-5}
$HCl + h\nu \rightarrow H + Cl$	0.0	1.7×10^{-9}	6.1×10^{-7}
$HOCl + h\nu \rightarrow OH + Cl$	1.3×10^{-4}	1.4×10^{-4}	2.3×10^{-4}
$CH_3Cl + h\nu \rightarrow Cl + CH_3$	0.0	1.1×10^{-9}	5.1×10^{-7}
$CH_2FCl + h\nu \rightarrow$ Products	0.0	1.2×10^{-10}	7.8×10^{-8}
$CHFCl_2 + h\nu \rightarrow CFCl_2 + H$	0.0	3.5×10^{-9}	1.4×10^{-6}
$CHF_2Cl + h\nu \rightarrow CF_2Cl + H$	0.0	1.1×10^{-11}	1.1×10^{-8}
$CCl_4 + h\nu \rightarrow CCl_3 + Cl$	0.0	6.7×10^{-8}	1.8×10^{-5}
$CFCl_3 + h\nu \rightarrow CFCl_2 + Cl$	0.0	3.8×10^{-8}	8.0×10^{-6}
$CF_2Cl_2 + h\nu \rightarrow CF_2Cl + Cl$	0.0	3.1×10^{-9}	1.9×10^{-6}

Appendix 10.1 (continued)

Reaction	Photodissociation coefficient[a] (s^{-1})		
	1 km	25 km	80 km
$CH_3CCl_3 + h\nu \rightarrow$ Products	0.0	5.7×10^{-8}	9.2×10^{-6}
$CH_3CFCl_2 + h\nu \rightarrow$ Products	0.0	9.8×10^{-9}	1.4×10^{-6}
$CH_3CF_2Cl + h\nu \rightarrow$ Products	0.0	8.6×10^{-11}	1.3×10^{-8}
$C_2Cl_3F_3 + h\nu \rightarrow$ Products	0.0	6.3×10^{-9}	1.6×10^{-6}
$C_2Cl_2F_4 + h\nu \rightarrow$ Products	0.0	3.5×10^{-10}	2.1×10^{-7}
$CF_3CHCl_2 + h\nu \rightarrow$ Products	0.0	7.5×10^{-9}	1.1×10^{-6}
$CF_3CHFCl + h\nu \rightarrow$ Products	0.0	7.9×10^{-11}	1.2×10^{-8}
$CF_3CCl_3 + h\nu \rightarrow$ Products	0.0	7.2×10^{-9}	2.6×10^{-6}
$CF_3CFCl_2 + h\nu \rightarrow$ Products	0.0	4.9×10^{-10}	2.1×10^{-7}
$CF_3CF_2CHCl_2 + h\nu \rightarrow$ Products	0.0	0.0	0.0
$CF_2ClCF_2CHFCl + h\nu \rightarrow$ Products	0.0	0.0	0.0
$CHClO + h\nu \rightarrow HCO + Cl$	2.8×10^{-4}	2.8×10^{-4}	2.8×10^{-4}
$COCl_2 + h\nu \rightarrow 2Cl + CO$	8.4×10^{-19}	2.5×10^{-8}	2.5×10^{-5}
$COFCl + h\nu \rightarrow$ Products	0.0	1.3×10^{-8}	3.2×10^{-6}
$HF + h\nu \rightarrow H + F$	0.0	0.0	1.1×10^{-7}
$CF_4 + h\nu \rightarrow CF_3 + F$	0.0	0.0	2.0×10^{-10}
$C_2F_6 + h\nu \rightarrow 2CF_3$	0.0	0.0	4.6×10^{-20}
$COF_2 + h\nu \rightarrow$ Products	0.0	9.9×10^{-10}	1.6×10^{-7}
$SF_6 + h\nu \rightarrow$ Products	0.0	0.0	0.0
$Br_2 + h\nu \rightarrow 2Br$	1.7×10^{-2}	1.6×10^{-2}	1.6×10^{-2}
$BrO + h\nu \rightarrow Br + O$	2.1×10^{-2}	2.3×10^{-2}	2.3×10^{-2}
$BrNO_2 + h\nu \rightarrow Br + NO_2$	0.0	0.0	0.0
$BrNO_3 + h\nu \rightarrow Br + NO_3$	5.2×10^{-4}	5.5×10^{-4}	1.1×10^{-3}
$HBr + h\nu \rightarrow H + Br$	0.0	1.1×10^{-7}	2.2×10^{-5}
$HOBr + h\nu \rightarrow OH + Br$	3.7×10^{-4}	3.8×10^{-4}	6.3×10^{-4}
$CH_3Br + h\nu \rightarrow CH_3 + Br$	0.0	8.9×10^{-8}	2.6×10^{-5}
$CHBr_3 + h\nu \rightarrow$ Products	1.6×10^{-7}	8.8×10^{-7}	6.2×10^{-4}
$BrCl + h\nu \rightarrow Br + Cl$	4.8×10^{-3}	4.8×10^{-3}	4.7×10^{-3}
$CF_2ClBr + h\nu \rightarrow CF_2Cl + Br$	1.5×10^{-15}	1.4×10^{-7}	5.9×10^{-5}
$CF_2Br_2 + h\nu \rightarrow$ Products	6.4×10^{-10}	1.2×10^{-7}	2.3×10^{-4}
$CF_3Br + h\nu \rightarrow CF_3 + Br$	0.0	1.4×10^{-8}	4.8×10^{-6}
$C_2F_4Br_2 + h\nu \rightarrow$ Products	0.0	1.4×10^{-7}	5.0×10^{-5}
$CH_4 + h\nu \rightarrow$ Products	0.0	0.0	6.6×10^{-7}
$CO_2 + h\nu \rightarrow CO + O$	0.0	5.3×10^{-14}	3.2×10^{-9}
$H_2CO + h\nu \rightarrow HCO + H$	1.5×10^{-5}	1.8×10^{-5}	3.8×10^{-5}
$H_2CO + h\nu \rightarrow H_2 + CO$	2.6×10^{-5}	3.8×10^{-5}	5.6×10^{-5}
$CH_3OOH + h\nu \rightarrow CH_3O + OH$	2.6×10^{-6}	3.0×10^{-6}	2.7×10^{-5}
$HCN + h\nu \rightarrow H + CN$	0.0	8.3×10^{-13}	1.8×10^{-6}
$CH_3CN + h\nu \rightarrow CH_3 + CN$	0.0	0.0	7.0×10^{-7}
$SO_2 + h\nu \rightarrow SO + O$	0.0	6.6×10^{-7}	9.4×10^{-5}
$H_2S + h\nu \rightarrow H + SH$	0.0	5.9×10^{-7}	1.6×10^{-4}
$OCS + h\nu \rightarrow CO + S$	5.4×10^{-13}	9.0×10^{-9}	1.3×10^{-5}

[a] The photodissociation coefficients are taken with updates from Michelangeli, D., Allen, M., and Yung, Y. L., 1989. *J. Geophys. Res.* **94**, 18429.

Appendix 10.2 Rate constants for second order reactions

Reaction	A-Factor[a]	E/R ± ΔE/R	k(298 K)[a]	f(298)[b]
O_x Reactions				
$O + O_3 \rightarrow O_2 + O_2$	8.0×10^{-12}	2060 ± 250	8.0×10^{-15}	1.15
$O(^1D)$ Reactions				
$O(^1D) + O_2 \rightarrow O + O_2$	3.2×10^{-11}	$-(70 \pm 100)$	4.0×10^{-11}	1.2
$O(^1D) + O_3 \rightarrow O_2 + O_2$	1.2×10^{-10}	0 ± 100	1.2×10^{-10}	1.3
$\rightarrow O_2 + O + O$	1.2×10^{-10}	0 ± 100	1.2×10^{-10}	1.3
$O(^1D) + H_2 \rightarrow OH + H$	1.1×10^{-10}	0 ± 100	1.1×10^{-10}	1.1
$O(^1D) + H_2O \rightarrow OH + OH$	2.2×10^{-10}	0 ± 100	2.2×10^{-10}	1.2
$O(^1D) + N_2 \rightarrow O + N_2$	1.8×10^{-11}	$-(110 \pm 100)$	2.6×10^{-11}	1.2
$O(^1D) + N_2O \rightarrow N_2 + O_2$	4.9×10^{-11}	0 ± 100	4.9×10^{-11}	1.3
$\rightarrow NO + NO$	6.7×10^{-11}	0 ± 100	6.7×10^{-11}	1.3
$O(^1D) + NH_3 \rightarrow OH + NH_2$	2.5×10^{-10}	0 ± 100	2.5×10^{-10}	1.3
$O(^1D) + CO_2 \rightarrow O + CO_2$	7.4×10^{-11}	$-(120 \pm 100)$	1.1×10^{-10}	1.2
$O(^1D) + CH_4 \rightarrow OH + CH_3$	1.2×10^{-10}	0 ± 100	1.2×10^{-10}	1.2
$\rightarrow H_2 + CH_2O$	3.0×10^{-11}	0 ± 100	3.0×10^{-11}	1.5
$O(^1D) + HCl \rightarrow$ products	1.5×10^{-10}	0 ± 100	1.5×10^{-10}	1.2
$O(^1D) + HF \rightarrow OH + F$	1.4×10^{-10}	0 ± 100	1.4×10^{-10}	2.0
$O(^1D) + HBr \rightarrow$ products	1.5×10^{-10}	0 ± 100	1.5×10^{-10}	2.0
$O(^1D) + SF_6 \rightarrow$ products	—	—	1.8×10^{-14}	1.5
Singlet O_2 Reactions				
$O_2(^1\Delta) + O \rightarrow$ products	—	—	$< 2 \times 10^{-16}$	—
$O_2(^1\Delta) + O_2 \rightarrow$ products	3.6×10^{-18}	220 ± 100	1.7×10^{-18}	1.2
$O_2(^1\Delta) + O_3 \rightarrow O + 2O_2$	5.2×10^{-11}	2840 ± 500	3.8×10^{-15}	1.2
$O_2(^1\Delta) + H_2O \rightarrow$ products	—	—	4.8×10^{-18}	1.5
$O_2(^1\Delta) + N_2 \rightarrow$ products	—	—	$< 10^{-20}$	—
$O_2(^1\Delta) + CO_2 \rightarrow$ products	—	—	$< 2 \times 10^{-20}$	—
$O_2(^1\Sigma) + O \rightarrow$ products	—	—	8×10^{-14}	5.0
$O_2(^1\Sigma) + O_2 \rightarrow$ products	—	—	3.9×10^{-17}	1.5
$O_2(^1\Sigma) + O_3 \rightarrow$ products	2.2×10^{-11}	0 ± 200	2.2×10^{-11}	1.2
$O_2(^1\Sigma) + H_2O \rightarrow$ products	—	—	5.4×10^{-12}	1.3
$O_2(^1\Sigma) + N \rightarrow$ products	—	—	$< 10^{-13}$	—
$O_2(^1\Sigma) + N_2 \rightarrow$ products	2.1×10^{-15}	0 ± 200	2.1×10^{-15}	1.2
$O_2(^1\Sigma) + CO_2 \rightarrow$ products	4.2×10^{-13}	0 ± 200	4.2×10^{-13}	1.2
HO_x Reactions				
$O + OH \rightarrow O_2 + H$	2.2×10^{-11}	$-(120 \pm 100)$	3.3×10^{-11}	1.2
$O + HO_2 \rightarrow OH + O_2$	3.0×10^{-11}	$-(200 \pm 100)$	5.9×10^{-11}	1.2
$O + H_2O_2 \rightarrow OH + HO_2$	1.4×10^{-12}	2000 ± 1000	1.7×10^{-15}	2.0
$H + O_3 \rightarrow OH + O_2$	1.4×10^{-10}	470 ± 200	2.9×10^{-11}	1.25
$H + HO_2 \rightarrow$ products	8.1×10^{-11}	0 ± 100	8.1×10^{-11}	1.3
$OH + O_3 \rightarrow HO_2 + O_2$	1.6×10^{-12}	940 ± 300	6.8×10^{-14}	1.3
$OH + H_2 \rightarrow H_2O + H$	5.5×10^{-12}	2000 ± 100	6.7×10^{-15}	1.1
$OH + HD \rightarrow$ products	5.0×10^{-12}	2130 ± 200	4.0×10^{-15}	1.2
$OH + OH \rightarrow H_2O + O$	4.2×10^{-12}	240 ± 240	1.9×10^{-12}	1.4
$OH + HO_2 \rightarrow H_2O + O_2$	4.8×10^{-11}	$-(250 \pm 200)$	1.1×10^{-10}	1.3
$OH + H_2O_2 \rightarrow H_2O + HO_2$	2.9×10^{-12}	160 ± 100	1.7×10^{-12}	1.2
$HO_2 + O_3 \rightarrow OH + 2O_2$	1.1×10^{-14}	$500 \pm ^{500}_{100}$	2.0×10^{-15}	1.3
$HO_2 + HO_2 \rightarrow H_2O_2 + O_2$	2.3×10^{-13}	$-(600 \pm 200)$	1.7×10^{-12}	1.3
NO_x Reactions				
$O + NO_2 \rightarrow NO + O_2$	6.5×10^{-12}	$-(120 \pm 120)$	9.7×10^{-12}	1.1
$O + NO_3 \rightarrow O_2 + NO_2$	1.0×10^{-11}	0 ± 150	1.0×10^{-11}	1.5
$O + N_2O_5 \rightarrow$ products			$< 3.0 \times 10^{-16}$	
$O + HNO_3 \rightarrow OH + NO_3$			$< 3.0 \times 10^{-17}$	
$O + HO_2NO_2 \rightarrow$ products	7.8×10^{-11}	3400 ± 750	8.6×10^{-16}	3.0
$H + NO_2 \rightarrow OH + NO$	4.0×10^{-10}	340 ± 300	1.3×10^{-10}	1.3

Appendix 10.2 (continued)

Reaction	A-Factor[a]	E/R ±ΔE/R	k(298 K)[a]	f(298)[b]
$OH + NO_3 \rightarrow$ products			2.2×10^{-11}	1.5
$OH + HONO \rightarrow H_2O + NO_2$	1.8×10^{-11}	$390 \pm ^{200}_{500}$	4.5×10^{-12}	1.5
$OH + HNO_3 \rightarrow H_2O + NO_3$	(See Note)			1.3
$OH + HO_2NO_2 \rightarrow$ products	1.3×10^{-12}	$-(380 \pm ^{270}_{500})$	4.6×10^{-12}	1.5
$OH + NH_3 \rightarrow H_2O + NH_2$	1.7×10^{-12}	710 ± 200	1.6×10^{-13}	1.2
$HO_2 + NO \rightarrow NO_2 + OH$	3.5×10^{-12}	$-(250 \pm 50)$	8.1×10^{-12}	1.15
$HO_2 + NO_3 \rightarrow HNO_3 + O_2$			3.5×10^{-12}	1.5
$N + O_2 \rightarrow NO + O$	1.5×10^{-11}	3600 ± 400	8.5×10^{-17}	1.25
$N + O_3 \rightarrow NO + O_2$			$< 2.0 \times 10^{-16}$	
$N + NO \rightarrow N_2 + O$	2.1×10^{-11}	$-(100 \pm 100)$	3.0×10^{-11}	1.3
$N + NO_2 \rightarrow N_2O + O$	5.8×10^{-12}	$-(220 \pm 100)$	1.2×10^{-11}	1.5
$NO + O_3 \rightarrow NO_2 + O_2$	2.0×10^{-12}	1400 ± 200	1.8×10^{-14}	1.1
$NO + NO_3 \rightarrow 2NO_2$	1.5×10^{-11}	$-(170 \pm 100)$	2.6×10^{-11}	1.3
$NO_2 + O_3 \rightarrow NO_3 + O_2$	1.2×10^{-13}	2450 ± 150	3.2×10^{-17}	1.15
$NO_3 + NO_3 \rightarrow 2NO_2 + O_2$	8.5×10^{-13}	2450 ± 500	2.3×10^{-16}	1.5
$NH_2 + O_2 \rightarrow$ products			$< 6.0 \times 10^{-21}$	
$NH_2 + O_3 \rightarrow$ products	4.3×10^{-12}	930 ± 500	1.9×10^{-13}	3.0
$NH_2 + NO \rightarrow$ products	4.0×10^{-12}	$-(450 \pm 150)$	1.8×10^{-11}	1.3
$NH_2 + NO_2 \rightarrow$ products	2.1×10^{-12}	$-(650 \pm 250)$	1.9×10^{-11}	3.0
$NH + NO \rightarrow$ products	4.9×10^{-11}	0 ± 300	4.9×10^{-11}	1.5
$NH + NO_2 \rightarrow$ products	3.5×10^{-13}	$-(1140 \pm 500)$	1.6×10^{-11}	2.0
$O_3 + HNO_2 \rightarrow O_2 + HNO_3$			$< 5.0 \times 10^{-19}$	
$N_2O_5 + H_2O \rightarrow 2HNO_3$			$< 2.0 \times 10^{-21}$	
Reactions of Hydrocarbons				
$O + CH_3 \rightarrow$ products	1.1×10^{-10}	0 ± 250	1.1×10^{-10}	1.3
$O + HCN \rightarrow$ products	1.0×10^{-11}	4000 ± 1000	1.5×10^{-17}	10
$O + C_2H_2 \rightarrow$ products	3.0×10^{-11}	1600 ± 250	1.4×10^{-13}	1.3
$O + H_2CO \rightarrow$ products	3.4×10^{-11}	1600 ± 250	1.6×10^{-13}	1.25
$O + CH_3CHO \rightarrow CH_3CO + OH$	1.8×10^{-11}	1100 ± 200	4.5×10^{-13}	1.25
$O_3 + C_2H_2 \rightarrow$ products	1.0×10^{-14}	4100 ± 500	1.0×10^{-20}	3
$O_3 + C_2H_4 \rightarrow$ products	1.2×10^{-14}	2630 ± 100	1.7×10^{-18}	1.25
$O_3 + C_3H_6 \rightarrow$ products	6.5×10^{-15}	1900 ± 200	1.1×10^{-17}	1.2
$OH + CO \rightarrow$ Products	1.5×10^{-13} $\times (1 + 0.6P_{atm}(air))$	0 ± 300	1.5×10^{-13} $\times (1 + 0.6P_{atm}(air))$	1.3
$OH + CH_4 \rightarrow CH_3 + H_2O$	4.6×10^{-12}	1965 ± 200	6.3×10^{-15}	1.1
$OH + CH_3D \rightarrow$ products	4.1×10^{-12}	2000 ± 200	5.0×10^{-15}	1.15
$OH + H_2CO \rightarrow H_2O + HCO$	2.0×10^{-12}	$-(480 \pm 200)$	1.0×10^{-11}	1.25
$OH + CH_3OH \rightarrow$ products	6.7×10^{-12}	600 ± 300	8.9×10^{-13}	1.2
$OH + CH_3OOH \rightarrow$ Products	3.8×10^{-12}	$-(200 \pm 200)$	7.4×10^{-12}	1.5
$OH + HC(O)OH \rightarrow$ products	4.0×10^{-13}	0 ± 200	4.0×10^{-13}	1.3
$OH + HCN \rightarrow$ products	1.2×10^{-13}	400 ± 150	3.1×10^{-14}	3
$OH + C_2H_6 \rightarrow H_2O + C_2H_5$	8.7×10^{-12}	1070 ± 100	2.4×10^{-13}	1.1
$OH + C_3H_8 \rightarrow H_2O + C_3H_7$	1.0×10^{-11}	660 ± 100	1.1×10^{-12}	1.2
$OH + CH_3CHO \rightarrow CH_3CO + H_2O$	5.6×10^{-12}	$-(270 \pm 200)$	1.4×10^{-11}	1.2
$OH + C_2H_5OH \rightarrow$ products	7.0×10^{-12}	235 ± 100	3.2×10^{-12}	1.3
$OH + CH_3C(O)OH \rightarrow$ products	4.0×10^{-13}	$-(200 \pm 400)$	8.0×10^{-13}	1.3
$OH + CH_3C(O)CH_3 \rightarrow CH_3C(O)CH_2 + H_2O$	2.2×10^{-12}	685 ± 300	2.2×10^{-13}	1.15
$OH + CH_3CN \rightarrow$ products	4.4×10^{-12}	1664 ± 200	1.7×10^{-14}	1.5
$OH + CH_3ONO_2 \rightarrow$ products	5.0×10^{-13}	890 ± 500	2.4×10^{-14}	3
$OH + CH_3C(O)O_2NO_2(PAN) \rightarrow$ products			$< 4 \times 10^{-14}$	
$OH + C_2H_5ONO_2 \rightarrow$ products	8.2×10^{-13}	450 ± 300	1.8×10^{-13}	3
$HO_2 + CH_2O \rightarrow$ adduct	6.7×10^{-15}	$-(600 \pm 600)$	5.0×10^{-14}	5
$HO_2 + CH_3O_2 \rightarrow CH_3OOH + O_2$	3.8×10^{-13}	$-(800 \pm 400)$	5.6×10^{-12}	2
$HO_2 + C_2H_5O_2 \rightarrow C_2H_5OOH + O_2$	7.5×10^{-13}	$-(700 \pm 250)$	8.0×10^{-12}	1.5
$HO_2 + CH_3C(O)O_2 \rightarrow$ products	4.5×10^{-13}	$-(1000 \pm 600)$	1.3×10^{-11}	2

(continued)

Appendix 10.2 (*continued*)

Reaction	A-Factor[a]	E/R ±ΔE/R	k(298 K)[a]	f(298)[b]
$NO_3 + CO \rightarrow$ products			$< 4.0 \times 10^{-19}$	
$NO_3 + CH_2O \rightarrow$ products			5.8×10^{-16}	1.3
$NO_3 + CH_3CHO \rightarrow$ products	1.4×10^{-12}	1900 ± 300	2.4×10^{-15}	1.3
$CH_3 + O_2 \rightarrow$ products			$< 3.0 \times 10^{-16}$	
$CH_3 + O_3 \rightarrow$ products	5.4×10^{-12}	220 ± 150	2.6×10^{-12}	2
$HCO + O_2 \rightarrow CO + HO_2$	3.5×10^{-12}	$-(140 \pm 140)$	5.5×10^{-12}	1.3
$CH_2OH + O_2 \rightarrow CH_2O + HO_2$	9.1×10^{-12}	0 ± 200	9.1×10^{-12}	1.3
$CH_3O + O_2 \rightarrow CH_2O + HO_2$	3.9×10^{-14}	900 ± 300	1.9×10^{-15}	1.5
$CH_3O + NO_2 \rightarrow CH_2O + HONO$	1.1×10^{-11}	1200 ± 600	2.0×10^{-13}	5
$CH_3O_2 + O_3 \rightarrow$ products			$< 3.0 \times 10^{-17}$	
$CH_3O_2 + CH_3O_2 \rightarrow$ products	2.5×10^{-13}	$-(190 \pm 190)$	4.7×10^{-13}	1.5
$CH_3O_2 + NO \rightarrow CH_3O + NO_2$	3.0×10^{-12}	$-(280 \pm 60)$	7.7×10^{-12}	1.15
$CH_3O_2 + CH_3C(O)O_2 \rightarrow$ products	1.3×10^{-12}	$-(640 \pm 200)$	1.1×10^{-11}	1.5
$C_2H_5 + O_2 \rightarrow C_2H_4 + HO_2$			$< 2.0 \times 10^{-14}$	
$C_2H_5O + O_2 \rightarrow CH_3CHO + HO_2$	6.3×10^{-14}	550 ± 200	1.0×10^{-14}	1.5
$C_2H_5O_2 + C_2H_5O_2 \rightarrow$ products	6.8×10^{-14}	0 ± 300	6.8×10^{-14}	2
$C_2H_5O_2 + NO \rightarrow$ products	2.6×10^{-12}	$-(365 \pm 150)$	8.7×10^{-12}	1.2
$CH_3C(O)O_2 + CH_3C(O)O_2 \rightarrow$ products	2.9×10^{-12}	$-(500 \pm 150)$	1.5×10^{-11}	1.5
$CH_3C(O)O_2 + NO \rightarrow$ products	5.3×10^{-12}	$-(360 \pm 150)$	1.8×10^{-11}	1.4
FO$_x$ Reactions				
$O + FO \rightarrow F + O_2$	2.7×10^{-11}	0 ± 250	2.7×10^{-11}	3.0
$O + FO_2 \rightarrow FO + O_2$	5.0×10^{-11}	0 ± 250	5.0×10^{-11}	5.0
$OH + CH_3F \rightarrow CH_2F + H_2O$ (HFC-41)	3.0×10^{-12}	1500 ± 300	2.0×10^{-14}	1.1
$OH + CH_2F_2 \rightarrow CHF_2 + H_2O$ (HFC-32)	1.9×10^{-12}	1550 ± 200	1.0×10^{-14}	1.2
$OH + CHF_3 \rightarrow CF_3 + H_2O$ (HFC-23)	1.0×10^{-12}	2440 ± 200	2.8×10^{-16}	1.3
$OH + CF_3OH \rightarrow CF_3O + H_2O$			$< 2 \times 10^{-17}$	
$OH + CH_3CH_2F \rightarrow$ products (HFC-161)	7.0×10^{-12}	1100 ± 300	1.7×10^{-13}	1.4
$OH + CH_3CHF_2 \rightarrow$ products (HFC-152a)	2.4×10^{-12}	1260 ± 200	3.5×10^{-14}	1.2
$OH + CH_2FCH_2F \rightarrow CHFCH_2F + H_2O$ (HFC-152)	4.0×10^{-12}	1070 ± 500	1.1×10^{-13}	2.0
$OH + CH_3CF_3 \rightarrow CH_2CF_3 + H_2O$ (HFC-143a)	1.8×10^{-12}	2170 ± 150	1.2×10^{-15}	1.1
$OH + CH_2FCHF_2 \rightarrow$ products (HFC-143)	4.0×10^{-12}	1650 ± 300	1.6×10^{-14}	1.5
$OH + CH_2FCF_3 \rightarrow CHFCF_3 + H_2O$ (HFC-134a)	1.5×10^{-12}	1750 ± 200	4.2×10^{-15}	1.1
$OH + CHF_2CHF_2 \rightarrow CF_2CHF_2 + H_2O$ (HFC-134)	1.6×10^{-12}	1680 ± 300	5.7×10^{-15}	2.0
$OH + CHF_2CF_3 \rightarrow CF_2CF_3 + H_2O$ (HFC-125)	5.6×10^{-13}	1700 ± 300	1.9×10^{-15}	1.3
$F + O_3 \rightarrow FO + O_2$	2.2×10^{-11}	230 ± 200	1.0×10^{-11}	1.5
$F + H_2 \rightarrow HF + H$	1.4×10^{-10}	500 ± 200	2.6×10^{-11}	1.2
$F + H_2O \rightarrow HF + OH$	1.4×10^{-11}	0 ± 200	1.4×10^{-11}	1.3
$F + HNO_3 \rightarrow HF + NO_3$	6.0×10^{-12}	$-(400 \pm 200)$	2.3×10^{-11}	1.3
$F + CH_4 \rightarrow HF + CH_3$	1.6×10^{-10}	260 ± 200	6.7×10^{-11}	1.4
$FO + O_3 \rightarrow$ products			$< 1 \times 10^{-14}$	
$FO + NO \rightarrow NO_2 + F$	8.2×10^{-12}	$-(300 \pm 200)$	2.2×10^{-11}	1.5
$FO + FO \rightarrow 2F + O_2$	1.0×10^{-11}	0 ± 250	1.0×10^{-11}	1.5
$FO_2 + O_3 \rightarrow$ products			$< 3.4 \times 10^{-16}$	
$FO_2 + NO \rightarrow FNO + O_2$	7.5×10^{-12}	690 ± 400	7.5×10^{-13}	2.0
$FO_2 + NO_2 \rightarrow$ products	3.8×10^{-11}	2040 ± 500	4.0×10^{-14}	2.0

Appendix 10.2 (*continued*)

Reaction	A-Factor[a]	E/R ±ΔE/R	k(298 K)[a]	f(298)[b]
$FO_2 + CO \to$ products			$< 5.1 \times 10^{-16}$	
$FO_2 + CH_4 \to$ products			$< 2 \times 10^{-16}$	
$CF_3O + O_2 \to FO_2 + CF_2O$	$< 3 \times 10^{-11}$	5000	$< 1.5 \times 10^{-18}$	
$CF_3O + O_3 \to CF_3O_2 + O_2$	2×10^{-12}	1400 ± 600	1.8×10^{-14}	1.3
$CF_3O + H_2O \to OH + CF_3OH$	3×10^{-12}	> 3600	$< 2 \times 10^{-17}$	
$CF_3O + NO \to CF_2O + FNO$	3.7×10^{-11}	$-(110 \pm 70)$	5.4×10^{-11}	1.2
$CF_3O + CO \to$ products			$< 2 \times 10^{-15}$	
$CF_3O + CH_4 \to CH_3 + CF_3OH$	2.6×10^{-12}	1420 ± 200	2.2×10^{-14}	1.1
$CF_3O + C_2H_6 \to C_2H_5 + CF_3OH$	4.9×10^{-12}	400 ± 100	1.3×10^{-12}	1.2
$CF_3O_2 + O_3 \to CF_3O + 2O_2$			$< 3 \times 10^{-15}$	
$CF_3O_2 + CO \to CF_3O + CO_2$			$< 5 \times 10^{-16}$	
$CF_3O_2 + NO \to CF_3O + NO_2$	5.4×10^{-12}	$-(320 \pm 150)$	1.6×10^{-11}	1.1
ClO_x Reactions				
$O + ClO \to Cl + O_2$	3.0×10^{-11}	$-(70 \pm 70)$	3.8×10^{-11}	1.2
$O + OClO \to ClO + O_2$	2.4×10^{-12}	960 ± 300	1.0×10^{-13}	2.0
$O + Cl_2O \to ClO + ClO$	2.7×10^{-11}	530 ± 150	4.5×10^{-12}	1.3
$O + HCl \to OH + Cl$	1.0×10^{-11}	3300 ± 350	1.5×10^{-16}	2.0
$O + HOCl \to OH + ClO$	1.7×10^{-13}	0 ± 300	1.7×10^{-13}	3.0
$O + ClONO_2 \to$ products	2.9×10^{-12}	800 ± 200	2.0×10^{-13}	1.5
$O_3 + OClO \to$ products	2.1×10^{-12}	4700 ± 1000	3.0×10^{-19}	2.5
$O_3 + Cl_2O_2 \to$ products	—	—	$< 1.0 \times 10^{-19}$	—
$OH + Cl_2 \to HOCl + Cl$	1.4×10^{-12}	900 ± 400	6.7×10^{-14}	1.2
$OH + ClO \to$ products	1.1×10^{-11}	$-(120 \pm 150)$	1.7×10^{-11}	1.5
$OH + OClO \to HOCl + O_2$	4.5×10^{-13}	$-(800 \pm 200)$	6.8×10^{-12}	2.0
$OH + HCl \to H_2O + Cl$	2.6×10^{-12}	350 ± 100	8.0×10^{-13}	1.2
$OH + HOCl \to H_2O + ClO$	3.0×10^{-12}	500 ± 500	5.0×10^{-13}	3.0
$OH + ClNO_2 \to HOCl + NO_2$	2.4×10^{-12}	1250 ± 300	3.6×10^{-14}	2.0
$OH + ClONO_2 \to$ products	1.2×10^{-12}	330 ± 200	3.9×10^{-13}	1.5
$OH + CH_3Cl \to CH_2Cl + H_2O$	4.0×10^{-12}	1400 ± 250	3.6×10^{-14}	1.2
$OH + CH_2Cl_2 \to CHCl_2 + H_2O$	2.1×10^{-12}	900 ± 100	1.0×10^{-13}	1.1
$OH + CHCl_3 \to CCl_3 + H_2O$	1.1×10^{-12}	715 ± 150	1.0×10^{-13}	1.2
$OH + CCl_4 \to$ products	$\sim 1.0 \times 10^{-12}$	> 2300	$< 5.0 \times 10^{-16}$	—
$OH + CFCl_3 \to$ products (CFC-11)	$\sim 1.0 \times 10^{-12}$	> 3700	$< 5.0 \times 10^{-18}$	—
$OH + CF_2Cl_2 \to$ products (CFC-12)	$\sim 1.0 \times 10^{-12}$	> 3600	$< 6.0 \times 10^{-18}$	—
$OH + CH_2ClF \to CHClF + H_2O$ (HCFC-31)	2.8×10^{-12}	1270 ± 200	3.9×10^{-14}	1.2
$OH + CHFCl_2 \to CFCl_2 + H_2O$ (HCFC-21)	1.7×10^{-12}	1250 ± 150	2.6×10^{-14}	1.2
$OH + CHF_2Cl \to CF_2Cl + H_2O$ (HCFC-22)	1.0×10^{-12}	1600 ± 150	4.7×10^{-15}	1.1
$OH + CH_3OCl \to$ products	2.4×10^{-12}	360 ± 200	7.2×10^{-13}	3.0
$OH + CH_3CCl_3 \to CH_2CCl_3 + H_2O$ (HCC-140)	1.8×10^{-12}	1550 ± 150	1.0×10^{-14}	1.1
$OH + C_2HCl_3 \to$ products	4.9×10^{-13}	$-(450 \pm 200)$	2.2×10^{-12}	1.25
$OH + C_2Cl_4 \to$ products	9.4×10^{-12}	1200 ± 200	1.7×10^{-13}	1.25
$OH + CCl_3CHO \to H_2O + CCl_3CO$	8.2×10^{-12}	600 ± 300	1.1×10^{-12}	1.5
$OH + CH_3CFCl_2 \to CH_2CFCl_2 + H_2O$ (HCFC-141b)	1.7×10^{-12}	1700 ± 150	5.7×10^{-15}	1.2
$OH + CH_3CF_2Cl \to CH_2CF_2Cl + H_2O$ (HCFC-142b)	1.3×10^{-12}	1800 ± 150	3.1×10^{-15}	1.2
$OH + CH_2ClCF_2Cl \to CHClCF_2Cl + H_2O$ (HCFC-132b)	3.6×10^{-12}	1600 ± 400	1.7×10^{-14}	2.0
$OH + CHCl_2CF_2Cl \to CCl_2CF_2Cl + H_2O$ (HCFC-122)	1.0×10^{-12}	900 ± 150	4.9×10^{-14}	1.2

(*continued*)

Appendix 10.2 (*continued*)

Reaction	A-Factor[a]	E/R ±ΔE/R	k(298 K)[a]	f(298)[b]
OH + CHFClCFCl$_2$ → CFClCFCl$_2$ + H$_2$O (HCFC-122a)	1.0×10^{-12}	1250 ± 150	1.5×10^{-14}	1.1
OH + CH$_2$ClCF$_3$ → CHClCF$_3$ + H$_2$O (HCFC-133a)	2.3×10^{-12}	1640 ± 200	9.4×10^{-15}	1.2
OH + CHCl$_2$CF$_3$ → CCl$_2$CF$_3$ + H$_2$O (HCFC-123)	7.0×10^{-13}	900 ± 150	3.4×10^{-14}	1.2
OH + CHFClCF$_2$Cl → CFClCF$_2$Cl + H$_2$O (HCFC-123a)	9.2×10^{-13}	1280 ± 150	1.3×10^{-14}	1.2
OH + CHFClCF$_3$ → CFClCF$_3$ + H$_2$O (HCFC-124)	8.0×10^{-13}	1350 ± 150	8.6×10^{-15}	1.2
HO$_2$ + Cl → HCl + O$_2$	1.8×10^{-11}	$-(170 \pm 200)$	3.2×10^{-11}	1.5
→ OH + ClO	4.1×10^{-11}	450 ± 200	9.1×10^{-12}	2.0
HO$_2$ + ClO → HOCl + O$_2$	4.8×10^{-13}	$-(700 \pm ^{250}_{700})$	5.0×10^{-12}	1.4
H$_2$O + ClONO$_2$ → products	—	—	$< 2.0 \times 10^{-21}$	—
NO + OClO → NO$_2$ + ClO	2.5×10^{-12}	600 ± 300	3.4×10^{-13}	2.0
NO + Cl$_2$O$_2$ → products	—	—	$< 2.0 \times 10^{-14}$	—
NO$_3$ + HCl → HNO$_3$ + Cl	—	—	$< 5.0 \times 10^{-17}$	—
HO$_2$NO$_2$ + HCl → products	—	—	$< 1.0 \times 10^{-21}$	—
Cl + O$_3$ → ClO + O$_2$	2.9×10^{-11}	260 ± 100	1.2×10^{-11}	1.15
Cl + H$_2$ → HCl + H	3.7×10^{-11}	2300 ± 200	1.6×10^{-14}	1.25
Cl + H$_2$O$_2$ → HCl + HO$_2$	1.1×10^{-11}	980 ± 500	4.1×10^{-13}	1.5
Cl + NO$_3$ → ClO + NO$_2$	2.4×10^{-11}	0 ± 400	2.4×10^{-11}	1.5
Cl + HNO$_3$ → products	—	—	$< 2.0 \times 10^{-16}$	—
Cl + CH$_4$ → HCl + CH$_3$	1.1×10^{-11}	1400 ± 150	1.0×10^{-13}	1.1
Cl + CH$_3$D → products	—	—	7.4×10^{-14}	2.0
Cl + H$_2$CO → HCl + HCO	8.1×10^{-11}	30 ± 100	7.3×10^{-11}	1.15
Cl + CH$_3$O$_2$ → products	—	—	1.6×10^{-10}	1.5
Cl + CH$_3$OH → CH$_2$OH + HCl	5.4×10^{-11}	0 ± 250	5.4×10^{-11}	1.5
Cl + C$_2$H$_6$ → HCl + C$_2$H$_5$	7.7×10^{-11}	90 ± 90	5.7×10^{-11}	1.1
Cl + C$_2$H$_5$O$_2$ → ClO + C$_2$H$_5$O	—	—	7.4×10^{-11}	2.0
→ HCl + C$_2$H$_4$O$_2$	—	—	7.7×10^{-11}	2.0
Cl + CH$_3$CN → products	1.6×10^{-11}	2140 ± 300	1.2×10^{-14}	2.0
Cl + CH$_3$CO$_3$NO$_2$ → products	—	—	$< 1 \times 10^{-14}$	—
Cl + C$_3$H$_8$ → HCl + C$_3$H$_7$	1.2×10^{-10}	$-(40 \pm 250)$	1.4×10^{-10}	1.3
Cl + OClO → ClO + ClO	3.4×10^{-11}	$-(160 \pm 200)$	5.8×10^{-11}	1.25
Cl + ClOO → Cl$_2$ + O$_2$	2.3×10^{-10}	0 ± 250	2.3×10^{-10}	3.0
→ ClO + ClO	1.2×10^{-11}	0 ± 250	1.2×10^{-11}	3.0
Cl + Cl$_2$O → Cl$_2$ + ClO	6.2×10^{-11}	$-(130 \pm 130)$	9.6×10^{-11}	1.2
Cl + Cl$_2$O$_2$ → products	—	—	1.0×10^{-10}	2.0
Cl + HOCl → products	2.5×10^{-12}	130 ± 250	1.6×10^{-12}	1.5
Cl + ClNO → NO + Cl$_2$	5.8×10^{-11}	$-(100 \pm 200)$	8.1×10^{-11}	1.5
Cl + ClONO$_2$ → products	6.5×10^{-12}	$-(135 \pm 50)$	1.0×10^{-11}	1.2
Cl + CH$_3$Cl → CH$_2$Cl + HCl	3.2×10^{-11}	1250 ± 200	4.8×10^{-13}	1.2
Cl + CH$_2$Cl$_2$ → HCl + CHCl$_2$	3.1×10^{-11}	1350 ± 500	3.3×10^{-13}	1.5
Cl + CHCl$_3$ → HCl + CCl$_3$	8.2×10^{-12}	1325 ± 300	9.6×10^{-14}	1.3
Cl + CH$_3$F → HCl + CH$_2$F (HFC-41)	2.0×10^{-11}	1200 ± 500	3.5×10^{-13}	1.3
Cl + CH$_2$F$_2$ → HCl + CHF$_2$ (HFC-32)	1.2×10^{-11}	1630 ± 500	5.0×10^{-14}	1.5
Cl + CF$_3$H → HCl + CF$_3$ (HFC-23)	—	—	3.0×10^{-18}	5.0
Cl + CH$_2$FCl → HCl + CHFCl (HCFC-31)	1.2×10^{-11}	1390 ± 500	1.1×10^{-13}	2.0
Cl + CHFCl$_2$ → HCl + CFCl$_2$ (HCFC-21)	5.5×10^{-12}	1675 ± 200	2.0×10^{-14}	1.3

Appendix 10.2 (continued)

Reaction	A-Factor[a]	E/R ±ΔE/R	k(298 K)[a]	f(298)[b]
Cl + CHF$_2$Cl → HCl + CF$_2$Cl (HCFC-22)	5.9×10^{-12}	2430 ± 200	1.7×10^{-15}	1.3
Cl + CH$_3$CCl$_3$ → CH$_2$CCl$_3$ + HCl	2.8×10^{-12}	1790 ± 400	7.0×10^{-15}	2.0
Cl + CH$_3$CH$_2$F → HCl + CH$_3$CHF (HCFC-161)	1.8×10^{-11}	290 ± 500	6.8×10^{-12}	3.0
→ HCl + CH$_2$CH$_2$F	1.4×10^{-11}	880 ± 500	7.3×10^{-13}	3.0
Cl + CH$_3$CHF$_2$ → HCl + CH$_3$CF$_2$ (HFC-152a)	6.4×10^{-12}	950 ± 500	2.6×10^{-13}	1.3
→ HCl + CH$_2$CHF$_2$	7.2×10^{-12}	2390 ± 500	2.4×10^{-15}	3.0
Cl + CH$_2$FCH$_2$F → HCl + CHFCH$_2$F (HFC-152)	2.6×10^{-11}	1060 ± 500	7.5×10^{-13}	3.0
Cl + CH$_3$CFCl$_2$ → HCl + CH$_2$CFCl$_2$ (HCFC-141b)	1.8×10^{-12}	2000 ± 300	2.2×10^{-15}	1.2
Cl + CH$_3$CF$_2$Cl → HCl + CH$_2$CF$_2$Cl (HCFC-142b)	1.4×10^{-12}	2420 ± 500	4.2×10^{-16}	1.2
Cl + CH$_3$CF$_3$ → HCl + CH$_2$CF$_3$ (HFC-143a)	1.2×10^{-11}	3880 ± 500	2.6×10^{-17}	5.0
Cl + CH$_2$FCHF$_2$ → HCl + CH$_2$FCF$_2$ (HFC-143)	5.5×10^{-12}	1610 ± 500	2.5×10^{-14}	3.0
→ HCl + CHFCHF$_2$	7.7×10^{-12}	1720 ± 500	2.4×10^{-14}	3.0
Cl + CH$_2$ClCF$_3$ → HCl + CHClCF$_3$ (HCFC-133a)	1.8×10^{-12}	1710 ± 500	5.9×10^{-15}	3.0
Cl + CH$_2$FCF$_3$ → HCl + CHFCF$_3$ (HFC-134a)	—	—	1.5×10^{-15}	1.2
Cl + CHF$_2$CHF$_2$ → HCl + CF$_2$CHF$_2$ (HCF-134)	7.5×10^{-12}	2430 ± 500	2.2×10^{-15}	1.5
Cl + CHCl$_2$CF$_3$ → HCl + CCl$_2$CF$_3$ (HCFC-123)	4.4×10^{-12}	1750 ± 500	1.2×10^{-14}	1.3
Cl + CHFClCF$_3$ → HCl + CFClCF$_3$ (HCFC-124)	1.1×10^{-12}	1800 ± 500	2.7×10^{-15}	1.3
Cl + CHF$_2$CF$_3$ → HCl + CF$_2$CF$_3$ (HFC-125)	—	—	2.4×10^{-16}	1.3
ClO + O$_3$ → ClOO + O$_2$	—	—	$< 1.4 \times 10^{-17}$	—
→ OClO + O$_2$	1.0×10^{-12}	> 4000	$< 1.0 \times 10^{-18}$	—
ClO + H$_2$ → products	$\sim 1.0 \times 10^{-12}$	> 4800	$< 1.0 \times 10^{-19}$	—
ClO + NO → NO$_2$ + Cl	6.4×10^{-12}	$-(290 \pm 100)$	1.7×10^{-11}	1.15
ClO + NO$_3$ → ClOO + NO$_2$	4.7×10^{-13}	0 ± 400	4.7×10^{-13}	1.5
ClO + N$_2$O → products	$\sim 1.0 \times 10^{-12}$	> 4300	$< 6.0 \times 10^{-19}$	—
ClO + CO → products	$\sim 1.0 \times 10^{-12}$	> 3700	$< 4.0 \times 10^{-18}$	—
ClO + CH$_4$ → products	$\sim 1.0 \times 10^{-12}$	> 3700	$< 4.0 \times 10^{-18}$	—
ClO + H$_2$CO → products	$\sim 1.0 \times 10^{-12}$	> 2100	$< 1.0 \times 10^{-15}$	—
ClO + CH$_3$O$_2$ → products	3.3×10^{-12}	115 ± 115	2.2×10^{-12}	1.5
ClO + ClO → Cl$_2$ + O$_2$	1.0×10^{-12}	1590 ± 300	4.8×10^{-15}	1.5
→ ClOO + Cl	3.0×10^{-11}	2450 ± 500	8.0×10^{-15}	1.5
→ OClO + Cl	3.5×10^{-13}	1370 ± 300	3.5×10^{-15}	1.5
HCl + ClONO$_2$ → products	—	—	$< 1.0 \times 10^{-20}$	—
BrO$_x$ Reactions				
O + BrO → Br + O$_2$	1.9×10^{-11}	$-(230 \pm 150)$	4.1×10^{-11}	1.5
O + HBr → OH + Br	5.8×10^{-12}	1500 ± 200	3.8×10^{-14}	1.3
O + HOBr → OH + BrO	1.2×10^{-10}	430 ± 300	2.8×10^{-11}	3.0
OH + Br$_2$ → HOBr + Br	4.2×10^{-11}	0 ± 600	4.2×10^{-11}	1.3
OH + BrO → products	—	—	7.5×10^{-11}	3.0
OH + HBr → H$_2$O + Br	1.1×10^{-11}	0 ± 250	1.1×10^{-11}	1.2
OH + CH$_3$Br → CH$_2$Br + H$_2$O	4.0×10^{-12}	1470 ± 150	2.9×10^{-14}	1.1
OH + CH$_2$Br$_2$ → CHBr$_2$ + H$_2$O	2.4×10^{-12}	900 ± 300	1.2×10^{-13}	1.1
OH + CHBr$_3$ → CBr$_3$ + H$_2$O	1.6×10^{-12}	710 ± 200	1.5×10^{-13}	2.0

(continued)

Appendix 10.2 (continued)

Reaction	A-Factor[a]	E/R ±ΔE/R	k(298 K)[a]	f(298)[b]
$OH + CHF_2Br \to CF_2Br + H_2O$	1.1×10^{-12}	1400 ± 200	1.0×10^{-14}	1.1
$OH + CH_2ClBr \to CHClBr + H_2O$	2.3×10^{-12}	930 ± 150	1.0×10^{-13}	1.2
$OH + CF_2ClBr \to$ products	—	—	$< 1.5 \times 10^{-16}$	—
$OH + CF_2Br_2 \to$ products	—	—	$< 5.0 \times 10^{-16}$	—
$OH + CF_3Br \to$ products	—	—	$< 1.2 \times 10^{-16}$	—
$OH + CH_2BrCF_3 \to CHBrCF_3 + H_2O$	1.4×10^{-12}	1340 ± 200	1.6×10^{-14}	1.3
$OH + CHFBrCF_3 \to CFBrCF_3$	7.2×10^{-13}	1110 ± 150	1.8×10^{-14}	1.5
$OH + CHClBrCF_3 \to CClBrCF_3 + H_2O$	1.3×10^{-12}	995 ± 150	4.5×10^{-14}	1.5
$OH + CF_2BrCHFCl \to CF_2BrCFCl + H_2O$	9.3×10^{-13}	1250 ± 150	1.4×10^{-14}	1.5
$OH + CF_2BrCF_2Br \to$ products	—	—	$< 1.5 \times 10^{-16}$	—
$HO_2 + Br \to HBr + O_2$	1.5×10^{-11}	600 ± 600	2.0×10^{-12}	2.0
$HO_2 + BrO \to$ products	3.4×10^{-12}	$-(540 \pm 200)$	2.1×10^{-11}	1.5
$NO_3 + HBr \to HNO_3 + Br$	—	—	$< 1.0 \times 10^{-16}$	—
$Cl + CH_2ClBr \to HCl + CHClBr$	4.3×10^{-11}	1370 ± 500	4.3×10^{-13}	3.0
$Cl + CH_3Br \to HCl + CH_2Br$	1.5×10^{-11}	1060 ± 100	4.3×10^{-13}	1.2
$Cl + CH_2Br_2 \to HCl + CHBr_2$	6.4×10^{-12}	810 ± 100	4.2×10^{-13}	1.2
$Br + O_3 \to BrO + O_2$	1.7×10^{-11}	800 ± 200	1.2×10^{-12}	1.2
$Br + H_2O_2 \to HBr + HO_2$	1.0×10^{-11}	> 3000	$< 5.0 \times 10^{-16}$	—
$Br + NO_3 \to BrO + NO_2$	—	—	1.6×10^{-11}	2.0
$Br + H_2CO \to HBr + HCO$	1.7×10^{-11}	800 ± 200	1.1×10^{-12}	1.3
$Br + OClO \to BrO + ClO$	2.6×10^{-11}	1300 ± 300	3.4×10^{-13}	2.0
$Br + Cl_2O \to BrCl + ClO$	2.1×10^{-11}	470 ± 150	4.3×10^{-12}	1.3
$Br + Cl_2O_2 \to$ products	—	—	3.0×10^{-12}	2.0
$BrO + O_3 \to$ products	$\sim 1.0 \times 10^{-12}$	> 3200	$< 2.0 \times 10^{-17}$	—
$BrO + NO \to NO_2 + Br$	8.8×10^{-12}	$-(260 \pm 130)$	2.1×10^{-11}	1.15
$BrO + NO_3 \to$ products	—	—	1.0×10^{-12}	3.0
$BrO + ClO \to Br + OClO$	1.6×10^{-12}	$-(430 \pm 200)$	6.8×10^{-12}	1.25
$\to Br + ClOO$	2.9×10^{-12}	$-(220 \pm 200)$	6.1×10^{-12}	1.25
$\to BrCl + O_2$	5.8×10^{-13}	$-(170 \pm 200)$	1.0×10^{-12}	1.25
$BrO + BrO \to$ products	1.5×10^{-12}	$-(230 \pm 150)$	3.2×10^{-12}	1.15
$CH_2BrO_2 + NO \to CH_2O + NO_2 + Br$	4×10^{-12}	$-(300 \pm 200)$	1.1×10^{-11}	1.5
IO$_x$ Reactions				
$O + I_2 \to IO + I$	1.4×10^{-10}	0 ± 250	1.4×10^{-10}	1.4
$O + IO \to O_2 + I$			1.2×10^{-10}	2.0
$OH + I_2 \to HOI + I$			1.8×10^{-10}	2.0
$OH + HI \to H_2O + I$			3.0×10^{-11}	2.0
$OH + CH_3I \to H_2O + CH_2I$	3.1×10^{-12}	1120 ± 500	7.2×10^{-14}	3.0
$OH + CF_3I \to HOI + CF_3$			3.1×10^{-14}	5.0
$HO_2 + I \to HI + O_2$	1.5×10^{-11}	1090 ± 500	3.8×10^{-13}	2.0
$HO_2 + IO \to HOI + O_2$			8.4×10^{-11}	1.5
$I + O_3 \to IO + O_2$	2.3×10^{-11}	870 ± 200	1.2×10^{-12}	1.2
$I + BrO \to IO + Br$			1.2×10^{-11}	2.0
$IO + NO \to I + NO_2$	9.1×10^{-12}	$-(240 \pm 150)$	2.0×10^{-11}	1.2
$IO + ClO \to$ products	5.1×10^{-12}	$-(280 \pm 200)$	1.3×10^{-11}	2.0
$IO + BrO \to$ products	—	—	6.9×10^{-11}	1.5
$IO + IO \to$ products	1.5×10^{-11}	$-(500 \pm 500)$	8.0×10^{-11}	1.5
$INO + INO \to I_2 + 2NO$	8.4×10^{-11}	2620 ± 600	1.3×10^{-14}	2.5
SO$_x$ Reactions				
$O + SH \to SO + H$	—	—	1.6×10^{-10}	5.0
$O + CS \to CO + S$	2.7×10^{-10}	760 ± 250	2.1×10^{-11}	1.1
$O + H_2S \to OH + SH$	9.2×10^{-12}	1800 ± 550	2.2×10^{-14}	1.7
$O + OCS \to CO + SO$	2.1×10^{-11}	2200 ± 150	1.3×10^{-14}	1.2
$O + CS_2 \to CS + SO$	3.2×10^{-11}	650 ± 150	3.6×10^{-12}	1.2
$O + CH_3SCH_3 \to CH_3SO + CH_3$	1.3×10^{-11}	$-(410 \pm 100)$	5.0×10^{-11}	1.1
$O + CH_3SSCH_3 \to CH_3SO + CH_3S$	5.5×10^{-11}	$-(250 \pm 100)$	1.3×10^{-10}	1.3
$O_3 + H_2S \to$ products	—	—	$< 2.0 \times 10^{-20}$	—

Appendix 10.2 (continued)

Reaction	A-Factor[a]	E/R ± ΔE/R	k(298 K)[a]	f(298)[b]
$O_3 + CH_3SCH_3 \to$ products	—	—	$< 1.0 \times 10^{-18}$	—
$O_3 + SO_2 \to SO_3 + O_2$	3.0×10^{-12}	> 7000	$< 2.0 \times 10^{-22}$	—
$OH + H_2S \to SH + H_2O$	6.0×10^{-12}	75 ± 75	4.7×10^{-12}	1.2
$OH + OCS \to$ products	1.1×10^{-13}	1200 ± 500	1.9×10^{-15}	2.0
$OH + CH_3SH \to CH_3S + H_2O$	9.9×10^{-12}	$-(360 \pm 100)$	3.3×10^{-11}	1.2
$OH + CH_3SCH_3 \to H_2O + CH_2SCH_3$	1.2×10^{-11}	260 ± 100	5.0×10^{-12}	1.15
$OH + CH_3SSCH_3 \to$ products	6.0×10^{-11}	$-(400 \pm 200)$	2.3×10^{-10}	1.2
$OH + S \to H + SO$	—	—	6.6×10^{-11}	3.0
$OH + SO \to H + SO_2$	—	—	8.6×10^{-11}	2.0
$HO_2 + H_2S \to$ products	—	—	$< 3.0 \times 10^{-15}$	—
$HO_2 + CH_3SH \to$ products	—	—	$< 4.0 \times 10^{-15}$	—
$HO_2 + CH_3SCH_3 \to$ products	—	—	$< 5.0 \times 10^{-15}$	—
$HO_2 + SO_2 \to$ products	—	—	$< 1.0 \times 10^{-18}$	—
$NO_2 + SO_2 \to$ products	—	—	$< 2.0 \times 10^{-26}$	—
$NO_3 + H_2S \to$ products	—	—	$< 8.0 \times 10^{-16}$	—
$NO_3 + OCS \to$ products	—	—	$< 1.0 \times 10^{-16}$	—
$NO_3 + CS_2 \to$ products	—	—	$< 4.0 \times 10^{-16}$	—
$NO_3 + CH_3SH \to$ products	4.4×10^{-13}	$-(210 \pm 210)$	8.9×10^{-13}	1.25
$NO_3 + CH_3SCH_3 \to CH_3SCH_2 + HNO_3$	1.9×10^{-13}	$-(500 \pm 200)$	1.0×10^{-12}	1.2
$NO_3 + CH_3SSCH_3 \to$ products	1.3×10^{-12}	270 ± 270	5.3×10^{-13}	1.4
$NO_3 + SO_2 \to$ products	—	—	$< 7.0 \times 10^{-21}$	—
$N_2O_5 + CH_3SCH_3 \to$ products	—	—	$< 1.0 \times 10^{-17}$	—
$CH_3O_2 + SO_2 \to$ products	—	—	$< 5.0 \times 10^{-17}$	—
$F + CH_3SCH_3 \to$ products	—	—	2.4×10^{-10}	2.0
$Cl + H_2S \to HCl + SH$	3.7×10^{-11}	$-(210 \pm 100)$	7.4×10^{-11}	1.25
$Cl + OCS \to$ products	—	—	$< 1.0 \times 10^{-16}$	—
$Cl + CS_2 \to$ products	—	—	$< 4.0 \times 10^{-15}$	—
$Cl + CH_3SH \to CH_3S + HCl$	1.2×10^{-10}	$-(150 \pm 50)$	2.0×10^{-10}	1.25
$ClO + OCS \to$ products	—	—	$< 2.0 \times 10^{-16}$	—
$ClO + CH_3SCH_3 \to$ products	—	—	9.5×10^{-15}	2.0
$ClO + SO \to Cl + SO_2$	2.8×10^{-11}	0 ± 50	2.8×10^{-11}	1.3
$ClO + SO_2 \to Cl + SO_3$	—	—	$< 4.0 \times 10^{-18}$	—
$Br + H_2S \to HBr + SH$	1.4×10^{-11}	2750 ± 300	1.4×10^{-15}	2.0
$Br + CH_3SH \to CH_3S + HBr$	9.2×10^{-12}	390 ± 100	2.5×10^{-12}	2.0
$BrO + CH_3SCH_3 \to$ products	1.5×10^{-14}	$-(850 \pm 200)$	2.6×10^{-13}	1.3
$BrO + SO \to Br + SO_2$	—	—	5.7×10^{-11}	1.4
$IO + CH_3SH \to$ products	—	—	6.6×10^{-16}	2.0
$IO + CH_3SCH_3 \to$ products	—	—	1.2×10^{-14}	1.5
$S + O_2 \to SO + O$	2.3×10^{-12}	0 ± 200	2.3×10^{-12}	1.2
$S + O_3 \to SO + O_2$	—	—	1.2×10^{-11}	2.0
$SO + O_2 \to SO_2 + O$	2.6×10^{-13}	2400 ± 500	8.4×10^{-17}	2.0
$SO + O_3 \to SO_2 + O_2$	3.6×10^{-12}	1100 ± 200	9.0×10^{-14}	1.2
$SO + NO_2 \to SO_2 + NO$	1.4×10^{-11}	0 ± 50	1.4×10^{-11}	1.2
$SO + OClO \to SO_2 + ClO$	—	—	1.9×10^{-12}	3.0
$SO_3 + NO_2 \to$ products	—	—	1.0×10^{-19}	10.0
$SH + O_2 \to OH + SO$	—	—	$< 4.0 \times 10^{-19}$	—
$SH + O_3 \to HSO + O_2$	9.0×10^{-12}	280 ± 200	3.5×10^{-12}	1.3
$SH + H_2O_2 \to$ products	—	—	$< 5.0 \times 10^{-15}$	—
$SH + NO_2 \to HSO + NO$	2.9×10^{-11}	$-(240 \pm 50)$	6.5×10^{-11}	1.2
$SH + Cl_2 \to ClSH + Cl$	1.7×10^{-11}	690 ± 200	1.7×10^{-12}	2.0
$SH + BrCl \to$ products	2.3×10^{-11}	$-(350 \pm 200)$	7.4×10^{-11}	2.0
$SH + Br_2 \to BrSH + Br$	6.0×10^{-11}	$-(160 \pm 160)$	1.0×10^{-10}	2.0
$SH + F_2 \to FSH + F$	4.3×10^{-11}	1390 ± 200	4.0×10^{-13}	2.0
$HSO + O_2 \to$ products	—	—	$< 2.0 \times 10^{-17}$	—
$HSO + O_3 \to$ products	—	—	1.0×10^{-13}	1.3
$HSO + NO \to$ products	—	—	$< 1.0 \times 10^{-15}$	—

(continued)

Appendix 10.2 (continued)

Reaction	A-Factor[a]	E/R ±ΔE/R	k(298 K)[a]	f(298)[b]
$HSO + NO_2 \rightarrow HSO_2 + NO$			9.6×10^{-12}	2.0
$HSO_2 + O_2 \rightarrow HO_2 + SO_2$			3.0×10^{-13}	3.0
$HOSO_2 + O_2 \rightarrow HO_2 + SO_3$	1.3×10^{-12}	330 ± 200	4.4×10^{-13}	1.2
$CS + O_2 \rightarrow OCS + O$			2.9×10^{-19}	2.0
$CS + O_3 \rightarrow OCS + O_2$			3.0×10^{-16}	3.0
$CS + NO_2 \rightarrow OCS + NO$			7.6×10^{-17}	3.0
$CH_3S + O_2 \rightarrow$ products			$< 3.0 \times 10^{-18}$	—
$CH_3S + O_3 \rightarrow$ products	2.0×10^{-12}	$-(290 \pm 100)$	5.3×10^{-12}	1.15
$CH_3S + NO \rightarrow$ products			$< 1.0 \times 10^{-13}$	—
$CH_3S + NO_2 \rightarrow CH_3SO + NO$	2.1×10^{-11}	$-(320 \pm 100)$	6.1×10^{-11}	1.15
$CH_2SH + O_2 \rightarrow$ products			6.5×10^{-12}	2.0
$CH_2SH + O_3 \rightarrow$ products			3.5×10^{-11}	2.0
$CH_2SH + NO \rightarrow$ products			1.9×10^{-11}	2.0
$CH_2SH + NO_2 \rightarrow$ products			5.2×10^{-11}	2.0
$CH_3SO + O_3 \rightarrow$ products			6.0×10^{-13}	1.5
$CH_3SO + NO_2 \rightarrow CH_3SO_2 + NO$			1.2×10^{-11}	1.4
$CH_3SOO + O_3 \rightarrow$ products			$< 8.0 \times 10^{-13}$	—
$CH_3SOO + NO \rightarrow$ products	1.1×10^{-11}	0 ± 100	1.1×10^{-11}	2.0
$CH_3SO_2 + NO_2 \rightarrow$ products	2.2×10^{-11}	0 ± 100	2.2×10^{-11}	2.0
$CH_3SCH_2 + NO_3 \rightarrow$ products			3.0×10^{-10}	2.0
$CH_3SCH_2O_2 + NO \rightarrow CH_3SCH_2O + NO_2$			1.9×10^{-11}	2.0
$CH_3SS + O_3 \rightarrow$ products			4.6×10^{-13}	2.0
$CH_3SS + NO_2 \rightarrow$ products			1.8×10^{-11}	2.0
$CH_3SSO + NO_2 \rightarrow$ products			4.5×10^{-12}	2.0
Metal Reactions				
$Na + O_3 \rightarrow NaO + O_2$	1.0×10^{-9}	95 ± 50	7.3×10^{-10}	1.2
$\rightarrow NaO_2 + O$	—	—	$< 4.0 \times 10^{-11}$	—
$Na + N_2O \rightarrow NaO + N_2$	2.8×10^{-10}	1600 ± 400	1.3×10^{-12}	1.2
$Na + Cl_2 \rightarrow NaCl + Cl$	7.3×10^{-10}	0 ± 200	7.3×10^{-10}	1.3
$NaO + O \rightarrow Na + O_2$	3.7×10^{-10}	0 ± 400	3.7×10^{-10}	3.0
$NaO + O_3 \rightarrow NaO_2 + O_2$	1.1×10^{-9}	570 ± 300	1.6×10^{-10}	1.5
$\rightarrow Na + 2O_2$	6.0×10^{-11}	0 ± 800	6.0×10^{-11}	3.0
$NaO + H_2 \rightarrow NaOH + H$	2.6×10^{-11}	0 ± 600	2.6×10^{-11}	2.0
$NaO + H_2O \rightarrow NaOH + OH$	2.2×10^{-10}	0 ± 400	2.2×10^{-10}	2.0
$NaO + NO \rightarrow Na + NO_2$	1.5×10^{-10}	0 ± 400	1.5×10^{-10}	4.0
$NaO + HCl \rightarrow$ products	2.8×10^{-10}	0 ± 400	2.8×10^{-10}	3.0
$NaO_2 + O \rightarrow NaO + O_2$	2.2×10^{-11}	0 ± 600	2.2×10^{-11}	5.0
$NaO_2 + NO \rightarrow NaO + NO_2$	—	—	$< 10^{-14}$	—
$NaO_2 + HCl \rightarrow$ products	2.3×10^{-10}	0 ± 400	2.3×10^{-10}	3.0
$NaOH + HCl \rightarrow NaCl + H_2O$	2.8×10^{-10}	0 ± 400	2.8×10^{-10}	3.0

The chemical kinetics data are based on an extensive compilation and evaluation by DeMore, W. B., et al., 1997, Chemical kinetics and photochemical data for use in stratospheric modeling, Evaluation number 12, JPL Publication 94-26.

[a] Units are cm^3/molecule-s.
[b] f(298) is the uncertainty factor at 298 K. To calculate the uncertainty at other temperatures, use the expression:

$$f(T) = f(298) \exp \left| \frac{\Delta E}{R} \left(\frac{1}{T} - \frac{1}{298} \right) \right|$$

Note that the exponent is absolute value.

Appendix 10.3 Rate constants for association reactions

Reaction	Low Pressure Limit[a] $k_o(T) = k_o^{300} (T/300)^{-n}$		High Pressure Limit[b] $k_\infty(T) = k_\infty^{300} (T/300)^{-m}$	
	k_o^{300}	n	k_∞^{300}	m
Ox Reactions				
$O + O_2 \xrightarrow{M} O_3$	$(6.0 \pm 0.5)(-34)$	2.3 ± 0.5	—	—
HOx Reactions				
$H + O_2 \xrightarrow{M} HO_2$	$(5.7 \pm 0.5)(-32)$	1.6 ± 0.5	$(7.5 \pm 4.0)(-11)$	0 ± 1.0
$OH + OH \xrightarrow{M} H_2O_2$	$(6.2 \pm 1.2)(-31)$	$1.0 \pm {}^{2.0}_{1.0}$	$(2.6 \pm 1.0)(-11)$	0 ± 0.5
NOx Reactions				
$O + NO \xrightarrow{M} NO_2$	$(9.0 \pm 2.0)(-32)$	1.5 ± 0.3	$(3.0 \pm 1.0)(-11)$	0 ± 1.0
$O + NO_2 \xrightarrow{M} NO_3$	$(9.0 \pm 1.0)(-32)$	2.0 ± 1.0	$(2.2 \pm 0.3)(-11)$	0 ± 1.0
$OH + NO \xrightarrow{M} HONO$	$(7.0 \pm 1.0)(-31)$	2.6 ± 0.3	$(3.6 \pm 1.0)(-11)$	0.1 ± 0.5
$OH + NO_2 \xrightarrow{M} HNO_3$	$(2.5 \pm 0.1)(-30)$	4.4 ± 0.3	$(1.6 \pm 0.2)(-11)$	1.7 ± 0.2
$HO_2 + NO_2 \rightarrow HO_2NO_2$	$(1.8 \pm 0.3)(-31)$	3.2 ± 0.4	$(4.7 \pm 1.0)(-12)$	1.4 ± 1.4
$NO_2 + NO_3 \xrightarrow{M} N_2O_5$	$(2.2 \pm 0.5)(-30)$	3.9 ± 1.0	$(1.5 \pm 0.8)(-12)$	0.7 ± 0.4
Hydrocarbon Reactions				
$CH_3 + O_2 \xrightarrow{M} CH_3O_2$	$(4.5 \pm 1.5)(-31)$	3.0 ± 1.0	$(1.8 \pm 0.2)(-12)$	1.7 ± 1.7
$C_2H_5 + O_2 \xrightarrow{M} C_2H_5O_2$	$(1.5 \pm 1.0)(-28)$	3.0 ± 1.0	$(8.0 \pm 1.0)(-12)$	0 ± 1.0
$OH + C_2H_2 \xrightarrow{M} HOCHCH$	$(5.5 \pm 2.0)(-30)$	0.0 ± 0.2	$(8.3 \pm 1.0)(-13)$	$-2 \pm {}^{2}_{1}$
$OH + C_2H_4 \xrightarrow{M} HOCH_2CH_2$	$(1.0 \pm 0.6)(-28)$	0.8 ± 2.0	$(8.8 \pm 0.9)(-12)$	$0 \pm {}^{1}_{2}$
$CH_3O + NO \xrightarrow{M} CH_3ONO$	$(1.4 \pm 0.5)(-29)$	3.8 ± 1.0	$(3.6 \pm 1.6)(-11)$	0.6 ± 1.0
$CH_3O + NO_2 \xrightarrow{M} CH_3ONO_2$	$(1.1 \pm 0.4)(-28)$	4.0 ± 2.0	$(1.6 \pm 0.5)(-11)$	1.0 ± 1.0
$C_2H_5O + NO \xrightarrow{M} C_2H_5ONO$	$(2.8 \pm 1.0)(-27)$	4.0 ± 2.0	$(5.0 \pm 1.0)(-11)$	1.0 ± 1.0
$C_2H_5O + NO_2 \xrightarrow{M} C_2H_5ONO_2$	$(2.0 \pm 1.0)(-27)$	4.0 ± 2.0	$(2.8 \pm 0.4)(-11)$	1.0 ± 1.0
$CH_3O_2 + NO_2 \xrightarrow{M} CH_3O_2NO_2$	$(1.5 \pm 0.8)(-30)$	4.0 ± 2.0	$(6.5 \pm 3.2)(-12)$	2.0 ± 2.0
$CH_3C(O)O_2 + NO_2 \xrightarrow{M} CH_3C(O)O_2NO_2$	$(9.7 \pm 3.8)(-29)$	5.6 ± 2.8	$(9.3 \pm 0.4)(-12)$	1.5 ± 0.3
FOx Reactions				
$F + O_2 \xrightarrow{M} FO_2$	$(4.4 \pm 0.4)(-33)$	1.2 ± 0.5	—	—
$F + NO \xrightarrow{M} FNO$	$(1.8 \pm 0.3)(-31)$	1.0 ± 10	$(2.8 \pm 1.4)(-10)$	0.0 ± 1.0
$F + NO_2 \xrightarrow{M} FNO_2$	$(6.3 \pm 3.0)(-32)$	2.0 ± 2.0	$(2.6 \pm 1.3)(-10)$	0.0 ± 1.0
$FO + NO_2 \xrightarrow{M} FONO_2$	$(2.6 \pm 2.0)(-31)$	1.3 ± 1.3	$(2.0 \pm 1.0)(-11)$	1.5 ± 1.5
ClOx Reactions				
$Cl + O_2 \xrightarrow{M} ClOO$	$(2.7 \pm 1.0)(-33)$	1.5 ± 0.5	—	—
$Cl + NO \xrightarrow{M} ClNO$	$(9.0 \pm 2.0)(-32)$	1.6 ± 0.5	—	—
$Cl + NO_2 \xrightarrow{M} ClONO$	$(1.3 \pm 0.2)(-30)$	2.0 ± 1.0	$(1.0 \pm 0.5)(-10)$	1.0 ± 1.0
$\xrightarrow{} ClNO_2$	$(1.8 \pm 0.3)(-31)$	2.0 ± 1.0	$(1.0 \pm 0.5)(-10)$	1.0 ± 1.0
$Cl + CO \xrightarrow{M} ClCO$	$(1.3 \pm 0.5)(-33)$	3.8 ± 0.5	—	—
$Cl + C_2H_2 \xrightarrow{M} ClC_2H_2$	$(5.9 \pm 1.0)(-30)$	2.1 ± 1.0	$(2.1 \pm 0.4)(-10)$	1.0 ± 0.5
$Cl + C_2H_4 \xrightarrow{M} ClC_2H_4$	$(1.6 \pm 1)(-29)$	3.3 ± 1.0	$(3.1 \pm 2)(-10)$	1.0 ± 0.5
$Cl + C_2Cl_4 \rightarrow C_2Cl_5$	$(1.4 \pm 0.6)(-28)$	8.5 ± 1.0	$(4.0 \pm 1.0)(-11)$	1.2 ± 0.5
$ClO + NO_2 \rightarrow ClONO_2$	$(1.8 \pm 0.3)(-31)$	3.4 ± 1.0	$(1.5 \pm 0.7)(-11)$	1.9 ± 1.9
$ClO + ClO \xrightarrow{M} Cl_2O_2$	$(2.2 \pm 0.4)(-32)$	3.1 ± 0.5	$(3.5 \pm 2)(-12)$	1.0 ± 1.0
$ClO + OClO \xrightarrow{M} Cl_2O_3$	$(6.2 \pm 1.0)(-32)$	4.7 ± 0.6	$(2.4 \pm 1.2)(-11)$	0 ± 1.0
$OClO + O \xrightarrow{M} ClO_3$	$(1.9 \pm 0.5)(-31)$	1.1 ± 1.0	$(3.1 \pm 0.8)(-11)$	0 ± 1.0
$CH_2Cl + O_2 \xrightarrow{M} CH_2ClO_2$	$(1.9 \pm 0.1)(-30)$	3.2 ± 0.2	$(2.9 \pm 0.2)(-12)$	1.2 ± 0.6
$CHCl_2 + O_2 \xrightarrow{M} CHCl_2O_2$	$(1.3 \pm 0.1)(-30)$	4.0 ± 0.2	$(2.8 \pm 0.2)(-12)$	1.4 ± 0.6
$CCl_3 + O_2 \xrightarrow{M} CCl_3O_2$	$(6.9 \pm 0.2)(-31)$	6.4 ± 0.3	$(2.4 \pm 0.2)(-12)$	2.1 ± 0.6
$CFCl_2 + O_2 \xrightarrow{M} CFCl_2O_2$	$(5.0 \pm 0.8)(-30)$	4.0 ± 2.0	$(6.0 \pm 1.0)(-12)$	1.0 ± 1.0
$CF_2Cl + O_2 \xrightarrow{M} CF_2ClO_2$	$(3.0 \pm 1.5)(-30)$	4.0 ± 2.0	$(3 \pm 2)(-12)$	1.0 ± 1.0
BrOx Reactions				
$Br + NO_2 \xrightarrow{M} BrNO_2$	$(4.2 \pm 0.8)(-31)$	2.4 ± 0.5	$(2.7 \pm 0.5)(-11)$	0 ± 1.0
$BrO + NO_2 \xrightarrow{M} BrONO_2$	$(5.2 \pm 0.6)(-31)$	3.2 ± 0.8	$(6.9 \pm 1.0)(-12)$	2.9 ± 1.0

(continued)

Appendix 10.3 (continued)

Reaction	Low Pressure Limit[a] $k_o(T) = k_o^{300} (T/300)^{-n}$		High Pressure Limit[b] $k_\infty(T) = k_\infty^{300} (T/300)^{-m}$	
	k_o^{300}	n	k_∞^{300}	m
IOx Reactions				
$I + NO \xrightarrow{M} INO$	$(1.8 \pm 0.5)(-32)$	1.0 ± 0.5	$(1.7 \pm 1.0)(-11)$	0 ± 1.0
$I + NO_2 \xrightarrow{M} INO_2$	$(3.0 \pm 1.5)(-31)$	1.0 ± 1.0	$(6.6 \pm 5.0)(-11)$	0 ± 1.0
$IO + NO_2 \xrightarrow{M} IONO_2$	$(5.9 \pm 2.0)(-31)$	3.5 ± 1.0	$(9.0 \pm 1.0)(-12)$	1.5 ± 1.0
SOx Reactions				
$O + SO_2 \xrightarrow{M} SO_3$	$(1.3 \pm ^{1.3}_{0.7})(-33)$	-3.6 ± 0.7		
$OH + SO_2 \xrightarrow{M} HOSO_2$	$(3.0 \pm 1.0)(-31)$	3.3 ± 1.5	$(1.5 \pm 0.5)(-12)$	$0 \pm ^0_2$
Metal Reactions				
$Na + O_2 \xrightarrow{M} NaO_2$	$(3.2 \pm 0.3)(-30)$	1.4 ± 0.3	$(6.0 \pm 2.0)(-10)$	0 ± 1.0
$NaO + O_2 \xrightarrow{M} NaO_3$	$(3.5 \pm 0.7)(-30)$	2.0 ± 2.0	$(5.7 \pm 3.0)(-10)$	0 ± 1.0
$NaO + CO_2 \xrightarrow{M} NaCO_3$	$(8.7 \pm 2.6)(-28)$	2.0 ± 2.0	$(6.5 \pm 3.0)(-10)$	0 ± 1.0
$NaOH + CO_2 \xrightarrow{M} NaHCO_3$	$(1.3 \pm 0.3)(-28)$	2.0 ± 2.0	$(6.8 \pm 4.0)(-10)$	0 ± 1.0

The chemical kinetics data are taken from the same source as Appendix 10.2.

Note: $k(Z) = k(M, T) = \left(\frac{k_o(T)[M]}{1 + (k_o(T)[M]/k_\infty(T))} \right) 0.6^{\{1+[\log_{10}(k_o(T)[M]/k_\infty(T))]^2\}^{-1}}$

The values quoted are suitable for air as the third body, M.
[a] Units are cm^6/molecule^2-sec.
[b] Units are cm^3/molecule-sec.

Appendix 10.4 Equilibrium constants

Reaction	A/cm^3 molecule^{-1}	B $\pm$ ΔB/°K	K_{eq}(298 K)	f(298 K)[a]
$HO_2 + NO_2 \leftrightarrow HO_2NO_2$	2.1×10^{-27}	10900 ± 1000	1.6×10^{-11}	5
$NO + NO_2 \leftrightarrow N_2O_3$	3.3×10^{-27}	4667 ± 100	2.1×10^{-20}	2
$NO_2 + NO_2 \leftrightarrow N_2O_4$	5.2×10^{-29}	6643 ± 250	2.5×10^{-19}	2
$NO_2 + NO_3 \leftrightarrow N_2O_5$	2.7×10^{-27}	11000 ± 500	2.9×10^{-11}	1.3
$CH_3O_2 + NO_2 \leftrightarrow CH_3O_2NO_2$	1.3×10^{-28}	11200 ± 1000	2.7×10^{-12}	2
$CH_3C(O)O_2 + NO_2$ $\leftrightarrow CH_3C(O)O_2NO_2$	9.0×10^{-29}	14000 ± 200	2.3×10^{-8}	2
$F + O_2 \leftrightarrow FOO$	3.2×10^{-25}	6100 ± 1200	2.5×10^{-16}	10
$Cl + O_2 \leftrightarrow ClOO$	5.7×10^{-25}	2500 ± 750	2.5×10^{-21}	2
$Cl + CO \leftrightarrow ClCO$	1.6×10^{-25}	4000 ± 500	1.1×10^{-19}	5
$ClO + O_2 \leftrightarrow ClO \cdot O_2$	2.9×10^{-26}	<3700	$<7.2 \times 10^{-21}$	—
$ClO + ClO \leftrightarrow Cl_2O_2$	1.3×10^{-27}	8744 ± 850	7.2×10^{-15}	1.5
$ClO + OClO \leftrightarrow Cl_2O_3$	1.1×10^{-24}	5455 ± 300	9.8×10^{-17}	3

The chemical kinetics data are taken from the same source as Appendix 10.2.
K/cm^3 molecule^{-1} = A exp (B/T) [200 < T/K < 300]
[a] f(298) is the uncertainty factor at 298 K. To calculate the uncertainty at other temperatures, use the expression:

$$f(T) = f(298K) \exp\left[\Delta B \left(\frac{1}{T} - \frac{1}{198} \right) \right]$$

Notes and Bibliography

Chapter 1

1.1 Nature of the Problem

There are several excellent books covering the subject matter in this chapter: Chamberlain, J. W., and Hunten, D. M., 1987, *Theory of Planetary Atmospheres* (New York: Academic Press); Lewis, J. S., and Prinn, R. G., 1984, *Planets and Their Atmospheres: Origin and Evolution* (New York: Academic Press); Atreya, S. K., Pollack, J. B., and Matthews, M. S., 1989, *Origin and Evolution of Planetary and Satellite Atmospheres* (Tucson: University of Arizona Press).

An earlier text giving a historical perspective is by Urey, H. C., 1959, "The Atmospheres of the Planets," in *Handbuch der Physik*, S. Flugge, ed. (Berlin: Springer-Verlag, 1959) pp. 363–418.

1.3 Chemical Composition of Planetary Atmospheres

There is a recent comprehensive compilation of chemical data for planetary atmospheres by Fegley, B., Jr., 1995, "Properties and Composition of the Terrestrial Oceans and of the Atmospheres of the Earth and Other Planets," in *AGU Handbook of Physical Constants*, T. Ahrens, ed. (Washington, DC: American Geophysical Union), pp. 320–345.

1.4 Stability of Planetary Atmospheres Against Escape

The fundamental theory of thermal escape is developed by Jeans, J. H., 1916, *The Dynamical Theory of Gases.* (Cambridge: Cambridge University Press) and Chamberlain, J. W., 1963, "Planetary Coronae and Atmospheric Evaporation." *Planet. Space Sci.* **11**, 901–960.

Chapter 2

2.1 Introduction

There are several excellent books covering the subject matter in this chapter: Okabe, H., 1978, *Photochemistry of Small Molecules* (New York: Wiley); Brasseur, G., and Solomon, S., 1984, *Aeronomy of the Middle Atmosphere* (Dordrecht: Reidel); Wayne, R. P., 1988, *Principles and Applications of Photochemistry* (New York: Oxford University Press).

2.2 Solar Flux

Much of the material in this chapter is based on Goody, R. M., and Yung, Y. L., 1989, *Atmospheric Radiation* (2nd ed.; New York: Oxford University Press).

A recent review of the sun's output and its evolution is by Sonnett, C. P., Giampapa, M. S., and Matthews, M. S., editors, 1991, *The Sun in Time* (Tucson: University of Arizona Press).

The usefulness of ^{14}C as a proxy for the variability of the sun in time is discussed by Damon, P. E., and Sonnett, C. P., 1991, "Solar and Terrestrial Components of the ^{14}C Variation Spectrum," in Sonett et al. (1991; cited above), p. 360.

The loss of heavy elements from the atmospheres of terrestrial planets via hydrodynamic escape in the early history of the solar system is discussed by Pepin, R. O., 1991, "On the Origin and Early Evolution of Terrestrial Planet Atmospheres and Meteoritic Volatiles." *Icarus* **92**, 2.

The discussion of entropy flux is taken from Peixoto, J. P. et al., 1991, "Entropy Budget of the Atmosphere." *J. Geophys. Res.* **96**, 10981.

2.3 Absorption in the Ultraviolet

The books by Okabe (1978) and Brasseur and Solomon (1984), cited in section 2.1, contain excellent summaries of absorption data of simple molecules.

Unless otherwise stated, the cross section data are taken from DeMore, W. B. et al., 1994, *Chemical Kinetics and Photochemical Data for Use in Stratospheric Modeling* (JPL Publ. No. 94-26; Pasadena, Calif.: Jet Propulsion Laboratory) p. 273.

The temperature dependence of CO_2 cross sections was measured by Lewis, B., and Carver, J. H., 1983, "Temperature Dependence of the Carbon Dioxide Photoabsorption Cross Section between 1200 and 1970 Å." *J. Quant. Spectrosc. Radiat. Transf.* **30**, 297.

Table 2.3 was compiled from Franklin, J. L. et al., 1969, *Ionization Potentials, Appearance Potentials, and Heats of Formation of Gaseous Ions* (National Bureau of Standards Ref. Publ. NSRDS-NBS 26; Washington, D.C.: U.S. Government Printing Office), p. 289.

2.4 Interaction with the Atmosphere

The fundamental theory of radiative transfer is developed in two classic texts by Chandrasekhar, S., 1960, *Radiative Transfer* (New York: Dover); and van de Hulst, H. C., 1980, *Multiple Light Scattering* (vols. 1 and 2; New York: Academic Press).

Applications of the theory of radiative transfer to the atmosphere are developed by Liou, K.-N., 1980, *An Introduction to Atmospheric Radiation* (New York: Academic Press); and Goody and Yung (1989; cited in section 2.2).

2.5 Sunlight and Life

The discussion on photosynthesis is based on material from Gregory, R. P. F., 1989, *Biochemistry of Photosynthesis* (New York: Wiley).

Chapter 3

3.1 Introduction

There are several excellent books covering the subject matter in this chapter: Johnston, H. S., 1966, *Gas Phase Reaction Rate Theory* (New York: Ronald Press); Benson, S. W., 1982, *Foundations of Chemical Kinetics* (New York: McGraw-Hill); Levine, R. D., and Bernstein, R. B., 1987, *Molecular Dynamics and Chemical Reactivity* (New York: Oxford University Press); Wayne, R. P., 1988, *Principles and Applications of Photochemistry* (New York: Oxford University Press); Bowers, M. T., editor, 1979, *Gas Phase Ion Chemistry* (vols. 1–3; New York: Academic Press).

Quote from Chuang Tsu is taken from *The Complete Work of Chuang Tsu*, translated by B. Watson (N.Y.: Columbia University Press), 1968.

3.2 Thermochemical Reactions

The standard reference for equilibrium chemistry is Lewis, G. N., and Randall, M., 1961, *Thermodynamics* (New York: McGraw-Hill).

Unless otherwise stated, the values of enthalpies and entropies are taken from DeMore, W. B. et al., 1994, *Chemical Kinetics and Photochemical Data for Use in Stratospheric Modeling* (Evaluation No., 11, JPL publication 94-26; Pasadena, Calif.: Jet Propulsion Laboratories).

3.3 Unimolecular Reactions

The material in this section is based on Robinson, P. J., and Holbrook, K. A., 1972, *Unimolecular Reactions* (New York: Wiley).

3.4 Bimolecular Reactions

The material on ion-molecule reactions is based on Levine and Bernstein (1989; cited in section 3.2).

3.5 Termolecular Reactions

Troe's theory of termolecular reactions is developed by Troe, J., 1977, "Theory of Thermal Unimolecular Reactions at Low Pressures. I. Solutions of the Master Equation." *J. Chem. Phys.* **66**, 4745; and "Theory of Thermal Unimolecular Reactions at Low Pressures. II. Strong Collision Rate Constants. Applications." *J. Chem. Phys.* **66**, 4758.

The importance of termolecular reactions for astrophysical and planetary applications was pointed out by Williams, D. A., 1971, "Association Reactions." *Astrophys. Lett.* **10**, 17; and Laufer, A. H. et al., 1983, "Computations and Estimates of Rate Constants for Hydrocarbon Reactions of Interest to the Atmospheres of the Outer Solar System." *Icarus* **56**, 560.

3.7 Miscellaneous Reactions

Data from table 3.3 are from Okabe, H., 1978, *Photochemistry of Small Molecules* (New York: Wiley).

Chapter 4

4.1 Introduction

There are several publications covering the subject matter in this chapter: Lewis, J. S., and Prinn, R. G., 1984, *Planets and Their Atmospheres* (New York: Academic Press); Black, D. C., and Matthews, M. S., editors, 1985, *Protostars and Planets II* (Tucson: University of Arizona Press); Levy, E. H., and Lunine, J. I., editors, 1993, *Protostars and Planets III* (Tucson: University of Arizona Press); Kerridge, J. F., and Matthews, M. S., editors, 1988, *Meteorites and the Early Solar System* (Tucson: University of Arizona Press); and Atreya, S. K., Pollack, J. B., and Matthews, M. S., editors, 1989, *Origin and Evolution of Planetary and Satellite Atmospheres* (Tucson: University of Arizona Press).

4.2 Cosmic Organization

The material on astronomy and astrophysics is based on Shu, F., 1981, *The Physical Universe: An Introduction to Astronomy* (Mill Valley, Calif.: University Science Books).

4.3 Elements of the Periodic Table

The chemical classification scheme used in this section is based on Larimer, J. W., 1987, "The Cosmochemical Classification of the Elements," in Kerridge and Matthews (1988; cited in section 4.1), pp. 375–389.

4.4 Molecular Clouds

The material in this section is based on Irvine, W. M., and Knacke, R. F., 1989, "Chemistry of Interstellar Gas and Grains," in Atreya et al. (1989; cited in section 4.1), pp. 3–34.

4.5 The Solar Nebula

The energy flux ratio is taken from Prinn, R. G., and Fegley, B., Jr., 1989, "Solar Nebula Chemistry: Origin of Planetary, Satellite and Cometary Volatiles," in Atreya et al. (1989; cited in section 4.1), pp. 78–136.

4.6 Meteorites

The material in this section is based on Wasserburg, G. J., 1985, "Short-lived Nuclei in the Early Solar System," in Black and Matthews (1985; cited in section 4.1), pp. 703–737. Table 4.3 was compiled from Swindle, T. D., 1993, "Extinct Radionuclides and Evolutionary Time Scales," in *Protostars and Planets III*, Levy, E. H., and Lunine, J. I., eds., (Tucson: University of Arizona Press), pp. 867–881. Table 4.4 was compiled from Cronin, J. R. et al., 1988, "Organic Matter in Carbonaceous Chondrites, Planetary Satellites, Asteroids, and Comets," in *Meteorites and the Early Solar System*, Kerridge, J. F., and Matthews, M. S., eds. (Tucson: University of Arizona Press), pp. 819–857.

4.7 Comets

The material in this section is based on Mumma, M. J., Weissman, P. R., and Stern, S. A., 1993, "Comets and the Origin of the Solar System: Reading the Rosetta Stone," in Levy and Lunine (1993; cited in section 4.1), pp. 1177–1252.

4.8 Planets

The material in this section is based on Pollack, J. B., 1985, "Formation of the Giant Planets and Their Satellite Ring Systems: An Overview," in Black and Matthews (1985; cited in section 4.1), pp. 791–831. Table 4.7 was compiled from Pollack, J. B., and Bodenheimer, P., 1989, "Theories of the Origin and Evolution of the Giant Planets," in Atreya et al. (1989; cited in section 4.1), pp. 564–602.

Appendix 4.2 was compiled from Sears, D. W. G., and Dodd, R. T., 1988, "Overview and Classification of Meteorites," in *Meteorites and the Early Solar System*, Kerridge, J. F., and Matthews, M. S., eds. (Tucson: University of Arizona Press), pp. 3–31.

Chapter 5

5.1 Introduction

There are several books that cover the subject matter of this chapter: Lewis, J. S., and Prinn, R. G., 1984, *Planets and Their Atmospheres: Origin and Evolution* (New York: Academic Press); Wayne, R. P., 1985, *Chemistry of Atmospheres* (New York: Oxford University Press); Atreya, S., 1986, *Atmospheres and Ionospheres of the Outer Planets and Their Satellites* (New York: Springer-Verlag); Chamberlain, J. W., and Hunten, D. M., 1987, *Theory of Planetary Atmospheres* (New York: Academic Press).

Several conference proceedings cover the subject matter in this chapter: Gehrels, T., editor, 1976, *Jupiter* (Tucson: University of Arizona Press); Gehrels, T., and Matthews, M. S., editors, 1982, *Saturn* (Tucson: University of Arizona Press); Bergstrahl, J. T., Miner, E. D., and Matthews, M. S., editors, 1991, *Uranus* (Tucson: University of Arizona Press); Atreya, S. K., Pollack, J. B., and Matthews, M. S., editors, 1989, *Origin and Evolution of Planetary and Satellite Atmospheres* (Tucson: University of Arizona Press); Black, D. C., and Matthews, M. S., editors, 1985, *Protostars and Planets II* (Tucson: University of Arizona Press); Levy, E. H., and Lunine, J. I., editors, 1993, *Protostars and Planets III* (Tucson: University of Arizona Press).

Pollack, J. B., and Bodenheimer, P., 1989, "Theories of the Origin and Evolution of Planetary and Satellite Atmospheres," cited in section 5.1.

5.1.1 Voyager Observations

The Voyager discoveries were reported in the following special issues.

Jupiter encounter: *Science* 1979, **204**, 945–1008 (Voyager 1); *Nature* 1979, **280**, 725–806 (Voyager 1); *Science* 1979, **206**, 925–995 (Voyager 1); *J. Geophys. Res.* 1981, **86**, 8123–8841 (Voyager 1 and 2).

Saturn encounter: *Science* 1981, **212**, 159–243 (Voyager 1); *Science* 1982, **215**, 499–594 (Voyager 2); *J. Geophys. Res.* 1983, **88**, 8639–9018 (Voyager 2).

Uranus encounter: *Science* 1986, **233**, 39–109 (Voyager 2); *J. Geophys. Res.* 1987, **92**, 14873–15375 (Voyager 2).

Neptune encounter: *Science* 1989, **246**, 1417–1501 (Voyager 2); *Geophys. Res. Lett.* 1990, **17**, 1643–1772 (Voyager 2); *J. Geophys. Res.* 1991, **96**, 18903–19268 (Voyager 2).

New results from the Galileo mission to Jupiter were reported in a special issue of *Science* **274**, p. 377–413, 1996.

5.1.2 Chemical Composition

There are several good reviews on the chemical composition of atmospheres in the outer solar system: Gautier, D., and Owen, T., 1989, *The Composition of Outer*

Planet Atmospheres, in Atreya et al. (1989; cited in section 5.1), p. 487; Pollack, J. B., and Bodenheimer, P., 1989, "Theories of the Origin and Evolution of Planetary and Satellite Atmospheres," cited in section 5.1.

The best He/H determinations were from the Voyager IRIS observations. See special issues of *Science* cited above.

A good review of the immiscibility of helium in hydrogen is by its principal proponent, Stevenson, D. J., 1982, "Interiors of the Giant Planets." *Ann. Rev. Earth Planet. Sci.* **10**, 257.

A good review on the history of gravitational contraction of the giant planets is by Pollack, J. B., and Bodenheimer, P., 1989, "Theories of the Origin and Evolution of the Giant Planets," in Atreya et al. (1989; cited in section 5.1), p. 564.

An alternative interpretation of He abundance in Neptune is by Marten, A. et al., 1993, "First Observations of CO and HCN on Neptune and Uranus at Millimeter Wavelengths and Their Implications for Atmospheric Chemistry." *Astrophys. J.* **406**, 285.

Table 5.1 reference: (1) Lindal, G. F. et al., 1992, *Astron. J.* **103**, 967.

Tables 5.2, 5.3, and 5.4 were compiled from Fegley, B. Jr., 1995, "Properties and Composition of the Terrestrial Oceans and of the Atmospheres of the Earth and the Other Planets," in *AGU Handbook of Physical Constants,* Ahrens, T., ed. (Washington, DC: American Geophysical Union), pp. 320–325.

Table 5.7 references: (1) Broadfoot, A. L. et al., 1981, *J. Geophys. Res.* **86**, 8259; (2) Broadfoot, A. L. et al., 1986, *Science* **233**, 74; (3) Strobel, D. F. et al., 1990, *J. Geophys. Res.* **95**, 10375.

5.1.3 The One-Dimensional Model

A detailed description of the numerical method for solving the continuity and diffusion equations can be found in Allen, M., Waters, J. W., and Yung, Y. L., 1991, "Vertical Transport and Photochemistry in the Terrestrial Mesosphere and the Lower Thermosphere (50–120 km)." *J. Geophys. Res.* **86**, 3617.

5.2.1 Energetics

A detailed discussion of the energetics of the thermosphere of the giant planets can be found in Atreya (1986; cited in section 5.1), section 6.3.

An early article that discusses a soft electron heat source for Jupiter's thermosphere is Hunten, D. M., and Dessler, A. J., 1977, "Soft Electrons as a Possible Heat Source for Jupiter's Thermosphere." *Planet. Space Sci.* **25**, 817.

A more recent discussion of the possibility of gravity waves as a heat source is by Yelle, R. V. et al., 1996, "Structure of Jupiter Upper Atmosphere—Predictions for Galileo." *J. Geophys. Res. Planets* **101**(E1), 2149.

Table 5.8 references: Atreya, S. K. et al., 1981, "Jupiter: Structure and Composition of the Upper Atmosphere." *Astrophys. J. Lett.* **247**, L43; McConnell, J. C. et al., 1982, "A New Look at the Ionosphere of Jupiter in Light of the UVS Occultation Results." *Planet. Space Sci.* **30**, 151; Smith, G. R. et al., 1983, "Saturn's Upper Atmosphere from the Voyager 2 EUV Solar and Stellar Occultations." *J. Geophys. Res.* **88**, 8667; Herbert, F. et al., 1987, "The Upper Atmosphere of Uranus—EUV

Occultations Observed by Voyager 2." *J. Geophys. Res.* **92**, 15093; Broadfoot, A. L. et al., 1989, "Ultraviolet Spectrometer Observations of Neptune and Triton." *Science* **246**, 1459; Waite, J. H. et al., 1988, "Superthermal Electron Processes in the Upper Atmosphere of Uranus—Aurora and Electroglow." *J. Geophys. Res.* **93**, 14295; Sandel, B. R. et al., 1982, "Extreme Ultraviolet Observations from the Voyager 2 Encounter with Saturn." *Science* **215**, 548.

5.2.2 Electroglow

The problem of the electroglow in the giant planets is discussed by Yelle, R. V. et al., 1987, "The Dependence of Electroglow on the Solar Flux." *J. Geophys. Res. Space Physics* **92**(A13), 15110; and Liu, W. H., and Dalgarno, A., 1996, "The Ultraviolet-Spectrum of the Jovian Dayglow." *Astrophys. J.* **462**(1), 502–518.

5.2.3 Ionosphere

The layered structures in the ionospheres of the giant planets are most likely caused by metals ions derived from ablation of meteorites, as suggested by Lyons, J. R., 1995 "Metal Ions in the Atmosphere of Neptune." *Science* **267**(5198), 648.

5.3 Hydrocarbon Chemistry

The foundation of hydrocarbon chemistry in the outer solar system was laid by Strobel, D. F., 1973, "The Photochemistry of Hydrocarbons in the Jovian Atmosphere." *J. Atmos. Sci.* **30**, 489.

A recent comprehensive model of hydrocarbon chemistry is by Gladstone, G. R., Allen, M., and Yung, Y. L., 1996, "Hydrocarbon Photochemistry in the Upper Atmosphere of Jupiter." *Icarus* **119**, 1.

Table 5.9 references: (1) Lias, S. G. et al., 1988, *J. Phys. Chem. Ref. Data* **17**, 1; Lias, S. G. et al., 1994, *NIST Stand Ref. Database* **25**, 1; (2) Domalski, E. S., and Hearing, E. D., 1993, *J. Phys. Chem. Ref. Data* **22**, 805; (3) Chase, M. W. et al., 1985, *J. Phys. Chem. Ref. Data* **14**, 1; (4) Wagman, D. D. et al., 1982, *J. Phys. Chem. Ref. Data* **11**, 1.

Table 5.10 references: (1) Mentall, J. E., and Gentieu, E. P., 1970, *J. Chem. Phys.* **52**, 5641; (2) van Dishoeck, E. F., 1989, personal communication; (3) Pilling, M. J. et al., 1971, *Chem. Phys. Lett.* **9**, 147; (4) Okabe, H., 1978, *Photochemistry of Small Molecules*, Wiley-Interscience, New York; (5) Parkes, D. A. et al., 1976, *J. Chem. Soc., Faraday Trans. 1* **72**, 1935; (6) Arthur, N. L., 1986a, *J. Chem. Soc., Faraday Trans. 2* **82**, 331; (7) Yu, H. T. et al., 1984, *J. Chem. Phys.* **80**, 2049; (8) Ye, H. T. et al., 1988, *J. Chem. Phys.* **89**, 2797; (9) Samson, J. A. R. et al., 1989, *J. Chem. Phys.* **90**, 6925; (10) Backx, C. et al., 1975, *J. Phys. B: Atom. Mol. Phys.* **8**, 3007; (11) Lee, L. C., and Chiang, C. C., 1983, *J. Chem. Phys.* **78**, 688; (12) Mount, G. H. et al., 1977, *Astrophys. J.* **214**, L47; (13) Mount, G. H., and Moos, H. W., 1978, *Astrophys. J.* **224**, L35; (14) Ditchburn, R. W., 1955, *Proc. R. Soc. London, Ser. A* **229**, 44; (15) Brion, C. E., and Thomson, J. P., 1984, *J. Elect. Spect. Related Phenom.* **33**, 301; (16) Rebbert, R. E., and Ausloos, P., 1972/1973, *J. Photochem.* **1**, 171; (17) Slanger, T. G., and Black, G., 1982, *J. Chem. Phys.* **77**, 2432; (18) Laufer, A. H., and McNesby, J. R., 1968, *J. Chem. Phys.* **49**, 2272; (19) Han, J. C. et al., 1989, *J. Chem. Phys.* **90**, 4000; (20) Suto, M., and Lee, L. C., 1984, *J. Chem. Phys.* **80**, 4824; (21) Wu, C. Y. R., 1990, personal communication; (22) Hamai, S., and Hirayama, F., 1979, *J. Chem. Phys.* **71**, 2934; (23) Cooper, G. et al., 1988, *Chem. Phys.* **125**, 307; (24) Okabe, H., 1981a, personal communication; Okabe, H., 1981b, *J. Chem. Phys.* **75**, 2772; Okabe, H., 1983a, *Can. J. Chem.* **61**, 850; Okabe, H., 1983b, *J. Chem. Phys.* **78**, 1312; (25) Fahr, A., and Laufer, A. H., 1986, *J. Photochem.* **34**, 261; (26) Wodtke, A. M., and Lee, Y. T., 1985, *J. Phys. Chem.* **89**, 4744; (27) Hunziker, H. E. et al., 1983, *Can. J. Chem.* **61**, 993; (28) Zelikoff, M., and Watanabe, K., 1953, *J. Opt. Soc. Amer.* **43**, 756; (29) Hara, H., and Tanaka, I., 1973, *Bull. Chem. Soc. Japan* **46**, 3012; (30) Adachi, H. et al., 1979, *Int. J. Chem. Kinet.* **11**, 995; (31) Blomberg, M. R. A., and Liu, B., 1985, *J. Chem. Phys.* **83**, 3995; (32) Calvert, J. G., and Pitts, J. N., 1966, *Photochemistry*, Wiley, New York; (33) Lias, S. G. et al., 1970, *J. Chem. Phys.* **52**, 1841; (34) Akimoto, H. et al., 1965, *J. Chem. Phys.* **42**, 3864; Akimoto, H. et al., 1973, *Bull. Chem. Soc. Japan* **46**, 2267; (35) Jacox, M. E., and Milligan, D. E., 1974, *Chem. Phys.* **4**, 45; (36) Nakayama, T., and Watanabe, K., 1964, *J. Chem. Phys.* **40**, 558; (37) Person, J. C., and Nicole, P. P., 1970, *J. Chem. Phys.* **53**, 1767; (38) Stief, L. J. et al., 1971, *J. Chem. Phys.* **54**, 1913; (39) Rabalais, J. W. et al., 1971, *Chem. Rev.* **71**, 73; (40) Fuke, K., and Schnepp, O., 1979, *Chem. Phys.* **38**, 211; (41) Shimo, N. et al., 1986, *J. Photochem.* **33**, 279; (42) Gierczak, T. et al., 1988, *J. Photochem. Photobiol. A* **43**, 1; (43) Samson, J. A. R. et al., 1962, *J. Chem. Phys.* **36**, 783; (44) Collin, G. J., 1988, *Adv. Photochem.* **14**, 135; (45) Johnston, G. R. et al., 1978, *J. Prog. React. Kin.* **8**, 231; (46) Glicker, S., and Okabe, H., 1987, *J. Phys. Chem.* **91**, 437; (47) Braude, E. A., 1945, *Ann. Rept. Progr. Chem.* **42**, 105; (48) Yung, Y. L. et al., 1984, *Astrophys. J. Suppl. Ser.* **55**, 465; (49) Hill, K. L., and Doepker, R. D., 1972, *J. Phys. Chem.* **76**, 1112; (50) Deslauriers, H. et al., 1980, *Can. J. Chem.* **58**, 2100; (51) Doepker, R. D., and Hill, K. L., 1969, *J. Phys. Chem.* **73**, 1313; (52) Collin, G. J., and Deslauriers, H., 1986, *J. Photochem.* **32**, 9; (53) Diaz, Z., and Doepker, R. D., 1977, *J. Phys. Chem.* **81**, 1442; (54) Doepker, R. D., 1968, *J. Phys. Chem.* **72**, 4037; (55) Bergmann, K., and Demtröder, W., 1968, *J. Chem. Phys.* **48**, 18; (56) Collin, G. J., and Wiceckowski, A., 1978, *J. Photochem.* **8**, 103; (57) Collin, G. J., 1973, *Can. J. Chem.* **51**, 2853; (58) Niedzielski, J. et al., 1978, 1979, *J. Photochem.* **10**, 287; (59) Jackson, J. A., and Lias, S. G., 1974, *J. Photochem.* **3**, 151; (60) Okabe, H., and Becker, D. A., 1963, *J. Chem. Phys.* **39**, 2549.

Table 5.11 references: (1) Tsang, W., and Hampson, R. F., 1986, *J. Phys. Chem. Ref. Data* **15**, 1087; (2) Zabarnick, S. J., Fleming, W., and Lin, M. C., 1986 estimate, *J. Chem. Phys.* **85**, 4373; (3) Allen, M., Yung, Y. L., and Gladstone, G. R., 1992, *Icarus* **100**, 527; (4) Yung, Y. L., Allen, M., and Pinto, J. P., 1984, *Astrophys. J. Suppl. Ser.* **55**, 465; (5) Sugawara, K., Okazaki, K., and Sato, S., 1981, *Bull. Chem. Soc. Jap.* **54**, 2872; (6) Lightfoot, P. D., and Pilling, M. J., 1987, *J. Phys. Chem.* **91**, 3373; (7) Pratt, G. L., and Wood, S. W., 1984, *J. Chem. Soc., Faraday Trans. 1* **80**, 3419; (8) Teng, L., and Jones, W. E., 1972, *J. Chem. Soc., Faraday Trans. 1* **68**, 1267; (9) Munk, J. et al., 1986, *J. Phys. Chem.* **90**, 2752; (10) Laufer, A. H. et al., 1983, *Icarus* **56**, 560; (11) Homann, K. H., and Schweinfurth, H., 1981, *Ber. Bunsenges. Phys. Chem.* **85**, 569; (12) Homann, K. H., and Wellmann, C., 1983, *Ber. Bunsenges. Phys. Chem.* **87**, 609; (13) Wagner, Von H. Gg., and Zellner, R., 1972b, *Ber. Bunsenges. Physik. Chem.* **76**, 518; (14) Allara, D. L., and Shaw, R., 1980, *J. Phys.*

Chem. Ref. Data **9**, 523; (15) Tsang, W., 1988, *J. Phys. Chem. Ref. Data* **17**, 887; (16) Nava, D. F., Mitchell, M. B., and Stief, L. J., 1986, *J. Geophys. Res.* **91**, 4585; (17) Schwanebeck, W., and Warnatz, J., 1975, *Ber. Bunsenges. Phys. Chem.* **79**, 530; (18) Berman, M. R., and Lin, M. C., 1984, *J. Chem. Phys.* **81**, 5743; (19) Berman, M. R., and Lin, M. C., 1983, *Chem. Phys.* **82**, 435; (20) Anderson, S. M., Freedman, A., and Kolb, C. E., 1987, *J. Phys. Chem.* **91**, 6272; (21) Berman, M. R. et al., 1982, *Chem. Phys.* **73**, 27; (22) Braun, W., Bass, A. M., and Pilling, M., 1970, *J. Chem. Phys.* **52**, 5131; (23) Langford, A. O., Petek, H., and Moore, C. B., 1983, *J. Chem. Phys.* **78**, 6650; (24) Bohland, T., Temps, F., and Wagner, H. Gg., 1985, *Ber. Bunsenges. Phys. Chem.* **89**, 1013; (25) Frank, P., Bhaskaran, K. A., and Just, Th., 1986, *J. Phys. Chem.* **90**, 2226; (26) Laufer, A. H., 1981, *Rev. Chem. Intermed.* **4**, 225; (27) Bohland, T., Temps, F., and Wagner, H. Gg., 1986, *Twenty-first Symposium International on Combustion*, The Combustion Institute, Pittsburgh, 841; (28) Möller, W., Mozzhukhin, E., and Wagner, H. Gg., 1986, *Ber. Bunsenges. Phys. Chem.* **90**, 854; (29) Slagle, I. R. et al., 1988, *J. Phys. Chem.* **92**, 2455; (30) Anastasi, C., and Arthur, N. L., 1987, *J. Chem. Soc., Faraday Trans. 2* **83**, 277; (31) Hautman, D. J. et al., 1981, *Int. J. Chem. Kinet.* **13**, 149; (32) Pitts, W. M., Pasternack, L., and McDonald, J. R., 1982, *Chem. Phys.* **68**, 417; (33) Stephens, J. W. et al., 1987, *J. Phys. Chem.* **91**, 5740; (34) Okabe, H., 1983a, *Can. J. Chem.* **61**, 850; Okabe, H., 1983b, *J. Chem. Phys.* **78**, 1312; (35) Tanzawa, T., and Gardiner, W. C., 1980, *J. Phys. Chem.* **84**, 236; (36) Callear, A. B., and Smith, G. B., 1986, *J. Phys. Chem.* **90**, 3229; (37) Fahr, A., and Laufer, A. H., 1990, *J. Photochem.* **34**, 261; (38) Pacey, P. D., and Wimalasena, J. H., 1984, *J. Phys. Chem.* **88**, 5657; (39) Arthur, N. L., 1986b, *J. Chem. Soc., Faraday Trans. 2* **82**, 1057; (40) Laufer, A. H., and Bass, A. M., 1979, *J. Phys. Chem.* **83**, 310; (41) Cole, J. A. et al., 1984, *Combust. Flame* **56**, 51.

5.3.1 Photochemistry and Chemical Kinetics

A thorough review of photochemistry and chemical kinetics in the outer solar system is by Yung, Y. L., Allen, M., and Pinto, J. P., 1984, "Photochemistry of the Atmosphere of Titan: Comparison between Model and Observations." *Astrophys. J. Suppl. Ser.* **55**, 465. Also, see Capone, L. A. et al., 1980, "Cosmic Ray Synthesis of Organic Molecules in Titan's Atmosphere." *Icarus* **44**, 72.

5.3.3 Comparison of Giant Planets

Modeling studies have been carried out for the chemistry of Uranus and Neptune by Summers, M. E., and Strobel, D. F., 1989, "Photochemistry of the Atmosphere of Uranus." *Astrophys. J.* **346**, 495; and Romani, P. N. et al., 1993, "Methane Photochemistry on Neptune: Ethane and Acetylene Mixing Ratios and Haze Production." *Icarus* **106**, 442.

5.4 Nitrogen Chemistry

The chemistry of nitrogen has been investigated by Strobel, D. F., 1973, "The Photochemistry of NH_3 in the Jovian Atmosphere." *J. Atmos. Sci.*, **30**, 1205; Atreya, S. K., Kuhn, W. R., and Donahue, T. M., 1980, "Saturn: Tropospheric Ammonia and Nitrogen." *Geophys. Res. Lett.* **7**, 474; Kaye, J. A., and Strobel, D. F., 1983, "HCN Formation on Jupiter: The Coupled Photochemistry of Ammonia and Acetylene." *Icarus* **54**, 417; and Kaye, J. A., and Strobel, D. F., 1983, "Formation and Photochemistry of Methylamine in Jupiter's Atmosphere." *Icarus* **55**, 399.

The detection of HCN in the atmosphere of Jupiter is still controversial, as discussed by Bezard, B., Griffith, C., Lacy, J., and Owen, T., 1995, "Nondetection of Hydrogen-Cyanide on Jupiter." *Icarus* **118**, 384.

5.5 Phosphorus Chemistry

The chemistry of phosphine has been studied by Prinn, R. G., and Lewis, J. S., 1975, "Phosphine on Jupiter and Implications for the Great Red Spot." *Science* **190**, 294.

5.6 Oxygen Chemistry

The chemistry of oxygen has been studied by Prinn, R. G., and Barshay, S. S., 1977, "Carbon Monoxide on Jupiter and Implications for Atmospheric Convection." *Science* **198**, 1031; Strobel, D. F., and Yung, Y. L., 1979, "The Galilean Satellites as a Source of CO in the Jovian Upper Atmosphere." *Icarus* **37**, 256; and Lodders, K., and Fegley, B., 1994, "The origin of Carbon-Monoxide in Neptune Atmosphere." *Icarus* **112**, 368.

5.7 Miscellaneous Topics

A comprehensive model of the deep atmosphere of Jupiter and Saturn is by Fegley, B., and Lodders, K., 1994, "Chemical-Models of the Deep Atmospheres of Jupiter and Saturn." *Icarus* **110**, 117.

Deuterium in the solar system is discussed by Yung, Y. L., and Dissly, R. W., 1992, "Deuterium in the Solar System." *Am. Chem. Soc. Symp. Ser.* **502**, 369.

Chapter 6

6.1 Introduction

Several conference proceedings cover the subject matter in this chapter: Atreya, S. K., Pollack, J. B., and Matthews, M. S., editors, 1989, *Origin and Evolution of Planetary and Satellite Atmospheres* (Tucson: University of Arizona Press); Black, D. C., and Matthews, M. S., editors, 1985, *Protostars and Planets II* (Tucson: University of Arizona Press); Levy, E. H., and Lunine, J. I., editors, 1993, *Protostars and Planets III* (Tucson: University of Arizona Press); Cruikshank, D. P., editor, 1995, *Neptune and Triton* (Tucson: University of Arizona Press).

The Voyager discoveries were reported in special issues of *Science* and *J. Geophys. Res.*, cited in section 5.1.1.

6.2 Io

The discoveries of optical emissions from Io were reported by Brown, R. A., 1974, *Optical Line Emission from Io.*, in *Exploration of the Planetary System*, A. Woszczyk and C. Iwaniszewska, editors (Hingham, Mass.: Reidel), pp. 527–531; and Trafton, L., 1975, "Detection of a Potassium Cloud Near Io." *Nature* **258**, 690.

The ionosphere of Io was discovered by Kliore, A. J. et al., 1974, "The Atmosphere of Io from Pioneer 10 Radio Occultation Measurements." *Icarus* **24**, 407–419.

The SO_2 atmosphere of Io and the SO_2 frost on the surface were discovered, respectively, by Pearl, J. C. et al., 1979, "Identification of Gaseous SO_2 and New Upper Limits for Other Gases on Io." *Nature* **280**, 755–758; and Fanale, F. P. et al., 1979, "Significance of Absorption Features in Io's IR Reflectance Spectrum." *Nature* **280**, 760–763.

The best constraints for the SO_2 atmosphere on Io were obtained by Lellouch, E. et al., 1990, "Io's Atmosphere from Microwave Detection of SO_2." *Nature* **346**, 639–641; Lellouch, E. et. al., 1996, "Detection of Sulfur Monoxide in Io's Atmosphere." *Astrophys. J.* **459**(2), L107–L110.

Modeling of the chemistry and dynamics of the SO_2 atmosphere of Io has been carried out by Kumar, S., 1985, "The SO_2 Atmosphere and Ionosphere of Io: Ion Chemistry, Atmospheric Escape, and Models Corresponding to the Pioneer 10 Radio Occultation Measurements." *Icarus* **61**, 101–120; Summers, M. E., and Strobel, D. F., 1996, "Photochemistry and Vertical Transport in Io's Atmosphere and Ionosphere." *Icarus* **120**(2), 290–316; Ingersoll, A. P., 1989, "Io Meterorology: How Atmospheric Pressure Is Controlled Locally by Volcanos and Surface Frosts." *Icarus* **81**, 298–313; and Wong, M. C., and Johnson, R. E., 1996, "A 3-Dimensional Azimuthally Symmetrical Model Atmosphere for Io. 2. Plasma Effect on the Surface." *J. Geophys. Res. Planets* **101**(E10), 23255–23259.

The interaction between Io's atmosphere and the plasma torus has been studied by Goertz, C. K., 1980, "Io's Interactions with the Plamsma Torus." *Geophys. Res. Lett.* **85**, 2949; Huang, T. S., and Siscoe, G. L., 1987, "Types of Planetary Tori." *Icarus* **70**, 366–378; Johnson, R. E., 1990, *Energetic Charged-Particle Interactions with Atmospheres and Surfaces* (Berlin: Springer-Verlag); Moos, H. W. et al., 1985, "Long-Term Stability of the Io High-Temperature Plasma Torus." *Astrophys. J.* **294**, 369–382; and Johnson, R. E., and Mcgrath, M., 1993, "Stability of the Io Plasma Torus Atmosphere Interaction." *Geophys. Res. Lett.* **20**(16), 1735–1738.

6.2.1 Neutral Atmosphere

Table 6.1 references: (1) Okabe, H., 1971, *J. Am. Chem. Soc.* **93**, 7095–7096; (2) Welge, K. H., 1974, *Can. J. Chem.* **52**, 1424–1435; (3) Driscoll, J. N., and Warneck, P., 1968, *J. Phys. Chem.* **72**, 3736–3740; (4) Phillips, L. F., 1981, *J. Phys. Chem.* **85**, 3994–4000; (5) Hudson, R. D., 1971, *Rev. Geophys. Sp. Phys.* **9** 305–406; (6) Brewer, L., and Brabson, G. D., 1966, *J. Chem. Phys.* **44**, 3274-3278; (7) Yung, Y. L., and DeMore, W. B., 1982, *Icarus* **51**, 199–247; (8) Herron, J. T., and Huie, R. E., 1980, *Chem. Phys. Lett.* **76**, 322–324; (9) Plane, J. M. C., 1991, *Int. Rev. Chem. Phys.* **10**, 55–106; (10) DeMore, W. B. et al., 1992, JPL92 Chemical Kinetics and

Photochemical Data for Use in Stratospheric Modeling; (11) Mallard, F. W. et al., 1994, NIST Chemical Kinetics Database: Version 6.0 National Institute of Standards and Technology, Gaithersburg, MD; (12) Baulch, D. L., Drysdale, D. D., and Horne, D. G., 1973, *Evaluated Kinetic Data for High Temperature Reactions, Vol. 2, Homogeneous Gas Phase Reactions of the H_2-N_2-O_2 System*, (London: Butterworths).

6.2.2 Ionosphere

Table 6.2 references: (1) Wu, C. Y. R., and Judge, D. L., 1981, *J. Chem. Phys.* **74**, 3804-3806; (2) McGuire, E. J., 1968, *Phys. Rev.* **175**, 20–30; (3) Prasad, S. S., and Huntress, W. T., 1980, *Astrophys. J. Suppl. Ser.* **43**, 1–35; (4) Anicich, V. G., and Huntress, W. T., 1986, *Astrophys. J. Suppl. Ser.* **62**, 553–672; (5) Anicich, V. G., 1993, *Astrophys. J. Suppl. Ser.* **84**, 215-315.

6.2.3 Torus

Table 6.3 references: (1) Kliore, A. J. et al., 1974, cited earlier; (2) Bagaenal, F., 1994, *J. Geophys. Res.* **99**, 11043; (3) Brown, R. A. et al., 1974, cited earlier.

6.3 Titan

The existence of an atmosphere on Titan was established by Kuiper, G. P., 1944, "Titan: A Satellite with an Atmosphere." *Astrophys. J.* **100**, 378.

The ionosphere on Titan was modeled by Keller, C. N. et al., 1992, "A Model of the Ionosphere of Titan." *J. Geophys. Res.* **97**, 12117; and Gan, L. et al., 1992, "Electrons in the Ionosphere of Titan. *J. Geophys. Res.* **97**, 12137.

Unless otherwise stated, all the modeling results of the neutral composition of Titan in this chapter are taken from Yung, Y. L. et al., 1984, "Photochemistry of the Atmosphere of Titan: Comparison between Model and Observations." *Astrophys. J. Suppl.* **55**, 465–506.

Updates and further comparisons between model and observations are in Coustenus, A. et al., 1993, "Titan's Atmosphere from Voyager Infrared Observations." *Icarus* **102**, 240.

The evolution the atmosphere of Titan was discussed by Lunine, J. I., 1983, "Ethane Ocean on Titan." *Science* **222**, 1229–1230; Lunine, J. I., and Nolan, M. C., 1992, "A Massive Early Atmosphere on Triton." *Icarus* **100**, 221–234; and Lunine, J. I., 1993, "Does Titan Have an Ocean? A Review of Current Understanding of Titan's Surface." *Rev. Geophys.* **31**, 133–149.

Table 6.5 references: (1) Strobel, D. F., and Shemansky, D. E., 1982, *J. Geophys. Res.* **87**, 1361; (2) Capone, L. A. et al., 1980, *Icarus* **44**, 72; (3) Lee, M. T., and McKay, V., 1982, *J. Phys. B.* **15**, 3971; (4) Parkes, D. A. et al., 1973, *Chem. Phys. Letters* **23**, 425; (5) Watanabe, K. et al., 1953, *Air Force Cambridge Res. Center Tech. Rept.* **53**, 57; (6) Strobel, D. F., 1973, *J. Atmos. Sci.* **30**, 839; (7) Mount, G. M., and Moos, H. W., 1978, *Astrophys. J.* **224**, 135; (8) Nakayama, T., and Watanabe, K., 1964, *J. Chem. Phys.* **40**, 558; (9) Okabe, H., 1981, *J. Chem. Phys.* **75**, 2772; (10) Okabe, H., 1983a, *J. Chem. Phys.* **78**, 1312; (11) Zelikoff, M., and Watanabe, K., 1953, *J. Opt.*

Soc. Am. **43**, 756; (12) Back, R. A., and Griffiths, D. W. L., 1967, *J. Chem. Phys.* **46**, 4839; (13) Akimoto, H. et al., 1965, *J. Chem. Phys.* **42**, 3864; (14) Hampson, R. F., Jr., and McNesby, J. R., 1965, *J. Chem. Phys.* **42**, 2200; (15) Lisa, S. G. et al., 1970, *J. Chem. Phys.* **52**, 1841; (16) Poole, C. P., Jr., and Anderson, R. S., 1959, *J. Chem. Phys.* **31**, 346; (17) Ramsay, D. A., and Thistlewaithe, P., 1966, *Canadian J. Phys.* **44**, 1381; (18) Jacox, M. E., and Milligan, D. E., 1974, *Chem Phys.* **4**, 45; (19) Stief, L. J. et al., 1971, *J. Chem. Phys.* **54**, 1913; (20) Hamai, S., and Hirayama, F., 1979, *J. Chem. Phys.* **71**, 2934; (21) Heller, S. R., and Milne, G. W. A., 1978, *EPA/NIH Mass Spectral Data Base*, **Vol. 1**, Molecular Weights 30-186 (NSRDS-NBS 63); (22) Sutcliffe, L. H., and Walsh, A. D., 1952, *J. Chem. Soc.* **1952**, 899; (23) Rabalais, J. W. et al., 1971, *Chem. Rev.* **71**, 73; (24) Calvert, J. G., and Pitts, J. N., Jr., 1966, *Photochemistry* (New York: Wiley); (25) Borrell, P. et al., 1971, *J. Chem. Soc. B.* **2**, 2293; (26) Collin, G. J. et al., 1979, *Canadian J. Chem.* **57**, 870; (27) Georgieff, K. K., and Richard, Y., 1958, *Canadian J. Chem.* **36**, 1280; (28) Kloster-Jensen, E. et al., 1974, *Hel. Chem. Acta* **57**, 1731; (29) West, G. A., 1975, *Ph.D. thesis, University of Wisconsin at Madison*; (30) Lee, L. C., 1980, *J. Chem. Phys.* **72**, 6414; (31) Connors, R. E. et al., 1974, *J. Chem. Phys.* **60**, 5011; (32) Nuth, J. A., and Glicker, S., 1982, *J. Quant. Spectrosc. Rad. Trans.* **28**, 223; (33) Shemansky, D. E., 1972, *J. Chem. Phys.* **56**, 1582; (34) DeMore, W. B., and Patapoff, M., 1972, *J. Geophys. Res.* **77**, 6291; (35) Allen, M. et al., 1981, *J. Geophys. Res.* **86**, 3717; (36) Pinto, J. P. et al., 1980, *Science* **210**, 183; (37) Okabe, H., 1978, *Photochemistry of Small Molecules* (New York: Wiley).

Table 6.6 references: (1) Payne, W. A., and Stief, L. J., 1976, *J. Chem. Phys.* **64**, 1150; (2) Michael, J. V. et al., 1973, *J. Chem. Phys.* **58**, 2800; (3) Lee, J. H. et al., 1978, *J. Chem. Phys.* **68d**, 1817; (4) Butler, J. E. et al., 1981, *Chem. Phys.* **56**, 355; (5) Ashfold, M. N. R. et al., 1980, *J. Photochem.* **12**, 75; (6) Laufer, A. H., 1981a, *Rev. Chem. Intermed.* **4**, 225; (7) Laufer, A. H. et al., 1983, *Icarus* **56**, 560; (8) Banyard, S. A. et al., 1980, *J. Chem. Soc. Chem. Comm.* **1980**, 1156; (9) Patrick, R. et al., 1980, *Chem. Phys.* **53**, 279; (10) Callear, A. B., and Metcalfe, M. P., 1976, *Chem. Phys.* **14**, 275; (11) Keil, D. G. et al., 1976, *Internat. J. Chem. Kinetics* **8**, 825; (12) MacFadden, K. O., and Currie, C. L., 1973, *J. Chem. Phys.* **58**, 1213; (13) Teng, L., and Jones, W. E., 1972, *J. Chem. Soc. Faraday Trans. I* **68**, 1267; (14) Whytock, D. A. et al., 1976, *J. Chem. Kinetics* **8**, 777; (15) von Wagner, H. Gg., and Zellner, R., 1972a, *Ber. Bunsenges. Phys. Chem.* **76**, 518; (16) von Wagner, H. Gg., and Zellner, R., 1972b, *Ber. Bunsenges. Phys. Chem.* **76** 667; (17) Pasternack, L., and McDonald, J. R., 1979, *Chem. Phys.* **43**, 173; (18) Brown, R. L., and Laufer, A. H, 1981, *J. Phys. Chem.* **58**, 3826; (19) Okabe, H., 1983b, *Canadian J. Chem.* **61**, 850; (20) Schwanebeck, W., and Warnatz, J., 1975, *Ber. Bunsenges. Phys. Chem.* **79**, 530.

Table 6.7 references: (1) Okabe, H., 1978, *Photochemistry of Small Molecules* (New York: Wiley); (2) Black, G. et al., 1969, *J. Chem. Phys.* **51**, 116; (3) Mayer, S. W. et al., 1966, *J. Chem. Phys.* **45**, 385; (4) Westley, F., 1980, NBS Pub. NSRDS-NBS67; (5) Schacke, H. et al., 1977, *Phys. Chem.* **81**, 670; (6) Becker, R. S., and Hong, J. H., 1983, *J. Phys. Chem.* **87**, 163; (7) Phillips, L. F., 1978, *Internat. J. Chem. Kinetics* **10**, 899; (8) Albers, E. A., 1969, Ph.D. thesis, University of Göttingen; (9) DeMore, W. B., 1982, JPL Pub. No. 82-57; (10) Homann, K. H., and Schweinfurth, H., 1981, *Ber. Bunsenges Phys. Chem.* **85**, 569; (11) Hampson, R. F., Jr., 1980, *US Dept. Transportations, Rept. No. FAA-EE* **80**, 17; (12) Baulch, D. L. et al., 1980, *J. Phys. Chem.*

Ref. Data **9**, 295; (13) Fenimore, C. P., 1969, *12th Internat. Symposium on Combustion.*, Pittsburgh: Combustion Institute, p. 463; (14) Perry, R. A., and Williamson, D., 1982, *Chem. Phys. Letters* **93**, 331; (15) Laufer, A. H., 1981a, *Rev. Chem. Intermed.* **4**, 225; (16) Pinto, J. P. et al., 1980, *Science* **210**, 183.

Table 6.8 reference: Yung, Y. L. et al., 1984, cited earlier.

Table 6.10 references: Lutz, B. L., de Bergh, C., and Owen, T., 1983, "Titan—Discovery of Carbon Monoxide in its Atmosphere." *Science* **200**, 1374; Muhleman, D. O. et al., 1984, "Microwave Measurements of Carbon-Monoxide on Titan." *Science* **223**, 393.

Table 6.11 reference: Strobel, D. F. and Shemansky, D. E., 1982, "EUV Emission from Titan's Upper-Atmosphere—Voyager-1 Encounter." *J. Geophys. Res.* **87**, 1361.

6.4 Triton

The airglow, composition, and thermal structure of Triton were modeled by Strobel, D. F. et al., 1991, "Nitrogen Airglow Sources—Comparison of Triton, Titan, and Earth." *Geophys. Res. Lett.* **18**(4), 689–692; Stevens, M. H. et al., 1992, "On the Thermal Structure of Triton's Thermosphere." *Geophys. Res. Lett.*" **19**, 669–672; and Krasnopolsky, V. A. et al., 1993, "On the Haze Model for Triton." *J. Geophys. Res.* **98** 17123.

The composition of the ionosphere of Triton was modeled by Lyons, J. R. et al., 1992, "Solar Control of the Upper Atmosphere of Triton." *Science* **256**, 204–206.

6.5 Pluto

Much of what is known about Pluto is summarized in the 1989 special issue on Pluto, "Pluto at Perihelion." *Geophys. Res. Lett.* **16**, 1203–1244; Stern, S. A., 1992, "The Pluto-Charon System." *Ann. Rev. Astron.* **30**, 185–233; Null, G. W. et al., 1993, "Masses and Densities of Pluto and Charon." *Astron. J.* **105**, 2319; and Stern, S. A. et al., 1991, "Rotationally Resolved Midultraviolet Studies of Triton and the Pluto/Charon System I: IUE Results." *Icarus* **92**, 332.

Stellar occultation by Pluto has been a major source of information about its atmosphere. These observations are reported by Elliot, J. L. et al., 1989, "Pluto's Atmosphere." *Icarus* **77**, 148–170.

Hydrodynamic escape of gases from Pluto is discussed by Trafton, L., 1990, "A Two-Component Volatile Atmosphere for Pluto. I. The Bulk Hydrodynamic Escape Regime." *Astrophys. J.* **359**, 512; and Clarke, J. T. et al., 1992, "Pluto's Extended Atmosphere: An Escape Model and Initial Observations." *Icarus* **95**, 173.

Chapter 7

7.1 Introduction

A recent conference proceedings publication discusses all important aspects of Mars: Kieffer, H. H., Jakosky, B. M., Snyder, C. W., and Matthews, M. S., editors, 1992, *Mars* (Tucson: University of Arizona Press).

The Viking discoveries were reported in special issues: *Science* 1976, **194**(4260), pp. 57–105; *J. Geophys. Res.* 1977, **82**(28), 3951–4684.

Other special issues on Mars include *J. Geophys. Res.* 1973, **78**(20), 4007–4440 (Mariner 9); *J. Geophys. Res.* 1979, **84**(B6), 2793–3007 (Mars volatiles); *J. Geophys. Res.* 1982, **87**(B12), 9715–10305 (3rd international); *J. Geophys. Res.* 1990, **95**(B9), 14087–14852 (4rd international); *J. Geophys. Res.* 1993, **98**(E2), 3091–3482 (MSATT); *J. Geophys. Res.* 1993, **98**(E6), 10897–11121 (MSATT).

A monograph covers the subject matter in this chapter: Krasnopolsky, V. A., 1986, *Photochemistry of the Atmospheres of Mars and Venus* (Berlin: Springer–Verlag).

A monograph covers the subject of interaction between solar wind and Mars: Luhmann, J. G., Tatrallyay, M., and Pepin, R. O., editors, 1992, *Venus and Mars: Atmospheres, Ionospheres, and Solar Wind Interactions* (AGU Monograph 66; Washington, DC: American Geophysical Union).

The seasonal behavior of the bulk atmosphere of Mars is described in Leighton, R. B., and Murray, B. C., 1966, "Behavior of Carbon Dioxide and Other Volatiles on Mars." *Science* **153**, 136.

Table 7.1 reference: Kieffer, H. H. et al., 1992, *Mars*, cited earlier.

Table 7.2 references: (1) Owen, T. et al., 1977, *J. Geophys. Res.* **82**, 4635; (2) Farmer, C. B., and Doms, P. E., 1979, *J. Geophys. Res.* **87**, 10215; (3) Barth, C. A., 1974, *Ann. Rev. Earth Planet. Sci.* **2**, 333.

Table 7.3 references: (1) Owen, T. et al., 1988, *Science* **240**, 1767. (2) Bjoraker, G. L. et al., 1989, *Bull. Amer. Astron. Soc.* **21**, 990; (3) Nier, A. O., and McElroy, M. B., 1977, *J. Geophys. Res.* **82**, 4341; (4) Biemann, K. et al., 1976, *Science* **194**, 76; (5) Owen, T. et al., 1977, *J. Geophys. Res.* **82**, 4635.

7.2 Photochemistry

The classic articles on the photochemistry and evolution of the Martian atmosphere are McElroy, M. B., and Donahue, T. M., 1972, "Stability of the Martian Atmosphere." *Science* **177**, 986; Parkinson, T. M., and Hunten, D. M., 1972, "Spectroscopy and Aeronomy of CO_2 on Mars." *J. Atmos. Sci.* **29**, 1380; McElroy, M. B., 1972, "Mars: An Evolving Atmosphere." *Science* **175**, 443.

Recent updates are by Fox, J. L., 1993, "Production and Escape of Nitrogen from Mars." *J. Geophys. Res.* **98**, 3297; and Nair, H., Allen, M., Anbar, A. D., Yung, Y. L., and Clancy, R. T., 1994, "A Photochemical Model of the Martian Atmosphere." *Icarus* **111**, 124.

7.3 Model Results

Unless otherwise stated, all model results are taken from the paper by Nair et al. (1994; cited in section 7.2). See also Fox (1993) and McElroy (1972), both cited in section 7.2.

The most important updating of the early models is the incorporation of temperature-dependent cross sections of CO_2: DeMore, W. B., and Patapoff, M., 1972, "Temperature and Pressure Dependence of CO_2 Extinction Coefficients." *J. Geophys. Res.* **77**, 6291; Lewis, B. R., and Carver, J. H., 1983, "Temperature Dependence of the

Carbon Dioxide Photoabsorption Cross Section between 1200 and 1970." *J. Quant. Spec. Radiat. Transf.* **30**, 297.

The effects of these reduced cross sections on the photochemistry of the atmosphere are reported by Nair et al. (1994) and Anbar, A. D., Allen, M., and Nair, H. A., 1993, "Photodissociation in the Atmosphere of Mars: Impact of High Resolution, Temperature-Dependent CO_2 Cross Section Measurements." *J. Geophys. Res.* **98**, 10925; Shimazaki, T., 1989, "Photochemical Stability of CO_2 in the Martian Atmosphere: Reevaluation of the Eddy Diffusion Coefficient and the Role of Water Vapor." *J. Geomag. Geoelectr.* **41**, 273.

The possible importance of heterogeneous chemistry on the surface of Martian dust is discussed by Anbar, A. D., Leu, M. T., Nair, H. A., and Yung, Y. L., 1993, "Adsorption of HOx on Aerosol Surfaces: Implications for the Atmosphere of Mars." *J. Geophys. Res.* **98**, 10933; and Atreya, S. K., and Gu, Z. G., 1994, "Stability of the Martian Atmosphere—Is Heterogeneous Catalysis Essential?" *J. Geophys. Res.* **99**, 13133.

Table 7.4 references: (1) Yung, Y. L. et al., 1988, *Icarus* **76**, 146; (2) Anbar, A. D. et al., 1993, *J. Geophys. Res.* **98**, 10925; (3) Nicolet, M., 1984, *Planet. Space Sci.* **32**, 1467; (4) Lee, L. C. et al., 1977, *J. Chem. Phys.* **67**, 5602; (5) Samson, J. A. R. et al., 1982, *J. Chem. Phys.* **76**, 393; (6) Taherian, M. R., and Slanger, T. G., 1985, *J. Chem. Phys.* **83**, 6246; (7) Turnipseed, A. A. et al., 1991, *J. Chem. Phys.* **95**, 3244; (8) Wine, P. H., and Ravishankara, A. R., 1982, *Chem. Phys.* **69**, 365; (9) Brock, J. C., and Watson, W. T., 1980, *Chem. Phys. Lett.* **71**, 371; (10) Sparks, R. K. et al., 1980, *J. Chem. Phys.* **72**, 1401; (11) Fairchild, C. E. et al., 1978, *J. Chem. Phys.* **69**, 3632; (12) Mentall, J. E., and Gentieu, E. P., 1970, *J. Chem. Phys.* **53**, 5641; (13) R. Gladstone, private communication; (14) Nee, J. B., and Lee, L. C., 1984, *J. Chem. Phys.* **81**, 31; (15) van Dishoeck, E. F., and Dalgarno, A., 1984, *Astrophys. J.* **277**, 576; (16) van Dishoeck, E. F. et al., 1984, *J. Chem. Phys.* **81**, 5709; (17) DeMore, W. B. et al., 1990, *Chemical Kinetics and Photochemical Data for Use in Stratospheric Modeling, Evaluation Number 7*, JPL Publication, 85-37; (18) Philips, E. et al., 1977, *J. Quant. Spectrosc. Radiat. Transf.* **18**, 309; (19) Schürgers, M., and Welge, K. H., 1968, *Naturforsch.* **23**, 1508; (20) Kronebusch, P. I., and Berkowitz, J., 1976, *Int. J. Mass Spectrom, Ion Process* **12**, 283; (21) Masuoka, T., and Samson, J. A. R., 1980, *J. Chem. Phys.* **77**, 623; (22) Lin, C. L., and Leu, M. T., 1982, *Int. J. Chem. Kinetics* **14**, 417; (23) Baulch, D. L. et al., 1976, *Evaluated Kinetic Data for High Temperature Reactions of The O_2-O_3 system, The CO_2-O_2-H_2 System and of Sulphur-Containing Species*, Butterworths, London/Boston; (24) Tsang, W., and Hampson, R. F., 1986, *J. Phys. Chem. Ref. Data* **15**, 1087; (25) Keyser, L. F., 1986, *J. Phys. Chem.* **90**, 2994; (26) Allen, M., and Frederick, J. E., 1982, *J. Atmos. Sci.* **39**, 2066; (27) Magnotta, F., and Johnston, H. S., 1980, *Geophys. Res. Lett.* **7**, 769; (28) Atkinson, R. et al., 1989, *J. Phys. Chem. Ref. Data* **18**, 881; (29) Brune, W. H. et al., 1983, *J. Phys. Chem.* **87**, 4503; (30) Fell, C. et al., 1990, *J. Chem. Phys.* **92**, 4768; (31) Piper, L. G. et al., 1987, *J. Phys. Chem.* **91**, 3883; (32) Schofield, K., 1979, *J. Phys. Chem. Ref. Data* **8**, 723; (33) Ko, T., and Fontijn, A., 1991, *J. Phys. Chem.* **95**, 3984; (34) Boodaghians, R. B. et al., 1988, *J. Chem. Soc., Faraday Trans.* **84**, 931; (35) Hall, I. W. et al., 1988, *J. Phys. Chem.* **92**, 5049; (36) Mellouki, A. et al., 1988, *J. Phys. Chem.* **92**, 2229; (37) Samson, J. A. R., and Pareek, P. N., 1985, *Phys. Rev.* **31**, 1470; (38) McElroy, M. B. et al., 1977, *J. Geophys. Res.* **82**, 4379; (39) Anicich, V. G., and Huntress,

W. T., 1986, *Astron. Astrophys. Suppl.* **62**, 553; (40) Anicich, V. G., 1993, *Astron. Astrophys. Suppl.* **84**, 215; (41) Kong, T. Y., and McElroy, M. B., 1977, *Planet. Space Sci.* **25**, 839.

7.4 Evolution

The seminal ideas on the evolution of the Martian atmosphere are due to McElroy (1972; cited in section 7.2), and Brinkmann, R. T., 1971, "Mars: Has Nitrogen Escaped?" *Science* **174**, 944.

The regulation of hydrogen and oxygen escape from Mars is demonstrated in Liu, S. C., and Donahue, T. M., 1976, "The Regulation of Hydrogen and Oxygen Escape from Mars." *Icarus* **28**, 231.

Solar wind–induced loss of the Martian atmosphere is modeled by Luhmann, J. G., Johnson, R. E., and Zhang, M. H. G., 1992, "Evolutionary Impact of Sputtering of the Martian Atmosphere by O^+ Pickup Ions." *Geophys. Res. Lett.* **19**, 2151; and Kass, D. M., and Yung, Y. L., 1995, "Loss of Atmosphere from Mars Due to Solar Wind Induced Sputtering." *Science* **268**, 697.

The effect of isotopic fractionation due to escape of gases is modeled by McElroy, M. B., and Yung, Y. L., 1976, "Oxygen Isotopes in the Martian Atmosphere: Implications for the Evolution of Volatiles." *Planet. Space Sci.* **24**, 1107; McElroy, M. B., Yung, Y. L., and Nier, A. O., 1976, "Isotopic Composition of Nitrogen: Implications for the Past History of Mars' Atmosphere." *Science* **194**, 70; and Yung, Y. L., Wen, J. S., Pinto, J. P., Allen, M., Pierce, K. K., and Paulson, S., 1988, "HDO in the Martian Atmosphere: Implications for the Abundance of Crustal Water." *Icarus* **76**, 146.

The lack of balance between the escape rates of oxygen and hydrogen is discussed in a recent article by Fox, J. L., 1993, "On the Escape of Oxygen and Hydrogen from Mars." *Geophys. Res. Lett.* **20**, 1847.

7.5 Terraforming Mars

The idea of terraforming Mars is explored by McKay, C. P., Toon, O. B., and Kasting, J. F., 1991, "Making Mars Habitable." *Nature* **352**, 489.

7.6 Unsolved Problems

The possible importance of heterogeneous chemistry on the surface of Martian dust is discussed by Anbar et al. (1993) and Atreya and Gu (1994), cited in section 7.3.

The lack of balance between the escape rates of oxygen and hydrogen is discussed in Fox (1993; cited in section 7.4).

The Milankovitch climatic cycles on Mars are known to the climatologists, but their effects on atmospheric chemistry are poorly quantified; see Ward, W. R., 1974, "Climatic Variations on Mars. I. Astronomical Theory of Insolation." *J. Geophys. Res.* **79**, 3375; Lindner, B. L., and Jakosky, B. M., 1985, "Martian Atmosphere Photochemistry and Composition during Periods of Low Obliquity." *J. Geophys. Res.* **909**, 3435.

Chapter 8

8.1 Introduction

Several conference proceedings cover the subject matter in this chapter: Hunten, D. M., Colin, L., Donahue, T. M., and Moroz, V., editors, 1983, *Venus* (Tucson: University of Arizona Press); Atreya, S. K., Pollack, J. B., and Matthews, M. S., editors, 1989, *Origin and Evolution of Planetary and Satellite Atmospheres* (Tucson: University of Arizona Press); Levine, J. S., editor, 1985, *The Photochemistry of Atmospheres* (New York: Academic Press).

A monograph on the photochemistry in the atmospheres of Mars and Venus is by Krasnopolsky, V. A., 1982, *Photochemistry of the Atmospheres of Mars and Venus* (Berlin: Springer-Verlag).

Mariner 10 encounter results were reported in a special issue: *Science* 1974, **183**, 1289–1321.

Pioneer Venus results were reported in special issues: *Science* 1979, **203**, 743–808; *Science* 1979, **205**, 41–121; *J. Geophys. Res.* 1980, **85**, 7573–8337.

An overview of the chemistry of Venus is by Prinn, R. G., 1985, "The Photochemistry of the Atmosphere of Venus," in Levine (1985; cited in section 8.1), p. 281; and Prinn, R. G., and Fegley, B., Jr., 1987, "The Atmospheres of Venus, Earth, and Mars: A Critical Comparison." *Ann. Rev. Earth Planet. Sci.* **15**, 171–212.

Some of the data in Table 8.1 are taken from the following sources: Davis, M. E. et al., 1992, *Celest. Mech. Dynam. Astron.* **53**, 377; Seiff, 1993, *Astronomical Almanac for Year 1994* (Washington, D.C.: US Government Printing Office); Beatty, J. K., and Chaikan, A., 1990, *The New Solar System* (Cambridge, Mass.: Sky Publishing).

Table 8.4 was compiled from DeMore, W. B., and Yung, Y. L., 1982, "Catalytic Processes in the Atmospheres of Earth and Venus." *Science* **217**, 1209–1213.

8.2 Photochemistry

The discussion on the possible importance of nitrogen chemistry in the production of sulfuric acid is based on Sill, G. T., 1983, "The Clouds of Venus: Sulfuric Acid by the Lead Chamber Process." *Icarus* **53**, 10.

The importance of thermal equilibrium chemistry between the atmosphere and the lithosphere is discussed by Lewis, J. S., 1970, "Venus: Atmospheric and Lithospheric Composition." *Earth Planet. Sci. Lett.* **10**, 73; and Fegley, B., Jr., and Prinn, R. G., 1989, "Estimation of the Rate of Volcanism on Venus from Reaction Rate Measurements." *Nature* **337**, 55.

Lightning on Venus is a controversial subject. The best optical detection places it at less than 0.1% of terrestrial lightning, as reported by Hansell, S. A., Wells, W. K., and Hunten, D. M., 1995, "Optical Detection of Lightning on Venus." *Icarus* **117**, 345.

8.3 Model Results

The yields of $O_2(^1\Delta)$ in the chemical reactions that are relevant for Venus are reported by Leu, M. T., and Yung, Y. L., 1987, "Determination of O_2 ($a^1\Delta_g$) and O_2

($b^1\Sigma^+g$) Yields in the Reaction O + ClO → Cl + O_2: Implications for Photochemistry in the Atmosphere of Venus." *Geophys. Res. Lett.* **14**, 949.

Table 8.5 references: (1) Shemansky, D. E., 1972, *J. Chem. Phys.* **56**, 1582; (2) DeMore, W. B., and Patapoff, M., 1972, *J. Geophys. Res.* **77**, 6291; (3) Watson, R. T., 1977, *J. Phys. Chem. Ref. Data.* **6**, 871; (4) Watanabe, K., 1958, *Advan. Geophys.* **5**, 153; (5) Hudson, R. D., 1971, *Rev. Geophys. Space Phys.* **9**, 305; (6) Prather, M. J., 1981, *J. Geophys. Res.* **86**, 5325; (7) Ackerman, M., 1971, *In Proceedings 4th ESRIN-ESLAB Symposium on Mesospheric Modeling and Related Experiments*, **p. 149**, Reidel, Hingham, Mass.; (8) Carver, J. H. et al., 1977, *J. Geophys. Res.* **82**, 1955; (9) Hudson, R. D., 1974, *Canad. J. Chem.* **52**, 1465; (10) Hudson, R. D., and Reed, E. I., 1979, *The Stratosphere: Present and Future*, **NASA Reference Publ. 1049**, Scientific and Technical Information Office, NASA; (11) Schügers, M., and Welge, K. W., 1968, *Naturforsch. A.* **23**, 1508; (12) Phillips, L., 1981, *J. Phys. Chem.* **85**, 3994; (13) Warneck, P. et al., 1964, *J. Chem. Phys.* **40**, 1132; (14) Bhatki, K. S. et al., 1982; (15) Okabe, H., 1971, *J. Amer. Chem. Soc.* **93**, 7095; (16) Hampson, R. F., Jr., and Garvin, D., 1978, *Nat. Bur. Stand. Spec. Publ.* **513**, Reaction rate and photochemical data for atmospheric chemistry—1977; (17) Baulch, D. L. et al., 1980, *J. Phys. Chem. Ref. Data* **9**, 295; (18) Demore, W. B. et al., 1981, *Chemical Kinetic and Photochemical Data for Use in Stratospheric Modeling. Evaluation Number 4: NASA Panel for Data Evaluation*, **JPL Publ. 81-3**, Jet Propulsion Laboratory; (19) Hampson, R. F., Jr., 1980, *Chemical Kinetic and Photochemical Data Sheets for Atmospheric Reactions*, **Report No. FAA-EE-80-17**, U.S. Dept. of Transportation. FAA Office of Environment and Energy, High Altitude Pollution Program, Washington, D.C.; (20) Value based on M = N_2, multiplied by 2; (21) Noxon, J. F. et al., 1976, *Astrophys. J.* **207**, 1025; (22) Traub, W. A. et al., 1979, *Astrophys. J.* **229**, 846; (23) Connes, P. J. et al., 1979, *Astrophys. J.* **233**, L29; (24) Okabe, H., 1978, Photochemistry of Small Molecules (New York: Wiley); (25) Trainor, D. W. et al., 1973, *J. Chem. Phys.* **58**, 4599; (26) Prather, M. J. et al., 1978, *Astrophys. J.* **223**, 1072; (27) Lee, Y-P., and Howard, C. J., 1982, *J. Chem. Phys.* **77**, 756; (28) Fair, R. W., and Thrush, B. A., 1969, *Trans. Faraday Soc.* **65**, 1208; (29) Clyne, M. A. A., and MacRobert, A. J., 1981, *Int. J. Chem. Kinet.* **13**, 187; (30) Herron, J. T., and Huie, R. E., 1980, *Chem. Phys. Lett.* **76**, 322; (31) Harris, G. W. et al., 1980, *Chem. Phys. Lett.* **69**, 378; (32) Leu, M. T., 1982, *J. Phys. Chem.* **86**, 4558; (33) Sander, S. P., and Watson, R. T., 1981, *Chem. Phys. Lett.* **77**, 473; (34) Strattan, L. W. et al., 1979, *Atmos. Environ.* **13**, 175; (35) Yung, Y. L., and McElroy, M. B., 1979, *Science* **203**, 1002; (36) Yung, Y. L. et al., 1980, *J. Atmos. Sci.* **37**, 339; (37) Stewart, A. I. et al., 1980, *J. Geophys. Res.* **85**, 7861; (38) Clark, T. C. et al., 1966, *Trans. Faraday Soc.* **62**, 3354; (39) Krasnopolsky, V. A., 1985, *Planet Spac.* **33**, 109.

Table 8.5 is taken from Yung and DeMore (1982), cited earlier.

8.4 Evolution

The origin and evolution of the atmosphere of Venus are discussed by Donahue, T. M., and Pollack, J. B., 1983, "Origin and Evolution of the Atmosphere of Venus," in Hunten et al. (1983; cited in section 8.1), p. 1003.

The significance of deuterium for understanding the evolution of Venus is pointed out by McElroy, M. B., Prather, M. J., and Rodriguez, J. M., 1982, "Escape of

Hydrogen from Venus." *Science* **215**, 1614; and Donahue, T. M. et al., 1982, "Venus Was Wet: A Measurement of the Ratio of Deuterium to Hydrogen." *Science* **216**, 630.

8.5 Mystery of the Missing Water

The geochemistry of the origin of water on Venus (and other terrestrial planets) is discussed by Liu, L. G., 1987, "Effects of H_2O on the Phase Behaviour of the Forsterite-Enstatite System at High Pressure and Temperatures and Implications for the Earth." *Phys. Earth Planet. Interiors* **49**, 142; Liu, L. G., 1988, "Water in the Terrestrial Planets and the Moon." *Icarus* **74**, 98; and Lange, M. A., and Ahrens, T. J., 1984, "FeO and H_2O and the Homogeneous Accretion of the Earth." *Earth Planet. Sci. Lett.* **71**, 111.

8.6 Unsolved Problems

Active volcanism on Venus at present is proposed by Esposito, L., 1984, "Sulfur Dioxide Shows Evidence for Venus Volcanism." *Science* **223**, 1072.

The unidentified UV absorber is most likely polysulfur. See discussion by Toon, O. B., Turco, R. P., and Pollack, J. B., 1982, "The Ultra-violet Absorber on Venus: Amorphous Sulfur." *Icarus* **51**, 358.

The problem of noble gases is discussed by Donahue and Pollack (1983; cited in section 8.4).

Chapter 9

9.1 Introduction

There is an enormous literature on the subject of Earth. The emphasis of this chapter is on the "planetary" aspects of the planet. A number of excellent texts are available: Holland, H. D., 1978, *The Chemistry of the Atmosphere and Oceans* (New York: Wiley); Holland, H. D., 1984, *The Chemical Evolution of the Atmosphere and Oceans* (Princeton, N.J.: Princeton University Press); Goody, R. M., and Yung, Y. L., 1989, *Atmospheric Radiation* (New York: Oxford University Press); Crowley, T. J., and North, G. R., 1991, *Paleoclimatology* (New York: Oxford University Press); Rambler, M. B., Margulis, L., and Fester, R., editors, 1989, *Global Ecology: Towards a Science of the Biosphere* (New York: Academic Press).

9.2 Gaia Hypothesis

The Gaia hypothesis was developed and elaborated by Lovelock, J. E., and Margulis, L., 1974, "Atmospheric Homeostasis, by and for the Biosphere: The Gaia Hypothesis." *Tellus* **26**, 1.

Table 9.1 is taken from Intergovernmental Panel on Climate Change, 1990, *Climate Change: The IPCC Scientific Assessment*, and 1992, *Climate Change 1992: Supplement to the IPCC Scientific Assessment* (Cambridge: Cambridge University Press).

Tables 9.2, 9.3, and 9.4 are taken from Fegley, B., Jr., "Properties and Composition of the Terrestrial Oceans and of the Atmospheres of the Earth and Other Planets," in *Global Earth Physics*, J. A. Thomas, ed. (AGU Reference Shelf 1; Washington, D.C.: American Geophysical Union), 320.

Table 9.5 is taken from Rambler, M. B. et al., eds., 1987, *Global Ecology: Towards a Science of the Biosphere* (New York: Academic Press).

Table 9.6 is taken from Holland, H. D., 1978, *The Chemistry of the Atmosphere and Oceans* (New York: Wiley).

Table 9.7 is taken from Intergovernmental Panel on Climate Change (1990, 1992), cited earlier.

Chapter 10

10.1 Introduction

There are many books and monographs covering the subject matter in this chapter: Intergovernmental Panel on Climate Change, 1990, *Climate Change: The IPCC Scientific Assessment* (Cambridge: Cambridge University Press), and 1992, *Climate Change 1992: Supplement to the IPCC Scientific Assessment* (Cambridge: Cambridge University Press); Houghton, J. T., editor, 1984, *The Global Climate* (Cambridge: Cambridge University Press); Schlesinger, W. H., 1991, *Biogeochemistry: An Analysis of Global Change* (New York: Academic Press); Hansen, J. E., and Takahashi, T., editors, 1984, *Climate Processes and Climate Sensitivity* (Geophysical Monograph 29; Washington, D.C.: American Geophysical Union); Takeuchi, K, and Yoshino, M., editors, 1991, *The Global Environment* (Berlin: Springer-Verlag); Brasseur, G., and Solomon, S., 1984, *Aeronomy of the Middle Atmosphere* (Dordrecht: Reidel); Holland, H. D., and Petersen, U., 1995, *Living Dangerously* (Princeton, N.J.: Princeton University Press); Singh, H. B., 1995, *Composition, Chemistry, and Climate of the Atmosphere* (New York: Van Nostrand Reinhold).

Much of the material in this chapter is based on International Panel on Climate Change (1990; cited above), and the following three fundamental reports: McElroy, M. B., 1975, "Chemical Processes in the Solar System: A kinetic Perspective," p. 127 in *Chemical Kinetics*, ed. D. R. Herschbach (London: Butterworths, 1975); Logan, J. A. et al., 1978, "Atmospheric Chemistry: Response to Human Influence." *Phil. Trans. Roy. Soc. Lond.* **290**, 187; Logan, J. A. et al., 1981, "Tropospheric Chemistry: A Global Perspective." *J. Geophys. Res.* **86**, 7210.

10.4 Stratospheric Chemistry

The Antarctic ozone hole was discovered by Farman, J. C., Gardiner, B. G., and Shanklin, J. D., 1985, "Large Losses of Total Ozone in Antarctica Reveal Seasonal ClOx/NOx Interaction." *Nature* **315**, 207.

Author and Proper Name Index

Agung, 343, 377
Ahrens, T.J., 104, 106
Allen, C.W., 19
Allende, 98, 110
Andreae, M.O., 352
Arrhenius, 65
Atreya, S.K., 133, 179
Axel-Danielson, 48, 177

Baliunas, S.L., 22
Barrow, J.D., 80
Barshay, S.S., 135, 185
Be, A.W.H., 334
Beer, 28
Beer, J., 27
Belcher, J.W., 200
Berner, R.A., 338
Black, G.A., 85, 87, 91
Blake, G., 63, 65
Bohren, C.F., 51
Bowers, M.T., 68, 69
Brasseur, G., 45
Broadfoot, A.L., 199, 200
Broadfoot, L., 124, 138
Broecker, W.S., 82, 354
Brown, 103

Canuto, V.M., 24
Chamberlain, J.W., 5, 47
Chapman, 45, 391
Chapman, C.R., 97
Chappuis, 35
Charon, 239
Chuang Tzu, 77
Clarke, A.C., 316
Clayton, R.N., 110, 112
Cogley, J.G., 105
Coloumb, 68
Columbus, 280

Dansgaard-Oeschger, 343
Demerjian, K.L., 49
DeMore, W.B., 295, 300–307
Dissly, R.W., 78, 108, 111
Donahue, R.A., 22
Donahue, T.M., 179, 248

El Chichon, 343, 352, 409
Evans-Polanyi, 67

Farmer, C.B., 248
Fegley, B. Jr., 93, 94
Festou, M.C., 125

Fischer-Tropsch, 98
Fox, J.L., 260, 262, 272, 279
Frederick, J.E., 41
Friedl, R.R., 72
Frohlich, C., 21
Fung, I., 355

Genesis, 362
Gibbs, 57
Giotto, 100
GISS, 379
Gladstone., 139, 163–177
Goldschmidt, 81
Gregory, R.P.F., 53
Grinspoon, 109
Gurwell, M.A., 312

Halley, 100
Hammer, C.U., 342
Hanel, R., 128
Henderson-Sellers, A., 105
Henry, 331
Herbert, F., 142
Herzberg, 34, 45
Herzberg, L., 46
Holbrook, K.A., 71
Holland, H.D., 350, 353
Houghton, J.T., 349
Hsu, K.J., 333
Hudson, R.D., 41
Huffman, D.R., 51
Huggins, 35
Hunten, D.M., 5, 47, 192, 248, 284

Jakosky, B.M., 248
Jeans, 15
Joseph, 361

Kant, 91
Kass, D.M., 276
Kasting, J.F., 336, 348
Kaye, J.A., 108, 111, 181
Keller, C.N., 208, 209, 210
Kliore, A.J., 126
Knollenberg, R., 284
Kozyra, J.U., 275
Krakatoa, 343
Krasnopolsky, V.A., 309, 310
Kuhn, W.R., 179
Kuiper, 99
Kumar, S., 192

Lange, M.A., 106
Langevin, 68, 69
Laplace, 91
Larimer, J.W., 81, 83
Lean, J., 23
Lee, T., 96
Leighton-Murray, 245
Levine, J.S., 346, 348
Lewis, J.S., 92, 109, 134, 182
Lindal, G.F., 127
Lindemann-Hinshelwood, 70
Liu, L.G., 314
Lovelock, J.E., 11, 322, 323, 326, 361
Luhmann, J.G., 275
Lunine, J.I., 87, 91
Lyons, J.R., 237, 238

Malicet, J., 37
Malthus, 361
Margulis, L., 322, 323, 326
Mariner, 282
Mathis, J.S., 26
Mauna Loa, 366
Maunder, 21
Maxwell-Boltzmann, 15
McConnell, J.C., 138
McElroy, 245, 248
McKenzie, J.A., 333
Mezger, P.G., 26
Michelangeli, D.V., 49
Mie, 48, 49
Milankovitch, 281, 340
Miller, 84, 98
Molina-Rowland, 305
Muckerman, J.T., 64
Murchison, 98

Nagy, A.F., 299
Nair, H., 251, 257, 261, 263–272, 274, 278
Najjar, R., 333
Nier, A.O., 246

O'Keefe, J.D., 106
Oort, 99
Oort, A.H., 328, 329
Owen, T.C., 240

Panagia, N., 26
Papanastassiou, D.A., 96
Parkinson, 248
Peixoto, J.P., 328, 329

Pepin, R.O., 107
Pinatubo, 343, 352, 409
Pioneer Venus, 282
Planck, 18
Pollack, J.B., 309, 310
Prinn, R.G., 92, 93, 94, 135, 185, 309

Raisbeck, G.M., 27
Raman, 17
Rayleigh, 17, 46, 277
Raymo, M.E., 337
Raynaud, D., 342
Rice-Ramsperger-Kassel-Marcus, 70
Robinson, P.J., 71
Ruddiman, W.F., 337

Sandel, B.R., 138
Sander, S.P., 72
Sarmiento, J.L., 333
Schumann-Runge, 35, 45, 273
Seiff, A., 283
Sharp, T.E., 31
Shemansky, D.E., 206
Solomon, S., 45
St.Helens, 343
Steinfeld, J.I., 36
Stevens, M.H., 142
Strobel, D.F., 142, 181, 192, 195, 198, 206, 237, 241

Su, T., 68, 69
Suess, 103
Summers, M.E., 192, 195, 198, 237

Tegen, I., 355
Tillman, J.E., 246
Tipler, F.J., 80
Toggweiler, J.R., 333

Urey, 83, 84, 98

van Dishoeck, E.F., 87, 91
Vega, 100
Venera, 282
Voyager, 123, 201

Walker, J.C.G., 105
Wasserburg, G.J., 96
Weidenschilling, S.J., 134, 182
Wetherill, G.W., 97
Wheeler, J.A., 14

Yelle, R.V., 235
Yiou, F., 27
Yung, Y.L., 72, 78, 108, 111, 216–233, 300–307

Zielinski, G.A., 344
Zinner, E., 109

Subject Index

Absorption, 28, 49
Absorption cross sections, 28
Accommodation coefficient, 72
Actinic flux, 48
Activated complex, 63
Activation energy, 66, 110
Adiabatic, 8
Aerosols, 48, 204, 377
Airglow, 17
Albedo, single scattering, 47
Albedo, surface, 11, 325
Alkalinity, 332
Allene, 156, 157
Anoxic environment, 347
Antarctic ozone hole, 358, 404
Arrhenius expression and parameters, 65
Asymmetry parameter, 50
Atmospheric cratering, 273
ATP, 54
Aurora, 17, 25, 141, 172
Axel-Danielson dust, 48, 177

Bacteria, 324
Beer's law, 28
Big Bang, 18
Bimolecular reactions, 56, 61

Biochemistry, 53, 331
Biogeochemical cycles, 317, 324
Biomass, 330
Biomass burning, 369
Biosphere, 317, 357
Blackbody radiation, 18
Bond energy, 29, 32, 33
Br (bromine), 355, 399, 407
Bromine compounds, 407

^{14}C (carbon isotope 14), 21
Callisto, 189
Carbohydrate, 53
Carbon cycle, 330
Carbonaceous chondrites, 104, 119
Carbonic acid, 331
Carbonyl sulfide (COS), 43
Catalyst, 249, 289, 404
Catalytic chemistry of
  Earth, 404
  Mars, 249
  Venus, 288
CFC, 40, 280, 358
$CFCl_3$ (CFC-11), 360, 375
CF_2Cl_2 (CFC-12), 360, 375, 397
CH_2, 153

Indexes 451

CH$_2$O (formaldehyde), 386
CH$_3$Br, 354, 370
CH$_3$Cl, 354, 370
CH$_3$D, 136, 186
CH$_3$I, 354
CH$_3$O$_2$ (methyl peroxy radical), 385
CH$_4$, 31, 144, 240, 385
  Earth, 369
  Jupiter, 129
  Neptune, 130
  Pluto, 240
  Saturn, 130
  Titan, 201
  Triton, 235
  Uranus, 130
C$_2$H (ethynyl radical), 145, 154, 155, 218
C$_2$H$_2$, 31, 154, 202
C$_2$H$_4$, 31, 153, 202
C$_2$H$_6$, 31, 153, 202
Chapman Chemistry, 391
Chapman function, 45
Chapman layer, 45
Chappuis bands, 35
Charge exchange reaction, 69, 196, 311
Charge transfer, 196
Chemical stability, 4, 52
Chemical time constant, 168
Chlorine compounds, 288
Chlorine dioxide (OClO), 407
Chlorine monoxide dimer, 40, 406
Chlorine nitrate (ClONO$_2$), 40, 399
Chlorofluorocarbon (CFC), 28, 40, 358
Chloroformyl radical, 288
Chlorophyll, 53
Chondrites, 98, 119
Chromophore, 183
Chuang Tzu, 77
Cl (chlorine), 404
ClO (chlorine monoxide), 40, 397, 400
ClO$_x$ (free chlorine), 402
Cl$_2$O$_2$, 40, 406
CLIMAP, 340
Climatic stability, 4, 52
ClONO$_2$, 40, 399
Clouds, 132
CN, 44, 224
C$_2$N$_2$, 43
CO (carbon monoxide), 37
  Earth, 385
  Jupiter, 135
  Mars, 247

Neptune, 135, 185
Pluto, 240
Saturn, 185, 135
Titan, 228
Uranus, 135, 185
Venus, 286
CO$_2$ (carbon dioxide), 37
  Earth, 330, 366, 377
  Mars, 247
  Titan, 228
  Venus, 286
CO$_2$ stability, 248, 285
Coloumb force, 68
Columbus, 280
Comparison of planets, 12, 203
Composition of atmospheres of planets, 12
  Earth, 318
  Io, 195
  Jupiter, 129
  Mars, 245
  Neptune, 130
  Pluto, 240
  Saturn, 130
  Titan, 202
  Triton, 237
  Uranus, 130
  Venus, 284
Conduction, 9
Convection, 9
Corona, 19
COS, 43, 292, 352, 403
Cosmic abundance, 82, 120
Cosmic rays, 25, 27, 224
Cosmology, 77
Critical level, 8, 15
Cross section, see Absorption cross section,
Cyano radical (CN), 44, 226
Cyanoacetylenes (HC$_3$N), 43, 228

Daisies, 325
Dansgaard-Oeschger events, 343
Degassing of gases, 103
Denitrification, 350, 372
Deuterium, 89, 108, 277, 312, 319
Devolatization, 313
Diffusion, molecular, 137, 257
  eddy, 137, 257, 295
Dissociation energy, 29
Dissociative recombination reactions, 74, 253
DMS, 403
DNA, 389

Dobson unit, 247
Dust, 256, 280, 355
Dust on Mars, 256, 280

Earth, 11
Eddy diffusion, 137, 257, 295
Efficiency, 49
Effusion velocity, 278
Einstein coefficient, 75
Electroglow, 123
Eleven-year solar cycle, 21, 189, 396
Energy barrier, 61
Enthalpies of simple molecules, 58
Entropies of simple molecules, 59
Entropy flux, 24
Equation of continuity, 136
Equation of radiative transfer, 47
Equilibrium chemistry, 122, 135, 310
   CO, 135
Equilibrium temperature, 325
Escape flux, 15, 106, 223
   Earth, 325, 346
   Io, 196
   Mars, 273, 277
   Pluto, 242
   Titan, 217
   Venus, 311
Escape of gas, 15
Escape velocity, 15, 198, 278, 311
Ethynyl radical (C_2H), 145, 155, 218, 310
Europa, 189
EUV, 22, 23, 199
Evans-Polanyi relationship, 67
Evaporation, 327
Evolution, 3, 4, 13, 110, 189, 273, 310
Excited species, 73
Exobase, 211, 252
Exosphere, 15
Exospheric temperature, 15, 141
   Earth, 15
   Io, 15
   Jupiter, 15, 141
   Mars, 15
   Saturn, 15, 141
   Uranus, 15, 141
   Venus, 15
Extinction, 49
Extreme ultraviolet (EUV), 22, 123, 199

Ferric iron, 349
Ferrous iron, 349
Fertilizers, 349
Fixation of N_2, 350
Foraminifera, 332
Formaldehyde (HCHO), 37, 86, 386

Gaia hypothesis, 317, 325
Galactic cosmic rays, 21, 25, 27, 226
Galaxy, 80
Ganymede, 189
Gas-kinetic rate coefficient, 62
Geometric optics, 50
Gibbs free energy, 57
Global warming, 363
Grain chemistry, 90
Grains, 76
Gravitational binding energies, 14
Greenhouse effect, 282, 363, 372
Greenhouse gases, 363

H (atomic hydrogen), 29
   Jupiter, 163
   Lyman-α airglow, 123
H_2 (molecular hydrogen), 39, 129, 179, 253, 263
   Earth, 386
   Jupiter, 129
   Mars, 253, 263
   Titan, 202
H_2CO (formaldehyde), 37, 86, 386
H_2O dissociation energy, 33
   Venus, 313
H_2O_2 (hydrogen peroxide), 33, 249, 383, 395
   Earth, 395
   Mars, 249
H_2S (hydrogen sulfide), 32, 129
H_2SO_4 (sulfuric acid),
   Earth, 351, 377, 410
   Venus, 290
Halocarbon, 369
Halogenated compounds, 369
Hartley bands, 35
HC_3N, 43, 224
HCl (hydrogen chloride), 40
   Earth, 399
   Venus, 287
HCN, 43, 132, 223
HD, 136, 186
He (helium), 29, 129, 130
   Jupiter, 129, 138
   Neptune, 130

Saturn, 130
Uranus, 130
Heat, 9
  conduction, 9
  convection, 9
Henry's Law, 331
Herzberg bands, 45
Herzberg continuum, 34
Heterogeneous reaction, 73
HF (hydrogen fluoride), 284, 397
  Earth, 397
  Venus, 284
HNO_3 (nitric acid), 38, 273, 396
HO (hydroxyl radical), 230, 248, 394
HO_2 (hydroperoxyl radical), 249, 394
HO_2NO_2, 38
Homeorrhesis, 325
Homeostasis, 325
Homeostat, 11
Homo Sapiens, 357
HO_x (odd hydrogen), 4, 250, 267, 384, 393
Huggins bands, 35
Hydrazine, 178
Hydrocarbon chemistry, 144, 210
Hydrodynamic escape, 104, 242
Hydrogen, 29
Hydrogen cyanide (HCN), 43, 223
Hydrogen peroxide (H_2O_2), 33, 383
Hydrogen sulfide (H_2S), 129
Hydrostatic equation, 6

Ice age, 342, 379
Ice core, 342, 378
Impact parameter, 68
Instability, 11
Internal heat of earth, 25
Interstellar medium, 73
Io (see composition), 10, 189, 191, 192
Ion-molecule reactions, 68, 69, 89
Ionosphere, 26, 142, 196, 204, 236
Irradiance, 20, 48
Isotopic composition, 319
Isotopic fractionation, 89, 279
Isotope enrichment, 279
Isotope fractionation resulting
  from escape of gases, 279
  from chemical reactions, 89

Jeans, 15
Jovian wind, 200
Junge layer, 49

Jupiter, 7, 120, 139, 144
Jupiter,
  composition, 129
  Great Red Spot, 183
  thermal structure, 7

K emission, 198
Kuiper Belt, 99

Langevin rate constant, 68, 69
Lapse rate, 7, 8, 10
  dry, 10
  wet, 10
Last Glacial Maximum (LGM), 342, 378
Leighton-Murray, 245
Lightning, 25, 27, 60, 349
Lindemann-Hinshelwood rate coefficient, 70
Lovelock, 361
Lyman α (see H), 123, 273
Lyman bands, 30

Magmatism, 339
Malthus, 361
Mariners, 244
Mars, 11
Mars,
  carbon dioxide, 247
  composition, 245
  escape of atmospheric compounds, 273
  isotope enrichment of D over H, 277
Maunder Minimum, 21
Maxwell-Boltzmann distribution, 15
Mean free path, 8
Mean temperature, 7
Metamorphism, 339
Methane (see CH_4)
Methylacetylene (propyne, CH_3CCH), 156, 157, 202
Methylene radical (CH_2), 145, 153
Mie scattering, 48, 49
Milankovitch theory of climate, 281, 340
Molecular diffusion, 137
Molina-Rowland, 305

N_2 (molecular nitrogen), 38, 223, 240, 318
  Earth, 318, 349
  Jupiter, 179
  Mars, 245
  Titan, 201
  Venus, 284
N_2O, 38, 256, 370

N_2O_5, 38, 256
Na emission, 198
Neptune, 7, 130, 181, 185
  composition, 130
  thermal structure, 7
NH_3 (ammonia), 32, 129, 130
  Jupiter, 129
  Saturn, 130
Nitrate radical (NO_3), 38, 255
Nitric acid (HNO_3), 38, 255
Nitric oxide (NO), 38, 255, 383
Nitrogen chemistry in the atmosphere of
  Earth, 349, 396
  Jupiter, 178
  Mars, 255
  Titan, 223
Nitrogen dioxide (NO_2), 38, 255
Nitrous oxide (N_2O), 38
  biological production, 374
  Mars, 256
NO (see nitric oxide)
NO_2 (see nitrogen dioxide)
NO_3 (see nitrate radical)
NO_x (odd nitrogen),
  catalytic cycle Earth, 402
  catalytic cycle Mars, 255
N_2O_5 (dinitrogen pentoxide), 38, 256

Ocean chemical composition, 320
$O(^1D)$ (first metastable state), 250
  Earth, 382
  Mars, 251
$O(^3P)$ (ground state), 34
  Earth, 402
  Mars, 247, 263
  Venus, 286
O_2 (molecular oxygen), 34, 191, 263, 348
  Earth, 348
  Mars, 245, 263
  Venus, 286
O_3 (ozone), 34
  biological screen, 348
  Earth abundance, 318, 393, 402
  Mars abundance, 247, 266
O_3 bands, 34
O_4, 34
OClO, 407
Odd nitrogen, 255
OH (hydroxyl), 230, 248, 267, 382, 403
  in HO_x cycle, 267, 384
OH, Earth, 382

Oort cloud, 99
Optical depth, 44
Organic compounds, 204
Organic synthesis, 84, 90
Origin, 3
Oxidants, 382
Oxygen, 34
Oxygen cycle, 347
Ozone hole, 404
Ozone hole (see Antarctic Ozone hole)

P_4, 182, 183
Paleosol, 335
Partition function, 65
Penultimate glacial maximum (PGM), 340
Perfect gas law, 6
Periodic table, 81
PH_3 (phosphine), 32, 129, 130, 182
Phase function, 47
Phosphine, 182
Phosphorus cycle, 353
Photochemistry, 247, 348
Photodissociation coefficient, 44
Photoelectrons, 143
Photosensitized dissociation, 32, 144, 155, 210
Photosphere, 19
Photosynthesis, 323, 330
Pioneer, 282
Planck function (see blackbody radiation)
Pluto, 240
  composition, 240
  thermal structure, 241
Polar stratospheric clouds, 408
Polarizability, 68
Polymerization, 177
Polysulfur, 291, 292
Polyynes, 144, 158, 173, 221
  Jupiter, 158
  Titan, 221
Population, 358
Potential energy, 13, 64
Precipitation, 327
Present atmospheric level (PAL), 345
Pressure, Venus, 283
Prokaryotes, 324
PSC, 408

Radiation entropy, 24
Radiative association reactions, 75, 90
Radiative forcing, 375

Radiative transfer, 47
Raman scattering, 17
Rate coefficient, 56
　gas kinetic, 62
　three-body, 56, 70
　two-body, 56, 62
Rates of reaction, 56
Rayleigh distillation, 277
Rayleigh scattering, 17, 46
Recombination reaction, 74, 202, 255
Resonance scattering, 17, 123, 202
Rice-Ramsperger-Kassel-Marcus, 70

S emission, 198
Saturn, 7, 130, 181, 185
　composition, 130
　thermal structure, 7
Scale height, 7
Scattering,
　Mie, 48
　Rayleigh, 46
Scattering by small particles, 48, 49
Scattering efficiency, 49
Schumann-Runge band, 35, 45
Sequestering, 273
Sidereal period, 6
Single scattering albedo, 47
SMOW, 110
SO, 43
SO_2 (Sulfur dioxide), 43, 191, 284, 377
　Earth, 377
　Io, 191
　Venus, 284
Solar constant, 18, 325
Solar cycle (*see* eleven year solar cycle), 22, 200, 396
Solar maximum, 27
Solar minimum, 27
Solar wind, 25, 275, 312
Stability, 4
　chemical, 4
　climatic, 4
Stability of CO_2, 248, 285
Starlight, 25, 26
Stratosphere, 134, 144, 202, 389
　Earth, 389
　Jupiter, 134
　Titan, 202
Structure, 205
Sulfate, 377
Sulfur, 186

Sulfur compounds, 32, 43, 193, 289
Sulfur cycle, 351
Sun, 20
Sun's variability, 21
Sunspots, 22, 340, 341
Synthesis of complex organic molecules, 71

T-Tauri phase, 23, 106
Temperature, thermal structure, 7
　Earth, 11
　Io, 10, 192
　Jupiter, 7, 139
　Mars, 11, 257
　Neptune, 7
　Pluto, 241
　Saturn, 7
　Titan, 10, 205
　Triton, 10, 237
　Uranus, 7
　Venus, 283
Termolecular reactions, 56, 70
thermal, 205
Thermal inversion, 11, 52
Thermal structure, 9
Thermochemical reactions, 56
Thermoequilbrium chemistry, 60
Three-body rate association, 70
Three-body rate coefficient, 71
Titan, 10, 205, 223
　composition, 202
　escape flux, 223
　ionosphere, 204
　photochemistry, 210
　thermal structure, 205
　Transition state, 63
Transition state, 63
Triton, 10, 234
Troe formula, 71
Troposphere of Earth, 381

Unimolecular reactions, 55, 61
Uranus, composition, 7, 130, 185
　temperature, 7
UV-A, UV-B, UV-C, 389

Venera, 282
Venus, 11, 282
　composition, 284
　deuterium, 312

Venus (*continued*)
   surface pressure, 283
   temperature, 283
   vertical structure, 283
Vibrationally excited H_2, 143
Viking, 244, 280
Vinyl radical, 160
Volcanic eruptions, 343

Volume absorption rate, 44
Voyager, 123, 201

Water, 313
Weathering, 328, 339
Werner bands, 30

Younger Dryas, 342